Series:

Advances in Agri-Genomics

Series editor: Chittaranjan Kole, Kolkata, West Bengal, India

1. **Potato Improvement in the Post-Genomics Era**
   Jagesh Kumar Tiwari, 2023

2. **Genetics, Genomics and Breeding of Bamboos**
   Malay Das, Liuyin Ma, Amita Pal, and *Chittaranjan Kole (eds.), 2023*

3. **The Avocado Genome: Genomics, Genetics and Breeding**
   J.I. Hormaza, Mary Lu Arpaia, Neena Mitter, and *Alice Hayward (eds.), Forthcoming*

4. **The Asparagus Genome: Genetics, Genomics and Breeding**
   Francesco Mercati, Jim Leebens-Mack, and *Francesco Sunseri (eds.), Forthcoming*

Advances in Agri-Genomics
Chittaranjan Kole (series editor)

Genetics, Genomics and Breeding of Bamboos

Editors:

Malay Das
Department of Life Science, Presidency University
Kolkata, India

Liuyin Ma
Associate Professor
Basic Forestry and Proteomics Research Center
Fujian Agriculture and Forestry University
Fuzhou, China

Amita Pal
Retired Senior Professor
Division of Plant Biology, Bose Institute
Kolkata, India

Chittaranjan Kole
Prof. Chittaranjan Kole Foundation for Science and Society

CRC Press
Taylor & Francis Group
Boca Raton London New York

CRC Press is an imprint of the
Taylor & Francis Group, an **informa** business

A SCIENCE PUBLISHERS BOOK

First edition published 2023
by CRC Press
6000 Broken Sound Parkway NW, Suite 300, Boca Raton, FL 33487-2742

and by CRC Press
4 Park Square, Milton Park, Abingdon, Oxon, OX14 4RN

Library of Congress Cataloging-in-Publication Data (applied for)

ISBN: 978-1-032-26305-2 (hbk)
ISBN: 978-1-032-26306-9 (pbk)
ISBN: 978-1-003-28760-5 (ebk)

DOI: 10.1201/9781003287605

Typeset in Times New Roman
by Radiant Productions

Preface to the Series

Agricultural genomics, popularly called as AgriGenomics, has emerged as one of the most frontier fields in life-sciences since the beginning of the 21st century. Whole genome sequencing of the leading crop plant rice in 2002 marked the begining of genome sequencing in an array of crop plant species. Elucidation of structural genomics of the nuclear genomes of the crop plants and their wild allied species included enumeration of the sequences and their assembly, repetitive sequences, gene annotation and genome duplication, synteny with other sequences and comparison of gene families. Sequencing of organellar genomes including chloroplasts and mitochondria enriched the structural genomics database. At the same time, researches on functional genomics such as transcriptomics, proteomics and metabolomics presented useful information. All these advances in genome sciences facilitated complementation of the erstwhile strategies of genome elucidation and improvement employing molecular breeding and genetic transformation with genomics-assisted breeding and most recent technique of genome editing. All these concepts, strategies, techniques and tools will be deliberated in the 'stand-alone' volumes of this book series.

Chittaranjan Kole
Kolkata, India

Preface

Bamboos are an important group of plants having enormous potential for commercial utilization. However, our current understanding on the physiology and genetics of the plant group is at a rudimentary state. One major reason is the over emphasis of the scientific community to understand the biology of reference plants. Studying bamboos poses multiple challenges such as polyploid genome, long generation time (perennialism), unpredictable flowering, low and asynchronous seed germination rate, recalcitrance to tissue culture and genetic transformation, absence of physical, genetic maps, and mutants. Recent genome/transcriptome sequencing of a couple of bamboo species provide an opportunity to address many fundamental questions at the genetic level.

The book entitled *Genetics, Genomics and Breeding of Bamboos* renders a comprehensive overview on fundamental as well as applied aspects of bamboo. It is aimed to provide the most recent advances in diverse areas of bamboo research that include taxonomy, genetic resources, traditional and molecular breeding, abiotic and biotic resistance, genome sequencing and genomic tools, tissue culture, genetic transformation, genome editing, and quality control of planting material. The different chapters have been contributed by internationally reputed experts, who have contributed enormously on these topics. This book will provide a good source of information for students, scientists, farmers, and bamboo resource management advisers. It will also help to prioritize future thrust areas for the bamboo research community. In conclusion, the Editors express thanks to the authors for their excellent contributions and Mr. Vijay Primlani of the Science Publishers for his enormous cooperation during the entire editing process.

Foreword

At least 55 million years ago, the ancestor of bamboos was already present on Earth. Since that time, that ancestor has diversified into over 1,700 species of both herbaceous and woody bamboos. Herbaceous bamboos first appeared around 42 million years ago, followed by woody bamboos approximately 17 million years later. Currently, bamboos are present on all continents, except for Europe and Antarctica. Taxonomically, bamboos are classified into three tribes: Olyreae or herbaceous bamboos (at least 135 species) and two tribes of woody bamboos, Arundinarieae or temperate woody bamboos (at least 580 species) and Bambuseae or tropical woody bamboos (over 1,000 species). The Bambuseae are divided into two large groups; the Paleotropical bamboos are exclusive to the Old World (Eastern hemisphere) and the Neotropical bamboos are endemic to the New World (Western hemisphere or the Americas).

It is highly possible that our ancestors used bamboo throughout their journey from Africa to the Americas through Asia. Archaeological records in Asia and America provide some corroboration for this statement, and bamboo cultures remain widespread in Asia and parts of South America. What makes woody bamboos so special? The answer, in part, is undoubtedly their strongly lignified culms and their fast growth rate. But woody bamboos also stand out from the rest of the grasses (Poaceae), the plant family to which they belong, as well as from the herbaceous bamboos, because of their long-life cycles that reach up to 120 years between each flowering and their monocarpy, in which most of the species reproduce only once and die.

The culms of many species of woody bamboos around the world are widely used in a variety of ways, from the very traditional (e.g., basketry) to the very advanced (e.g., laminate flooring for exteriors). And, of course, they are used in the construction of traditional residential houses built by local ethnic groups in Asia and the Americas, and now even in the construction of modern houses and touristic centers around the world. Notably, the shoots of many bamboo species are also consumed for food, especially in several Asian countries (Chapter 2).

The use of a resource by humans always leads us to learn more about it. In this sense, the same thing has happened with the bamboos. Although the study of bamboo properties for engineering is now common, probably the earliest scientific studies were in taxonomy, that is, the description and classification of the world's bamboos. Over the years and with advances in molecular technologies, we now have a broader picture of the phylogenetic relationships of bamboos at the tribal and generic levels.

However, knowledge about the genetic diversity of bamboo species, even widely cultivated ones, or the genetic bases of culm lignification or the flowering cycles of woody bamboos, has been lacking.

To try to better understand these fabulous plants, the bamboo community has developed *ex situ* collections and germplasm resources (Chapter 3). We have learned of their phylogenetic relationships, using various molecular markers and techniques, and have begun to apply these to understanding the genetic diversity of bamboo species and populations (Chapters 4 and 5). We are learning about the breeding system of woody bamboos (Chapter 6), the role of photoperiod genes in flowering (Chapter 7), and the genetic basis of abiotic resistance to stress (Chapter 8). We are studying bamboo diseases and the genes that control pathogenic interactions (Chapter 9). In the age of biotechnology and genomics, we are learning about genetic transformation and *in vitro* culture. The sequencing and annotation of complete genomes is underway (Chapters 10 and 11), as is the application of CRISPR for the technological development of bamboo and the editing of genes and genomes to obtain better culms for quality production (Chapters 12 to 16). These topics are treated in the 16 chapters included in this book, especially as they may apply to widely grown or potentially useful woody bamboo species.

The current book is intended to gather the most recent advances in the study of bamboos from genetic and genomic perspectives (Chapter 1). The intention is that this book will serve both as a handy reference for bamboo researchers and a solid foundation for the genomic study of bamboos soon.

Lynn G. Clark
Iowa State University, USA

Eduardo Ruiz-Sanchez
Universidad de Guadalajara, Mexico

Contents

Preface to the Series   v

Preface   vi

Foreword   viii

Unit 1

1. **Introduction**   1
   Malay Das, Liuyin Ma, Amita Pal and *Chittaranjan Kole*

2. **Taxonomy, Diversity, Distribution, and Use of Bamboos: Special**   4
   Importance to Selected Four Northeast States of India
   Evanylla Kharlyngdoh, Bharat G Somkuwar and *Pulok Kumar Mukherjee*

   2.1   Introduction   4
   2.2   Taxonomy   5
   2.3   Distribution and Diversity   6
         2.3.1   Diversity of Edible Bamboo Species   9
         2.3.2   Diversity of Commercial Bamboo Species   11
   2.4   Area and Resource Stock   12
   2.5   Uses of Bamboo   13
         2.5.1   Bamboo as Food   13
         2.5.2   Bamboo in Medicinal, Cosmetic, and Agricultural Uses   16
         2.5.3   Bamboo in Environmental Conservation and Carbon   20
                 Sequestration
   2.6   Conclusion   20

3. **Germplasm Resources of Bamboos**   27
   KC Koshy

   3.1   Introduction   27
   3.2   Bamboo Gardens and Their Coordinating Organizations   28
   3.3   Botanic Gardens having Living Bamboo Collections   28
   3.4   Earliest Living Bamboo Collections   43
   3.5   Largest Living Bamboo Collections   43

3.6   Bamboos Conserved in Botanic Gardens and Bambuseta   44
3.7   Dominant Bamboo Genera under Living Collections   108
3.8   What are the Benefits of Bamboo Conservation?   108
3.9   How Should a Modern Bambusetum Function?   109
3.10   Seed and Rhizome Banks   109
3.11   Summary and Conclusion   110

4. Genetic Diversity Assessment and Molecular Markers in Bamboos   117
      Eduardo Ruiz-Sanchez, Jessica Pérez-Alquicira,
      María de la Luz Perez-Garcia and *Miguel Angel García-Martínez*

4.1   Introduction   117
4.2   Molecular Markers Used to Infer Phylogenetic Relationships   118
       in Bamboos
       4.2.1   Bambusoideae   118
       4.2.2   Arundinarieae   126
       4.2.3   Bambuseae   127
       4.2.4   Paleotropical Woody Bamboos   128
       4.2.5   Neotropical Woody Bamboos   129
       4.2.6   Olyreae   130
4.3   Molecular Markers Used to Analyze Genetic Diversity   131
       in Bamboos
       4.3.1   AFLP Markers   134
       4.3.2   ISSR Markers   139
       4.3.3   RAPD Markers   143
       4.3.4   SSR Markers   144
       4.3.5   SNP (Single Nucleotide Polymorphism) Markers   146
4.4   Transcriptomes   147

5. Genetic Diversity Assessment and Molecular Markers in   154
      African Bamboos
      Oumer Abdie and *Muhamed Adem*

5.1   Introduction   154
5.2   Bamboo Cytogenetics, Ploidy Level, and Genome Size   156
5.3   Geographical Distribution and Area Coverage of Bamboo   157
5.4   Importance of Genetic Diversity and Population Structure Study   158
5.5   Molecular Markers Used for the Genetic Diversity and Population 159
       Structure Study of Bamboos Found in Africa
5.6   Simple Sequence Repeat (SSR) Markers for the Genetic Diversity   160
       and Population Structure Study of *Arundinaria alpina* (K. Schum)
5.7   Inter Simple Sequence Repeat (ISSR) Markers for the Genetic   162
       Diversity and Population Structure Study of
       Oxytenanthera abyssinica (A. Rich.)
5.8   Chloroplast (cpDNA) Genes for the Population Genetic Diversity   166
       and Structure Study of *Oxytenanthera abyssinica* (A. Rich.)
5.9   Conclusions   170

Unit 2

6. **Breeding System, Molecular Genetic Map, and Artificial**   **176**
   Hybridization in Woody Bamboos
   PeiTong Dou, TiZe Xia and *HanQi Yang*

   6.1   Introduction   176
   6.2   The Breeding System of Woody Bamboo   176
         6.2.1   Flowering and Fruiting Phenomena of Woody Bamboos   177
                 6.2.1.1   Flowering Cycle   177
                 6.2.1.2   Flowering Types   177
         6.2.2   Breeding System of Woody Bamboos   185
                 6.2.2.1   Inflorescence   185
                 6.2.2.2   Floret   187
                 6.2.2.3   Pollen   187
                 6.2.2.4   Pollination   188
                 6.2.2.5   Mating System   191
                 6.2.2.6   Fruits or Seeds   192
                 6.2.2.7   Seeds Setting Rate   192
   6.3   Genetic Map of Woody Bamboo   193
   6.4   Artificial Hybridization and Application to Woody Bamboo   194
   6.5   Research Prospects   195

7. **Circadian Clock Genes and their Role in Bamboo Flowering**   **202**
   Smritikana Dutta, Sukanya Chakraborty and *Malay Das*

   7.1   Introduction   202
   7.2   Bamboo Flowering: Types and their Ecological Impact   203
   7.3   Photoperiodism in Flowering: What is Known about Bamboo?   204
   7.4   Early Background on Important Photoreceptors Involved   205
         in Flowering
         7.4.1   Red Light Photoreceptors   206
         7.4.2   Blue Light Photoreceptors   206
   7.5   Early Background on Molecular Mechanisms Controlling   206
         the Circadian Clock Regulation of Flowering
         7.5.1   The Central Loop   208
         7.5.2   The Morning Loop   208
         7.5.3   The Evening Loop   208
   7.6   *LHY/CCA1, TOC1/PRR, ZTL, GI* Gene Homologs Identified from   209
         Diverse Angiosperm Plants
   7.7   Circadian Clock in Bamboo Flowering   211
   7.8   Circadian Clock Integrator Gene *CONSTANS* (*CO*) in Bamboo   216
         and other Plants
   7.9   Transcriptional Regulation of *CO*: Model Plants vs. Bamboo 217
   7.10 Post Translational Regulation of *CO*: What is Happening   218
         in Bamboo?
   7.11 Conclusions and Future Perspective   219

8. Genetics of Abiotic Stress Resistance in Bamboos   **228**
   Muhamed Adem and *Oumer Abdie*

   8.1   Introduction   228
   8.2   Bamboo and the Environment   230
   8.3   Bamboo Genome   230
   8.4   Development of Abiotic Stress Resistance in Bamboo   233
   8.5   Stress and Bamboo Shoot Organogenesis   234
   8.6   Abiotic Stresses   236
          8.6.1   Drought and High Temperature   236
          8.6.2   Salinity   238
          8.6.3   Cold/Chilling   239
   8.7   Transgenic Approaches for Enhancing Abiotic Stress Tolerance   240
          in Bamboos
   8.8   Role of Transcription Factors in Abiotic Stress Response   245
   8.9   Effect of Abiotic Stresses on Physiology of Transgenic Lines with   247
          Bamboo Originated Genes
   8.10   Conclusions   247

9. Current Understanding on Major Bamboo Diseases, Pathogenicity,   **256**
   and Resistance Genes
   Sonali Dey, Subhadeep Biswas, Anirban Kundu, Amita Pal and *Malay Das*

   9.1   Introduction   256
   9.2   Distribution of Major Bamboo Diseases and Pests in Asia   257
   9.3   Bamboo Diseases Caused by Fungi   258
   9.4   Bamboo Diseases Caused by Bacteria   266
   9.5   Bamboo Diseases Caused by Virus   267
   9.6   Bamboo Diseases Caused by Insects   270
   9.7   Physical and Chemical Methods of Bamboo Preservation   271
   9.8   Biochemical and Genetic Defence in Bamboo   272
   9.9   The Bamboo Resistance (R)-gene Family and their Potential   273
          in Marker Assisted Selection
   9.10   Conclusion and Future Perspective   274

10. Genome Annotation, *In silico* Tools and Databases with Special   **279**
   Reference to Moso Bamboo
   Hansheng Zhao, Lianfu Chen, Yu Wang, Yinguang Hou, Lei Sun,
   Junwei Gan, Zeyu Fan and *Shanying Li*

   10.1   Introduction   279
   10.2   Genome Annotation   280
          10.2.1   Evaluation of Genome Assembly Quality   280
          10.2.2   Genome Structure Annotation   281
                    10.2.2.1   Repetition Sequence Identification   281
                    10.2.2.2   Annotation of the Coding Gene   281
                    10.2.2.3   Release of Data   282
          10.2.3   Annotation of Gene Functions   282
   10.3   Moso Bamboo Genome Annotation   283

11. Functional Genomics Study of Bamboo Shoot and Rhizome   **286**
Development An Overview
Jiaxiang Zhang, Han Li, Huiming Xu, Xingyan Fang, Gangjian Cao
and *Liuyin Ma*

11.1	Introduction	286
11.2	The Rapid Growth of Bamboo Culms	287
11.3	The Characteristics of Rapid Growth in Bamboos	287
	11.3.1 Rapid-growth Stages	287
	11.3.2 Cell Division or Cell Elongation	288
	11.3.3 Dynamics of Cell Division and Cell Elongation	288
11.4	Transcriptional Regulation of Rapid Growth in Bamboos	288
	11.4.1 Auxin and BR Responses	288
	11.4.2 Hormones in SAM Region	289
	11.4.3 Crosstalk Between Hormones	290
	11.4.4 Transcription Factors	290
	11.4.5 Crosstalk Between TF and Hormones	290
11.5	Post-transcriptional Regulation of Rapid Growth in Bamboos	291
	11.5.1 Kinase	291
	11.5.2 Small RNAs	291
	11.5.3 Alternative Splicing, cis-NATs, and Circular RNAs	292
11.6	Rhizome Development of Bamboos	292
	11.6.1 Characteristics of Bamboo Rhizome	292
	11.6.2 Transcriptional and Post-transcriptional Regulation of Rhizome Development	292
	11.6.2.1 Transcriptional Regulation	292
	11.6.2.2 Post-transcriptional Regulation	293
11.7	Nitrate Signaling in Bamboo Development	293

Unit 3

12. *In vitro* Propagation and Genetic Transformation in Bamboo: **296**
Present Status and Future Prospective
Adla Wasi, Zishan Ahmad, Anwar Shahzad, Sabaha Tahseen, Yulong Ding
Abolghassem Emamverdian and *Muthusamy Ramakrishnan*

12.1	Introduction	296
12.2	Micropropagation	298
	12.2.1 Explant Selection and Disinfection	298
	12.2.2 Nutrient Medium	298
	12.2.3 Regeneration Process	304
	12.2.3.1 Direct Organogenesis	304
	12.2.3.2 Indirect Organogenesis	305
	12.2.4 Somatic Embryogenesis	307
	12.2.5 Acclimatization and Field Transfer	307
12.3	Genetic Transformation and Editing of the Genome in Bamboos	310
12.4	Conclusion and Future Prospective	311

13. Current Status of Bamboo Tissue Culture and Genetic   **320**
   Transformation Technology
   Vidya R Sankar and *Muralidharan Enarth Maviton*

   13.1 Introduction   320
   13.2 Propagation   321
         13.2.1 Vegetative Propagation   321
         13.2.2 Micropropagation   322
   13.3 Limitations   324
         13.3.1 Tissue Browning   324
         13.3.2 Microbial Contamination   325
   13.4 Conventional Strategies for Control of Contamination   325
   13.5 Latent Contamination   326
   13.6 Systemic Acquired Resistance to Control Latent Contamination   327
   13.7 Rooting   327
   13.8 *In vitro* Rhizome Induction   328
   13.9 *In vitro* Flowering   328
   13.10 Application of Modern Technology   329
   13.11 Potential for Photoautotrophic Micropropagation (PAM)   329
   13.12 Liquid Cultures/Bioreactors   330
   13.13 Genetic Transformation for Bamboo Improvement   330
   13.14 Conclusions   332

14. CRISPR/Cas Based Genome Editing and its Possible Implication in   **338**
    Bamboo Research
    Tsheten Sherpa, Khushbu Kumari, Deepak Kumar Jha,
    Manas Kumar Tripathy and *Nrisingha Dey*

   14.1 Introduction   338
   14.2 Different Genome Editing Techniques and Advantages of   340
         CRISPR/Cas9
         14.2.1 Zinc Finger Nuclease (ZFN)   340
         14.2.2 Transcription Activator-like Effector Nucleases (TALEN)   341
         14.2.3 Clustered Regularly Interspaced Short Palindromic   342
                 Repeats (CRISPR)-associated Protein (CRISPR/Cas)
   14.3 Different Methods in CRISPR-mediated Gene Editing   343
         14.3.1 Base Editing through CRISPR   343
         14.3.2 Prime Editing   345
         14.3.3 Gene Knock-out using CRISPR   345
         14.3.4 Gene Knock-in using CRISPR   346
         14.3.5 Transcriptional Control Models   346
         14.3.6 CRISPR interference (CRISPRi)   346
         14.3.7 CRISPR activation (CRISPRa)   347
   14.4 Applications of CRISPR/Cas System in Bamboo   347
         Improvement
   14.5 Challenges in Employing CRISPR/Cas System in Bamboo   348
   14.6 Conclusion   350

15. Genome Editing and Future Scope in Bamboos   **354**
Muthusamy Ramakrishnan, Theivanayagam Maharajan, Zishan Ahmad
Stanislaus Antony Ceasar and *Qiang Wei*

15.1 Introduction   354
    15.1.1   ZFNs   354
    15.1.2   TALENs   355
    15.1.3   CRISPR/Cas   356
15.2 Mechanism and Classifications of CRISPR/Cas System   357
15.3 Characterization of Various Genes by CRISPR/Cas9 System   358
15.4 Genome Editing and Future Scope in Bamboo   359

16. Production of Quality Planting Material in Bamboo   **364**
Muralidharan Enarth Maviton

16.1 Introduction   364
16.2 Flowering Behavior of Bamboo   365
16.3 Bamboo Propagation   365
    16.3.1   Natural Propagation in Bamboo   365
    16.3.2   Vegetative Propagation Methods   366
    16.3.3   Micropropagation   366
16.4 Limitations to Conventional Breeding in Bamboo   366
16.5 Genetic Improvement in Bamboo   367
16.6 Hurdles in Species and Clonal Identification   368
16.7 DNA Barcoding   368
16.8 DNA Fingerprinting   369
16.9 Modern Analytical Techniques   369
16.10 Certification of Quality Planting Material in Bamboo   369
16.11 Identification of Species and Origin of Planting Material   370
16.12 Large Scale Multiplication of Planting Material   371
16.13 Traceability   371
16.14 Morphological and Plant Health Indicators of Planting
    Stock Quality   371
16.15 Accreditation of Bamboo Nurseries   372
    16.15.1   Criteria for Accreditation of Bamboo Nurseries   372
16.16 Quality Parameters   373
16.17 Conclusions   373

Index   **375**

Chapter 1
Introduction

Malay Das,[1,*] *Liuyin Ma,*[2] *Amita Pal*[3] and *Chittaranjan Kole*[4]

> *One would rather omit meat from his diet than omit bamboo from his life.*
> —Su Shi, Chinese poet, AD 1073

Bamboos are a group of plants, which have been adopted to diverse environmental conditions and have enormous ecological as well as commercial utilities. The last decade has observed enormous progress in different areas of bamboo biology. Many of these studies have been published as original research or as a review article in different journals. However, not much effort has been undertaken to compile current knowledge on fundamental (e.g., taxonomy, genetic diversity, and breeding) as well as advanced aspects of bamboo biology (e.g., genetics and genomics). Therefore, this book is designed to help people understand the status of bamboo research worldwide, which ranges from macroscopic bamboo classification and distribution to microscopic molecular biology analyses. This book is best for college students, teachers, and forestry station staff who are interested in bamboo research. This is a theoretical and investigative book designed to provide researchers and scientists with a quick overview regarding bamboo distribution, morphology, genetics, genomics, and breeding.

A major goal of the bamboo research community is to understand the characteristics of bamboo and to translate the research results into the practice of bamboo breeding, cultivation, and management. This book aims to introduce readers to the distribution and preservation of bamboo; what is the molecular basis of bamboo development and stress resistance; and what is the status of knowledge on bamboo breeding. With the help of this book, one can quickly and easily understand

[1] Department of Life Sciences, Presidency University, Kolkata.
[2] Fujian Agriculture and Forestry University, China.
[3] Bose Institute, Kolkata, India.
[4] Prof. Chittaranjan Kole Foundation for Science and Society.
* Corresponding author: malay.dbs@presiuniv.ac.in

the intricacies and high-level on-going scientific research in the field of bamboo. The authors of this book are in Asia, America, and Africa. Contributions of approximately ~ 50 authors from nearly 20 Academic Institutes are presented here. The objective of this book is to invite scientists to write investigative reports or outline the molecular mechanisms of bamboo growth and development, stress resistance, and molecular breeding strategies. Therefore, this book has been compiled based on the important survey resources in bamboo distribution areas or by authors who have significantly contributed to the basic research on bamboo.

There are 16 chapters in this book. The first chapter is an introduction, and the remaining 15 chapters can be divided into three units, each of which consists of several articles. Unit 1 deals with bamboo distribution, germplasm, and genetic diversity; Unit 2 deals with genomics and functional genomics of bamboo development and response to environmental stress; and Unit 3 is on bamboo breeding. As the purpose of this book is to expose the reader to different aspects of bamboo biology, each unit is designed to help in understanding the current state of bamboo research globally.

Unit 1: Distribution, Germplasm and Genetic Diversity of Bamboo. Unit 1 consists of four chapters: Chapters 2–5. The chapters in this unit are mainly differentiated according to the bamboo distribution area. Chapter 2 presents the taxonomy, diversity, distribution, and use of bamboo in northeastern India by authors from the Institute of Bioresources and Sustainable Development (IBSD) in India. Chapter 5 deals with the genetic diversity assessment and molecular markers of African bamboo by authors from Asosa University and University of Madda Walabu in Ethiopia. Chapter 4 summarizes the molecular marker technologies such as restriction site associated DNA sequencing (RAD-seq), Inter-simple sequence repeats (ISSRs), simple-sequence repeats (SSRs), amplified fragment length polymorphisms (AFLPs) and restriction fragment length polymorphism (RFLPs) for the genetic diversity assessment of bamboo by authors at Centro Universitario de Ciencias Biológicas y Agropecuarias, Universidad de Guadalajara in Mexico. Chapter 3 systematically analyses the world's bamboo germplasm resources and highlights the benefits of bamboo conservation by the authors of the Jawaharlal Nehru Tropical Botanical Garden and Research Institute in India. After reading these four chapters, readers will understand the status and uses of molecular marker technologies to assess global bamboo distribution, germplasm conservation and genetic diversity.

Unit 2: Genomic and functional genomics regulation of bamboo development and response to environmental stress. Unit 2 has six chapters: Chapters 6–11. The chapters of this unit are differentiated from each other according to the research topic. The chapters of the second unit are divided into three subsections according to their research themes: Development, Environmental Tolerance, and Genomes. The first part is development, and the book focuses on two of the most fascinating scientific questions in Moso bamboo or tortoise-shell bamboo: the enormous, long flowering cycle (about 60 years) and rapid growth (the top-ten fastest growing species of land plants). Chapters 6 and 7 describe bamboo flowering types, key factors of the circadian clock, and transcriptional and post-transcriptional regulation of CO in bamboo by authors from Chinese Academy of Forestry, Kunming, Yunnan, China and the Presidency University and Jawaharlal Nehru University, New Delhi of India. Chapter 11 mainly introduces the definition of the rapid growth stages of bamboo, the contribution of cell elongation and cell division to the rapid growth process, the transcriptome and post-transcriptional reprogramming during rapid growth of bamboos by the authors from the Fujian Agriculture and Forestry University, China. The second part is the environmental response of bamboo. Giant bamboo with commercial value is limited to tropical and subtropical regions, and it is important to enhance the stress resistance to expand the distribution range of giant bamboo in temperate regions. Chapter 8 reveals the molecular mechanisms of how bamboo responds to abiotic stresses such as drought, heat, salt, and cold, by authors from the Asosa University and the University of Madda Walabu in Ethiopia. Chapter 9 comprises of the distribution of pests and diseases in bamboo, as well as bamboo resistance genes in marker-assisted selection of disease-resistant bamboo by authors from the Presidency University, Ramakrishna Mission Vivekananda Centenary College, and Bose Institute, Kolkata, India. The last part is on genome analyses. Chapter 10 describes genome assembly and annotation tools as well as their application in bamboo genomes by authors from the International Center for Bamboo and Rattan, China. These six chapters will enable the readers to understand the current state of bamboo genome, development, and environmental tolerance at the molecular level.

Unit 3: Bamboo breeding. It covers five chapters: Chapters 12–16. The chapters in this unit are primarily applied in nature. Chapter 12 discusses micropropagation and genetic transformation of bamboo by authors from the Aligarh Muslim University of India and Nanjing Forestry University of China. Chapter 13 describes the tissue culture and genetic transformation of bamboo by authors at the Kerala Forest Research Institute, India. Chapter 14 deals with genome editing techniques such as Zinc Finger Nuclease (ZFN), transcription activator-like effector nucleases (TALEN) and advantage of clustered regularly interspaced short palindromic repeats (CRISPR)-associated protein (CRISPR/Cas) in bamboo by authors from the Institute of Life Sciences, Odisha, India. Chapter 15 also focuses on genome editing methods, but mainly describes the current state of CRISPR/Cas genome editing in Ma bamboo (*Dendrocalamus latoflorus* Munro) by authors from the Nanjing Forestry University and Rajagiri School of Social Sciences. Finally, Chapter 16 discusses on the utility and strategy of producing quality planting material in bamboo and has been contributed by an author from the Kerala Forest Research Institute, India.

Collectively, all these chapters will give the readers an understanding of the current state of bamboo breeding, from traditional hybridization to advanced genome editing for molecular breeding.

The editors hope that the people involved in bamboo research, breeding, cultivation, and policy making will be benifitted by this book. We hope that current gaps that exist in many areas of bamboo biology will be fulfilled soon. We thank all the reviewers, series editor, and the Publisher for constantly cooperating with us to make this Project a reality.

Chapter 2

Taxonomy, Diversity, Distribution, and Use of Bamboos
Special Importance to Selected Four Northeast States of India

Evanylla Kharlyngdoh,[1,2,*] *Bharat G Somkuwar*[1,3]
and *Pulok Kumar Mukherjee*[1]

2.1 Introduction

The selected Northeast region of India is highly rich and diverse in biological diversity and represents the Indo-Burma biodiversity hotspot, which ranks sixth among the 25 global biodiversity hotspots. Among the diverse plant bioresources that the region has, bamboo forms one of the major components of the bioresources in the entire region, many of which are still untapped. Bamboo is an important non timber forest product (NTFP) and meets several commercial, social, environmental, and economic needs (Blowfield et al., 1996; Rao, 1998; Sundriyal et al., 2002). It is also known as a "poor man's timber" and "green gold"' due to its wide range of uses and versatile nature. The present study aims at providing a broad description on taxonomy, distribution, diversity, and uses of bamboos from four selected Northeast states of India, viz., Assam, Manipur, Meghalaya, and Nagaland, following a thorough literature and field surveys.

---

[1] Institute of Bioresources and Sustainable Development (IBSD), Manipur, Takyelpat, Imphal-795001, Manipur, India.

[2] Institute of Bioresources and Sustainable Development (IBSD), Meghalaya Node, 6th Mile, Upper Shillong-793009, Meghalaya, India.

[3] Institute of Bioresources and Sustainable Development (IBSD), Chawanga Road, Nursery Veng, Aizawl, Mizoram, India.

Emails: director.ibsd@nic.in; bgs.ibsd@nic.in

* Corresponding author: evanylla@gmail.com

2.2 Taxonomy

Bamboo belongs to the subfamily *Bambusoideae* under the grass family (*Poaceae*). Its distribution ranges from 46°N– 47°S latitude (Dransfield 1992), and from 0–4000 m a.s.l. altitude in the Himalayas and parts of China (McNeely, 1995). Bamboo thrives in the temperature range from 8.8°C to 36°C and annual precipitation of 1,020–6,350 mm (Huberman, 1959) but some species like *Fargesia frigidis* and *Phyllostachys bissetii* (Zhengyi et al., 2006) can even grow in low temperature up to –20°C (Wang and Shen, 1987). Worldwide, over 70 genera and 1,642 woody and herbaceous bamboo species are classified under three tribes, i.e., Bambuseae (tropical woody bamboo), Arundinarieae (temperate woody bamboo) and Olyreae (herbaceous bamboo) (Liese and Kohl, 2015; Maria et al., 2016). About 39 genera and 500 species are distributed in Asia primarily China (Zhu et al., 1994; Fu, 2000). Over 20 genera and 200 species are found in Southeast Asia (Dransfield and Widjaja, 1995), 20 genera and 429 species from Latin America (Judziewicz et al., 1999), six genera and 20 species from Madagascar, two genera and three species from Australia (Dransfield, 2000; Tran, 2010) and one native species from North America has been reported (Shor, 2002). Africa has three genera with three species (Dransfield and Widjaja, 1995; Fu et al., 2000). Bamboos are naturally distributed in all continents except Antarctica and Europe (McClure, 1993; Liese, 2001). However, since the 19th century, bamboos from China and Japan have been introduced to Europe. At present, over 400 species of bamboo are grown or cultivated in Europe (Gielis and Oprins, 2000).

The first comprehensive work on the bamboos of India was prepared by Gamble (1896) in which he reported 15 genera and 115 species in his book *The Bambuseae of British India*. Camus (1913), Parker (1929), and Blatter (1929) carried out additional taxonomical works on bamboo. Other important publications on bamboo taxonomy in this region are that of Bor (1938, 1940), McClure (1936, 1954, 1966), Bahadur (1979), Dransfield (1980), Soderstrom (1985), Chao and Renvoize (1989), Bennet and Gaur (1990), Tewari (1992), Kumar (1995) and Dransfield and Widjaja (1995). Morphological identification and classification of Indian bamboos were also updated by many researchers, viz., Bahadur and Naithani (1976, 1983), Majumdar (1983), Naithani (1986, 1990a, 1990b, 1993), Bennet (1988, 1989), Seethalakshmi and Kumar (1998) and Muktesh et al. (2001). As per their reports, there are about 23 genera and 136 bamboo species in India. The Northeast region of India harbours 68% of the total bamboo species reported from India. However, bamboos of Northeast region have received very little attention by the workers in the past with few sporadic works published by Shukla (1996). Recent bamboo exploration works in Manipur, Meghalaya, and Nagaland states have led to addition of new species, which were published by Naithani (2007, 2008, 2011), Kumari and Singh (2014) and Naithani et al. (2015).

Scientific euphoria has now centered their attention towards molecular taxonomy of important bamboos, which are extensively on the verge of extinction. Currently, the IUCN Red List of Threatened Plants contains 23 species of bamboo (IUCN, 2015), and all of them are from the Asia Pacific region. Earlier, 25–30 bamboo species from India were believed to be potentially threatened (Tewari, 1992; Banik, 1995;

Biswas, 1995; Subramanian, 1995; Seethalakshmi and Kumar, 1998). However, with changing land-use patterns and human disturbances, this information needs to be scientifically updated and verified to comprehend their current distribution and threat status.

The main purpose of the molecular taxonomy is (a) phylogenetic annotation of the taxa and (b) development of taxonomic key. Two types of molecular taxonomic methods have been widely used: (a) DNA fingerprinting-based methods like RFLP, RAPD, SCARs, AFLP, SSRs, ESTs, Transposon and (b) DNA sequence-based methods-organellar genes, nuclear genes, comparative genomics (Das et al., 2008; Ghosh et al., 2017).

Recent advances in molecular prospecting and mapping of bamboo resources have gained momentum due to the sophisticated tools and techniques of omics (genomics, transcriptomics, proteomics, and metabolomics). Certainly, in the future, bambomics (bamboo grass omics studies) will provide new insights towards understanding key genes that anticipate in regulatory mechanisms of some of the important traits like fast growth, high tensile strength, long fiber, and flowering cycle complexity.

2.3 Distribution and Diversity

A thorough evaluation and analysis of bamboo species from Assam, Manipur, Meghalaya, and Nagaland showed that there are 23 genera and 85 bamboo species, accounting to 62.5% of the country's total bamboo species (Gamble, 1896, 1978; Bor, 1938, 1940; Tewari, 1992; Shukla, 1996; Seethalakshmi and Kumar, 1998; Naithani, 2008, 2011; Kumari and Singh, 2014; Naithani et al., 2015). These species are distributed in the tropical, subtropical, and temperate regions either in the understorey of the natural forests or as pure stands called bamboo brakes. The genera include *Ampelocalamus, Arundinaria, Bambusa, Cephalostachyum, Chimonobambusa, Dendrocalamus, Dinochloa, Drepanostachyum, Gigantochloa, Himalayacalamus, Melocalamus, Melocanna, Neohouzeaua, Neomicrocalamus, Phyllostachys, Pleioblastus, Pseudosasa, Pseudostachyum, Schizostachyum, Sinarundinaria, Thamnocalamus, Thyrsostachys* and *Yushania.*

Regarding growth habit, bamboo is categorized as (i) sympodial or clumping and (ii) monopodial or non-clumping bamboo. Sympodial bamboos contributed over 67% of the total growing stock with *Dendrocalamus strictus* (45%) being the highest followed by *Bambusa bamboos* (13%), *D. hamiltonii* (7.0%), *B. tulda* (5.0%), and *B. pallida* (4.0%). The remaining sympodial bamboo species contributes only 6.0%. Monopodial bamboos such as *Melocanna baccifera, Phyllostachys bambusoides* and *P. mannii*, accounts for nearly 20% of the growing stock and most of it are found in the Northeast region (Naithani, 1993; Tewari et al., 2019).

Among the four selected states, Manipur has the maximum number of 61 species followed by 50 species from Meghalaya (53.8%) (Tables 2.1 and 2.2).

The distribution pattern of bamboo species in these four states is highly variable. Of the total 85 bamboo species, about 38.8% of the species are found only in either one state indicating their restricted distribution, whereas 21.2% within two states, 18.8% within three states, and 21.2% are distributed in all four states (Table 2.2).

Table 2.1. Number of bamboo genera and species in Northeast region of India.

Place	Genera	Species	% Contribution to Total No. of Bamboo Species in the Country	% Contribution to Total No. of Bamboo Species in Regions under Study
Assam	16	45	33.1	52.9
Manipur	18	61	44.9	71.8
Meghalaya	18	50	36.8	58.8
Nagaland	14	36	26.5	42.4
Regions under study	**23**	**85**	**62.5**	
India	**23**	**136**		

Sources: Gamble, 1896; Tewari, 1992; Naithani, 2011; Kumari and Singh, 2014; Naithani et al., 2015.

Table 2.2. Distribution of bamboo species in Northeast states of India.

Sl. No.	Species Name	AS	MN	ME	NG	Total States
1.	*Ampelocalamus patellaris* (Gamble) Stapleton	■			■	2
2.	*Arundinaria racemosa* Munro		■			1
3.	*Bambusa balcooa* Roxb.	■	■	■	■	4
4.	*Bambusa bambos* (L.) Voss	■	■		■	3
5.	*Bambusa binghamii* Gamble		■			1
6.	*Bambusa burmanica* Gamble	■	■			2
7.	*Bambusa cacharensis* R.B. Majumdar	■	■	■		3
8.	*Bambusa griffithiana* Munro	■				1
9.	*Bambusa jaintiana* R.B. Majumdar	■	■	■	■	4
10.	*Bambusa khasiana* Munro	■	■	■		3
11.	*Bambusa kingiana* Gamble		■			1
12.	*Bambusa manipureana* H.B. Naithani & N.S. Bisht		■			1
13.	*Bambusa mizorameana* H.B. Naithani		■			1
14.	*Bambusa multiplex* (Lour.) Raeusch. ex Schults	■	■	■	■	4
15.	*Bambusa multiplex* var. *rivieroerum* Maire			■		1
16.	*Bambusa myanmarnica* Gamble	■	■			2
17.	*Bambusa nutans* Wall. ex Munro	■	■	■	■	4
18.	*Bambusa oliveriana* Gamble		■			1
19.	*Bambusa pallida* Munro	■	■	■	■	4
20.	*Bambusa polymorpha* Munro	■	■		■	3
21.	*Bambusa rangaensis* Borthakur & Barooah	■				1
22.	*Bambusa schizostachyoides* Kurz ex Gamble		■			1
23.	*Bambusa teres* Munro	■	■	■	■	4
24.	*Bambusa tulda* Roxb.	■	■	■	■	4
25.	*Bambusa vulgaris* Schrad.	■	■	■	■	4
26.	*Cephalostachyum capitatum* Munro		■	■	■	3

Table 2.2 contd. ...

...Table 2.2 contd.

Sl. No.	Species Name	AS	MN	ME	NG	Total States
27.	*Cephalostachyum latifolium* Munro		■		■	2
28.	*Cephalostachyum mannii* (Gamble) Stapleton		■	■		3
29.	*Cephalostachyum pallidum* Munro	■				3
30.	*Cephalostachyum pergracile* Munro			■	■	3
31.	*Chimonobambusa callosa* (Munro) Nakai		■			4
32.	*Chimonocalamus griffithianus* (Munro) Hsueh & T.P. Yi		■			4
33.	*Dendrocalamus asper* (Schult.) Backer					1
34.	*Dendrocalamus brandisii* (Munro) Kurz	■				2
35.	*Dendrocalamus calostachyus* (Kurz) Kurz			■		2
36.	*Dendrocalamus giganteus* Munro	■				4
37.	*Dendrocalamus hamiltonii* Nees and Arn. ex Munro	■				4
38.	*Dendrocalamus hookeri* Munro	■				4
39.	*Dendrocalamus latiflorus* Munro	■				2
40.	*Dendrocalamus longispathus* (Kurz) Kurz	■				3
41.	*Dendrocalamus manipureanus* Naithani & Bisht		■			1
42.	*Dendrocalamus membranaceus* Munro	■				2
43.	*Dendrocalamus sahnii* H.B. Naithani Bahadur	■				1
44.	*Dendrocalamus sericeus* Munro	■				1
45.	*Dendrocalamus sikkimensis* ex Oliv.	■				3
46.	*Dendrocalamus strictus* (Roxb.) Nees	■				3
47.	*Dinochloa macclellandii* (Munro) Kurz		■			2
48.	*Drepanostachyum intermedium* (Munro) Keng f.		■			1
49.	*Drepanostachyum khasianum* (Munro) Keng f.			■		2
50.	*Drepanostachyum kurzii* (Gamble) Pandey ex D.N. Tewari	■				3
51.	*Drepanostachyum polystachyum* (Kurz ex Gamble) R.B. Majumdar					1
52.	*Gigantochloa albociliata* (Munro) Kurz	■				2
53.	*Gigantochloa apus* (Schult.) Kurz					1
54.	*Gigantochloa macrostachya* Kurz	■				2
55.	*Gigantochloa nigrociliata* (Buse) Kurz	■				2
56.	*Gigantochloa parvifolia* (Brandis ex Gamble) T.Q. Nguyen					1
57.	*Himalayacalamus falconeri* (Hook. f. ex Munro) Keng f.		■			1
58.	*Himalayacalamus hookerianus* (Munro) Stapleton					1
59.	*Melocalamus compactiflorus* (Kurz) Benth.	■	■			4
60.	*Melocalamus indicus* R.B. Majumdar	■				2
61.	*Melocalamus mastersii* (Munro) R.B. Majumdar	■				1
62.	*Melocanna arundina* C.E. Parkinson	■				1
63.	*Melocanna baccifera* (Roxb.) Kurz	■	■	■	■	1

Table 2.2 contd. ...

...Table 2.2 contd.

Sl. No.	Species Name	AS	MN	ME	NG	Total States
64.	*Melocanna clarkei* (Gamble ex Brandis) P. Kumari & P. Singh					1
65.	*Neohouzeaua helferi* (Munro) Gamble					3
66.	*Neomicrocalamus prainii* (Gamble) Keng f.					3
67.	*Phyllostachys bambusoides* Siebold & Zucc.					2
68.	*Phyllostachys mannii* Gamble					4
69.	*Phyllostachys nigra* (Lodd. ex Lindl.) Munro					1
70.	*Pseudosasa japonica* (Steud.) Makino					1
71.	*Pseudostachyum polymorphum* Munro					4
72.	*Schizostachyum beddomei* (C.E.C. Fisch.) R.B. Majumdar					1
73.	*Schizostachyum dullooa* (Gamble) R.B. Majumdar					4
74.	*Schizostachyum griffithii* (Munro) R.B. Majumdar					3
75.	*Schizostachyum latifolium* Gamble					2
76.	*Schizostachyum mannii* R.B. Majumdar					4
77.	*Sinarundinaria debilis* (Thwaites) C.S. Chao & Renvoize					1
78.	*Thamnocalamus spathiflorus* (Trin.) Munro					1
79.	*Thyrsostachys oliveri* Gamble					1
80.	*Thyrsostachys siamensis* Gamble					1
81.	*Yushania elegans* (Kurz) R.B. Majumdar					3
82.	*Yushania hirsuta* (Munro) R.B. Majumdar					2
83.	*Yushania maling* (Gamble) R.B. Majumdar & Karthik					1
84.	*Yushania microphylla* (Munro) R.B. Majumdar					1
85.	*Yushania rolloana* (Gamble) T.P. Yi					2
Total	**85**	**45**	**61**	**50**	**36**	

Abbreviations: AS: Assam, **MN:** Manipur, **ME:** Meghalaya, **NG:** Nagaland. A colored cell represents presence of species whereas an uncolored cell represents absence of species.

Sources: Gamble, 1896; Tewari, 1992; Naithani, 2011; Kumari and Singh, 2014; Naithani et al., 2015.

2.3.1 Diversity of Edible Bamboo Species

About 35 edible bamboo shoot species are found in the four selected states and these belongs from the genera *Bambusa*, *Cephalostachyum*, *Dendrocalamus*, *Gigantochloa*, *Himalayacalamus*, *Melocanna*, *Phyllostachys*, *Pleioblastus*, *Schizostachyum*, and *Yushania* (Table 2.3). Manipur has the maximum number of edible bamboo species (30) followed by Assam (23), Meghalaya (23), and Nagaland (21) (Table 2.3).

Although the Northeast region has a high diversity of edible bamboo shoot species, only 15 species are available in local markets as young shoots (Table 2.4). The availability of these edible shoots in the local markets of the Northeast region is also highly variable due to species distribution pattern and seasonality. For instance, among the 15 species, the most popular edible shoots available in most of the Northeast states local markets and consumed by communities are *Bambusa tulda*, *Dendrocalamus hamiltonii*, and *Melocanna baccifera*. About nine

Table 2.3. Distribution of edible bamboo species in Northeast states of India.

Sl. No.	Edible Bamboo Species	AS	MN	ME	NG	Total
1.	*Bambusa balcooa*					4
2.	*Bambusa bambos*					3
3.	*Bambusa burmanica*					2
4.	*Bambusa cacharensis*					3
5.	*Bambusa jaintiana*					4
6.	*Bambusa kingiana*					1
7.	*Bambusa mizorameana*					1
8.	*Bambusa nutans*					4
9.	*Bambusa pallida*					4
10.	*Bambusa polymorpha*					2
11.	*Bambusa tulda*					4
12.	*Bambusa vulgaris*					4
13.	*Cephalostachyum capitatum*					3
14.	*Cephalostachyum pergracile*					3
15.	*Chimonobambusa callosa*					4
16.	*Dendrocalamus asper*					1
17.	*Dendrocalamus calostachyus*					2
18.	*Dendrocalamus giganteus*					4
19.	*Dendrocalamus hamiltonii*					4
20.	*Dendrocalamus hookeri*					4
21.	*Dendrocalamus latiflorus*					2
22.	*Dendrocalamus longispathus*					3
23.	*Dendrocalamus manipureanus*					1
24.	*Dendrocalamus membranaceus*					2
25.	*Dendrocalamus sikkimensis*					3
26.	*Dendrocalamus strictus*					3
27.	*Gigantochloa albociliata*					2
28.	*Gigantochloa apus*					1
29.	*Himalayacalamus hookerianus*					1
30.	*Melocanna baccifera*					4
31.	*Phyllostachys bambusoides*					2
32.	*Phyllostachys mannii*					4
33.	*Schizostachyum beddomei*					1
34.	*Schizostachyum dullooa*					4
35.	*Yushania elegans*					3
	Total	**23**	**30**	**23**	**21**	

Abbreviations: AS: Assam, **MN:** Manipur, **ME:** Meghalaya, **NG:** Nagaland. A colored cell represents presence of edible bamboo species whereas an uncolored cell represents absence of species.

Sources: Gamble, 1896; Tewari, 1992; Naithani, 2011; Kumari and Singh, 2014; Naithani et al., 2015; Premlata et al., 2020.

Table 2.4. Availability of edible bamboo shoots in local markets in different Northeast states of India.

Sl. No.	Edible Bamboo Shoot Species Available in Local Markets of NE Region	AS	MN	ME	NG	Total
1.	*Bambusa balcooa*		■	■		2
2.	*Bambusa cacharensis*			■		1
3.	*Bambusa nutans*		■			1
4.	*Bambusa tulda*	■	■	■	■	4
5.	*Cephalostachyum capitatum*		■			1
6.	*Cephalostachyum pergracile*		■			1
7.	*Chimonobambusa callosa*		■			1
8.	*Dendrocalamus giganteus*		■		■	2
9.	*Dendrocalamus hamiltonii*	■	■	■	■	4
10.	*Dendrocalamus hookeri*		■	■		2
11.	*Dendrocalamus latiflorus*		■			1
12.	*Dendrocalamus longispathus*		■			1
13.	*Dendrocalamus membranaceus*				■	1
14.	*Dendrocalamus sikkimensis*		■			1
15.	*Melocanna baccifera*	■	■	■		3
Total		3	13	6	4	
Annual consumption (in tonnes) (Bhatt et al., 2004)		NA	2188	442	442	

Abbreviations: AS: Assam, **MN:** Manipur, **ME:** Meghalaya, **NG:** Nagaland. A colored cell represents the availability of edible bamboo shoots in local markets whereas an uncolored cell represents absence of edible shoots.

Sources: Gamble, 1978; Tewari, 1992; Kanjilal, 1997; Bhatt et al., 2003; Chongtham et al., 2011; Naithani, 2011; Kumari and Singh, 2014; Naithani et al., 2015; Premlata et al., 2020.

edible species, viz., *Bambusa cacharensis, B. nutans, Cephalostachyum capitatum, C. pergracile, Chimonobambusa callosa, Dendrocalamus latiflorus, D. longispathus, D. membranaceus,* and *D. sikkimensis* are sold only in either one Northeast state local markets as young shoots. The remaining three species, viz., *Bambusa balcooa, Dendrocalamus giganteus,* and *D. hookeri* are available only in only two Northeast states local markets (Table 2.4). In terms of species richness and consumption, Manipur sells maximum number of edible shoots (13 species) with an annual consumption of 2,188 tonnes (Bhatt et al., 2004; Chongtham et al., 2011, Nirmala et al., 2014, 2018; Waikhom et al., 2013).

2.3.2 *Diversity of Commercial Bamboo Species*

Tribal communities in the Northeast region have been using over 84 bamboo species for various traditional and commercial applications. Currently only 15 bamboo species have attained commercial importance mainly due to its woody characteristics and these include: *Bambusa balcooa, B. bambos, B. nutans, B. polymorpha, B. tulda, B. vulgaris, Dendrocalamus brandisii, D. hamiltonii, D. hookeri, D. latiflorus, D. membranaceus, D. sikkimensis, D. strictus, Gigantochloa giganteus,* and *Melocanna baccifera* (Table 2.5). These woody bamboo species are mostly used

Table 2.5. Matured culm characteristics of commercial bamboo species available in the Northeast states.

Sl. No.	Species Name	Culm Height (m)	Culm Diam. (cm)	Internode Length (cm)
1.	*Bambusa balcooa*	16.0–24.0	8.0–15.0	20.0–45.0
2.	*Bambusa bambos*	15.0–30.0	10.0–18.0	20.0–45.0
3.	*Bambusa nutans*	6.0–15.0	5.0–10.0	25.0–45.0
4.	*Bambusa polymorpha*	15.0–25.0	7.0–15.0	40.0–60.0
5.	*Bambusa tulda*	7.0–25.0	5.0–10.0	40.0–70.0
6.	*Bambusa vulgaris*	6.0–20.0	5.0–10.0	24.0–45.0
7.	*Dendrocalamus brandisii*	15.0–30.0	5.0–20.0	20.0–40.0
8.	*Dendrocalamus giganteus*	25.0–35.0	15.0–30.0	35.0–45.0
9.	*Dendrocalamus hamiltonii*	10.0–25.0	10.0–20.0	30.0–50.0
10.	*Dendrocalamus hookeri*	15.0–20.0	10.0–15.0	40.0–50.0
11.	*Dendrocalamus latiflorus*	14.0–25.0	8.0–20.0	20.0–70.0
12.	*Dendrocalamus membranaceus*	20.0–24.0	6.0–10.0	20.0–40.0
13.	*Dendrocalamus sikkimensis*	17.0–22.0	12.0–20.0	20.0–45.0
14.	*Dendrocalamus strictus*	6.0–20.0	2.5–8.0	30.0–45.0
15.	*Melocanna baccifera*	10.0–20.0	3.0–7.0	20.0–50.0

Sources: Tewari, 1992; Shukla, 1996; Seethalakshmi and Kumar, 1998; Naithani, 2011; Kumari and Singh, 2014; Naithani et al., 2015.

for construction material, paper and pulp, charcoal, prefabricated bamboo products, furniture making, matting and basketry, handicrafts, household and agricultural utilities, lifestyle products, sports, musical items, and for food consumption and are available throughout the region.

2.4 Area and Resource Stock

Worldwide, the total bamboo forests cover was estimated at 37 million hectares accounting for 3.2% of the world's total forest area (INBAR, 2019). India is second to China in terms of resource stock and varieties of bamboo it has. According to the India State of Forest Report (ISFR, 2021), India holds about 15.0 million hectares of bamboo area. However, this area has decreased by 0.11 million hectares as compared to ISFR 2019. In the Northeast region of India, bamboo covers about 53,485 km² that constitutes about 36% of the country's total bamboo area. In terms of bamboo area in these four selected states, Assam has the highest bamboo area (10,659 km²) followed by Manipur (8,377 km²), Meghalaya (5,007 km²), and Nagaland (3,947 km²).

As per the ISFR (2021) report, the area under pure bamboo stands in the country was recorded as 5,516 km²; while area under dense and scattered bamboo was 29,208 km² and 89,648 km², respectively. The culm density at the national level is 53,336 million out of which green culms contributed 73.4%, dry culms, 17.5%, and decayed culms, 9.1%. The total culm biomass at the national level was 402.0 million tonnes, where green and dry culms contributed 65.6% and 34.4%, respectively.

In these four states, the total area under pure bamboo stands was recorded at 1,266 km² while the area under dense and scattered category was 5,272 km² and

21,157 km², respectively. Pure bamboo stands are largely distributed in Meghalaya (484 km²) and Assam (449 km²), whereas dense and scattered bamboo area is maximum in Assam (2,300 and 7,798 km²) and Manipur (1,450 and 6,766 km²), respectively. The total culm density in these four states is 12,128 million out of which green culms contributed 82.4%, dry culms, 9.9%, and decayed culms 7.8%. The total culm biomass was recorded at 107.1 million tonnes with green culms contributing to 79.3% and dry culms to 20.7%.

2.5 Uses of Bamboo

In the Northeast region of India, over 84 bamboo species including commercial and non-commercial ones reported to have one or more uses for various applications. Bamboos are harvested directly from the private- or community-owned forests, farmlands, and homesteads. Based on the diversity and quality of species available in the region, different bamboo species have similar or diverse utilities in the region (Figs. 2.1–2.5). Different bamboo species used for various traditional and modern utilities in the Northeast states are tabulated in Table 2.6. Although a variety of traditional and modern bamboo products are produced from the region, most of these products still lack value addition concepts. Except for a few engineered bamboo products manufactured in the bamboo-based industries, most of the handicraft items, daily household utilities, and woven products are still produced by artisans at the individual or household level. The harvesting and processing technique continues to be traditional, raw, and unscientific. Nearly 62% of the Northeast Tribal population lack appropriate knowledge of bamboo felling rules and 92% of these artisans used traditional hand tools for primary processing (NCDPD and BCDI, 2010). The value chain of bamboo in the Northeast region suffers a huge challenge starting from harvesting till marketing of the finished products. Therefore, bamboo products with less or no finishing remains low paying and are sold only locally with no systematic marketing channels.

2.5.1 Bamboo as Food

As food, bamboo shoots are being used as traditional food item for centuries in many bamboo-growing countries. However, at present bamboo shoots have become a high value food crop in the global market as well. In the Northeast region, bamboo shoots are traditionally consumed as vegetables either as fresh or fermented. It is also used as a flavoring agent usually as dried or roasted form. The fresh and fermented shoots are also prepared as pickle with chillies or meat. In recent times, through innovation and value addition techniques, new edible bamboo products like bamboo juice, candies, cookies, and bamboo shoot powder are also prepared in Tripura, Assam, and Nagaland (Singhal et al., 2013). About 200 species of bamboo are edible at the global level and 37 of these found in the Northeast region of India. Currently, there is a high demand for processed bamboo shoots both at national and global markets, and China is by far the leading exporter of bamboo shoot products, earning around USD 259 million (INBAR, 2021). In India, particularly the Northeast region, the marketing channel is still very weak and unorganized, and there are not enough processed shoots to even supply to the domestic market despite being the second

Table 2.6. Number of bamboo species used for various traditional and modern utilities in the Northeast Region.

Sl. No.	Category	Sub-category	No. of Bamboo Species Used
1.	Bamboo raw materials for construction	Housing, bridges	43
		Fencing	34
		Roofing/thatching	27
		Scaffolding	24
		Walling	22
		Flooring	8
		Hedge	7
		Raft	6
		Floating timber	5
		Boat Mast	1
		Boat outriggers	1
2.	Bamboo pulp and paper articles	Paper and pulp	22
		Rayon	6
3.	Bamboo furniture and seats	Shelf/racks, chair, stool, tables, beds, desks, computer chair, sofa	25
4.	Bamboo-based panels	Bamboo particle boards, corrugated sheets, roofs, door and wall Panels, plywood, tiles, laminated bamboo lumber	9
5.	Woven bamboo products	Baskets, Cones	52
		Table mats, floor mats, wall mats, roof mats	25
		Ropes	15
		Trays	8
		Handloom	5
		Buckets	2
		Hats	2
		Bamboo screens	1
		Shoes, Sandals	1
6.	Agricultural implements	Farm tools and accessories	39
		Fishing rod/trap	19
		Water pipes	11
7.	Fuel	Firewood, charcoal, briquettes	14
8.	Food	Canned and pickled bamboo shoots, dried bamboo shoots, fermented bamboo shoots, bamboo shoot juice/flavoring agent, bamboo shoot powder	39
		Fodder	19
		Biscuits/cookies	1
9.	Medicine	Ethnomedicinal	12

Table 2.6 contd. ...

...Table 2.6 contd.

Sl. No.	Category	Sub-category	No. of Bamboo Species Used
10.	Bamboo tableware and kitchenware	Cutleries, Utensils, containers	37
		Chopstick	4
		Toothpick	3
		Knife handle	3
11.	Bamboo articles of daily use	Rituals/flag poles	9
		Broom	8
		Umbrella stick	6
		Walking stick	5
		Toothbrush	4
		Incense stick	4
		Comb	4
		Thread spindle	2
		Smoking pipes	2
		Cigarette cover	1
12.	Lifestyle products	Handicrafts, decorative items, toys, bags, wallets, chandeliers, office stationery items like pen, pen-stand, notebook cover, cycle, watch, table lamps, packaging box, jewelery box, hangers, flower vases, photo frames, water bottle, hand fans, bamboo art, sculptures and paintings, statutes, carvings, cycles	58
13.	Ornamental	Planting, landscaping	12
14.	Musical instrument	Flute, wind instrument, guitar, drums	18
15.	Sports items	Bows and arrows	11
		Kite	2
		Quivers	1
		Hockey stick	1
		Javelin	1
		Spear	1

Sources: Gamble, 1978; Tewari, 1992; Shukla, 1996; Kanjilal, 1997; Rao, 1998; Seethalakshmi and Kumar, 1998; Laha, 2000; Bhatt et al., 2003; Singh et al., 2003; Haque and Karmaka, 2004; Bhardwaj and Gahkar, 2005; Kharlyngdoh and Barik, 2008; Sharma and Borthakur, 2008; Singh et al., 2010; Chongtham et al., 2011; Naithani, 2011; Tynsong et al., 2012a, 2012b; Kumari and Singh, 2014; Naithani et al., 2015; Sharma et al., 2018; Singh et al., 2018; Singh, 2019; Manna and Debnath, 2020; Premlata et al., 2020.

largest bamboo resources in the world. In the Northeast region, the annual harvest of bamboo shoot was estimated at 5,685 tonnes (excluding Assam and Sikkim states) and these bamboo shoots are consumed and sold at the local level only with an annual income of 28 million rupees (Bhatt et al., 2004).

Bamboo shoots are a popular source of food and nutrition in most Asian countries including the northeastern and eastern parts of India (Chongtham et al., 2011; Nirmala et al., 2014). They are rich in various macro-nutrients, vitamins, minerals, cellulose,

Fig. 2.1. Traditional bamboo houses of various tribes in Meghalaya (a, c, d, e, f, i), Assam (b), and Nagaland (g, h, j).

essential amino acids, fibees, and bioactive compounds which have nutraceutical benefits (Akao et al., 2004; Choudhury et al., 2011, 2012; Chongtham et al., 2011, Nirmala et al., 2014; Singhal et al., 2013; Waikhom et al., 2013).

2.5.2 *Bamboo in Medicinal, Cosmetic, and Agricultural Uses*

Bamboo is a high value plant, and it has enormous potentiality to enter in the nutraceutical, medicinal, and cosmetic markets due to its reported rich source of natural antioxidants and bioactive compounds (Nirmala et al., 2018). Every part of the plant has different applications that are being used traditionally through many generations in many parts of the Asian countries (Bensky and Borolet, 1990; Piper,

Fig. 2.2. A variety of traditional woven bamboo products processed and crafted by tribal communities in Assam, Manipur, Meghalaya, and Nagaland for daily household and agricultural applications.

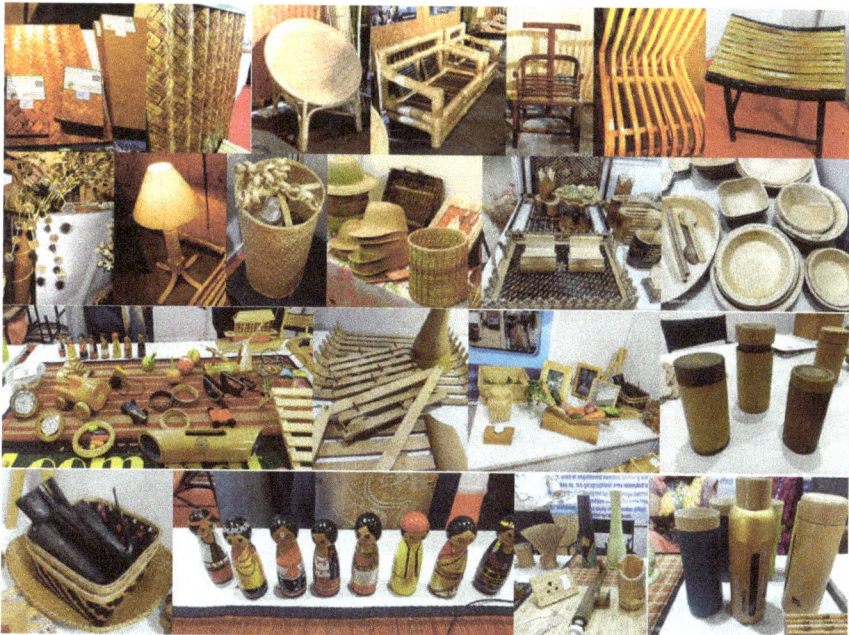

Fig. 2.3. A variety of engineered and modern lifestyle bamboo products processed and manufactured in the Northeast region.

Fig. 2.4. Construction of tree house (a) local bridge (b), heritage village (g–h), and beautification of tourist places (e–f) in different parts of Meghalaya and hosting of festivals (c–d) in Manipur using bamboo as raw material.

1992; Huang and Wang, 1993; Zhang et al., 2005; Tripathi et al., 2009; Liese and Kohl, 2015). Even the siliceous secretion, also known as *Tabasheer/Banslochan/ Bamboo mana* or bamboo biosilica that is found in the hollow internodes of bamboo culms, is being used in traditional medicine for centuries in many parts of the world (Nirmala and Bisht, 2017; Nirmala et al., 2018). Bamboo biosilica has been used since ancient times as a stimulant, astringent, febrifuge, cooling tonic, aphrodisiac, treatment of asthma, cough, and other antispasmodic diseases (Nadkarni, 2005; Prajapati and Kumar, 2005; Khare, 2007; Prajapati et al., 2009; Anjum et al., 2019). *Vanshlochan* is one of the important ingredients of Ayurvedic immune booster recipe *Chyawanprash*. In India, the *Chyawanprash*, an Ayurvedic immune booster, is prepared from a combination of various medicinal herbs, including bamboo biosilica (*Vanshlochan*; Sharma et al., 2019). Currently, Indian *Vanshlochan* is on high demand in the international market (Nirmala et al., 2018). During the recent COVID-19 pandemic, the importance of immune-boosters significantly escalated.

Fig. 2.5. Various fresh and processed (fermented and dried) edible bamboo shoot products of the Northeast region.

Studies have shown that regular consumption of bamboo-based foods may help fight various chronic diseases such as cancer, diabetes, cardiovascular, Alzheimer's, and Parkinson's disease (Muniappan and Sundararaj, 2003; Panee, 2008; Aakruti et al., 2013; Nirmala et al., 2018). Bamboo leaves contains 2–5% flavone and phenolic compounds that help removal of active oxy–free radicals and blood fat (Zhang and Ding, 1997; Lu et al., 2005; Ogunjinmi et al., 2009). Bamboo roots, sprouts and grains are used in the Ayurvedic system of medicine for treatment of many diseases (Chatterjee and Pakrashi, 2009). Burnt bamboo roots are used for the treatment of ring worm, bleeding gums, painful joints, and wounds (Seethalakshmi and Kumar, 1998). During pyrolysis, bamboo vinegar can be obtained from bamboo culms and leaves as by-products along with charcoal (Jiang, 2004). Since it contains > 200 organic substances, bamboo vinegar is widely used for pharmaceutical, cosmetics, beverage, and agricultural applications (Zhang, 1997; Lobovikov et al., 2007; Liese and Kohl, 2015). In Mizoram and Meghalaya, the green glossy surface of *Melocanna baccifera* culm is heated and applied directly on the infected skin to cure wounds (Bhardwaj and Gahkar, 2005). In Meghalaya, leaves of *Bambusa jaintiana* are boiled with other herbs and a decoction of it is prescribed for treatment of Tonsil (Khongsit, 1987). In Manipur, the juice of young *B. tulda* shoots is applied on injured nails and the boiled decoction of fermented shoos is used for treatment of tumors, ringworm, and meningitis, whereas the fruit is believed to enhance fertility issues (Singh et al., 2010). The various bamboo species exhibit medicinal properties are *Ampelocalamus callosa* (sgin disorder), *Bambusa arundinacea* (anti-inflammatory, antiulcer), *B. bambos* (gynecological disorder), *B. vulgaris* (antioxidant, anti-hepatitis, and measles), *B. balcooa* (anti-diabetic), *B. nana* (weed poison), *B. tulda* (antioxidant), *B. blumeana* (antiulcer), *B. oliveriana* (ring worm infection and alopecia), *Cephalostachyum capitatum* (anti-cancer), *Dendrocalamus asper* (antioxidant), *D. latiflorus* (anti-diabetic), *D. strictus* (astringent, Calcium supplement), *Phyllostachys bambusoides* (paracitidial, antioxidant, and skin disease), *P. nigra* (COVID-19 treatment, anti-cancer, anti-inflammatory, cardio-protective), *P. edulis* (antitumor, lipid lowering), *P. heterocycla* (anti-cancer), and *P. pubescens* (skin disorder) (Rastogi and Mehrotra, 1994; Devi and Th. Sobita, 1995; Sinha, 1996; Said, 1997; Maheshwari, 2000; Kimura et al., 2002; Mu et al., 2004; Singh, 2006; Rathour et al., 2022).

2.5.3 Bamboo in Environmental Conservation and Carbon Sequestration

Bamboo plays a crucial role in environmental cleaning, biodiversity conservation, soil and water conservation, and waste purification (Embaye, 2000; Liese, 2001; Hunter and Wu, 2002). It is also useful for climate change mitigation due to its carbon sequestration capabilities whereby it can sequester > 5 tonnes of CO_2 per acre while releasing 35% oxygen to the atmosphere (Atanda, 2015; Nath et al., 2015). Studies on Life-Cycle Assessment (LCA) and carbon footprint analysis of bamboo products from cradle to coffin have shown that items made from bamboo can be carbon-neutral or even carbon-negative if production parameters were optimized (Van der Lugt et al., 2015). As climate issues rises, many companies are pledging zero deforestation to increase green cover and are opting bamboo as a wood replacement for fuel, fiber, furniture, and other products. In 2017, the Parliament of India has amended the "clause (7) of Section 2 of the Indian Forest Act (1927)" to exempt 'bamboo' grown in non-forest areas from the definition of tree. This was a key initiative to promote bamboo cultivation on non-forested land by farmers and private landowners to increase green cover of the country. Bamboo also plays a great role in rehabilitation of degraded land (Embaye et al., 2003). In China, India, and Thailand, bamboo agroforestry models for cultivation on degraded and marginal lands have been successfully developed (Christanty et al., 1996; Takamatsu et al., 1997; Fu et al., 2000; Nath et al., 2009). In the Northeast region, vast degraded unutilized lands such as mining areas can be utilized efficiently for bamboo plantation. This will not only help in rehabilitation of degraded lands but also increase green cover and carbon sequestration.

2.6 Conclusion

The above study reveals the diverse distribution and utility of bamboo species of the Northeast region. Despite its richness only 10–15 woody species are recognized as commercial species that are widely used for various bamboo-based applications. The potential and prospects of the remaining woody and herbaceous bamboo species remains highly unexplored in terms of their distribution and utility. Further research to understand their growth and biomass stock, propagation and management strategies for commercial utilization and conservation needs imminent attention. Commercial bamboo species are mostly harvested from the wild regions, without any sustainable cultivation practices. Therefore, immediate attention for cultivation, harvesting, and proper management of all priority species is needed for continuous supply of high quality, authentic raw material for product development. Except for a few engineered bamboo-based products, most of the bamboo processing and crafts making in the entire region continues to be traditional and hand-made with only a rough finish. Therefore, efforts to improve processing and value addition techniques through scientific interventions need augmentation for economic improvements of the region. The global export and import markets for bamboo-based products are vast and growing rapidly, and the Northeast region of India has the potential for sustainable utilization of bamboo to meet several commercial, social, environmental, and economic needs.

Acknowledgements

The authors are thankful to the Director of the Institute of Bioresources and Sustainable Development (IBSD), Imphal, Manipur for his support and encouragement in publishing the work. The above work was funded by the Department of Biotechnology, Government of India under the IBSD Core Fund.

References

Aakruti, A.K., Swati, R.D. and Vilasrao, J.K. (2013). Overview of Indian medicinal tree: *Bambusa bambos* (Druce). *International Research Journal of Pharmacy*, 4: 52–56.

Akao, Y., Seki, N., Nakagawa, Y., Yi, H., Matsumoto, K., Ito, Y., Ito, K., Funaoka, M., Maruyama, W., Naoi, M. and Nozawa, Y. (2004). A highly bioactive lignophenol derivative from bamboo lignin exhibit a potent activity to suppress apoptosis induced by oxidative stress in human neuroblastoma SH-SY5Y cells. *Bioorganic and Medicinal Chemistry*, 12: 4791–4801.

Anjum, A.A., Tabassum, K. and Ambar, S. (2019). Tabasheer (*Bambusa arundinaceae* Retz.), a plant origin drug of Unani medicine: A review. *Journal of Ayurvedic and Herbal Medicine*, 5: 31–34.

Atanda, J. (2015). Environmental impacts of bamboo as a substitute constructional material in Nigeria. *Case Studies in Construction Materials*, 3: 33–39.

Bahadur, K.N. (1979). Taxonomy of bamboos. *Indian Journal of Forestry*, 2: 222–241.

Bahadur, K.N. and Naithani, H.B. (1976). On a rare Himalayan Bamboo. *Indian Journal of Forestry*, 1: 39–43.

Bahadur, K.N. and Naithani, H.B. (1983). On the identity, nomenclature, flowering, and utility of the climbing bamboo *Melocalamus compactiflorus*. *Indian Forester*, 109: 566–568.

Banik, R.L. (1995). Diversities, reproductive biology and strategies for germplasm conservation of bamboos. pp. 1–22. *In*: *Bamboo and Rattan Genetic Resources and Use*. Proceedings of the First INBAR Biodiversity, Genetic Resources and Conservation Working Group, Singapore, 7–9 November, 1994. International Plant Genetic Resources Institute and International Network for Bamboo and Rattan.

Bennet, S.S.R. (1988). Notes on an exotic bamboo, *Thyrsostachys siamensis* Gamble. *Indian Forester*, 114: 711–713.

Bennet, S.S.R. (1989). The climbing bamboos. *Melocalamus* in India. *Van Vigyan Patrika*, 27: 119.

Bennet, S.S.R. and Gaur, R.C. (1990). *Thirty-seven Bamboos Growing in India*. Forest Research Institute, Dehra Dun, 100 pp.

Bensky, D. and Barolet, R. (1990). *Chinese Herbal Medicine: Formulas and Strategies* (Rev. Edn.), 337 Eastland Press, Seattle, WA.

Bhardwaj, S. and Gahkar, S.K. (2005). Ethnomedicinal plants used by the tribals of Mizoram to cure cuts and wounds. *Indian Journal of Traditional Knowledge*, 4: 7–80.

Bhatt, B.P., Singha, L.B., Singh, K. and Sachan, M.S. (2003). Some commercial edible bamboo species of northeast India: Production, indigenous uses, cost–benefit and management strategies. *Bamboo Science Culture*, 17: 4–20.

Bhatt, B.P., Singha, L.B., Sachan, M.S. and Singh, K. (2004). Commercial edible bamboo species of the North-Eastern Himalayan region, India. Part I: Young shoot sales. *Journal of Bamboo and Rattan*, 3: 337–364.

Blatter, E. (1929). Indian bamboos brought up to date. *Indian Forester*, 55: 541–562; 586–612.

Blowfield, M., Boa, E. and Chandrashekara, U.M. (1996). The role of bamboos in village-based enterprise. pp. 10–21. *In*: Belcher, B., Karki, M. and Williams, T. (eds.). *Bamboo, People, and the Environment*, vol. 4. Socioeconomics and Culture, International Network for Bamboo and Rattan (INBAR). New Delhi.

Bor, N.L. (1938). A list of grasses of Assam. *Indian Forest Record* (n.s. Bot.), 1: 47–102.

Bor, N.L. (1940). *Flora of Assam*. Reprinted 1982. A Von Book Company, Delhi, 480 pp.

Camus, E.G. (1913). *Les Bambusées Monographie Biologie Culture Principaux Usages*. Paris: Paul Lechevalier.

Chao, C.S. and Renvoize, S.A. (1989). A revision of the species described under *Arundinaria* (Graminae) in Southeast Asia and Africa. *Kew Bulletin*, 44: 349–367.

Chatterjee, A. and Pakrashi, S.C. (2009). *The Treatise on Indian Medicinal Plants*. Vol. 6, New Delhi: NISCAIR, CSIR, pp. 50–51.

Chongtham, N., Bisht, M.S. and Haorongbam, S. (2011). Nutritional properties of bamboo shoots: Potential and prospects for utilization as a health food. *Comprehensive Reviews in Food Science and Food Safety*, 10: 15368.

Choudhury, D., Sahu, J.K. and Sharma, G.D. (2011). Bamboo based fermented food products: A review. *Journal of Science and Industrial Research*, 70: 199–203.

Choudhury, D., Sahu, J.K. and Sharma, G.D. (2012). Value addition to bamboo shoots: A review. *Journal of Food Science and Technology*, 49: 407–414.

Christanty, L., Mailly, D. and Kimmins, J.P. (1996). Without bamboo, the land dies: Biomass, litterfall, and soil organic matter dynamics of a Javanese bamboo talun–kebun system. *Forest Ecology and Management*, 87: 75–88.

Das, M., Bhattacharya, S., Singh, P., Filgueiras, T.S. and Pal, A. (2008). Bamboo taxonomy and diversity in the era of molecular markers. *Advances in Botanical Research*, 47: 225–268.

Devi, Dhanapati, L. and Devi, Th. S. (1995). *Ethnobotanical Study of Bamboos in Manipur*. Keishamthong, Irom Pukhri Mapal, Imphal, Manipur, 69 pp.

Dransfield, S. (1980). Bamboo taxonomy in the Indo-Malayan region. pp. 121–130. *In*: Lessard, G. and Chouinard, A. (eds.). *Bamboo Research in Asia*. International Development Research Centre, Canada.

Dransfield, S. (1992). *The Bamboo of Sabah Sabah Forest Records*. Forestry Department, Sabah Malaysia, pp. 1–10.

Dransfield, S. (2000). Woody bamboos (Gramineae–Bambusoideae) of Madagascar. pp. 43–50. *In*: Jacobs, S.W.L. and Everett, J. (eds.). *Grasses Systematics and Evolution*. CSIRO, Melbourne.

Dransfield, S. and Widjaja, E.A. (1995). *Plant Resources of Southeast Asia, No. 7. Bamboos*. Backhuys Publishers, Leiden, Netherlands, 189 pp.

Embaye, K. (2000). The indigenous bamboo forests of Ethiopia: An overview. *Ambio*, 29: 518–521.

Embaye, K., Christersson, L., Ledin, S. and Weih, M. (2003). Bamboo as bioresource in Ethiopia: Management strategy to improve seedling performance (*Oxytenanthera abyssinica*). *Bioresource Technology*, 88: 33–39.

Fu, J. (2000). Moso bamboo in China. *Bamboo Science and Culture: The American Bamboo Society Magazine*, 21: 12–17.

Fu, J. (2001). Chinese Moso bamboo: Its importance. *American Bamboo Society Magazine*, 22: 5.

Fu, M., Xiao, J. and Lou, Y. (2000). *Cultivation and Utilization on Bamboo*. Beijing: China Forestry Publishing House.

Gamble, J.S. (1896). *The Bambuseae of British India*. Annals of Royal Botanic Garden, Calcutta, Vol. 7. 133 pp.

Gamble, J.S. (1978). *The Bambuseae of British India*. Vol. VII. Royal Botanic Garden, Calcutta. M/S Bishen Singh Mahendra Pal Singh Publications, Dehradun.

Gielis, J. and Oprins, J. (2000). *Bamboo for Europe*. EU-FAIR CT1747 Final Individual Report, 65 pp.

Ghosh, J.S., Bhattacharya, S. and Pal, A. (2017). Molecular phylogeny of 21 tropical bamboo species reconstructed by integrating non-coding internal transcribed spacer (ITS1 and 2) sequences and their consensus secondary structure. *Genetica*, 145: 319–333. Doi: 10.1007/s10709-017-9967-9.

Haque, M.S. and Karmakar, K.G. (2004). Potential and economics of Kanak Kaich bamboo (*Bambusa affinis*) cultivation in Tripura. *Indian Forester*, 130: 867–872.

Huang, B. and Wang, Y. (1993). *Thousand Formulas and Thousand Herbs of Traditional Chinese Medicine*, Vol. 2. Heilongjiang Education Press, Harbin.

Hossain, M.F., Islam, M.A. and Numan, S.M. (2015). Multipurpose uses of bamboo plants: A review. *Int. Res. J. Biol. Sci.*, 4: 57–60.

Huberman, M.A. (1959). Bamboo silviculture. *Unasylva*, 13: 36–48.

Hunter, I.R. and Wu, J.Q. (2002). *Bamboo Biomass*. International Network for Bamboo and Rattan Working Paper No. 36.

[INBAR] International Network of Bamboo and Rattan. (2019). *Trade Overview 2017. Bamboo and Rattan Commodities in the International Market*. INBAR: Beijing, China.

[INBAR] International Network of Bamboo and Rattan. (2021). *Trade Overview 2019: Bamboo and Rattan Commodities in the International Market*. INBAR: Beijing, China.

[ISFR] India State of Forest Report. (2019). *Forest Survey of India*, Vol. I. Ministry of Environment Forest and Climate Change, Kaulagarh Road, P.O. IPE Dehradun – 248195, Uttarakhand, India.

IUCN. (2015). *The IUCN Red List of Threatened Species*. Version 2015-4. <https://www.iucnredlist.org>.

Jiang, S. (2004). *Training Manual of Bamboo Charcoal for Producers and Consumers*. Bamboo Engineering Research Center, Nanjing Forestry University, 54 pp.

Judziewicz, E.J., Clark, L.G., Londoño, X. and Stern, M.J. (1999). *American Bamboos*. Smithsonian Institution Press, Washington, D.C., USA.

Kanjilal, U.N. (1997). *Flora of Assam*. Volume V, Gramineae, Omsons publications, New Delhi, pp. 1–57.

Kimura, Y., Suto, S. and Tatsuka, M. (2002). Evaluation of carcinogenic/co-carcinogenic activity of chikusaku-eki, a bamboo charcoal by-product used as a folk remedy, in BALB/c 3T3 cells. *Biological and Pharmaceutical Bulletin*, 25: 1026–1029.

Khare, C.P. (2007). *Indian Medicinal Plants*. Rajkamal Electronic Press, Delhi, 8081 pp.

Kharlyngdoh, E. and Barik, S.K. (2008). Diversity, distribution pattern and use of bamboos in Meghalaya. *Journal of Bamboo and Rattan*, 7: 73–90.

Khongsit, S. (1987). *Hangne Tangia U Siej. Swila Khongngain*, Shillong.

Kleinhenz, V. and Midmore, D.J. (2001). Aspects of bamboo agronomy. *Advances in Agronomy*, 74: 99–145. Academic Press, New York, USA.

Kumar, M. (1988). Reed bamboos (*Ochlandra*) in Kerala: Distribution and management. pp. 39–43. *In*: Rao, I.V.R., Gnanaharan, R. and Sastry, C.B. (eds.). *Bamboos Current Research*. Proceedings of the International Bamboo Workshop, 14–18, November, Cochin. Kerala Forest Research Institute, Peechi and International Development Research Centre, Canada.

Kumar, M. (1995). A re-investigation on the taxonomy of the genus *Ochlandra* (Thw.) (Poaceae - Bambusoidea). *Rheedea*, 5: 63–89.

Kumari, P. and Singh, P. (2014). *Bamboos of Meghalaya*. BSI publications, 302 pp. ISBN: 9788181770561.

Laha, R. (2000). Bamboo uses for housing by the different tribes of Northeast India. *Journal of the American Bamboo Society*, 14: 10–14.

Liese, W. (2001). Advances in bamboo research. *Journal of Nanjing Forestry University*, 25: 1–6.

Liese, W. and Kohl, M. (2015). *Bamboo: The Plant and its Uses*. Springer International Publishing, 362 pp. Doi: 10.1007/978-3-31914133-6. ISSN 1614-9785.

Lobovikov, M., Paudel, S., Piazza, M., Ren, H. and Wu, J. (2007). *Non–wood Forest Products: A Thematic Study Prepared in the Framework of the Global Forest Resources Assessment 2005* (18. World bamboo resources). Food and Agriculture Organization of the United Nations, Rome, Italy, 73 pp.

Lu, B., Wu, X., Tie, X., Zhang, Y. and Zhang, Y. (2005). Toxicology and safety of antioxidant of bamboo leaves. Part I: Acute and sub-chronic toxicity studies on antioxidant of bamboo leaves. *Food and Chemical Toxicology*, 43: 783–792.

Majumdar, R. (1983). Three new taxa of Indian bamboos. *Bulletin Botanical Survey of India*, 25: 235–238.

Maheshwari, J.K. (2000). *Ethnobotany and Medicinal Plants of Indian Subcontinent*. Scientific Publishers, Jodhpur, India, pp. 1–672.

Manna, K. and Debnath, B. (2020). Phytochemicals, nutritional properties and anti-oxidant activity in young shoots of *Melocanna baccifera*, a traditional vegetable of Northeast India. *Journal of Food, Agriculture and Environment*, 18: 59–65.

Maria, S.V., Lynn, G.C., Dransfield, J., Govaerts, R. and Baker, W.J. (2016). *World Checklist of Bamboo and Rattans*. INBAR Technical Report No. 37, 454 pp.

McClure, F.A. (1936). The bamboo genera, *Dinochloa* and *Melocalamus*. *Kew Bulletin*, 251–254.

McClure, F.A. (1954). The taxonomic conquest of the bamboos with notes on their silviculture at status in America. *Proceedings of the IV World Forestry Conference*, Dehra Dun, 3: 410–414.

McClure, F.A. (1966). *The Bamboos: A Fresh Perspective.* Harvard University Press, Cambridge, Massachusetts, 347 pp.

McClure, F.A. (1993). *The Bamboos.* Smithsonian Institution Press, Washington and London, 368 pp.

McNeely, J.A. (1995). Bamboo biodiversity and conservation in Asia: Biodiversity and genetic conservation. INBAR. Vol. 2. *In*: Rao, V.R. and Ramanuja Rao, I.V.R (eds.). *Bamboo, People, and the Environment.* Proceedings of the 5th International Bamboo Workshop and the 4th International Bamboo Congress, Ubud, Bali, Indonesia.

Mu, J., Uehara, T., Li, J. and Furuno, T. (2004). Identification and evaluation of antioxidant activities of bamboo extracts. *Forestry Studies in China*, 6: 1–5.

Muktesh, K., Remesh, M. and Sequiera, S. (2001). Field identification key to the native bamboos of Kerala, India. Bamboo Science & Culture. *The Journal of the American Bamboo Society*, 15: 35–47.

Muniappan, M. and Sundararaj, T. (2003). Anti-inflammatory and antiulcer activities of *Bambusa arundinacea. Journal of Ethnopharmacology*, 88: 161–167.

Naithani, H.B. (1986). *Pleioblastus simonii* (Carriere) Nakai: A bamboo new to India from Arunachal Pradesh. *Indian Forester*, 112: 85–87.

Naithani, H.B. (1990a). Two new combinations of bamboos. *Indian Forester*, 116: 990–991.

Naithani, H.B. (1990b). Nomenclature of Indian species of *Oxytenanthera Munro. Journal of the Bombay Natural History Society*, 87: 439–440.

Naithani, H.B. (1993). *Dendrocalamus somdevai*: A new species of bamboo from Uttar Pradesh, India. *Indian Forester*, 119: 504–506.

Naithani, H.B. (1993). *Contributions to the Taxonomic Studies of Indian Bamboos.* PhD. Thesis, H.N.B. Garhwal University, Srinagar, Garhwal.

Naithani, H.B. (2007). *Survey Report on the Distribution of Bamboo Species in Meghalaya, India.* Study sponsored by Department of Forest and Environment, Government of Meghalaya, Shillong – 22, India.

Naithani, H.B. (2008). Diversity of Indian bamboos with special reference to Northeast India. *The Indian Forester*, 134: 756–788.

Naithani, H.B. (2011). *Bamboos of Nagaland.* NEPED and NBDA, ISBN: 978-81-904539-1-2, 206 pp.

Naithani, H.B., Bisht, N.S. and Singsit, S. (2015). *Bamboo Species of Manipur.* National Bamboo Mission (NBM), Manipur, Forest Department, Government of Manipur, 88 pp.

Nadkarni, K.M. (2005). *Indian Plants and Drugs.* New Delhi, Srishti Book Distributors, 50 pp.

Nath, A.J., Lal, R. and Das, A.K. (2015). Managing woody bamboos for carbon farming and carbon trading. *Global Ecology and Conservation*, 3: 654–663.

Nath, S., Das, R., Chandra, R. and Sinha, A. (2009). *Bamboo Based Agroforestry for Marginal Lands with Special Reference to Productivity, Market Trend and Economy.* Institute of Forest Productivity, Ranchi, Jharkhand, 26 pp.

National Bamboo Mission (NBM). (2021). *National Consultation on Opportunities and Challenges for Bamboo in India.* Concept Note, 56 pp.

NCDPD (National Centre for Design and Product Development) and BCDI (Bamboo and Cane Development Institute). (2010). *Feasibility Study for the Identification and Promotion of Commercially Viable Technologies for Product Development of the Value Addition of Bamboo & Cane-Based Products from North-east Region of India.* Study conducted under the R&D Scheme of the Development Commissioner (Handicrafts).

Nirmala, C. and Bisht, M.S. (2017). 10th WBC Reports: Bamboo: A prospective ingredient for functional food and nutraceuticals. *Bamboo Journal*, 30: 82–99.

Nirmala, C., Sheena, H. and David, E. (2011). Bamboo shoots: A rich source of dietary fibres. pp. 173–187. *In*: Klein, F. and Moller, G. (eds.). *Dietary Fibres, Fruit and Vegetable Consumption and Health.* Nova Science Publishers, USA.

Nirmala, C., Bisht, M.S. and Laishram, M. (2014). Bioactive compounds in bamboo shoots: Health benefits and prospects for developing functional foods. *International Journal of Food Science & Technology*, 49: 1425–1431.

Nirmala, C., Bisht, M.S., Bajwa, H.K. and Santosh, O. (2018). Bamboo: A rich source of natural antioxidants and its applications in the food and pharmaceutical industry. *Trends in Food Science & Technology*, 77: 91–99.

Ogunjinmi, A.A., Ijeomah, H.M. and Ayieloja, A.A. (2009). Socio-economic importance of bamboo (*Bambusa vulgaris*) in Borgu local government area of Niger state, Nigeria. *Journal of Sustainable Development in Africa*, 10: 284–298.

Panee, J. (2008). Bamboo extract in the prevention of diabetes and breast cancer. pp. 159–177. *In*: Watson, R.R. (ed.). *Complementary and Alternative Therapies and the Aging Population: An Evidence Based Approach*. Elsevier, San Diego.

Parker, R.N. (1929). The Indian bamboos brought up to date (additional notes). *Indian Forester*, 55: 612–613.

Piper, J.M. (1992). *Bamboo and Rattan: Traditional Uses and Beliefs*. Oxford University Press, Oxford.

Prajapati, N.D. and Kumar, U. (2005). *Agro's Dictionary of Medicinal Plants*. Shyam Printing Press, Jodhpur, 44 pp.

Prajapati, N.D., Purohit, S.S. and Sharma, A.K. (2009). *A Handbook of Medicinal Plants*. 1st Edition. Agrobios, Jodhpur, 82 pp.

Premlata, T., Sharma, V., Bisht, M.S. and Nirmala, C. (2020). Edible bamboo resources of Manipur: Consumption pattern of young shoots, processing techniques and their commercial status in the local market. *Indian Journal of Traditional Knowledge*, 19: 73–82.

Radhapyri Devi Oinum and Singh, I. (1990). *Activity of Bamboo Shoot Ferment on Dimethyl-Amino-Azobenzene-induced Carcinogenesis in Fish*. PhD Thesis, Dept. of Life Sciences, Manipur University.

Rao, A.N. (1998). Genetic diversity of woody bamboos: Their conservation and improvement. *In*: Rao, A.N. and Rao, R. (eds.). *Bamboo Conservation, Diversity, Ecogeography, Resource Utilization and Taxonomy*. Proceedings of a workshop held in May 1998, Yunnan, China.

Rastogi, R.P. and Mehrotra, B.N. (1990). *Compendium of Indian Medicinal Plants*. Central Drug Research Institute.

Rathour, R., Kumar, H., Prasad, K., Anerao, P., Kumar, M., Kapley, A., Pandey, A., Kumar Awasthi, M. and Singh, L. (2022). Multifunctional applications of bamboo crop beyond environmental management: An Indian prospective. *Bioengineered*, 13: 8893–8914.

Said, M. (1997). *Hamdard Pharmacopeia of Eastern Medicine*. Sri Satguru Publication, Delhi, 65 pp.

Scurlock, J., Dayton, D. and Hames, B. (2000). Bamboo: An overlooked biomass resource? *Biomass Bioenergy*, 19: 229–244.

Seethalakshmi, K.K. and Kumar, M.S.M. (1998). *Bamboos of India: A Compendium*. Kerala Forest Research Institute, Peechi and International Network for Bamboo and Rattan, Beijing, People's Republic of China, 342 pp.

Sharma, B.K., Bhat, M.H., Fayaz, M., Kumar, A. and Jain, A.K. (2018). Diversity, distribution pattern and utilization of bamboos in Sikkim. *NeBIO*, 9: 18–27.

Sharma, R., Martins, N., Kuca, K., Chaudhary, A., Kabra, A., Rao, M.M. and Prajapati, P.K. (2019). Chyawanprash: A traditional Indian bioactive health supplement. *Biomolecules*, 9: 161. Doi:10.3390/biom9050161.

Sharma, T.P. and Borthakur, S.K. (2008). Ethnobotanical observations on Bamboos among Adi tribes in Arunachal Pradesh. *Indian Journal of Traditional Knowledge*, 7: 594–597.

Shor, G. (2002). *Species Source List, No. 22*. Portland: American Bamboo Society.

Shukla, U. (1996). *The Grasses of North-Eastern India*. Scientific Publishers, Jodhpur, pp. 165–229.

Sinha, S.C. (1996). *Medicinal Plants of Manipur*. Manipur Association for Science and Society, Imphal, pp. 1–202.

Singh, H.B., Kumar, B. and Singh, R.S. (2003). Bamboo resources of Manipur: An overview for management and conservation. *Journal of Bamboo and Rattan*, 2: 43–55.

Singh, K.P., Devi, S.P., Devi, K.K., Ningombam, D.S. and Athokpam, P. (2010). *Bambusa tulda* Roxb. in Manipur State, India: Exploring the local values and commercial implications. *Notulae Scientia Biologicae*, 2: 35–40.

Singh, M.K. (2019). Ethno-medico-botanical observations of bamboos among Indigenous people of Manipur. *Journal of Medicinal Plants Studies*, 7: 159–162.

Singh, M.K., Ibrahim, M., Bordoloi, S., Meena, D. and Kakoti, S. (2018). Bamboo diversity, distribution and utility in forest fringe villages of Manipur (India). *The Pharma Innovation Journal*, 7: 503–507.

Singh, S.S. (2006). *The Economic Plants of Manipur and their uses*. S. Chandani Devi, Moirangkhomp, Imphal, Manipur, 96 pp.

Singhal, P., Bal, L.M., Satya, S., Sudhakar, P. and Naik, S.N. (2013). Bamboo shoots: A novel source for food and medicine. *Critical Reviews in Food Science Nutrition*, 53: 517–534.

Soderstrom, T.R. (1985). Bamboo systematics: Yesterday, today, and tomorrow. *Journal of American Bamboo Society*, 6: 4–16.

Soumya, V., Muzib, Y.I. and Venkatesh, P. (2016). A novel method of extraction of bamboo seed oil (Bambusa bambos Druce) and its promising effect on metabolic symptoms of experimentally induced polycystic ovarian disease. *Indian Journal of Pharmacology*, 48: 162–167.

Subramaniam, K.N. (1995). Bamboo genetic resources in India. *In*: Vivekanandan, K., Rao, A.N. and Ramanatha, R.V. (eds.). *Bamboo and Rattan Genetic Resources in Certain Asian Countries*. Serdang, IPGRI-APO, 229 p.

Sundriyal, R.C., Upreti, T.C. and Varuni, R. (2002). Bamboo and cane resource utilization and conservation in the Apatani plateau, Arunachal Pradesh, India: Implications for management. *Journal of Bamboo and Rattan*, 1: 205–246.

Takamatsu, T., Kohno, T. and Ishida, K. (1997). Role of the dwarf bamboo (*Sasa*) community in retaining basic cations in soils and preventing soil acidification in mountainous area of Japan. *Plant and Soil*, 192: 167–179.

Tewari, D.N. (1992). *A Monograph on Bamboo*. International Book Distributors, Indian Council of Forestry Research and Education, Dehradun, India. 495 pp.

Tewari, S., Negi, H. and Kaushal, R. (2019). Status of Bamboo in India. *International Journal of Economic Plants*, 6: 30–39.

Tran, V.H. (2010). *Growth and Quality of Indigenous Bamboo Species in the Mountainous Regions of Northern Vietnam*. PhD dissertation, Faculty of Forest Science and Forest Ecology, Georg–August–University, Göttingen.

Tripathi, Y.C., Khawlhring, L. and Vasu, N.K. (2009). Traditional and contemporary medicinal applications of bamboo. pp. 242–232. *In*: Nath, S., Singh, S., Sinha, A., Das, R. and Krishnamurty, R. (eds.). *Conservation and Management of Bamboo Resources IFP*, Ranchi.

Tynsong, H., Dkhar, M. and Tiwari, B.K. (2012a). Traditional knowledge-based management and utilization of bio-resources by War Khasi tribe of Meghalaya, North-east India. *Indian Journal of Innovations and Developments*, 1: 162–174.

Tynsong, H., Tiwari, B.K. and Dkhar, M. (2012b). Traditional knowledge associated with traditional harvesting bird harvesting of War Khasi Meghalaya. *Indian Journal of Traditional Knowledge*, 11: 334–341.

Van Der Lugt, P., Vogtländer, J.G., Van Der Vegte, J.H. and Brezet, J.C. 2015. Environmental assessment of industrial bamboo products: Life cycle assessment and carbon sequestration. *Proceedings in 10th World Bamboo Congress*, pp. 1–16.

Waikhom, S.D., Louis, B., Sharma, C.K., Kumari, P., Somkuwar, B.G., Singh, M.W. and Talukdar, N.C. (2013). Grappling the high altitude for safe edible bamboo shoots with rich nutritional attributes and escaping cyanogenic toxicity. *BioMed Research International*, ID 289285, 11 pp. https://doi.org/10.1155/2013/289285.

Wang, D. and Shen, S.J. (1987). *Bamboos of China*. Timber Press, Portland, Oregon. 176 pp.

Zhang, Y. (1997). Bio-antioxidative activity of functional factors in bamboo leaf. *In*: Fu, M.Y. and Lou, Y.P. (eds.). *Paper Summaries. International Bamboo Workshop, Bamboo Towards 21st Century*, 7–11 September 1997. Research Institute of Subtropical Forestry, Chinese Academy of Forestry, Anji, Zhejiang, People's Republic of China.

Zhang, Y. and Ding, X.L. (1997). Existence of specific amino acid in bamboo leaves and its biological significance. *Journal of Wuxi University of Light Industry*, 16: 29–32.

Zhang, Y., Bao, B.L., Lu, B.Y., Ren, Y.P., Tie, X.W. and Zhang, Y. (2005). Determination of flavone C glucosides in antioxidant of bamboo leaves (AOB) fortified foods by reversed-phase high performance liquid chromatography with ultraviolet diode array detection. *Journal of Chromatography*, 1065: 177–185.

Zhengyi, W., Raven, P.H. and Deyuan, H. (2006). *Flora of China, Volume 22: Poaceae*. Beijing and St. Louis, MO. Science Press and Missouri Botanical Garden.

Zhu, S., Ma, N. and Fu, M. (1994). *A Compendium of Chinese Bamboo*. China Forestry Publication House, Beijing, pp. 241.

Chapter 3
Germplasm Resources of Bamboos

KC Koshy

3.1 Introduction

Bamboos enjoy a unique position among the flowering plants because of their fast growth, predominantly monocarpic nature, versatile use, economic potential, and aesthetic appeal. All parts of the bamboo plant are used, viz., leaves for fodder, shoots for human consumption, and culms for construction, textiles, musical and agricultural instruments, paper, and pulp making. They are widely employed in ornamental gardening. Bamboo has become very popular in the last four decades, largely due to its potential in meeting present-day economic, environmental and social needs as a sustainable and renewable resource (INBAR, 2014; Liese and Köhl, 2015). Nowadays, the number of bamboo products has increased immensely with the backing of cutting-edge techniques. Bamboo-based industry employs about 2.5 billion people in the

> Definitions:
>
> **Germplasm**
> The genetic material of an individual that may be transmitted, sexually or somatically, from one generation to another, and is represented by a species, population, landrace, hybrid, or cultivar (Offord, 2017).
>
> **Botanic gardens**
> Botanic gardens are institutions holding documented collections of living plants for the purposes of scientific research, conservation, display, and education (Wyse Jackson, 1999).
>
> **Bambusetum (*pl.* bambuseta)**
> Any area, small or large, holding documented collections of living bamboo plants for scientific research, conservation, display and educational purpose. It may form either part of a botanic garden or an arboretum.

production and allied sectors (Scurlock et al., 2000). All these impelled wide cultivation of bamboos on plantation basis and in public and private gardens. As a result, there is

Formerly Senior Principal Scientist and Head, Plant Genetic Resources Division, Jawaharlal Nehru Tropical Botanic Garden and Research Institute, Palode, Thiruvananthapuram, 695562, Kerala, India.
Email: koshykc1@gmail.com

an upsurge of bamboo gardens, especially in North America, Europe and Southeast Asian countries. Most of them are intended for commercial purposes and not meant for scientific research. However, these gardens have utilitarian goals such as improvement of air quality, control of erosion or replenishment of soil (Volis, 2019). These establishments inadvertently indulge in bamboo conservation, introducing many species to new areas outside their native domain (Hunter, 2003; Canavan et al., 2017).

Bamboos belong to the grass family Poaceae under the subfamily Bambusoideae which comprise three tribes: Arundinarieae (temperate woody bamboos, 31 genera and 581 species), Bambuseae (tropical woody bamboos, 73 genera and 966 species) and Olyreae (herbaceous bamboos, 21 genera and 123 species) totalling to 1,670 species in 125 genera (Soreng et al., 2017). They are distributed across 122 countries (Canavan et al., 2017). Conservation of germplasm of bamboos, as in the case of other plants, materializes in their natural locations (*in situ*) or in botanic gardens (*ex situ*), seedbanks, and tissue repositories.

3.2 Bamboo Gardens and Their Coordinating Organizations

Bamboos are normally conserved as living collections in botanic gardens. There are two main organizations working for the promotion of bamboo sector, the International Bamboo and Rattan Organisation (INBAR) and the World Bamboo Organization (WBO). INBAR is an intergovernmental development organization and its mission is to improve the well-being of producers and users of bamboo and rattan. Since its founding in 1997, with headquarters in China, it has the presence of 48 Member States with Regional Offices in Cameroon, Ecuador, Ethiopia, Ghana, and India (https://www.inbar.int/about-inbar/). WBO, founded in 2005, with headquarters in Antwerp, Belgium, is a diverse group consisting of individual people, commercial businesses, non-profit associations, institutions, and allied trade corporations that share a common interest 'Bamboo'. Originally founded as the International Bamboo Association (IBA) in 1992, its primary responsibility was to conduct the International Bamboo Congress and International Bamboo Workshop. Currently, these two meetings are combined into one: the World Bamboo Congress, which is organized by the World Bamboo Organization (https://worldbamboo.net/). The World Bamboo Day celebrations on 18 September every year since 2009 has helped in the popularization and cultivation of bamboos in a big way. The activities of both these organizations have become the driving force to promote conservation, propagation, and industrial application of bamboos in many countries.

This chapter addresses the following issues: (1). How many gardens are involved in conservation of bamboos? (2). How many species out of the estimated 1,670 species occurring in the world have been conserved? (3). What are the dominant bamboo genera under the living collections? (4). What are the benefits of bamboo conservation? (5). How should a modern bambusetum function?

3.3 Botanic Gardens having Living Bamboo Collections

Among the 3,026 known botanic gardens spread over 175 countries in the world, 139 gardens distributed in 37 countries have established bambuseta (Table 3.1;

Table 3.1. Bambuseta in the World.

Continent	Country	Name of Garden/Institution	Details	References
AFRICA	Algeria	Jardin Botanique du Hamma Rue Hassiba Ben Bouali - B.P. 141 Hamma District, Algiers	1832 Collections include bamboos	BGCIa
ASIA	Bangladesh	Bangladesh Agricultural University Botanic Garden Department of Crop Botany, Mymensingh 2202	1962 Special collections include bamboos	BGCIa
ASIA	Bangladesh	Forest Research Institute Chittagong (BFRI)	1983 1.5 ha bambusetum holds 27 taxa (BFRI, Table 3.2)	Islam, 2003; Wong, 2004
ASIA	Bangladesh	Mirpur National Botanic Garden Dhaka	1961 The live collection includes bamboos	BGCIa; Wong, 2004
ASIA	Brunei	Brunei Forestry Centre, Simpang 50, Jalan Labi, Sungai Liang, Belait District, Brunei Darussalam	1993 Bambusetum contains 17 species	URL, 1; Wong, 2004
ASIA	China	Anji Bamboo Botanic Garden (ABG) HMQ6+F8G, Anji County Huzhou, Zhejiang, China	1974 Area 80 ha., the largest bamboo garden in Asia with 211 species (ABG, Table 3.2)	Maoyi, 1999; URL, 2, 3
ASIA	China	Arboretum of Shandong College of Forestry (Shangdong) Forestry Department of Shandong Province, Northern Gate, Taian Shandong	1956 Special collection includes bamboos	BGCIa
ASIA	China	Fairy Lake Botanical Garden (SZBG) Shenzhen & Chinese Academy of Sciences 160 Xianhu Road, Liantang, Luohu District Shenzhen, Guangdong 518004	1988 In SZBG, 131 bamboo species are conserved (SZBG, Table 3.2)	BGCIa; Huang, 2018
ASIA	China	Fushan Botanic garden P.O. Box 132, Yilan, 260 Taiwan	1990 Living collection include bamboos	BGCIa; Lucas, 2013
ASIA	China	Guangxi Institute of Botany (GXIB) Chinese Academy of Sciences Yanshan, Guilin	1935 GXIB bambusetum conserves129 species (GXIB, Table 3.2)	Huang, 2018

Table 3.1 contd. ...

...*Table 3.1 contd.*

Continent	Country	Name of Garden/Institution	Details	References
ASIA	China	Hangzhou Botanical Garden (HBG) Taoyuanling, Yuquan Hangzhou City, Zhejiang Province	1956 HBG conserves over 100 bamboo species in 14.5 ha. of which 7 species are listed (HBG, Table 3.2)	BGCIa
ASIA	China	Hainan Botanical Garden of Tropical Economic Plants, Institute of Tropical Horticulture South China Academy of Tropical Crops, Baodao Xincun, Danxian County, Hainan	1958 Special collection include bamboos	BGCIa; Maoyi, 1999
ASIA	China	Hong Kong Zoological and Botanical Gardens, Albany Road, Central Hong Kong	Living collection include bamboos	BGCIa; Lucas, 2013
ASIA	China	Hunan Botanic Garden, (*Hunan Forest Botanical Garden*), Dongjing Town, Yuhua District, Changsha, Hunan 410116	1985 Well-known because of large number of bamboo species grown in this garden	BGCIa; Maoyi, 1999
ASIA	China	Jiangsu Institute of Botany (CNBG) No. 39, East Beijing Road, Nanjing 210008	1929 CNBG has a specialized bamboo garden, 10 species are listed (CNBG, Table 3.2)	Huang, 2018
ASIA	China	Jinan Quancheng Park, Parks and Landscape Bureau of Jinan, No. 106 Jingshi Road, Jinan Shandong	1986 Special collection includes bamboos	BGCIa
ASIA	China	Jinyunshan Botanical Garden (Chongqing) Jinyunshan, Jialing River, Beibei District Chongqing, Sichuan, 630702, China	1985 Special collection includes bamboos	BGCIa
ASIA	China	Institute of Botany (IBCAS), the Chinese Academy of Sciences Add: No. 20 Nanxincun, Xiangshan, Beijing	IBCAS, one of the oldest comprehensive research institutes in China, also holds a bamboo collection (IBCAS, Table 3.2)	Huang, 2018
ASIA	China	Kunming Institute of Botany (KIB) Chinese Academy of Sciences No. 132, Lanhei Road, Kunming 650201 Yunnan	KIB bambusetum harbours 20 species (KIB, Table 3.2)	Huang, 2018

ASIA	China	Lushan Botanical Garden (LSBG) Jiangxi Province and Chinese Academy of Sciences, P.O. No. 9 Zhiqing Road, Lushan, Jiangxi	Bambusetum of LSBG holds 13 species (LSBG, Table 3.2)	Huang, 2018
ASIA	China	Shanghai Botanical Garden No. 1111, Longwu Road Shanghai 200231	1974 Special collection includes bamboos	BGCIa; Lucas, 2013
ASIA	China	South China Botanical Garden (SCBG), No. 723, Xingke Road, Tianhe District Guangzhou Guangdong 510650	1929 The largest Chinese bamboo collection, bambusetum area 4.2 ha, holding 222 species (SCBG, Table 3.2)	Huang, 2011; URL, 4, 5
ASIA	China	Taipei Botanical Garden Botanical Garden Division, Taiwan Forestry Research Institute, 53, Nanhai Road Taipei City 10066 Taiwan	1896 Botanical Garden contains a beautiful collection of bamboos	Lucas, 2013
ASIA	China	Taihu Botanic Garden, Wuxi (Jiangsu) Bureau of Gardens and Forestry of Wuxi, Yuantouzhu Wuxi, Jiangsu, 214086	Special collection includes bamboos	BGCIa
ASIA	China	Wenzhou Botanical Garden (Zhejiang) Institute of Subtropical Crops, Zhejiang Academy of Sciences, Jinshan, Western Suburbs, Wenzhou, Zhejiang, 325005	Special collection includes bamboos	BGCIa
ASIA	China	World Horti-Expo Garden, 10 Shibo Rd, Panlong District, Kunming, Yunnan	1999 Harbours a 'Theme bamboo garden'	Lucas, 2013
ASIA	China	Wuhan Botanic Garden (WHIOB), Wuhan Institute of Botany, Chinese Academy of Sciences, Moshan, Donghu Lake, Hongshan, Wuchang, Hubei 430074	In Wuhan Botanic Garden 59 bamboo species are conserved (WHIOB, Table 3.2)	Huang, 2018; URL, 6
ASIA	China	Xishuangbanna Tropical Botanical Garden (XTBG), CAS, Menglun, Mengla Yunnan 666303	1959 Bamboo garden with an area 7 ha holds over 200 species. Regarded as one of the largest display bamboo gardens in China. 125 species are listed (XTBG, Table 3.2)	BGCIa; Huang, 2018; URL, 7; Yang et al., 2008; Wang and Li, 2019

Table 3.1 contd. ...

...Table 3.1 contd.

Continent	Country	Name of Garden/Institution	Details	References
ASIA	China	Xiamen Botanical Garden (XMBG), Parks and Landscape Bureau of Xiamen, 25 Huyuan Road, Siming District Xiamen, Fujian 361003	1959 Xiamen Botanical Garden conserves 39 bamboo species (XMBG, Table 3.2)	BGCIa; Huang, 2018
ASIA	Hong Kong SAR China	Shing Mun Arboretum (SMA) Flora Conservation Section, AFCD 303, Cheung Sha Wan Road, Hong Kong	1970 Special collection include bamboos with 22 species (SMA, Table 3.2)	URL, 8
ASIA	India	Acharya Jagadish Chandra Bose Indian Botanic Garden (AJCBIBG), (*Indian Botanic Garden, Calcutta Botanic Garden*), Botanical Survey of India, P.O. Botanic Garden, Howrah 711103.	1787 One of the world's earliest established bambusetum which has 36 species (Table 3.2)	Roxburgh, 1814; Biswas, 1999; Wong, 2004; Koshy, 2010; Kumari, 2017
ASIA	India	Bamboorium, Basar, Sinang Disrtict	One of the largest collection of bamboos in India with 57 species	Subramaniam, 1998; URL, 9; Wong, 2004
ASIA	India	Bambusetum, Forest Research Institute (FRI), Chakarata Rd, New Forest P.O., Dehradun, Uttarakhand	1932 A well-known bambusetum, harbours 45 species (FRI, Table 3.2)	Biswas, 1999, 2008; Wong, 2004; Koshy, 2010
ASIA	India	Bambusetum, Kerala Forest Research Institute (KFRI) Peechi, Thrissur, Kerala	1988–95 Conserves 63 species in two locations, 55 species in Velupadam and eight in Devikulam (KFRI, Table 3.2)	Wong, 2004; KFRI, 2005; URL, 10
ASIA	India	Bambusetum, Van Vigyan Kendra (VVK), Chessa	1981 The largest bambusetum in the North-east region of India, holds 40 species (VVK, Table 3.2)	Beniwal and Haridasan, 1988; Wong, 2004; Koshy, 2010, 2017
ASIA	India	Botanical Garden, Guru Nanak Dev University Department of Botanical and Environmental Sciences, Amritsar, Punjab	1975 Special collection includes bamboos	BGCIa

	Country	Institution / Address	Details	Reference
	India	Botanical garden of Punjab University, Department of Botany, Punjab University, Chandigarh, U.T.	Known botanical garden attached to an Indian university. Bambusetum contains over 24 species	Subramaniam, 1998; URL, 11
ASIA	India	Experimental Gardens Dehra Dun, Botanical Survey of India (Northern Circle), P.O. Khirsu, Garhwal 246 001 Uttar Pradesh	Special collection includes bamboos	BGCIa; URL, 12
ASIA	India	Government Botanic Gardens, Dept. of Museum & Zoos, Thiruvananthapuram 695 001, Kerala	1858 Live collection includes bamboos	Koshy K.C., personal observation
ASIA	India	Jawaharlal Nehru Tropical Botanic Garden and Research Institute (JNTBGRI), Palode, Thiruvananthapuram, Kerala	1987 Largest bamboo collection in India with 82 Species and 14 Hybrids (JNTBGRI, Table 3.2)	BGCIa; Koshy, 2010, 2017; Koshy and Gopakumar, 2016
ASIA	India	Institute of Forest Genetics and Tree Breeding (Indian Council of Forestry Research and Education) Coimbatore 641002	1988 The Institute garden holds 26 bamboo species	Subramaniam, 1998; Koshy, 2010
ASIA	India	Kerala Forest Department, Begur P.O., Wayanad District, Kerala	Harbours 12 bamboo species	Subramaniam, 1998
ASIA	India	Lalbagh Botanical Garden Bangalore 560 004, Karnataka	1856 Collection includes bamboos	URL, 12
ASIA	India	Rain Forest Research Institute, A.T. Road (East), Jorhat- 785 001, Assam	1988 RFRI bambusetum contains 49 species	ICFRI, 2015; URL, 13
ASIA	India	Advanced Research Centre for Bamboo and Rattans (ARCBR), P.O. Box 171, Kulikawn Aizwal 796001, Mizoram	2004 ARCBR bambusetum holds 28 species	ICFRI, 2015; URL, 13
ASIA	India	Regional Plant Resource Centre (RPRC), Forests and Environment Department, Government of Odisha, Nayapalli, Bhubaneswar 751015, Odisha	1985 Special collection includes 28 bamboo species (RPRC, Table 3.2)	BGCIa; URL, 14
ASIA	India	State Forest Research Institute, Polipathar Jabalpur 482008, Madhya Pradesh	1963 Bambusetum area 1 ha, consists of 47 species	SFRI, 2020

Table 3.1 contd. …

...Table 3.1 contd.

Continent	Country	Name of Garden/Institution	Details	References
ASIA	India	Sri Chamarajendra Park (formerly Cubbon Park), Bangalore, Karnataka	19th century Collections include bamboos	URL, 12
ASIA	India	Pilikula Development Authority, Mangaluru in Karnataka State, India	1996 The bambusetum holds 27 species belonging to 8 genera	URL, 15
ASIA	Indonesia	Arboretum, Agricultural University (IPB), Bogor	1963 Collections include bamboos	Wong, 2004
ASIA	Indonesia	Bali Botanic Garden (BBG) (*Ekakarya Botanic Garden*), Bedugul Bali, Indonesia	1980s. Bambusetum area 2 ha, holds 52 species (BBG, Table 3.2)	Wong, 2004; Sujarwo, 2018
ASIA	Indonesia	Bamboo collection, PT. Great Giant Pineapple Co., Terbanggi Besar, Lampung Tengah, Sumatra	Contains various bamboos	Wong, 2004
ASIA	Indonesia	Bogor Botanic Gardens (*Kebun Raya Bogor*) (Center for Plant Conservation - Botanic Gardens), Jalan Ir. H. Juanda No. 13, P.O. Box 309, Bogor, West Java 16003	1817 Collections include bamboos	BGCIa; Wong, 2004; Lucas, 2013
ASIA	Indonesia	Cibodas Botanic Gardens Cimacan, Cipanas Cianjur, West Java	1852 Collections include bamboos	Wong, 2004
ASIA	Indonesia	Bali Botanic Garden (Ekakarya Botanic Garden), Bedugul, Bali	1959 Collections include bamboos	Wong, 2004
ASIA	Indonesia	Perhutani Bamboo Germplasm Collection Garden, Huar Bentes, Jasinga, West Java	1990s Collections include bamboos	Wong, 2004; URL, 16
ASIA	Indonesia	Purwodadi Botanic Garden (PBG) Jl. Surabaya-Malang, Purwodadi Pasuruan East Java	1941 PBG bambusetum holds 30 species. 7 species were listed (PBG, Table 3.2)	BGCIa; Wong, 2004
ASIA	Israel	Botanical Garden 'Mikveh-Israel' Holon	1930 Special collection includes bamboos	BGCIa

ASIA	Japan	Botanic Gardens of Toyama 42 Kamikutsuwada, Fuchu-machi, Toyama City, Toyama 939-2713	1989 Collections include bamboos	Lucas, 2013
ASIA	Japan	Fuji Bamboo Garden 885 Minami Isshiki, Nagaizumi-cho, Sunto-gun, Shizuoka	1955 One of the largest bambuseta in the world containing 450 species and c. 100,000 clumps	BGCIa; Lucas, 2013
ASIA	Japan	Koishikawa Botanical Garden, 3 Chome-7-1 Hakusan, Bunkyo City, Tokyo 112-0001	Collections include bamboos	Lucas, 2013
ASIA	Japan	Minamata memorial garden Minamata	Holds a 'scenic bamboo garden'	Lucas, 2013
ASIA	Japan	Rakusai Chikurin (Bamboo Forest) Park Kyoto City, Japan	Holds c. 110 bamboo taxa	Lucas, 2013
ASIA	Malaysia	Agricultural Research Station Ulu Dusun, Sabah	Collections include bamboos	Wong, 2004
ASIA	Malaysia	Bambusetum, Rimba Ilmu Botanic Garden University of Malaya, Kuala Lumpur	Bambusetum was completed in 2003	Wong, 2004; URL, 17
ASIA	Malaysia	Bambusetum, Forest Research Institute Malaysia (FRIM), Kepong	1994 Bambusetum holds 45 species	Wong, 2004; URL, 18
ASIA	Malaysia	Bamboo collection, Putra Jaya Wetlands, Putra Jaya	The bambusetum is a tourist spot	Wong, 2004
ASIA	Malaysia	Taman Botani, Putra Jaya	Bambusetum is one among 8 botanical themes of the park	Wong, 2004
ASIA	Malaysia	Botanical Garden, Pulau Pinang	Bambusetum is a key feature of this garden	Wong, 2004
ASIA	Malaysia	Penang Botanic Gardens Jabatan Taman Botani, Kompleks Pentadbiran, Bangunan Pavilion, Jalan Kenbun Bunga, Penang 10350	1884 Bambusetum contains c. 36 species	BGCIa; URL, 19
ASIA	Malaysia	Taman Kiara Arboretum Federal Territory of Kuala Lumpur, Kuala Lumpur	1982 Special collection includes bamboos	BGCIa
ASIA	Myanmar	Forest Research Institute, Yezin Forest Research Institute, Forest Department Ministry of Natural Resources and Environmental Conservation, Myanmar	1983 The Institute has a small bambusetum with 11 species	Wong, 2004; URL, 20

Table 3.1 contd....

...Table 3.1 contd.

Continent	Country	Name of Garden/Institution	Details	References
ASIA	Myanmar	National Kandawgyi Botanical Gardens (Maymyo Botanical Garden), Nandar Road, Pyin Oo Lwin, Mandalay Division	1915 It also has a bamboo garden	BGCIa; URL, 21; Wong, 2004
ASIA	Philippines	Makiling Botanic Gardens, College of Forestry and Natural Resources, University of the Philippines, Los Baños, College, Los Baños, Laguna 4031, Philippines	1963 Special collection includes bamboos	BGCIa; Wong, 2004
ASIA	Philippines	Los Baños Bambusetum, Los Baños, Laguna	Bambusetum area is 2.2 ha., has 34 species	Roxas, 1999; URL, 22
ASIA	Philippines	Davao Bambusetum Nabunturan, Davao del Norte	Bambusetum area is 2 ha., contains 33 species	Roxas 1999; URL, 22
ASIA	Philippines	Bukidnon Bambusetum Malaybalay, Bukidnon	Bambusetum area is 10 ha., holds 31 species	Roxas, 1999; URL, 22
ASIA	Philippines	ERDB Bambusetum, Los Baños, Laguna	The bambusetum holds 20 species	Roxas, 1999; URL, 22
ASIA	Philippines	Philippine Bambusetum, ERDS Loakan Rd., Baguoi City 2600	Special collections include bamboos	BGCIa; Roxas, 1999; Tañgau and Veracion, 1989; Wong, 2004; URL, 22
ASIA	Singapore	Singapore Botanic Garden (SBG), 1 Cluny Rd., Singapore 259569	1859 SBG bambusetum holds 32 species (SBG, Table 3.2)	Wong, 2004, 2022; URL, 23
ASIA	South Korea	Damyang Bamboo Park, 148 Binaedong-gil, Geumseong-myeon, Damyang-gun, Jeollanam-do	Special collection includes bamboos	Lucas, 2013
ASIA	South Korea	Juknokwon (Bamboo Forest), 119 Jungnogwon-ro, Damyang-eup, Damyang-gun, Jeollanam-do	2003 The bamboo forest is spread over 16 ha.	Lucas, 2013
ASIA	Sri Lanka	Gampaha Botanic Gardens (GBG), Heneratgoda, Gampaha, Sri Lanka	1876 GBG contains 12 bamboo species (GBG, Table 3.2)	Kariyawasam, 1999; URL, 24
ASIA	Sri Lanka	Hakgala Botanic Gardens (HBGH), Hakgala, Sri Lanka	1861 HBGH holds 5 bamboo species (HBGH, Table 3.2)	Kariyawasam, 1999; URL, 24
ASIA	Sri Lanka	Royal Botanic Gardens (RBGPS), P.O. Box 14 Peradeniya, 20400 Sri Lanka	1821 RBGPS harbours 18 bamboo species (RBGPS, Table 3.2)	Kariyawasam, 1999; Wong, 2004; URL, 24

Region	Country	Location	Description	Reference
ASIA	Thailand	Kanchanaburi Research Station, Agro-Ecological System Research and Development Institute, Kasetsart University Bangkhen Campus, Bangkok 10900	The research station maintains 33 bamboo species	Wong, 2004; Henpithaksa, 2008
ASIA	Thailand	Peninsular (Thung Khai) Botanic Garden, Trang-Palien Road, Thung Khai Arboretum and Forest Reserve, Yan Ta Khao district, Trang province 92140	1993 Taxonomic garden of Thung Khai Arboretum contains bamboos	BGCIa
ASIA	Thailand	The Queen Sirikit Botanical Garden, Mae Rim District, Chiang Mai Province	Arboretum contains bamboos	Wong, 2004
ASIA	Thailand	Walai Rukhavej Botanical Research Institute, Srinakharinwirot University, Mahasarakham	1993 Special collection includes bamboos	BGCIa
AUSTRALIA/ OCEANIA	Australia	Mount Mirinjo Farm, Imisfail, Queensland	*Ex situ* collection includes bamboos (*Gigantochloa ridleyi*)	Muller, 1998; Wong, 2004
AUSTRALIA/ OCEANIA	Australia	Perth Zoo (PZ), 20 Labouchere Road, South Perth 6951 South Perth, Western Australia 6151	1898 PZ contains 37 bamboo species, two of them mentioned (PZ, Table 3.2)	BGCIa; URL, 25
AUSTRALIA/ OCEANIA	Papua New Guinea	Lae Botanic Gardens, Bumneng, Eriku, Lae City, Morobe Province, Papua New Guinea	1900s The first botanic garden in Papua New Guinea Plant collection include bamboos	Wong, 2004; URL, 26
AUSTRALIA/ OCEANIA	Papua New Guinea	Agricultural Experimental Station, Laloki Central Province, Port Moresby	Bambusarium has been established at Lae with species from S.E. Asia, South and North America	Sharma, 1980; Wong, 2004
EUROPE	Belgium	Bokrijk Arboretum, Provinciaal Domein Bokrijk Genk (Bokrijk), B-3600	1965 The arboretum houses the national bamboo collection	BGCIa; Lucas, 2013
EUROPE	Belgium	Jardin des Plantes Medicinales Universite Catholique de Louvain, Avenue Mounier, 51 1200 Bruxelles, Brussels	1981 Special collection includes bamboos	BGCIa
EUROPE	France	Jardin d'Oiseaux Tropicaux, RD 559, Quartier Saint-Honoré, La Londe-les-Maures 83250	1989 Living collection contains bamboos	BGCIa

Table 3.1 contd....

...Table 3.1 contd.

Continent	Country	Name of Garden/Institution	Details	References
EUROPE	France	La Bambouseraie (Maurice Negre Parc, Exotique de Prafrance) (LBMG), Generargues, F-30140 Anduze France	1956 LBMG is a private botanical garden specializing in bamboos and represents one of Europe's oldest bamboo collections. It contains 106 species including 59 *Phyllostachys* spp.(LBMG, Table 3.2)	BGCIa; Wong, 2004; URL, 27
EUROPE	Italy	Arboreto di Arco - Parco Arciducale, c/o MUSE Museo delle scienze, Corso della scienza, 3 Trento I-38123	Special collection includes bamboos	BGCIa
EUROPE	Italy	Orto Botanico dell'Univerita della Tuscia, Via San Camillo de Lellis, 01100 Viterbo	1991 Special collection includes bamboos	BGCIa
EUROPE	Italy	Die Gärten von Schloss Trauttmansdorff / I Giardini di Castel Trauttmansdorff / The Gardens of Traut St. Valentin Str. 51a Via San Valentino, Meran/Merano 39012	2001 The garden maintains a 'Bamboo Forest'	BGCIa
EUROPE	Germany	Palmengarten der Stadt Frankfurt am Main Siesmayerstrasse 61, Frankfurt am Main, D-60323 Germany	1868 *Ex situ* collection include bamboos	BGCIa; Lucas, 2013
EUROPE	Netherlands	Hortus Botanicus Vrije Universiteit Van der Boechorststraat 8 1081 BT Amsterdam	1951 The garden has collections of winter hardy bamboos	BGCIa
EUROPE	Netherlands	Hortus Haren, P.O. Box 179, Haren	Special collection include bamboos	BGCIa
EUROPE	Rome	The Botanical Garden of Rome, Sapienza University Rome	Bamboo collection includes over 60 taxa. In 2019, *Phyllostachys bambusoides* flowered here, which flowers at intervals of 60 to 100 years	URL, 28
EUROPE	Spain	Jardí Botànic Marimurtra Passeig Carles (JBMPC) Faust No. 9 Box 112, Blanes Girona 17300	1918 Special collection includes bamboos (JBMPC, Table 3.2)	BGCIa; URL, 29

EUROPE	Spain	Jardin Botanico-Historico 'La Concepcion' de Malaga (JBHLCM), Patronato Botanico Municipal 'Ciudad de Malaga', Camino del Jardin Botánico, 3 Malaga 29014	1943 — The 'Historical-artistic garden' harbours 100-year-old plants including 22 bamboo species (JBHLCM, Table 3.2)	BGCIa; URL, 30
EUROPE	United Kingdom	Royal Botanic Gardens (RBG), Kew, Richmond, London TW9 3AE	One of the earliest and largest *ex situ* bamboo collection in UK holding 130 species introduced from China, Japan, the Himalaya, and America. First bamboo planted in 1826	Lucas, 2013; Townsend, 2018; URL, 31
EUROPE	United Kingdom	Benmore Botanic Garden, Benmore Argyll, Scotland PA23 8QU	1850 — *Ex situ* collection include bamboos	Lucas, 2013
EUROPE	United Kingdom	Carwinion House and Bamboo Garden Carwinion Rd, Mawnan Smith, Falmouth TR11 5JA	The garden, famous for its collection of over 200 bamboos species	URL, 32
EUROPE	United Kingdom	Logan Botanic Garden Port Logan, Nr. Stranraer DG9 9ND, Scotland	A satellite garden of RBG Edinburgh. It is leading an international project to investigate the bamboos in giant panda habitat and the diet of giant panda.	Lucas, 2013; URL, 33
EUROPE	United Kingdom	The Lost Gardens of Heligan, B3273, Pentewan, Saint Austell PL26 6EN	1992 — *Ex situ* collection contains bamboos	URL, 34
EUROPE	United Kingdom	Ness Botanic Gardens, University of Liverpool, Neston Rd, Little Neston Ness CH64 4AY	1898 — *Ex situ* collection comprises bamboos	Lucas, 2013; BGCIa
EUROPE	United Kingdom	Royal Botanic Garden Edinburgh (RBGE), 20A Inverleith, Row Edinburgh City of Edinburgh EH3 5LR	Home to one of the world's richest collections of living plants (over 13,500 species) including 47 bamboo taxa (RBGE, Table 3.2)	Lucas, 2013; URL, 35
EUROPE	United Kingdom	Royal Horticultural Society's Garden, Wisley, Surrey GU23 6QB	1904 — *Ex situ* collection includes bamboos	BGCIa; Lucas, 2013
EUROPE	United Kingdom	Trebah Garden Trust (TGT), Mawnan Smith, Falmouth TR11 5JZ	1826 — Trebah garden holds 47 bamboos (TGT, Table 3.2)	URL, 36

Table 3.1 contd...

...Table 3.1 contd.

Continent	Country	Name of Garden/Institution	Details	References
EUROPE	United Kingdom	Tresco Abbey Gardens, Tresco, Isles of Scilly, TR24 0QQ	*Ex situ* collection includes bamboos	URL, 37
NORTH AMERICA	Bermuda	The Bermuda Botanical Gardens, P.O. Box HM 20, Hamilton, HM AX Bermuda	1898 Special collection includes bamboos	BGCIa
NORTH AMERICA	British Virgin Islands	Joseph Reynold O'Neal Botanic Gardens, P.O. Box 878, Road Town, Tortola, British Virgin Islands	1986 Special collection includes bamboos	BGCIa
NORTH AMERICA	Costa Rica	Jardin Botánico Lankester, Universidad de Costa Rica Cartago 302-7050	1973 Collection include bamboos, mainly *Phyllostachys aurea*	BGCIa
NORTH AMERICA	Cuba	Jardin Botanico de Cienfuegos C.A.I. - Pepito Tey, Cienfuegos C.P. 59290	1902 *Ex situ* collections includes 23 bamboo taxa	BGCIa
NORTH AMERICA	Honduras	Lancetilla Botanic Garden & Research Center, Apartado Postal 49, Tela 00000	1925 The only the botanical Garden in Honduras, holds a special collection of 16 bamboo species	BGCIa; URL, 38
NORTH AMERICA	Puerto Rico	Tropical Agriculture Research Station (MAY) Mayaguez, Puerto Rico	1901 Natl. Germplasm Repository - Mayaguez, Puerto Rico has 48 bamboo species (MAY, Table 3.2)	Wong, 2004; USDA, 2015; URL, 39
NORTH AMERICA	United States of America	Coastal Georgia Botanical Gardens at the Historic Bamboo Farm, 2 Canebrake Road, Savannah, Georgia 31419	The historic bamboo plantings, from China in the 1920s, '30s and '40s. The living collection has over 70 bamboo species	Lucas, 2013; BGCIa; URL, 40
NORTH AMERICA	United States of America	Fairchild Tropical Botanic Garden, 10901 Old Cutler Road, Coral Gables, Florida 33156	1938 Special collection includes bamboos	BGCIa; Lucas, 2013
NORTH AMERICA	United States of America	Huntington Library, Art Museum and Botanical Gardens, The1151 Oxford Road, San Marino, California	Special collection includes bamboos	BGCIa; Lucas, 2013

NORTH AMERICA	United States of America	Kanapaha Botanical Gardens, 4700 SW 58th Drive Gainesville, Florida 32608	1978 Florida state's largest public display of bamboos	URL, 41
NORTH AMERICA	United States of America	Los Angeles County Arboretum and Botanic Garden, 301 North Baldwin Avenue, Arcadia, California 91007	1948 Special collection includes bamboos	BGCIa
NORTH AMERICA	United States of America	Missouri Botanical Garden (MBG) 4344 Shaw Boulevard, St. Louis, Missouri	1859 One of the top three botanical gardens in the world. It has a bamboo collection of which 8 species are listed (MBG, Table 3.2)	URL, 42
NORTH AMERICA	United States of America	Plant Genetic Resources Conservation Unit Griffin, Georgia (GRIFFIN)	Maintains a well-documented collection of bamboo species (GRIFFIN, Table 3.2)	USDA, 2022; URL, 39
NORTH AMERICA	United States of America	National Germplasm Repository (MIA), Miami, Florida	1980 (formation of the National Plant Germplasm System). Holds 48 bamboo species (MIA, Table 3.2)	USDA, 2022; URL, 39
NORTH AMERICA	United States of America	Mercer Botanic Gardens, The Mercer Society 22306, Aldine Westfield Road, Humble, Texas 77338	1974 This garden houses 40 bamboo taxa	BGCIa
NORTH AMERICA	United States of America	The Palomar College Gardens (PCG), 1140 West Mission Road, San Marcos, California 92069-1415	1973 PCG holds a bamboo collection, and 23 taxa are listed (PCG, Table 3.2)	BGCIa; URL, 43
NORTH AMERICA	United States of America	Rutgers Gardens Cook College, Rutgers University, 112 Ryders Lane New Brunswick, New Jersey 08901-8519	1934 Special collection includes bamboos	BGCIa
NORTH AMERICA	United States of America	San Diego Botanic Garden (SDBG) San Diego, Encinitas, California	1970 The largest bamboo collection in a North American public garden, consists of 145 taxa (SDBG, Table 3.2)	URL, 44
NORTH AMERICA	United States of America	Rip Van Winkle House and Gardens, Jefferson Island, Iberia Parish, Louisiana	*Ex situ* collection includes bamboos	Lucas, 2013

Table 3.1 contd. ....

...Table 3.1 contd.

Continent	Country	Name of Garden/Institution	Details	References
SOUTH AMERICA	Brazil	Jardim Botânico do Instituto Agronômico de Campinas, Av. Theodureto de Almeida Camargo, no.1500, Vila Nova, Campinas, São Paulo	1998 Maintains a bamboo germplasm bank	Lucas, 2013
SOUTH AMERICA	Brazil	Jardim Botânico de São Paulo Instituto de Botânica, Av. Miguel Estefano 3031, Caixa Postal 400501051, Sao Paulo	*Ex situ* collection includes bamboos	Lucas, 2013
SOUTH AMERICA	Colombia	Jardin Botânico "Juan Maria Cespedes" Instituto Vallecaucano de Investigaciones Científicas (INCIVO), Av. Rooselvet No. 24–80, Tulua	1968 Special collections include bamboos	BGCIa
SOUTH AMERICA	Mexico	The Francisco Javier Clavijero Botanical Garden at the Instituto de Ecología (FJCBG) Congregación El Haya P.O. Box 63, 91000 Xalapa, Veracruz 91073	1977 Houses the Mexican national living bamboo collection of 56 species including native species of *Aulonemia, Guadua, Chusquea, Rhipidocladum, Olmeca and Otatea*. 47 species are listed (FJCBG, Table 3.2)	BGCIa; Mejia-Saulés and Ordóñez, 2018; URL, 45

BGCIa, b; Wyse Jackson and Sutherland, 2013; URL, 1–45). Many private gardens and institutions also conserve bamboos in an informal way (Benfield, 2013) but have not surfaced in the websites/literature. They are not included in Table 3.1. Asia tops first with 88 bambuseta distributed in 14 countries followed by Europe with 23 in eight countries, North America with 19 in seven countries, Australia/Oceania with four in two countries, South America with four in three countries, and Africa having one (Table 3.1). China stands out in bamboo conservation with 25 bambuseta followed by India with 19 bambuseta. The Chinese living bamboo collections are well documented (Huang, 2018).

3.4 Earliest Living Bamboo Collections

The earliest known cultivation of bamboos took place in India. It was in Indian Botanic Garden, Calcutta (presently known as Acharya Jagadish Chandra Bose Indian Botanic Garden), established under the leadership of Robert Kyd, during 1787–1793. In 1794, it had seven bamboo species (Roxburgh, 1814; Biswas, 1999; Koshy, 2010, 2017; Kumari, 2017). The black bamboo *Phyllostachys nigra* was introduced to Royal Botanic Gardens, Kew in 1826, though a full-fledged Bamboo Garden took shape only in 1891 (Townsend, 2018). Incidentally, in 1826 itself, the same bamboo species was introduced to Trebah Garden Trust, Falmouth, UK. Species such as *Indocalamus tessellatus* was introduced to UK from China in 1845 and *Yushania anceps* from Northwest Himalayas in 1865 (URL, 35). The Bambouseraie de Prafrance, a private botanical garden specializing in bamboos, established in 1856 by an amateur botanist Eugène Mazel (BGCIa; URL, 26) is reportedly having the oldest bamboo collections among the European countries. The Missouri Botanical Garden in St. Louis, USA has a good bamboo collection (URL, 41). The historic bamboo plantings at Coastal Georgia Botanical Gardens at the Historic Bamboo Farm, Georgia ensued through plant collecting trips to eastern China in the 1920s, 1930s, and 1940s (URL, 39).

3.5 Largest Living Bamboo Collections

The Fuji Bamboo Garden, established in 1955, is Japan's only botanical garden specializing in bamboo. It is said to be the world's largest collection of bamboo, holding more than 450 taxa from all around the world. The garden contains approximately 100,000 bamboo plants (BGCIa). The bamboo garden at South China Botanical Garden (SCBG) under the Chinese Academy of Sciences takes a total area of 4.2 ha with an initial collection of 222 species. Currently, SCBG is holding the largest Chinese collection of bamboos (URL, 4, 5; Huang, 2011). Anji Bamboo Garden in China, built in 1974, takes a total area of 80 ha with 211 bamboo species. It has the most comprehensive collection of *Phyllostachys* and in 2015, c. 100 bamboo species were added (Maoyi, 1999; URL, 2, 3). Xishuangbanna Tropical Botanical Garden, China, founded in 1959, has a large collection of more than 200 bamboo species in 7 ha. It is said to be one of the largest display bamboo gardens in China, functioning for scientific research, science popularization, and eco-tourism (URL, 7; Yang et al., 2008; Wang and Li, 2019). The bamboo garden at Hangzhou Botanical

Garden, Hangzhou, China, founded in 1962, covers an area of 14.5 ha with more than 100 species (BGCIa). The Royal Botanic Gardens, Kew have one of the largest bamboo collections in the UK with 130 bamboo species introduced from China, Japan, the Himalaya, and Americas (Townsend, 2018). The Bambouseraie de Prafrance, France harbours more than 100 species (BGCIa). Coastal Georgia Botanical Gardens at the Historic Bamboo Farm, Georgia has a bamboo collection of over 70 species (URL, 39). San Diego Botanic Garden's (SDBG) bamboo collection, the largest among the North American public gardens, has its origin in 1979. It consists of 145 taxa including species from Asia, South America, and Africa.

As for India, the living bamboo collection at Acharya Jagadish Chandra Bose Indian Botanic Garden (AJCBIBG), Kolkata has now 36 species (Kumari, 2017). The Forest Research Institute (FRI) in Dehra Dun, Uttarakhand, has been the pioneer institution in bamboo research. FRI maintains an excellent bamboo collection, established in 1932, consisting of 45 species (Biswas, 1999, 2008). Van Vignan Kendra, Chessa (VVK), established in 1981, is credited with a live collection of 40 species (Beniwal and Haridasan, 1988). The two important bambuseta in southern India occur in Kerala state. Kerala Forest Research Institute (KFRI), Peechi holds 63 species (KFRI, 2005; URL, 10). The bambusetum at Jawaharlal Nehru Tropical Botanic Garden and Research Institute (JNTBGRI), Thiruvananthapuram (formerly Tropical Botanic Garden and Research Institute) is the largest and scientifically well-managed one in India (Alam, 2011; Christenhusz, 2011; Vorontosova, 2011; Wong, 2011; Xia, 2011). Founded in 1987, the collection rose to 82 species and 14 hybrids, each species having many accessions taking the total to 1,168 (Koshy, 2010; Koshy and Gopakumar, 2016).

3.6 Bamboos Conserved in Botanic Gardens and Bambuseta

Globally, a total of 560 species under 60 genera have been conserved in 139 bambuseta (Tables 3.1, 3.2). Nomenclature of the species is updated with IPNI (2022) and POWO (2022) to assess the correct number of species as the same taxon has been given under different synonyms in many garden publications. The taxa in Table 3.2 are classified following BPG (2012) and Soreng et al. (2017) to determine the number of species conserved under Arundinarieae, Bambuseae, and Olyreae. The first group is represented by 28 genera and 228 species, the second by 28 genera and 327 species and third by four genera and five species.

The following is the genus/species status of conservation (genera with number of species in bracket): Arundinarieae: *Acidosasa* (6), *Ampelocalamus* (8), *Arundinaria* (4), *Bashania* (2), *Bergbambos* (1), *Brachystachyum* (1), *Chimonobambusa* (13), *Chimonocalamus* (3), *Drepanostachyum* (3), *Fargesia* (19), *Ferrocalamus* (1), *Gelidocalamus* (2), *Himalayacalamus* (6), *Indocalamus* (13), *Indosasa* (10), *Oligostachyum* (8), x*Phyllosasa* (1), *Phyllostachys* (55), *Pleioblastus* (14), *Pseudosasa* (9), *Sasa* (11), *Sasaella* (4), *Sasamorpha* (1), *Semiarundinaria* (8), *Shibataea* (6), *Sinobambusa* (7), *Thamnocalamus* (1), and *Yushania* (11).

Bambuseae: The Neotropical Woody Bamboos are represented by eight genera and 62 species, viz., *Arthrostylidium* (1), Aulonemia (1), *Chusquea* (28), *Guadua* (9),

Table 3.2. Bamboo taxa conserved in various bambuseta in the world.

Names in bold italics indicate correct names and those in regular italics indicate synonyms as given in the respective garden publication. d = ditto, unresolved = names of which current status unknown.

No.	Species	Tribe	Native Range	Repository	Country
1.	***Acidosasa chinensis* C.D. Chu & C.S. Chao**	Arundinarieae	China	ABG, SCBG	China
2.	***Acidosasa edulis* (T.H. Wen) T.H. Wen**	Arundinarieae	China	ABG, GXIB	China
3.	***Acidosasa lingchuanensis* (C.D. Chu et. C.S. Chao) Q.Z. Xie et X.Y. Chen**	Arundinarieae	China	GXIB	China
4.	***Acidosasa nanunica* (McClure) C.S. Chao & G.Y. Yang** *Pseudosasa nanunica* var. *angustifolia*	Arundinarieae	China	SCBG	China
5.	***Acidosasa notata* (Z.P. Wang & G.H. Ye) S.S. You** *Acidosasa longiligula*	Arundinarieae	China	ABG	China
	Pseudosasa notata	d	d	SDBG	USA
	Pleioblastus maculosoides	d	d	ABG	China
	Pleioblastus intermedius S.Y. Chen	d	d	GXIB	China
6.	***Acidosasa venusta* (McClure) Z.P. Wang & G.H. Ye ex Ohrnb. & Goerrings**	Arundinarieae	China	ABG, SZBG	China
7.	***Ampelocalamus actinotrichus* (Merr. & Chun) S.L. Chen T.H. Wen & G.Y. Sheng**	Arundinarieae	China	SCBG, XMBG	China
8.	***Ampelocalamus hirsutissimus* (W.D. Li & Yuan C. Zhong) Stapleton & D.Z. Li** *Drepanostachyum hirsutissimus*	Arundinarieae	China	ABG	China
9.	***Ampelocalamus luodianensis* T.P. Yi & R.S. Wang** *Drepanostachyum luodianense*	Arundinarieae	China	ABG	China
10.	***Ampelocalamus melicoideus* (Keng f.) D.Z. Li & Stapleton** *Drepanostachyum melicoideum*	Arundinarieae	China	ABG	China

Table 3.2 contd. ....

...Table 3.2 contd.

No.	Species	Tribe	Native Range	Repository	Country
11.	**Ampelocalamus microphyllus (Hsueh & T.P. Yi) Hsueh & T.P. Yi** / *Drepanostachyum microphyllum (Hsueh et. Yi) Keng f.*	Arundinarieae	China	WHIOB	China
12.	**Ampelocalamus patellaris (Gamble) Stapleton**	Arundinarieae	Nepal to China	JNTBGRI	India
13.	**Ampelocalamus scandens Hsueh & W.D. Li** / *Drepanostachyum scandens (Hsueh et. W.D. Li) Keng f. ex Yi*	Arundinarieae	China	ABG, GXIB, SCBG	China
14.	**Ampelocalamus stoloniformis (S.H. Chen & Zhen Z. Wang) C.H. Zheng, N.H. Xia & Y.F. Deng** / *Drepanostachyum stoloniforme S.H. Chen et Z.Z. Wang*	Arundinarieae	China	XMBG	China
15.	**Arthrostylidium excelsum Griseb.**	Bambuseae	Mexico to America	FJCBG	Mexico
16.	**Arundinaria gigantea (Walter) Muhl.**	Arundinarieae	U.S.A.	SDBG, MBG	USA
17.	**Arundinaria oleosa (T.H. Wen) Demoly**	Arundinarieae	China	GRIFFIN, MAY	USA
18.	**Arundinaria sp.**	Arundinarieae	0	GBG	Sri Lanka
19.	**Arundinaria sp.**	Arundinarieae	0	RBGPS	Sri Lanka
20.	**Aulonemia laxa (F. Maek.) McClure**	Bambuseae	Mexico	FJCBG	Mexico
21.	**Bambusa affinis Munro**	Bambuseae	Indo-China	AJCBIBG, KFRI	India
22.	**Bambusa albociliata** = (unresolved)	Bambuseae	0	KFRI	India
23.	**Bambusa albolineata L.C. Chia**	Bambuseae	China, Taiwan	SMA	Hong Kong
	Bambusa albo-lineata	d	d	SCBG, WHIOB, XTBG, SZBG, GXIB, XMBG	China
	Bambusa albostriata	d	d	KFRI	India
24.	**Bambusa balcooa Roxb.**	Bambuseae	India to Indo-China	FRI, KFRI, AJCBIBG, RPRC VVK, JNTBGRI	India
		d	d	PZ	Australia
		d	d	BFRI	Bangladesh

No.	Species	Bambuseae	India to Indo-China	Institutions	Country
25.	*Bambusa bambos* (**L.**) **Voss**	d	d	AJCBIBG, JNTBGRI, KFRI	India
		d	d	MIA	USA
		d	d	GBG, RBGPS	Sri Lanka
	Bambusa arundinacea (Retz.) Willd	d	d	SCBG, XTBG	China
		d	d	VVK, FRI, RPRC	India
		d	d	LBMG	France
	Gigantochloa maxima	d	d	SCBG	China
		d	d	SDBG	USA
	Bambusa bambos var. *gigantea* Bennet & Gaur	d	d	JNTBGRI	India
	Bambusa arundinacea var. *gigantea*	d	d	FRI	India
		d	d	KFRI	India
26.	**Bambusa basihirsuta McClure** / *Dendrocalamopsis basihirsuta* (McClure) Q.F. Zheng ex W.T.Lin	Bambuseae	China	SCBG, XTBG, SZBG, GXIB, ABG	China
27.	**Bambusa beecheyana Munro**	Bambuseae	China to Indo-China, Taiwan	PCG, SDBG	USA
	Dendrocalamopsis beecheyana (Munro) Keng f.		d	SCBG, XTBG, SZBG, GXIB, XMBG	China
	Dendrocalamopsis beecheyana var. *pubescens* (P.F. Li) Keng f.	d	d	SCBG, SZBG	China
28.	**Bambusa bicicatricata (W.T. Lin) L.C. Chia & H.L. Fung** / *Dendrocalamopsis bicicatricata* (W.T. Lin) Keng f.	Bambuseae	Hainan	SCBG, XTBG, GXIB	China
29.	**Bambusa boniopsis McClure**	Bambuseae	China	ABG, SCBG, XTBG, XMBG	China
	Bambusa boniopsis 'Christian Lydick Form'	d	d	SDBG	USA
30.	**Bambusa burmanica Gamble**	Bambuseae	Bangladesh to China and Malaysia.	AJCBIBG, VVK, FRI, JNTBGRI	India

Table 3.2 contd. ....

...Table 3.2 contd.

No.	Species	Tribe	Native Range	Repository	Country
	d	d	d	SCBG	China
	d	d	d	SDBG	USA
	d	d	d	BFRI	Bangladesh
31.	*Bambusa cacharensis* R.B. Majumdar	Bambuseae	India to Bangladesh	AJCBIBG, JNTBGRI	India
	d	d	d	BFRI	Bangladesh
32.	*Bambusa cerosissima* McClure *Bambusa cerosissima*	Bambuseae	China to Vietnam	SCBG, GXIB	China
33.	*Bambusa cerosissima* McClure var. *glabra* R. S. Lin et J. B. Ni = (unresolved)	Bambuseae	0	SCBG	China
34.	*Bambusa chungii* McClure *Bambusa chungii* var. *barbelatta*	Bambuseae	S. China to Vietnam	SDBG	USA
	Bambusa chungii McClure var. *velatina* T.P. Yi et. J.Y. Shi = (unresolved)	d	d	SCBG	China
	Bambusa chungii McClure	d	d	SCBG, WHIOB, XTBG, SZBG, GXIB, XMBG	China
	d	d	d	SDBG, KBG	USA
	d	d	d	SMA	Hong Kong
	Lingnania chungii	d	d	ABG	China
35.	*Bambusa clavata* Stapleton	Bambuseae	Bhutan	RBGE	UK
36.	*Bambusa comillensis* Alam	Bambuseae	Bangladesh	BFRI	Bangladesh
37.	*Bambusa contracta* L.C. Chia & H.L. Fung	Bambuseae	China	ABG, SZBG, GXIB, SCBG	China
38.	*Bambusa copelandii* Gamble *Bambusa copelandi*	Bambuseae	Myanmar	FRI	India
	Dendrocalamus copelandii (Gamble) N.H. Xia & Stapleton	d	d	JNTBGRI	India
39.	*Bambusa corniculata* L.C. Chia & H.L. Fung	Bambuseae	China	SCBG, SZBG	China
40.	*Bambusa cornigera* McClure	Bambuseae	China	SCBG, GXIB	China

No.	Name	Tribe		Location	Gardens	Country
41.	**Bambusa diaoluoshanensis L.C. Chia & H.L. Fung**	Bambuseae		Hainan	SCBG, SZBG, XMBG	China
42.	**Bambusa diffusa Blanco** *Schizostachyum diffusum*	Bambuseae		Philippines, Taiwan	SDBG	USA
43.	**Bambusa dissimulator McClure** *Bambusa dissimulator*	Bambuseae		China	KBG, MAY	USA
		d			KFRI	India
	Bambusa dissimulator var. hispida McClure	d			SCBG, GXIB	China
44.	**Bambusa distegia (Keng & Keng f.) L.C. Chia & H.L. Fung**	Bambuseae		China	SCBG, XTBG, SZBG, GXIB	China
		d			SDBG	USA
45.	**Bambusa dolichoclada Hayata**	d			SCBG, XTBG	China
	Bambusa dolichoclada 'Stripe'	Bambuseae		China, Taiwan	SDBG	USA
	Bambusa dolichoclada Hayata 'stripe'	d			PCG	USA
46.	**Bambusa duriuscula W.T. Lin**	d			SCBG	China
47.	**Bambusa emeiensis L.C. Chia & H.L. Fung** *Neosinocalamus affinis* cv. Viridiflavus	Bambuseae		S. China	SCBG, ABG	China
	Neosinocalamus affinis cv. Flavidorivens	d			SCBG	China
	Neosinocalamus affinis cv. Chrysotrichus	d			SCBG, ABG	China
	Neosinocalamus affinis	d			SCBG, ABG	China
	Bambusa emeiensis Chia et. H.L. Fung	d			SCBG, WHIOB, KIB, SZBG, GXIB	China
	Bambusa emeiensis 'Flavidovirens'	d			SDBG	USA
	Bambusa emeiensis 'Viridiflavus'	d			SDBG	USA
48.	**Bambusa eutuldoides McClure**	Bambuseae		China	ABG, SCBG, XTBG, SZBG, GXIB, XMBG	China
		d			KBG, SDBG	USA
	Bambusa eutuldoides McClure var. viridi-vittata (W.T. Lin) L.C. Chia	d			SCBG, SCBG, XTBG, SZBG	China

Table 3.2 contd. ...

...Table 3.2 contd.

No.	Species	Tribe	Native Range	Repository	Country
	Bambusa eutuldoides var. viridivittata	d	d	SDBG	USA
49.	**Bambusa eutuldoides var. basistriata** = (unresolved)	Bambuseae	0	SCBG, LSBG, SZBG, GXIB	China
	d	d	d	SMA	Hong Kong
50.	**Bambusa farinacea K.M. Wong**	Bambuseae	Thailand to Malaysia	SBG	Singapore
51.	**Bambusa flexuosa Munro**	Bambuseae		SCBG, XTBG, SZBG, XMBG	China
52.	**Bambusa funghomii McClure**		China	SCBG	China
53.	**Bambusa gibba McClure**	Bambuseae	China to Vietnam	SMA	Hong Kong
	d	d	d	SCBG, XTBG, SZBG, GXIB	China
54.	**Bambusa gibboides W.T. Lin**	Bambuseae	China	SCBG, SZBG, GXIB, XMBG	China
55.	**Bambusa glaucophylla Widjaja**	Bambuseae	Jawa	SBG	Singapore
56.	**Bambusa grandis (Q.H. Dai & X.L. Tao) Ohrnb.** *Dendrocalamopsis daii* Keng f.	Bambuseae	China	SCBG, SZBG, GXIB	China
57.	**Bambusa guangxiensis Chia et. H. L. Fung**	Bambuseae	China	GXIB	China
58.	**Bambusa hainanensis L.C. Chia & H.L. Fung**	Bambuseae	Hainan	SCBG	China
59.	**Bambusa hirticaulis R.S. Lin**	Bambuseae	China	SCBG	China
60.	**Bambusa indigena L.C. Chia & H.L. Fung**	Bambuseae	China	SCBG, SZBG, GXIB	China
61.	**Bambusa insularis L.C. Chia & H.L. Fung**	Bambuseae	Hainan	SMA	Hong Kong
	d	d	d	SZBG	China
62.	**Bambusa intermedia Hsueh & T.P. Yi**	Bambuseae	China	ABG, SCBG, XTBG, SZBG, GXIB, XMBG	China
63.	**Bambusa jaintiana R.B. Majumdar**	Bambuseae	Nepal to Myanmar	AJCBIBG, JNTBGRI	India
	d	d	d	BFRI	Bangladesh

No.	Species	Tribe	Distribution	Codes	Country
64.	**Bambusa lako Widjaja**	Bambuseae	Lesser Sunda Islands (E. Timor)	MAY, PCG, SDBG	USA
	d		d	SBG	Singapore
65.	**Bambusa lapidea McClure**	Bambuseae	China	SCBG, XTBG, SZBG, GXIB	China
	d		d	SMA	Hong Kong
66.	**Bambusa longispiculata Gamble** *Bambusa longispiculata*	Bambuseae	Bangladesh to Myanmar	ABG, SCBG, XTBG, SZBG, GXIB	China
	d		d	KFRI, VVK, FRI	India
	Bambusa longispiculata Gamble ex Brandis	d		d	MAY	USA
67.	**Bambusa maculata Widjaja**	Bambuseae	Lesser Sunda Islands to Maluku	BBG	Indonesia
68.	**Bambusa malingensis McClure**	Bambuseae	Hainan	PCG, KBG	USA
	d		d	SMA	Hong Kong
69.	**Bambusa multiplex (Lour.) Raeusch. ex Schult. & Schult. f.**	Bambuseae	Nepal to Taiwan	MAY, GRIFFIN	USA
	Bambusa multiplex (Lour.) Raeusch. ex Schult.f.	d		d	AJCBIBG, KFRI, RPRC, JNTBGRI	India
	d		d	HBGH, RBGPS, GBG	Sri Lanka
	d		d	KBG, SDBG	USA
	d		d	LBMG	France
	d		d	SBG	Singapore
	d		d	SMA	Hong Kong
	d		d	BBG	Indonesia
	d		d	ABG, SCBG, WHIOB, KIB XTBG, CNBG, SZBG, GXIB XMBG	China

Table 3.2 contd. ...

...*Table 3.2 contd.*

No.	Species	Tribe	Native Range	Repository	Country
	d	d	d	BFRI	Bangladesh
	Bambusa glaucesence	d	d	KFRI, VVK	India
	Bambusa nana	d	d	KFRI	India
	Bambusa multiplex (Lour.) Raeusch. ex Schult. var. *incana* B.M. Yang	d	d	SCBG, SZBG, XMBG	China
	Bambusa multiplex (Lour.) Raeusch. ex Schult. var. *riviereorum* R. Maire	d	d	ABG, KIB, XTBG, SZBG, GXIB	China
	Bambusa multiplex var. *riviereorum*	d	d	SCBG	China
	d	d	d	SMA	Hong Kong
	d	d	d	SBG	Singapore
	Bambusa multiplex cv. alphonse-karr	d	d	SCBG	China
	Bambusa multiplex cv. Fernleaf	d	d	SCBG	China
	Bambusa multiplex cv. Silverstripe	d	d	SCBG	China
	Bambusa multiplex cv. stripestem fernleaf	d	d	SCBG	China
	Bambusa dolichomerithalla 'Silverstripe'	d	d	SDBG	USA
	Bambusa dolichomerithalla 'Green Stripe'	d	d	SDBG	USA
	Bambusa multiplex 'alphonse karr'	d	d	KBG	USA
	Bambusa multiplex 'Fernleaf'	d	d	KBG	USA
	Bambusa multiplex 'Golden goddess'	d	d	KBG	USA
	Bambusa multiplex 'Rivieorum'	d	d	KBG	USA
	Bambusa multiplex 'Silverstripe'	d	d	KBG	USA
	Bambusa multiplex 'Alphonse Karr'	d	d	SDBG	USA
	d	d	d	SMA	Hong Kong

#	Name				Country
	d			LBMG	France
	d			SDBG	USA
	Bambusa multiplex 'Alponse Carr' & 'Silver Stripe'			PCG	USA
	Bambusa multiplex cv. Willowy			SCBG	China
	Bambusa multiplex cv. Yellowstripe			SCBG	China
	Bambusa multiplex 'Elegans'			LBMG	France
	Bambusa multiplex 'Fernleaf'			SDBG	USA
	d			HBG	China
	Bambusa multiplex 'Gold Stripe'			SDBG	USA
	Bambusa multiplex 'Golden goddess'			LBMG	France
	d			SDBG	USA
	Bambusa multiplex 'Riviereorum'			SDBG	USA
	Bambusa multiplex 'Silverstripe'			SDBG	USA
	Bambusa multiplex 'Stripestem Fernleaf'			SDBG	USA
	Bambusa multiplex 'Fernleaf'			SDBG	USA
	d			ABG	China
	Bambusa multiplex 'Tiny Fern'			SDBG	USA
	Bambusa multiplex (Lour.) Raeusch. (cultivar)			RBGE	UK
	Bambusa multiplex 'Alphonse-Karri'			ABG	China
	Bambusa multiplex 'Silverstripe'			ABG	China
70.	**Bambusa mutabilis McClure**	Bambuseae	Hainan	SCBG, SZBG	China
71.	**Bambusa nutans Wall. ex Munro**	Bambuseae	Himalaya to Indo-China	SCBG, XTBG	China
	d			SDBG	USA

Table 3.2 contd. ...

...Table 3.2 contd.

No.	Species	Tribe	Native Range	Repository	Country
	d	d	d	KFRI, VVK, RPRC, FRI, JNTBGRI	India
	d	d	d	BFRI	Bangladesh
72.	***Bambusa odashimae* Hatus. ex Ohrnb.** *Dendrocalamopsis edulis* (Odash.) Keng f.	Bambuseae	Taiwan	SCBG	China
73.	***Bambusa oldhamii* Munro**	Bambuseae	China to Hainan	MAY	USA
	d	d	d	SDBG, KBG, PCG	USA
	d	d	d	LBMG	France
	d	d	d	KFRI	India
	Dendrocalamopsis oldhamii (Munro) Keng f.	d	d	SCBG, XTBG, SZBG, GXIB, ABG	China
	Dendrocalamopsis oldhamii (Munro) Keng f. f. *revoluta* (W.T. Lin & J.Y. Lin) W.T. Lin	d	d	SCBG	China
74.	***Bambusa oliveriana* Gamble**	Bambuseae	Myanmar	AJCBIBG	India
75.	***Bambusa ooh* Widjaja & Astuti**	Bambuseae	Lesser Sunda Islands (Bali)	BBG	Indonesia
76.	***Bambusa pachinensis* Hayata** *Bambusa pachinensis*	Bambuseae	China to Taiwan	SDBG	USA
	d	d	d	SCBG	China
77.	***Bambusa pachinensis* Hayata var. *hirsutissima* (Odashima) W. C. Lin** =(unresolved)	d	0	SCBG	China
78.	***Bambusa pallida* Munro** *Bambusa pallida*	Bambuseae	Sikkim, China to Malaysia	KFRI, VVK, FRI, JNTBGRI	India
	d	d	d	XTBG, SZBG	China
79.	***Bambusa papillata* (Q.H. Dai) K.M. Lan**	d	d	SCBG, XTBG	China

No.	Species	Tribe	Distribution	Institution	Country
80.	*Bambusa pervariabilis* × *Dendrocalamopsis daii* No 3	Bambuseae	0	ABG	China
81.	*Bambusa pervariabilis* × *Dendrocalamus latiflorus* No 7	Bambuseae	0	ABG	China
82.	*Bambusa pervariabilis* McClure	Bambuseae	China	ABG, SCBG, WHIOB, XTBG GXIB, XMBG	China
	d	d	d	SDBG	USA
	d	d	d	SMA	Hong Kong
	Bambusa pervariabilis 'Viridistriatus'		d	SDBG	USA
	Bambusa pervariabilisiar 'Viridistriata'		d	ABG	China
	Bambusa pervariabilis McClure var. *viridistriata* Q.H. Dai et. X.C. Liu	d	d	SCBG, GXIB	China
83.	*Bambusa piscatorum* McClure	Bambuseae	Hainan	SCBG, SZBG	China
84.	*Bambusa polymorpha* Munro	Bambuseae	Bangladesh to China and Indo-China	AJCBIBG, VVK, FRI, RPRC JNTBGRI, KFRI	India
	d	d	d	MAY	USA
	d	d	d	ABG, SCBG, XTBG, SZBG	China
	d	d	d	BFRI	Bangladesh
	d	d	d	GBG, RBGPS	Sri Lanka
85.	*Bambusa prominens* H.L. Fung & C.Y. Sia	Bambuseae	China	SCBG, XTBG	China
86.	*Bambusa rectocuneata* (W.T. Lin) N.H. Xia, R.S. Lin & R.H. Wang *Neosinocalamus recto-cuneatus* W.T.Lin	Bambuseae	China	SCBG	China
87.	*Bambusa remotiflora* (Kuntze) L.C. Chia & H.L. Fung	Bambuseae	China to Vietnam	SCBG, XTBG, SZBG, GXIB	China
	Lingnania remotiflora	d	d	ABG	China
88.	*Bambusa rigida* Keng & Keng f.	Bambuseae	China	ABG, SCBG, WHIOB, XTBG GXIB	China
89.	*Bambusa rongchengensis* (T.P. Yi & C.Y. Sia) D.Z.Li	Bambuseae	China	SCBG, XTBG	China

Table 3.2 contd. ...

...Table 3.2 contd.

No.	Species	Tribe	Native Range	Repository	Country
90.	*Bambusa rutila* **McClure**	Bambuseae	China	SCBG	China
91.	*Bambusa salarkhanii* **Alam**	Bambuseae	Nepal to Bangladesh	BFRI	Bangladesh
92.	*Bambusa sinospinosa* **McClure**	Bambuseae	China to Hainan	SDBG	USA
	d	d	d	SCBG, XTBG, SZBG, GXIB	China
93.	*Bambusa* **sp**	Bambuseae	0	ABG	China
94.	*Bambusa* **sp.**	Bambuseae	0	SDBG	USA
95.	*Bambusa* **sp.**	Bambuseae	India	VVK	India
96.	*Bambusa* **sp.**	Bambuseae	India (Assam)	VVK	India
97.	*Bambusa* **sp.**	Bambuseae	India	VVK	India
98.	*Bambusa* **sp.**	Bambuseae	India	JNTBGRI	India
99.	*Bambusa* **sp.** (Hijo)	Bambuseae	India	VVK	India
100.	*Bambusa* **sp.** (Nangal)	Bambuseae	India	RPRC	India
101.	*Bambusa* **sp.** (Tapi)	Bambuseae	India	RPRC	India
102.	*Bambusa* **sp.** (Tapii)	Bambuseae	India	VVK	India
103.	*Bambusa* **sp.**- 2 accessions	Bambuseae	India	JNTBGRI	India
104.	*Bambusa* **sp.**- 2 accessions	Bambuseae	India	JNTBGRI	India
105.	*Bambusa* **sp.** 'Nana'	Bambuseae	0	SDBG	USA
106.	*Bambusa* **sp.** (Maithang)	Bambuseae	India	VVK	India
107.	*Bambusa* **sp.** (Routa)	Bambuseae	India	VVK	India
108.	*Bambusa* **sp.**1	Bambuseae	0	BBG	Indonesia
109.	*Bambusa* **sp.**	Bambuseae	India (Kerala)	JNTBGRI	India
110.	*Bambusa* **sp.**2	Bambuseae	0	BBG	Indonesia

111.	**Bambusa sp.3**	Bambuseae	0	BBG	Indonesia
112.	**Bambusa sp.'Nal'**	Bambuseae	India (Assam)	VVK	India
113.	**Bambusa sp.'Nal'**	Bambuseae	India	VVK	India
114.	**Bambusa sp.'Nangal'**	Bambuseae	India (Assam)	VVK	India
115.	**Bambusa spinosa Roxb.**	Bambuseae	Jawa to Maluku	FRI	India
	Bambusa blumeana Schult.f.	d	d	BBG	Indonesia
	d	d	d	SCBG, XTBG, GXIB, XMBG	China
	Bambusa bambos var. spinosa	d	d	BFRI	Bangladesh
116.	**Bambusa stenoaurita (W.T. Lin) T.H. Wen** *Dendrocalamopsis stenoaurita* (W.T. Lin) Keng f. ex W.T. Lin	Bambuseae	China	SCBG, SCBG, XTBG, GXIB	China
117.	**Bambusa subaequalis H.L. Fung & C.Y. Sia**	Bambuseae	China	SCBG	China
118.	**Bambusa subtruncata L.C. Chia & H.L. Fung**	Bambuseae	China	SDBG, SCBG, SZBG, GXIB	USA
	d	Bambuseae	China	SCBG	China
119.	**Bambusa surrecta (Q.H. Dai) Q.H. Dai**	Bambuseae	China	SCBG, GXIB	China
120.	**Bambusa teres Buch. Ham ex Munro**	Bambuseae	Nepal to China	AJCBIBG	India
	d	d	d	SMA	Hong Kong
121.	**Bambusa textilis McClure**	Bambuseae	China to Vietnam	GRIFFIN, MAY, SDBG	USA
	d	d	d	JBHLCM	Spain
	d	d	d	ABG, SCBG, WHIOB, KIB XTBG, SZBG, GXIB	China
	d	d	d	KFRI	India
	d	d	d	SMA	Hong Kong
	d	d	d	LBMG	France
	Bambusa textilis var. fusca	d	d	SCBG	China

Table 3.2 contd. ...

...*Table 3.2 contd.*

No.	Species	Tribe	Native Range	Repository	Country
	Bambusa textilis cv. Purpurascens	d	d	SCBG	China
	Bambusa textilis 'Albolineata'	d	d	SDBG	USA
	Bambusa textilis 'Dwarf form'	d	d	SDBG	USA
	Bambusa textilis 'Gracilis'	d	d	SDBG, KBG	USA
	Bambusa textilis 'Kanapaha'	d	d	SDBG, KBG	USA
	Bambusa textilis 'Mutabilis'	d	d	KBG	USA
	Bambusa textilis 'Scranton'	d	d	SDBG	USA
122.	**Bambusa textilis McClure var. glabra McClure** = (unresolved)	Bambuseae	0	SCBG, GXIB	
123.	**Bambusa textilis McClure var. gracilis McClure** = (unresolved)	Bambuseae	0	SCBG, WHIOB, XTBG, SZBG GXIB, XMBG	China
	Bambusa textilis var *gracilis*	d	d	SBG	Singapore
124.	**Bambusa tulda Roxb.**	Bambuseae	Himalaya to China and Indo-China	AJCBIBG, JNTBGRI, FRI, KFRI, RPRC, VVK	India
	d	d	d	SCBG, XTBG, SZBG, GXIB	China
	d	d	d	MAY, SDBG	USA
	d	d	d	BFRI	Bangladesh
	B. tulda cv. 'striata'	d	d	PCG	USA
125.	**Bambusa tulda var. gamblei P. Kumari & P.Singh**	Bambuseae	India (Meghalaya)	AJCBIBG	India
126.	**Bambusa tuldoides Munro**	Bambuseae	China to Malaysia	BBG	Indonesia
	d	d	d	MIA, MAY, SDBG	USA
	d	d	d	ABG, SCBG, XTBG, GXIB, XMBG	China
	d	d	d	SMA	Hong Kong

	Bambusa blumeana		d	PBG	Indonesia
	d		d	KFRI	India
	d		d	SDBG	USA
	d		d	SCBG	China
	d		d	SBG	Singapore
	Bambusa tuldoides 'Swolleninternode'		d	ABG, SCBG	China
	Bambusa tuldoides 'Ventricosa Kimmei'		d	SDBG	USA
127.	**Bambusa valida (Q.H. Dai) W.T. Lin** / *Dendrocalamopsis validus* Q.H. Dai	Bambuseae	China	ABG, SCBG, XTBG	China
128.	**Bambusa variostriata (W.T. Lin) L.C. Chia & H.L. Fung** / *Dendrocalamopsis vario-striata*	Bambuseae	China (Guangdong)	SCBG	China
129.	**Bambusa ventricosa McClure**	Bambuseae	China to Vietnam	BGR	Italy
	d		d	MAY, KBG, MBG, SDBG	USA
	d		d	JBHLCM	Spain
	d		d	PZ	Australia
	d		d	GBG, RBGPS	Sri Lanka
	d		d	FRI, RPRC	India
	d		d	SMA	Hong Kong
	d		d	LBMG	France
	d		d	SCBG, IBCAS, WHIOB, KIB CNBG, SZBG, GXIB, XMBG HBG, ABG	China
	d		d	BFRI	Bangladesh
	Bambusa ventricosa 'Kimmei'		d	KBG, SDBG	USA

Table 3.2 contd. ...

...*Table 3.2 contd.*

No.	Species	Tribe	Native Range	Repository	Country
	d	d	d	LBMG	France
130.	**Bambusa vulgaris Schrad. ex J.C. Wendl.**	Bambuseae	China to Indo-China	MAY	USA
	d	d	d	BBG	Indonesia
	d	d	d	AJCBIBG, VVK, FRI, RPRC JNTBGRI, KFRI	India
	d	d	d	RBGPS	Sri Lanka
	d	d	d	JBHLCM	Spain
	d	d	d	SCBG, XTBG, GXIB	China
	d	d	d	BFRI	Bangladesh
	d	d	d	SMA	Hong Kong
	d	d	d	RBGE	UK
	d				
	Bambusa vulgaris Schrad. ex J.C. Wendl. 'Vittata'	d	d	JNTBGRI	India
	d	d	d	SDBG	USA
	Bambusa vulgaris 'Vittata'	d	d	HBGH, GBG, RBGPS	Sri Lanka
	d	d	d	ABG, SCBG	China
	d	d	d	RPRC	India
	d	d	d	PCG	USA
	Bambusa vulgaris Schrad. ex J.C. Wendl. 'Wamin'	d	d	JNTBGRI	India
	d	d	d	FRI, KFRI	India
	Bambusa wamin E.G. Camus	d	d	AJCBIBG	India
	d	d	d	FRI	India

No.	Name				
	d	d		RPRC	India
	Bambusa vulgaris var. *wamin*	d		KFRI	India
	Bambusa vulgaris var.*vittata* A.& C. Rivierea			AJCBIBG	India
	Bambusa vulgaris var. *striata* (Lodd. ex Lindl.) Gamble	d		BBG	Indonesia
	Bambusa vulgaris 'Striata'	d		LBMG	France
	d	d		SMA	Hong Kong
	Bambusa vulgaris 'Wamin'	d		SMA	Hong Kong
	d	d		ABG, SCBG	China
	d	d		SBG	Singapore
	d	d		SDBG, PCG	USA
	Bambusa vulgaris 'Wamin Striata'	d		SDBG	USA
	Bambusa vulgaris 'Buddha's belly'	d		XTBG	China
	Bambusa vulgaris f. *waminii* T.H. Wen	d		BBG	Indonesia
	Bambusa vulgaris 'green form'	d		FRI	India
	Bambusa ventricosa 'Buddha's Belly Bamboo'	d		PCG	USA
	Gigantochloa auriculata (Kurz) Kurz	d		AJCBIBG	India
	Phyllostachys mitis Rivière & C. Rivière	d		RBGE	UK
131.	**Bambusa xiashanensis L.C. Chia & H.L. Fung**	Bambuseae	China	SCBG, SZBG	China
132.	**Bambusa xueliniana R.S. Lin et C.H. Zheng**	Bambuseae	SCBG		
133.	**Bashania fargesii (E.G. Camus) Keng f. & T.P. Yi**	Arundinarieae	China	ABG, HBG, WHIOB, SZBG	China
	d	d		RBGE	UK
134.	**Bashania qingchengshanensis Keng f. & T.P. Yi**	Arundinarieae	China	ABG	China
135.	**Bergbambos tessellata (Nees) Stapleton** *Thamnocalamus tessellatus* (Nees) Soderstr. & Ellis	Arundinarieae	S. Africa	RBGE, TGT	UK

Table 3.2 contd. ...

...Table 3.2 contd.

No.	Species	Tribe	Native Range	Repository	Country
	d	d	d	SDBG	USA
	d	d	d	LBMG	France
136.	**Bonia saxatilis (L.C. Chia et al.) N.H. Xia var. solida (C.D. Chu & C.S. Chao) D. Z. Li**	Bambuseae	China	GRIFFIN, MAY	USA
	Monocladus saxatilis	d	d	SCBG	China
137.	**Brachystachyum densiflorum (Rendle) Keng**	Arundinarieae	S. China	GRIFFIN, MAY	USA
	d	d	d	SZBG, GXIB	China
	Brachystachyum densiflorum var. villosum	d	d	ABG	China
138.	**Cephalostachyum chinense (Rendle) D.Z. Li & H.Q. Yang** Schizostachyum chinense Rendle	Bambuseae	China	RBGE	UK
	Schizostachyum chinense Rendle	d	d	XTBG	China
139.	**Cephalostachyum latifolium Munro** Cephalostachyum fussianum	Bambuseae	Nepal to China	KFRI, VVK	India
	Cephalostachyum fuchsianum Gamble	d	d	XTBG	China
140.	**Chimonobambusa angustifolia C.D. Chu & C.S. Chao**	Arundinarieae	China	ABG, SCBG, WHIOB	China
141.	**Chimonobambusa communis (Hsueh & T.P. Yi) T.H. Wen & Ohrnb.** Qiongzhuea communis Hsueh	Arundinarieae	China	WHIOB, SZBG, ABG	China
142.	**Chimonobambusa convoluta Q.H. Dai & X.L. Tao**	Arundinarieae	China	XTBG, GXIB	China
143.	**Chimonobambusa damingshanensis Hsueh & W.P. Zhang**	Arundinarieae	China	GXIB	China
144.	**Chimonobambusa sp.** Qiongzhuea sp.	Arundinarieae	E. Himalaya to China, Indo-China, Taiwan, Japan	SCBG	China
145.	**Chimonobambusa marmorea (Mitford) Makino**	Arundinarieae	China, Japan	ABG, SCBG, XTBG, SZBG	China
	d	d	d	LBMG	France
	Chimonobambusa marmorea 'Variegata'	d	d	LBMG	France

No.	Name				
146.	**Chimonobambusa ningnanica Hsueh f. & L.Z. Gao** *Chimonobambusa yunnanensis Hsueh & W.P. Zhang*	d	China	KIB, XTBG	China
147.	**Chimonobambusa purpurea Hsueh & T.P. Yi** *Chimonobambusa neopurpurea T.P. Yi*	d	China	WHIOB, SZBG, XMBG	China
148.	**Chimonobambusa quadrangularis (Franceschi) Makino**	Arundinariinae	China to Vietnam, Taiwan	JBHLCM	Spain
	d	d	d	LBMG	France
	d	d	d	SDBG	USA
	d	d	d	TGT, RBGE	UK
	d	d	d	SCBG, IBCAS, WHIOB, LSBG, CNBG, SZBG, GXIB	China
	Chimonobambusa quadrangularis 'Tatejima'	d	d	LBMG	France
149.	**Chimonobambusa sichuanensis (T.P. Yi) T.H. Wen** *Menstruocalamus sichuanensis (Yi) Yi*	Arundinarieae	China	ABG, SZBG	China
150.	**Chimonobambusa sp.**	Arundinarieae	0	HBGH	Sri Lanka
151.	**Chimonobambusa tumidissinoda Ohrnb.**	Arundinarieae	China	TGT	UK
	d	d	d	SDBG	USA
	Qiongzhuea tumidinoda Hsueh et. Yi	d	d	KIB, GXIB, SCBG, ABC	China
152.	**Chimonobambusa utilis (Keng) Keng f.**	Arundinarieae	China	ABG, KIB	China
153.	**Chimonocalamus delicatus Hsueh & T.P. Yi**	Arundinarieae	China	SDBG	USA
	d	d	d	KIB, GXIB	China
154.	**Chimonocalamus griffithianus (Munro) Hsueh & T.P. Yi** *Arundinaria griffithiana*	Arundinarieae	E. Himalaya to China and Indo-China	KFRI	India
155.	**Chimonocalamus lushaiensis Ohrnb.** *Sinarundinaria longispiculata*	Arundinarieae	India	RPRC	India

Table 3.2 contd. ...

...Table 3.2 contd.

No.	Species	Tribe	Native Range	Repository	Country
156.	**Chusquea aperta L.G. Clark** *Chusquea aperta*	Bambuseae	Mexico	FJCBG	Mexico
157.	**Chusquea bilimekii E. Fourn.** *Chusquea bilimekii*	Bambuseae	Mexico	FJCBG	Mexico
158.	**Chusquea circinata Soderstr. & C.E. Calderón**	Bambuseae	Mexico	SDBG	USA
	Chusquea circinata 'Teague'	d	Mexico to Central America	SDBG	USA
	Chusquea circinata	d	Central & S. Mexico to Guatemala	FJCBG	Mexico
159.	**Chusquea coronalis Soderstr. & C.E. Calderón**	Bambuseae	Mexico	LBMG	France
	d		Mexico	SDBG	USA
	d		Mexico	FJCBG	Mexico
	d		Mexico	SCBG	China
160.	**Chusquea cortesii L.G. Clark & Ruiz-Sanchez**	Bambuseae	Mexico to Honduras	FJCBG	Mexico
161.	**Chusquea culeou É. Desv.**	Bambuseae	Mexico to Central America	TGT, RBGE	UK
	d		Mexico (Chiapas) to Guatemala	SDBG	USA
162.	**Chusquea cumingii Nees**	Bambuseae	Mexico	RBGE	UK
163.	**Chusquea enigmatica Ruiz-Sanchez, Mejía-Saulés & L.G. Clark**	Bambuseae	Mexico	FJCBG	Mexico
164.	**Chusquea foliosa L.G. Clark**	Bambuseae	Mexico	SDBG	USA
165.	**Chusquea galeottiana Rupr. ex Munro**	Bambuseae	Mexico	FJCBG	Mexico
166.	**Chusquea gibcooperi Ruiz-Sanchez, Mejía-Saulés, G.Cortés & L.G. Clark**	Bambuseae	Mexico	FJCBG	Mexico
167.	**Chusquea gigantea Demoly**	Bambuseae	Mexico	TGT	UK
168.	**Chusquea glauca L.G. Clark**	Bambuseae	Mexico	FJCBG	Mexico

169.	*Chusquea lanceolata* Hitchc.	Bambuseae	Mexico	FJCBG	Mexico
	Chusquea 'Las Vigas'	Bambuseae	Mexico to Ecuador and Venezuela	SDBG	USA
170.	*Chusquea liebmannii* E. Fourn.	Bambuseae	Mexico	FJCBG	Mexico
	d	d	Mexico	SCBG	China
171.	*Chusquea longifolia* Swallen	Bambuseae	Mexico	FJCBG	Mexico
172.	*Chusquea mallatzinca* L.G. Clark & Ruiz-Sanchez	Bambuseae	Mexico	FJCBG	Mexico
173.	*Chusquea mimosa* 'Australis'	Bambuseae	Mexico	SDBG	USA
174.	*Chusquea mulleri* Munro	Bambuseae	Mexico	FJCBG	Mexico
175.	*Chusquea nedjaquithii* Ruiz-Sanchez, Mejia-Saulés & L.G. Clark	Bambuseae	Mexico to Trop. America	FJCBG	Mexico
176.	*Chusquea nelsonii* Scribn. & J.G. Sm.	Bambuseae	Mexico (Veracruz)	FJCBG	Mexico
177.	*Chusquea perotensis* L.G. Clark, G. Cortes & Cházaro	Bambuseae	Mexico	FJCBG	Mexico
178.	*Chusquea pittieri* Hack.	Bambuseae	Mexico	SDBG, PCG	USA
	d	d	Mexico	FJCBG	Mexico
179.	*Chusquea quila* Kunth	Bambuseae	Mexico	TGT	UK
	d	d	Mexico	RBGE	UK
180.	*Chusquea repens* L. G. Clark & Londoño *Chusquea repens* subsp. *repens* & subsp. *oaxacacensis*	Bambuseae	Mexico	FJCBG	Mexico
181.	*Chusquea septentrionalis* Ruiz-Sanchez, Art. Castro & L.G. Clark	Bambuseae	Mexico	FJCBG	Mexico
182.	*Chusquea simpliciflora* Munro	Bambuseae	Mexico	FJCBG	Mexico
183.	*Chusquea sulcata* Swallen	Bambuseae	S. Mexico to Honduras, Peru to Brazil and Paraguay	FJCBG	Mexico

Table 3.2 contd. ...

...*Table 3.2 contd.*

No.	Species	Tribe	Native Range	Repository	Country
184.	**Cryptochloa strictiflora (E. Fourn.) Swallen** *Cryptochloa strictiflora*	Olyreae	Mexico to Tropical America	FJCBG	Mexico
185.	**Davidsea attenuata (Thwaites) Soderstr. & R.P. Ellis** *Teinostachyum attenuatum*	Bambuseae	Mexico	KFRI	India
186.	**Dendrocalamus asper (Schult. & Schult. f.) Backer ex K. Heyne**	Bambuseae	Mexico	MAY	USA
	d	d	d	BBG, PBG	Indonesia
	d	d	d	RBGPS	Sri Lanka
	d	d	d	FRI, KFRI, JNTBGRI	India
	d	d	d	SBG	Singapore
	d	d	d	SDBG, PCG	USA
	d	d	d	SCBG, XTBG, SZBG	China
	Dendrocalamus asper 'Betung Hitam'	d	d	SDBG	USA
187.	**Dendrocalamus bambusoides Hsueh & D.Z. Li**	Bambuseae	Mexico	SCBG, XTBG	China
188.	**Dendrocalamus barbatus Hsueh & D.Z. Li**	Bambuseae	Mexico	SCBG, XTBG, SZBG	China
	d	d	d	RBGE	UK
	Dendrocalamus barbatus Hsueh et D.Z. Li var. *internodiiradicatus* Hsueh et. D.Z. Li	d	Mexico	SCBG, XTBG, SZBG	China
189.	**Dendrocalamus birmanicus A. Camus**	Bambuseae	Mexico to Central America	SCBG, XTBG	China
190.	**Dendrocalamus brandisii (Munro) Kurz**	Bambuseae	Mexico to Surinam and Argentina	FRI, VVK, KFRI, JNTBGRI	India
	d	d	d	ABG, SCBG, XTBG, GXIB	China
	d	d	d	BFRI	Bangladesh
	d	d	d	SDBG	USA

			Distribution	Institutions	Country
	Dendrocalamus brandisii 'Maroochy'	♂		SDBG	USA
	Dendrocalamus brandisii 'Silver'	♂		SDBG	USA
	Dendrocalamus brandisii 'Teddy Bear'	♂		SDBG	USA
191.	**Dendrocalamus calostachyus (Kurz) Kurz**	Bambuseae	India to China	SCBG, XTBG	China
		♂		FRI, JNTBGRI	India
192.	**Dendrocalamus elegans (Ridl.) Holttum**	Bambuseae	Thailand to Malaysia	SBG	Singapore
193.	**Dendrocalamus exauritus (W.T. Lin) N.H. Xia & Y.B. Guo** *Drepanostachyum exauritum*	Bambuseae	China	ABG	China
194.	**Dendrocalamus farinosus (Keng & Keng f.) L.C. Chia & H.L. Fung**	Bambuseae	China to Vietnam	SCBG, XTBG, GXIB, ABG	China
	Dendrocalamus ovatus	♂		SCBG	China
195.	**Dendrocalamus fugongensis Hsueh & D.Z. Li**	♂	China	XTBG	China
196.	**Dendrocalamus giganteus Munro**	Bambuseae	India to China	GBG, RBGPS	Sri Lanka
		♂		FRI, RPRC, KFRI, JNTBGRI AJCBIBG	India
		♂		SBG	Singapore
		♂		SCBG, XTBG, SZBG, GXIB	China
		♂		SDBG	USA
		♂		BFRI	Bangladesh
197.	**Dendrocalamus hamiltonii Nees & Arn. ex Munro**	Bambuseae	Nepal to China and Indo-China	AJCBIBG, FRI, KFRI, VVK, RPRC, JNTBGRI	India
		♂		RBGPS	Sri Lanka
		♂		SCBG, XTBG, SZBG, ABG	China
		♂		SDBG	USA
		♂		BFRI	Bangladesh

Table 3.2 contd. ...

...*Table 3.2 contd.*

No.	Species	Tribe	Native Range	Repository	Country
	Dendrocalamus hamiltonii var. *hamiltonii* *Dendrocalamus semiscandens* Hsueh et. D.Z. Li	d	d	SCBG, XTBG	China
	d	d	d	SCBG	China
	Dendrocalamus hamiltonii 'Hastings'	d	d	SDBG	USA
198.	**Dendrocalamus hookeri Munro**	Bambuseae	Nepal to Myanmar	JNTBGRI	India
	d	d	d	XTBG	China
199.	**Dendrocalamus jianshuiensis Hsueh & D.Z. Li**	Bambuseae	China	SDBG	USA
200.	**Dendrocalamus latiflorus Munro**	Bambuseae	Myanmar to China and Taiwan	SCBG, XTBG, SZBG, GXIB	China
	Dendrocalamus latiflorus 'Mei-nung'	d	d	SDBG	USA
201.	**Dendrocalamus longispathus (Kurz) Kurz**	Bambuseae	India to Malaysia	FRI, JNTBGRI	India
	d	d	d	GBG, RBGPS, HBGH	Sri Lanka
	d	d	d	BFRI	Bangladesh
202.	**Dendrocalamus membranaceus Munro**	Bambuseae	Bangladesh to China and Indo-China	AJCBIBG	India
	d	d	d	VVK	India
	d	d	d	PCG	USA
	d	d	d	SCBG, XTBG, SZBG	China
	d	d	d	RBGE	UK
	Dendrocalamus membranaceus Munro f. *fimbriligulatus* Hsueh & D.Z. Li	d	d	XTBG	China
	Dendrocalamus membranaceus Munro f. *pilosus* Hsueh & D.Z. Li	d	d	XTBG	China
	Bambusa membranacea (Munro) Stapleton & N.H. Xia)	d	d	JNTBGRI, FRI, RPRC, KFRI	India

	d	d		d	SDBG	USA
203.	**Dendrocalamus menglongensis Hsueh & K.L. Wang ex N.H. Xia, R.S. Lin et Y.B. Guo**	Bambuseae		China	SCBG, XTBG	China
204.	**Dendrocalamus minor (McClure) L.C. Chia et. H.L. Fung**	Bambuseae		China	SCBG, XTBG, SZBG, GXIB	China
	Dendrocalamus sapidus Q.H. Dai et D.Y. Huang	d		China	SDBG, XTBG, GXIB	USA
	Dendrocalamus minor McClure L.C. Chia et. H.L. Fung var. *amoenus* (Q.H.Dai et C.F. Huang) Hsueh et. D.Z. Li	d			SCBG, GXIB, WHIOB, SZBG, XMBG	China
	Dendrocalamus minor 'Amoenus'	d		d	SDBG	USA
205.	**Dendrocalamus pachystachyus Hsueh et D.Z. Li**	Bambuseae		China	SCBG, XTBG	China
206.	**Dendrocalamus peculiaris Hsueh & D.Z. Li**	Bambuseae		China	XTBG, SZBG	China
207.	**Dendrocalamus pulverulentus L.C. Chia & But**	Bambuseae		China	SCBG	China
208.	**Dendrocalamus sahnii H.B. Naithani & Bahadur**	Bambuseae		India	JNTBGRI	India
209.	**Dendrocalamus sikkimensis Gamble ex Oliv.**	Bambuseae		E. Himalaya to China	FRI, VVK, JNTBGRI, KFRI	India
	d		d	d	XTBG	China
	d		d	d	SDBG	USA
210.	**Dendrocalamus sikkimensis Gamble ex Oliv. var. tumidus K.L. Wang** =(unresolved)	Bambuseae		0	XTBG	China
211.	**Dendrocalamus sinicus L.C. Chia & J.L. Sun**	Bambuseae		China to Laos	SMA	Hong Kong
	d		d	d	SCBG, KIB, XTBG, SZBG, XMBG	China
	Dendrocalamus sinicus L.C. Chia et J.L. Sun f. *aequatus* K.L. Wang	d		d	XTBG, SZBG	China
	Dendrocalamus sinicus L.C.Chia et. J.L. Sun var. *fimbriligulatus* K.L. Wang	d		d	XTBG	China
212.	**Dendrocalamus somdevae H.B. Naithani** *Dendrocalamus somdevai*	Bambuseae		India	FRI	India
213.	**Dendrocalamus sp.**	Bambuseae		0	BBG	Indonesia

Table 3.2 contd. ...

...*Table 3.2 contd.*

No.	Species	Tribe	Native Range	Repository	Country
214.	*Dendrocalamus* sp.	Bambuseae	0	SDBG	USA
215.	*Dendrocalamus* sp.	Bambuseae	India	VVK	India
216.	*Dendrocalamus* sp.	Bambuseae	0	JBMPC	Spain
217.	*Dendrocalamus* sp.	Bambuseae	0	SBG	Singapore
218.	*Dendrocalamus* sp.-	Bambuseae	India (Kerala)	JNTBGRI	India
219.	*Dendrocalamus* sp.	Bambuseae	India (Manipur)	JNTBGRI	India
220.	*Dendrocalamus* sp.	Bambuseae	India (Manipur)	JNTBGRI	India
221.	*Dendrocalamus* sp.	Bambuseae	India	JNTBGRI	India
222.	*Dendrocalamus* sp.	Bambuseae	India (Kerala)	JNTBGRI	India
223.	*Dendrocalamus* sp.	Bambuseae	India (Kerala)	JNTBGRI	India
224.	*Dendrocalamus* sp.	Bambuseae	India (Kerala)	JNTBGRI	India
225.	*Dendrocalamus strictus* (Roxb.) Nees	Bambuseae	Indian Subcontinent to Indo-China	AJCBIBG, FRI, RPRC, VVK, JNTBGRI, KFRI	India
	d	d	d	MIA, MAY	USA
	d	d	d	RBGPS	Sri Lanka
	d	d	d	SCBG, XTBG, GXIB	China
	d	d	d	BFRI	Bangladesh
226.	*Dendrocalamus tsiangii* (McClure) L.C. Chia & H.L. Fung	Bambuseae	China	SCBG	China
227.	*Dendrocalamus yunnanicus* Hsueh & D.Z. Li	Bambuseae	China to Vietnam	SCBG, XTBG, SZBG, GXIB	China
228.	*Dinochloa andamanica* Kurz	Bambuseae	Andaman & Nicobar Islands, Myanmar to Thailand	AJCBIBG, JNTBGRI	India
229.	*Dinochloa kostermansiana* S. Dransf.	Bambuseae	Lesser Sunda Islands	BBG	Indonesia

230.	*Dinochloa macclellandii* (Munro) Kurz	Bambuseae	E. Himalaya to Indo-China	AJCBIBG, KFRI, VVK, FRI, RPRC, JNTBGRI	India
231.	*Dinochloa matmat* **S. Dransf. & Widjaja**	Bambuseae	Jawa	PBG	Indonesia
232.	*Dinochloa puberula* **McClure** *Neohouzeaua puberula* (McClure) T.H. Wen	Bambuseae	Hainan	XMBG	China
233.	*Dinochloa scabrida* **S. Dransf.**	Bambuseae	N. & E. Borneo	SBG	Singapore
234.	*Dinochloa sepang* **Widjaja & Astuti**	Bambuseae	Lesser Sunda Islands (Bali)	BBG	Indonesia
235.	*Dinochloa* sp.	Bambuseae	0	BBG	Indonesia
236.	*Dinochloa* sp.	Bambuseae	Little Andamans	JNTBGRI	India
237.	*Drepanostachyum annulatum* **Stapleton**	Arundinarieae	S. Bhutan	RBGE	UK
238.	*Drepanostachyum falcatum* **(Nees) Keng f.**	Arundinarieae	W. & Central Himalaya to Indo-China	TGT	UK
	d	d	d	FRI	India
239.	*Drepanostachyum khasianum* **(Munro) Keng f.**	Arundinarieae	Nepal to N. Myanmar	TGT	UK
	d	d	d	SDBG	USA
240.	*Fargesia altior* **T.P. Yi**	Arundinarieae	China	XTBG	China
241.	*Fargesia angustissima* **T.P. Yi** *Borinda angustissima*	Arundinarieae	China	SDBG	USA
242.	*Fargesia contracta* **T.P. Yi**	Arundinarieae	China	TGT	UK
243.	*Fargesia denudata* **T.P. Yi**	Arundinarieae	China	SDBG	USA
244.	*Fargesia dracocephala* **T.P. Yi**	Arundinarieae	China	TGT	UK
	d	d	d	SZBG	China
	Fargesia dracocephala 'Rufa'	d	d	SDBG	USA
245.	*Fargesia frigidis* **T.P. Yi** *Fargesia frigidus*	Arundinarieae	China	TGT	UK

Table 3.2 contd. ....

...*Table 3.2 contd.*

No.	Species	Tribe	Native Range	Repository	Country
246.	**Fargesia fungosa T.P. Yi** *Borinda fungosa* 'Chocolate'	Arundinarieae	China	SDBG	USA
	Borinda fungosa	d	d	SDBG	USA
247.	**Fargesia lushuiensis Hsueh & T.P. Yi** *Fargesia lushiensis*	Arundinarieae	China	TGT	UK
248.	**Fargesia macclureana (Bor) Stapleton**	Arundinarieae	SE. Tibet	TGT	UK
	Borinda macclureana	d	d	SDBG	USA
249.	**Fargesia murielae (Gamble) T.P. Yi**	Arundinarieae	China	RBGE	UK
	d	d	d	LBMG	France
	d	d	d	TGT	UK
	d	d	d	SDBG	USA
	Fargesia murielae 'Jumbo'	d	d	LBMG	France
	Fargesia murielae 'Simba'	d	d	LBMG	France
	Fargesia murielae 'Harewood'	d	d	LBMG	France
250.	**Fargesia nitida (Mitford) Keng f. ex T.P. Yi**	Arundinarieae	China	LBMG	France
	d	d	d	WHIOB	China
	d	d	d	TGT, RBGE	UK
	d	d	d	SDBG	USA
	Fargesia nitida 'Jiuzhaigou'	d	d	SDBG	USA
251.	**Fargesia papyrifera T.P. Yi**	Arundinarieae	China	TGT	UK
	d	d	d	RBGE	UK
	d	d	d	KIB	China
252.	**Fargesia plurisetosa T.H. Wen**		China	XTBG	China

253.	*Fargesia robusta* **T.P. Yi**	Arundinarieae	China	TGT	UK
	d	d	d	SDBG	USA
	d	d	d	LBMG	France
	d	d	d	SDBG	USA
	Fargesia robusta 'Wolong'	d	0	SDBG	USA
254.	*Fargesia rufa* **T.P. Yi**	Arundinarieae	China	TGT	UK
	d	d	d	MBG	USA
	Fargesia rufa 'Dracocephala'	d	d	SDBG	USA
255.	*Fargesia* **sp.** 'Scabrida'	Arundinarieae	0	SDBG	USA
256.	*Fargesia spathacea* **Franch.** *Fargesia spathaceus*	Arundinarieae	China	SDBG	USA
	d	d	d	SZBG	China
257.	*Fargesia yulongshanensis* **T.P. Yi**	Arundinarieae	China	TGT	UK
258.	*Fargesia yunnanensis* **Hsueh & T.P. Yi**	Arundinarieae	China	KIB, GXIB	China
	Ferrocalamus rimosivaginus T.H. Wen	d	d	SCBG	China
259.	*Ferrocalamus strictus* **Hsueh & Keng f.**	Arundinarieae	China	SCBG	China
260.	*Gelidocalamus kunishii* **(Hayata) Keng f. & T.H. Wen** *Arundinaria kunishii*	Arundinarieae	Nansei-shoto to N. & Central Taiwan	LBMG	France
261.	*Gelidocalamus stellatus* **T.H. Wen**	Arundinarieae	China	ABG	China
262.	*Gigantochloa albociliata* **(Munro) Kurz**	Bambuseae	India to China and Indo-China	SDBG	USA
	d	d	d	FRI, KFRI, VVK, JNTBGRI	India
	d	d	d	SCBG, XTBG	China
263.	*Gigantochloa apus* **(Blume ex Schult.f) Kurz**	Bambuseae	India to China and Malesia	AJCBIBG	India

Table 3.2 contd. ...

...Table 3.2 contd.

No.	Species	Tribe	Native Range	Repository	Country
	d	d	d	MAY, SDBG	USA
	d	d	d	BBG, PGB	Indonesia
	d	d	d	BFRI	Bangladesh
	d	d	d	SCBG	China
264.	*Gigantochloa atroviolacea* Widjaja	Bambuseae	China, Jawa to Lesser Sunda Islands (Bali)	BBG	Indonesia
	d	d	d	AJCBIBG, KFRI, FRI, JNTBGRI	India
	d	d	d	PCG, SDBG	USA
	d	d	d	BFRI	Bangladesh
265.	*Gigantochloa atter* (Hassk.) Kurz ex Munro	Bambuseae	Indo-China to New Guinea	BBG	Indonesia
	d	d	d	RBGPS	Sri Lanka
	d	d	d	FRI, KFRI	India
266.	*Gigantochloa aya* Widjaja & Astuti	Bambuseae	Lesser Sunda Islands (Bali)	BBG	Indonesia
267.	*Gigantochloa baliana* Widjaja & Astuti	Bambuseae	Lesser Sunda Islands (Bali)	BBG	Indonesia
268.	*Gigantochloa balui* K.M. Wong	Bambuseae	Borneo	SBG	Singapore
269.	*Gigantochloa felix* (Keng) Keng f.		China	SCBG, XTBG, SZBG	China
270.	*Gigantochloa hasskarliana* (Kurz) Backer	Bambuseae	Indo-China to W. Malesia and Lesser Sunda Islands (Bali)	BBG	Indonesia
271.	*Gigantochloa kuring* Widjaja	Bambuseae	Sumatera	BBG	Indonesia

	Gigantochloa kuring 'Sumatra 3751'	d	d	SDBG	USA
272.	**Gigantochloa latifolia Ridl.**	Bambuseae	Laos, Malaysia	SBG	Singapore
273.	**Gigantochloa levis (Blanco) Merr.**	Bambuseae	China to Vietnam, Malesia	SCBG, XTBG, SZBG	China
274.	**Gigantochloa ligulata Gamble**	Bambuseae	Thailand to Malaya	SBG	Singapore
	d	d	d	XTBG	China
275.	**Gigantochloa luteostriata Widjaja**	Bambuseae	Borneo, Lesser Sunda Islands (Bali)	BBG	Indonesia
276.	**Gigantochloa macrostachya Kurz**	Bambuseae	India to Myanmar	VVK, KFRI	India
277.	**Gigantochloa magentea Widjaja**	Bambuseae	Sumatera	BBG	Indonesia
278.	**Gigantochloa manggong Widjaja**	Bambuseae	Myanmar, E. Jawa to Bali	BBG, PBG	Indonesia
	d	d	d	KFRI	India
279.	**Gigantochloa nigrociliata (Buse) Kurz**	Bambuseae	India to Malesia	SCBG, XTBG	China
	d	d	d	JNTBGRI, AJCBIBG	India
	d	d	d	BBG	Indonesia
	Gigantochloa andamanica	d	d	BFRI	Bangladesh
	Oxytenanthera nigrociliata	d	d	RPRC	India
280.	**Gigantochloa parviflora (Keng f.) Keng f.**	Bambuseae	China	SCBG, XTBG, SZBG	China
	Oxytenanthera parvifolia Brandis ex Gamble	d	d	AJCBIBG	India
281.	**Gigantochloa pubinervis Widjaja**	Bambuseae	Sumatera, Lesser Sunda Islands (Bali)	BBG	Indonesia
	d	d	d	SDBG	USA
282.	**Gigantochloa pubipetiolata Widjaja**	Bambuseae	Sumatera, Lesser Sunda Islands (Bali)	BBG	Indonesia

Table 3.2 contd. ...

...Table 3.2 contd.

No.	Species	Tribe	Native Range	Repository	Country
283.	*Gigantochloa ridleyi* Holttum	Bambuseae	Malaysia	SBG	Singapore
284.	*Gigantochloa robusta* Kurz	Bambuseae	Sumatera to W. Jawa, Lesser Sunda Islands	BBG	Indonesia
285.	*Gigantochloa rostrata* K.M. Wong	Bambuseae	Malaysia	AJCBIBG, FRI, RPRC	India
	d	d	d	XTBG	China
286.	*Gigantochloa scortechinii* Gamble	Bambuseae	Malaysia to Sumatera	SBG	Singapore
	d	d	d	XTBG	China
287.	*Gigantochloa serik* Widjaja	Bambuseae	Sumatera, Lesser Sunda Islands	BBG	Indonesia
288.	*Gigantochloa* sp.	Bambuseae	0	SDBG	USA
289.	*Gigantochloa* sp.	Bambuseae	0	SBG	Singapore
290.	*Gigantochloa* sp. 'Malaysian Black'	Bambuseae	0	SDBG	USA
291.	*Gigantochloa* sp.1	Bambuseae	0	BBG	Indonesia
292.	*Gigantochloa* sp.2	Bambuseae	0	BBG	Indonesia
293.	*Gigantochloa* sp.	Bambuseae	Andaman Islands	JNTBGRI	India
294.	*Gigantochloa* sp.	Bambuseae	0	JNTBGRI	India
295.	*Gigantochloa* sp.	Bambuseae	0	JNTBGRI	India
296.	*Gigantochloa taluh* Widjaja & Astuti	Bambuseae	Sumatera, Lesser Sunda Islands	BBG	Indonesia
297.	*Gigantochloa thoi* K.M. Wong	Bambuseae	Malaysia to Sumatera, Lesser Sunda Islands (Bali)	BBG	Indonesia
	d	d	d	SBG	Singapore

298.	*Gigantochloa velutina* **Widjaja**	Bambuseae	Sumatera, Lesser Sunda Islands (Bali)	BBG	Indonesia
299.	*Gigantochloa verticillata* **(Willd.) Munro**	Bambuseae	Indo-China to W. Malesia	SDBG	USA
	Gigantochloa pseudoarundinacea	d	d	SCBG, XTBG, SZBG	China
		d	d	FRI	India
300.	*Gigantochloa wrayi* **Gamble**	Bambuseae	Myanmar to Malaysia	SBG	Singapore
		d	d	SDBG	USA
301.	*Guadua aculeata* **E. Fourn.**	Bambuseae	Mexico to America	FJCBG	Mexico
302.	*Guadua amplexifolia* **J. Presl**	Bambuseae	Mexico to Colombia and Trinidad	FJCBG	Mexico
	Guadua amplexifolia 'Velutina'	d	d	SDBG	USA
303.	*Guadua angustifolia* **Kunth**	Bambuseae	Trinidad, Venezuela to N. Peru	MAY	USA
		d	d	KFRI, JNTBGRI	India
		d	d	SDBG	USA
		d	d	XTBG	China
304.	*Guadua chacoensis* **(Rojas Acosta) Londoño & P.M.Peterson**	Bambuseae	Bolivia to Brazil and Argentina	BBG	Indonesia
	Guadua angustifolia Kunth ssp. *chacoensis* (Rojas Acosta) S.M. Young et. W.S. Judd	d	d	XTBG	China
305.	*Guadua inermis* **E. Fourn.**	Bambuseae	Mexico	FJCBG	Mexico
306.	*Guadua longifolia* **(E. Fourn.) R.W. Pohl**	Bambuseae	Mexico to America	FJCBG	Mexico
307.	*Guadua paniculata* **Munro**	Bambuseae	Mexico to tropical America	FJCBG	Mexico

Table 3.2 contd. ...

...*Table 3.2 contd.*

No.	Species	Tribe	Native Range	Repository	Country
308.	*Guadua tuxtlensis* Londoño & Ruiz-Sanchez	Bambuseae	Mexico	FJCBG	Mexico
309.	*Guadua velutina* Londoño & L.G. Clark	Bambuseae	Mexico	FJCBG	Mexico
310.	*Himalayacalamus asper* Stapleton	Arundinarieae	Nepal	LBMG	France
311.	*Himalayacalamus cupreus* Stapleton	Arundinarieae	Nepal	RBGE	UK
312.	*Himalayacalamus falconeri* (Hook.f. ex Munro) Keng. f.	Arundinarieae	Himalaya to Tibet	RBGE	UK
	Himalayacalamus falconeri 'Damarapa'	d	d	TGT	UK
	d	d	d	SDBG	USA
313.	*Himalayacalamus hookerianus* (Munro) Stapleton	Arundinarieae	Nepal to Assam	TGT	UK
314.	*Himalayacalamus planatus* Stapleton	Arundinarieae	Nepal	SDBG	USA
315.	*Himalayacalamus porcatus* Stapleton	Arundinarieae	Nepal	TGT	UK
	d	d	d	SDBG	USA
316.	*Indocalamus barbatus* McClure	Arundinarieae	China	ABG, GXIB	China
317.	*Indocalamus decorus* Q.H. Dai	Arundinarieae	China	SCBG, SZBG, GXIB	China
318.	*Indocalamus guangdongensis* H.R. Zhao & Y.L. Yang	Arundinarieae	S. China	ABG	China
	d	d	d	SCBG, SZBG	China
319.	*Indocalamus herklotsii* McClure	Arundinarieae	China (Hong Kong)	ABG	China
	d	d	d	SCBG, XTBG	China
320.	*Indocalamus hirsutissimus* Z.P. Wang & P.X. Zhang *Indocalamus hirsutissimus* var. *glabrifolius*	Arundinarieae	China	SCBG	China
321.	*Indocalamus hunanensis* B.M.Yang *Indocalamus wuxiensis* Yi	Arundinarieae	China	ABG, SCBG, WHIOB	China
322.	*Indocalamus latifolius* (Keng) McClure	Arundinarieae	China	SCBG, KIB, WHIOB, XTBG LSBG, CNBG, ABG, SZBG XMBG	China

No.	Name	Tribe	Distribution	Garden	Country
	Indocalamus lacunosus		d	ABG	China
	Indocalamus migoi		d	ABG	China
	Pseudosasa hirta S.L. Chen		d	LSBG	China
323.	**Indocalamus longiauritus Hand.-Mazz. Pseudosasa guanxianensis**	Arundinarieae	China	SCBG, SZBG, GXIB	China
324.	**Indocalamus pedalis (Keng) Keng f.**	Arundinarieae	China	ABG, SZBG	China
325.	**Indocalamus pumilus Q.H. Dai & C.F. Huang**	d	China	XTBG, GXIB	China
326.	**Indocalamus sinicus (Hance) Nakai**	Arundinarieae	China to Hainan	SBG	Singapore
	d	d	d	SCBG	China
327.	**Indocalamus tessellatus (Munro) Keng f.**	Arundinarieae	China	GRIFFIN	USA
	d	d	d	JBHLCM	Spain
	d	d	d	TGT, RBGE	UK
	d	d	d	SCBG, IBCAS, HIOB, CNBG, ABG	China
	Bambusa tesselata	d	d	HBGH	Sri Lanka
	Sasa tessellata	d	d	LBMG	France
328.	**Indocalamus victorialis Keng f.**	Arundinarieae	China	ABG	China
329.	**Indosasa angustata McClure**	Arundinarieae	China to Vietnam	SCBG, SZBG, XTBG	China
330.	**Indosasa gigantea (T.H. Wen) T.H. Wen Acidosasa gigantea**	Arundinarieae	China	ABG, XMBG	China
331.	**Indosasa glabrata C.D. Chu & C.S. Chao**	Arundinarieae	China	ABG, GXIB	China
	Indosasa glabrata var. *albo-hispidula*				
	Indosasa glabrata var. *albohispidula* (Dai et. Huang) Chao et Chu			GXIB	China
332.	**Indosasa hispida McClure**	Arundinarieae	China	XTBG, SZBG	China
333.	**Indosasa ingens Hsueh & T.P. Yi**	Arundinarieae	China	XTBG	China
334.	**Indosasa longispicata W.Y.Hsiung & C.S. Chao Sinobambusa striata**	Arundinarieae	China	ABG	China

Table 3.2 contd. ...

...Table 3.2 contd.

No.	Species	Tribe	Native Range	Repository	Country
335.	**Indosasa shibataeoides McClure**	Arundinarieae	China	ABG, XTBG, GXIB	China
	Indosasa acutiligulata	d	d	ABG	China
	Indosasa levigata	d	d	ABG	China
336.	**Indosasa singulispicula T.H. Wen**	Arundinarieae	China	XTBG	China
337.	**Indosasa sinica C.D. Chu & C.S. Chao**	Arundinarieae	China to Laos	SCBG, XTBG, SZBG, GXIB	China
	d	d	d	ABG	China
338.	**Indosasa sp.**	Arundinarieae	0	ABG	China
339.	*Lingnania chungii var. velutina* = (unresolved)	Bambuseae	China to Vietnam	ABG	China
340.	**Lithachne pauciflora (Sw.) P. Beauv.**	Olyreae	America	FJCBG	Mexico
341.	**Melocalamus arrectus T.P. Yi**	Bambuseae	China	SCBG, XTBG	China
342.	**Melocalamus compactiflorus (Kurz) Benth.**	Bambuseae	India to China and Indo-China	JNTBGRI, VVK	India
	d	d	d	BFRI	Bangladesh
	d	d	d	RBGE	UK
	d	d	d	XTBG	China
	Melocalamus compactiflorus (Kurz) Benth. Et. Hook. f. var. fimbriatus (J.R. Xue et. C.M. Hui) D.Z. Li et Z.H. Guo	d	d	XTBG	China
	Dinochloa compactiflora	d	d	FRI	India
343.	**Melocanna arundina C.E. Parkinson** *Melocanna humilis* Kurz	Bambuseae	Bangladesh to Thailand	SCBG, XTBG	China
344.	**Melocanna baccifera (Roxb.) Kurz**	Bambuseae	India to Myanmar	AJCBIBG, FRI, RPRC, KFRI, VVK, JNTBGRI	India
	d	d	d	MAY	USA

	♂		RBGPS	Sri Lanka
	♂		SBG	Singapore
	♂		SCBG, WHIOB, XTBG, XMBG	China
	♂		BFRI	Bangladesh
345.	*Merostachys mexicana* Ruiz-Sanchez & L.G. Clark	Bambuseae	SE. Mexico	FJCBG	Mexico
346.	*Neololeba atra* (Lindl.) Widjaja	Bambuseae	Malesia to Queensland, Caroline Islands (Palau)	BBG	Indonesia
	Bambusa atra		♂	JNTBGRI, AJCBIBG	India
			♂	GBG, RBGPS	Sri Lanka
347.	*Neomicrocalamus prainii* (Gamble) Keng f.	Bambuseae	India to China	SCBG, ABG	China
348.	*Ochlandra ebracteata* Raizada & Chatterji	Bambuseae	SW. India	JNTBGRI, KFRI	India
349.	*Ochlandra scriptoria* (Dennst.) C.E.C. Fisch.	Bambuseae	India	AJCBIBG, KFRI, JNTBGRI	India
350.	*Ochlandra setigera* Gamble- 10 accessions	Bambuseae	India	JNTBGRI	India
351.	*Ochlandra* sp.	Bambuseae	India	RBGPS	Sri Lanka
352.	*Ochlandra* sp.- 2 accessions	Bambuseae	India (Kerala)	JNTBGRI	India
353.	*Ochlandra* sp.- 5 accessions	Bambuseae	India (Kerala)	JNTBGRI	India
354.	*Ochlandra* sp.- 9 accessions	Bambuseae	India (Kerala)	JNTBGRI	India
355.	*Ochlandra* sp.- accessions	Bambuseae	India (Kerala)	JNTBGRI	India
356.	*Ochlandra* sp.-11 accessions	Bambuseae	India (Kerala)	JNTBGRI	India
357.	*Ochlandra* sp.-12 accessions	Bambuseae	India (Kerala)	JNTBGRI	India
358.	*Ochlandra* sp.-6 accessions	Bambuseae	India (Kerala)	JNTBGRI	India
359.	*Ochlandra talbotii* Brandis	Bambuseae	India	KFRI	India
360.	*Ochlandra travancorica* (Bedd.) Gamble	Bambuseae	S. India	JNTBGRI, KFRI, AJCBIBG, FRI	India

Table 3.2 contd. ...

...*Table 3.2 contd.*

No.	Species	Tribe	Native Range	Repository	Country
	Ochlandra sivagiriana (Gamble) E.G. Camus	d	d	JNTBGRI	India
361.	**Ochlandra wightii (Munro) C.E.C. Fisch.**	Bambuseae	S. India	JNTBGRI, KFRI	India
362.	**Oligostachyum glabrescens (T.H. Wen) Q.F. Zheng & Z.P. Wang**	Arundinarieae	China	SDBG	USA
363.	**Oligostachyum hupehense (J.L. Lu) Z.P. Wang & G.H.Ye** *Sinobambusa anaurita*	Arundinarieae	China	ABG	China
364.	**Oligostachyum lubricum (T.H. Wen) Keng f.**	Arundinarieae	China	ABG, SZBG, GXIB, XMBG	China
365.	**Oligostachyum nuspiculum (McClure) Z.P. Wang & G.H. Ye**	Arundinarieae	Hainan	SCBG	China
366.	**Oligostachyum oedogonatum (Z.P. Wang & G.H. Ye) Q.F. Zhang & K.F. Huang**	Arundinarieae	China	ABG, SZBG, GXIB	China
367.	**Oligostachyum shiuyingianum (L.C. Chia & But) G.H. Ye & Z.P. Wang**	Arundinarieae	China (Hong Kong) to Hainan	SCBG	China
	Arundinaria shiuyingiana	d	d	SMA	Hong Kong
368.	**Oligostachyum spongiosum (C.D. Chu & C.S. Chao) Q.F. Zheng & Y.M. Lin** *Sinobambusa dushanensis*	Arundinarieae	China	ABG, SZBG, GXIB	China
369.	**Oligostachyum sulcatum Z.P. Wang & G.H. Ye**	Arundinarieae	China	ABG, SZBG, GXIB, XMBG	China
370.	**Olmeca clarkiae (Davidse & R.W. Pohl) Ruiz-Sanchez, Sosa & Mejia-Saulés**	Bambuseae	Mexico to Honduras	FJCBG	Mexico
371.	**Olmeca fulgor (Soderstr.) Ruiz-Sanchez, Sosa & Mejia-Saulés**	Bambuseae	Mexico	FJCBG	Mexico
	Aulonemia fulgor	d	d	SDBG	USA
372.	**Olmeca recta Soderstr.**	Bambuseae	Mexico	FJCBG	Mexico
373.	**Olmeca reflexa Soderstr.**	Bambuseae	Mexico	FJCBG	Mexico
374.	**Olmeca zapotecorum Ruiz-Sanchez, Sosa & Mejia-Saulés**	Bambuseae	Mexico	FJCBG	Mexico

375.	Olyra glaberrima Raddi	Olyreae	Mexico to Honduras, Peru, Brazil and Paraguay	FJCBG	Mexico
376.	Olyra latifolia L. / Olyra latifolia	Olyreae	Mexico to trop. America and Madagascar	FJCBG	Mexico
377.	Otatea acuminata (Munro) C.E. Calderon & Soderstr.	Bambuseae	Mexico	JBHLCM	Spain
	d	d	d	LBMG	France
	d	d	d	FJCBG	Mexico
	d	d	d	BBG	Indonesia
	Otatea acuminata ssp. Aztecorum	d	d	PCG, SDBG	USA
	Otatea aztecorum (McClure et E.W. Smith) C.E. Calderón ex Soderstr.	d	d	SCBG	China
378.	Otatea carrilloi Ruiz-Sanchez, Sosa & Mejia-Saulés	Bambuseae	Mexico	FJCBG	Mexico
379.	Otatea fimbriata Soderstr.	Bambuseae	Mexico to Honduras, Colombia	SDBG	USA
	d	d	d	SDBG	USA
	d	d	d	FJCBG	Mexico
380.	Otatea glauca L.G. Clark & G. Cortés	Bambuseae	Mexico	SDBG, PCG	USA
	d	d	d	FJCBG	Mexico
381.	Otatea nayeeri Ruiz-Sanchez	Bambuseae	Mexico	FJCBG	Mexico
382.	Otatea ramirezii Ruiz-Sanchez	Bambuseae	Mexico	FJCBG	Mexico
383.	Otatea reynosoana Ruiz-Sanchez & L.G. Clark	Bambuseae	Mexico	FJCBG	Mexico
384.	Otatea rzedowskiorum Ruiz-Sanchez	Bambuseae	Mexico	FJCBG	Mexico
385.	Otatea sp.	Bambuseae	0	SDBG	USA

Table 3.2 contd. ...

...*Table 3.2 contd.*

No.	Species	Tribe	Native Range	Repository	Country
386.	*Otatea* sp. Western Mexico 'clone WFZ'	Bambuseae	0	SDBG	USA
387.	*Otatea transvolcanica* Ruiz-Sanchez & L.G. Clark	Bambuseae	Mexico	FJCBG	Mexico
388.	*Otatea victoriae* Ruiz-Sanchez	Bambuseae	Mexico	FJCBG	Mexico
389.	*Otatea ximenae* Ruiz-Sanchez & L.G. Clark	Bambuseae	Mexico	FJCBG	Mexico
390.	*Oxytenanthera abyssinica* (A. Rich.) Munro	Bambuseae	Trop. & S. Africa	FRI, KFRI, VVK	India
391.	*Oxytenanthera* sp.	d	0	KFRI	India
392.	*Oxytenanthera* sp. 'Medang'	Bambuseae	India	VVK	India
393.	×*Phyllosasa tranquillans* (Koidz.) Demoly	Arundinarieae	Occuring between the native and naturalized parents	TGT	UK
	d	d	d	GRIFFIN	USA
	Phyllosasa tranquillans (Koidz.) Demoly 'Shiroshima'	d	d	JBHLCM	Spain
	Hibanobambusa tranquillans	d	d	LBMG	France
	Hibanobambusa tranquillans 'Shiroshima'	d	d	LBMG	France
	Hibanobambusa tranquillans f. 'Shiroshima'	d	d	ABG	China
	Hibanobambusa tranquillans 'Shiroshima'	d	d	SDBG	USA
	× *Hibanobambusa tranquillans* (Koidz.) Maruy. Et. H. Okamura f. shiroshima H. Okamura	d	d	SCBG, SZBG, XMBG	China
	Hibanobambusa tranquillans 'Shiroshima'	d	d	AJCBIBG	India
394.	*Phyllostachys acuta* C.D. Chu & C.S. Chao	Arundinarieae	China	ABG	China
395.	*Phyllostachys angusta* McClure	Arundinarieae	China	GRIFFIN, MAY	USA
	d	d	d	ABG, WHIOB, GXIB	China
396.	*Phyllostachys arcana* McClure	Arundinarieae	China	GRIFFIN, MAY	USA

No.	Name				Country
		d		ABG, WHIOB, XTBG	China
	Phyllostachys arcana 'Luteosulcata'	d		ABG	China
		d		LBMG	France
397.	**Phyllostachys atrovaginata** C.S. Chao & H.Y. Chou	Arundinarieae	China	ABG	China
398.	**Phyllostachys aurea** (André) Rivière & C. Rivière	Arundinarieae	China to Vietnam	BBG	Indonesia
		d		MAY, MIA	USA
		d		RBGE, TGT	UK
		d		JBHLCM	Spain
		d		FRI, RPRC	India
		d		LBMG	France
		d		PCG	USA
		d		SCBG, WHIOB, KIB, SZBG, GXIB, ABG	China
	Phyllostachys aurea 'Holochrysa'	d		TGT	UK
		d		LBMG	France
	Phyllostachys aurea 'Flavescens inversa'	d		LBMG	France
	Phyllostachys aurea 'Koi'	d		LBMG	France
399.	**Phyllostachys aureosulcata** McClure	Arundinarieae	China	GRIFFIN, MAY, MBG, PCG, SDBG	USA
		d		JBHLCM	Spain
	Phyllostachys aureosulcata	d		LBMG	France
		d		TGT	UK
		d		ABG, GXIB	China
	Phyllostachys aureosulcata 'Spectabilis'	d		LBMG	France

Table 3.2 contd. ...

...*Table 3.2 contd.*

No.	Species	Tribe	Native Range	Repository	Country
	d	d	d	ABG	China
	Phyllostachys aureosulcata 'Aureocaulis'	d	d	ABG	China
	d	d	d	LBMG	France
	Phyllostachys aureosulcata 'Pekinensis'	d	d	ABG	China
400.	**Phyllostachys bissetii McClure**	Arundinarieae	China	GRIFFIN, MAY	USA
	d	d	d	LBMG	France
	d	d	d	ABG, WHIOB, SZBG	China
401.	**Phyllostachys dulcis McClure**	Arundinarieae	China	GRIFFIN, MAY	USA
	d	d	d	LBMG	France
	d	d	d	ABG	China
402.	**Phyllostachys circumpilis C.Y. Yao & S.Y. Chen**	Arundinarieae	China	WHIOB, XTBG, SZBG	China
403.	**Phyllostachys edulis (Carrière) J. Houz.**	Arundinarieae	China, Taiwan	GRIFFIN, MBG	USA
	d	d	d	TGT, RBGE	UK
	d	d	d	JNTBGRI	India
	Phyllostachys edulis 'Moso'	d	d	LBMG	France
	Phyllostachys heterocycla	d	d	ABG	China
	Phyllostachys heterocycla ' Luteosulcata'	d	d	ABG	China
	Phyllostachys heterocycla 'Pachyloen'	d	d	ABG	China
	Phyllostachys heterocycla 'Tao Kiang'	d	d	ABG	China
	Phyllostachys heterocycla 'Ventricosa'	d	d	ABG	China
	Phyllostachys heterocycla 'Gracilis'	d	d	ABG	China
	Phyllostachys heterocycla 'Pubescens'	d	d	ABG	China

	Phyllostachys heterocycla 'Tubaeformis'		d	ABG	China
	Phyllostachys heterocycla 'Viridisulcata'		d	ABG	China
	Phyllostachys heterocycla 'Obliquinoda'		d	ABG	China
	Phyllostachys pubescens (Pradelle) Mazel ex Lehaie		d	BGR	Italy
	d		d	SCBG	China
	d		d	KFRI	India
	d		d	LBMG	France
	Phyllostachys pubescens 'Bicolor'		0	LBMG	France
	Phyllostachys pubescens 'Heterocycla'		0	LBMG	France
404.	***Phyllostachys elegans* McClure**	Arundinarieae	China	GRIFFIN, MAY	USA
	d		d	ABG	China
405.	***Phyllostachys fimbriligula* T.H. Wen**	Arundinarieae	China	GXIB, ABG	China
406.	***Phyllostachys flexuosa* Rivière & C. Rivière**	Arundinarieae	China	GRIFFIN, MAY, PCG	USA
	d		d	LBMG	France
	d		d	RBGE	UK
	d		d	GXIB, ABG	China
407.	***Phyllostachys glabrata* S.Y. Chen & C.Y. Yao**	Arundinarieae	China	ABG	China
408.	***Phyllostachys glauca* McClure**	Arundinarieae	China	GRIFFIN, MAY	USA
	d		d	LBMG	France
	d		d	XTBG, SZBG, ABG	China
	Phyllostachys glauca var. *variabilis* J.L. Lu		d	WHIOB, GXIB, ABG	China
	Phyllostachys glauca 'Yunzhu'		d	ABG, SCBG	China
409.	***Phyllostachys heteroclada* Oliv.**	Arundinarieae	China	GRIFFIN	USA

Table 3.2 contd. ...

...*Table 3.2 contd.*

No.	Species	Tribe	Native Range	Repository	Country
	d	d	d	LBMG	France
	d	d	d	SCBG, WHIOB, XTBG, SZBG GXIB, ABG	China
	Phyllostachys heteroclada f. *purpurata*	d	d	ABG	China
	Phyllostachys heteroclada f. *solida* (S.L. Chen) Z.P. Wang et. Z.H. Yu	d	d	WHIOB, ABG	China
410.	**Phyllostachys humilis** =(unresolved)	Arundinarieae	0	LBMG	France
411.	**Phyllostachys incarnata T.H. Wen**	Arundinarieae	China	SZBG, GXIB, ABG	China
412.	**Phyllostachys iridescens C.Y. Yao & S.Y. Chen**	Arundinarieae	China	SCBG, WHIOB, GXIB, ABG	China
413.	**Phyllostachys kwangsiensis W.Y. Hsiung, Q.H. Dai & J.K. Liu**	Arundinarieae	China	GXIB, XMBG, ABG, SCBG	China
414.	**Phyllostachys makinoi Hayata**	Arundinarieae	China, Taiwan	GRIFFIN, MAY, SDBG	USA
	d	d	d	LBMG	France
	d	d	d	SZBG, ABG	China
415.	**Phyllostachys mannii Gamble**	Arundinarieae	Himalaya to Myanmar	GRIFFIN, MAY	USA
	d	d	d	WHIOB, XTBG, SZBG, GXIB, ABG	China
	d	d	d	JNTBGRI	India
	d	d	d	LBMG	France
	Phyllostachys assamica	d	d	KFRI	India
	Phyllostachys decora	d	d	SCBG	China
416.	**Phyllostachys meyeri McClure**	Arundinarieae	China	GRIFFIN, MAY	USA
	d	d	d	SCBG, ABG	China
	d	d	d	LBMG	France
417.	**Phyllostachys nidularia Munro**	Arundinarieae	China	GRIFFIN, MAY	USA

		d	JBHLCM	Spain
		d	LBMG	France
		d	SCBG, WHIOB, XTBG, LSBG SZBG, GXIB, ABG	China
	Phyllostachys nidularia f. *farcata* H.R. Chao et. A.T. Liu	d	GXIB, ABG	China
	Phyllostachys nidularia f. *glabrovagina* (McClure) Wen	d	GXIB, ABG	China
	Phyllostachys nidularia Munro f. *mirabilis* Yi et. C.G. Chen	d	SZBG	China
	Phyllostachys nidularia Munro f. *speciosa* Yi et. C.G. Chen	d	SZBG	China
418.	**Phyllostachys nigella T.H. Wen**	Arundinarieae	SZBG, ABG	China
419.	**Phyllostachys nigra (Lodd. ex Lindl.) Munro**	Arundinarieae	GRIFFIN, MAY, KBG, SDBG, PCG	USA
		d	JBMPC, JBHLCM	Spain
		d	LBMG	France
		d	SBG	Singapore
		d	JNTBGRI, RPRC	India
		d	TGT	UK
		d	SCBG, IBCAS, WHIOB, KIB, XTBG, LSBG, SZBG, GXIB, XMBG, ABG	China
420.	**Phyllostachys nigra (Lodd. ex Lindl.) Munro var. *henonis* (Mitford) Stapf ex Rendle**	Arundinarieae	GRIFFIN, MAY	USA
		d	WHIOB, KIB, XTBG, LSBG, SZBG, GXIB, ABG	China
	Phyllostachys nigra 'Henonis'	d	LBMG	France
		d	KBG	USA

Table 3.2 contd. ....

...Table 3.2 contd.

No.	Species	Tribe	Native Range	Repository	Country
	Phyllostachys nigra (Lodd. ex Lindl.) Munro	d	d	BBG	Indonesia
	d	d	d	RBGE	UK
	Phyllostachys nigra 'Boryana'	d	d	TGT	UK
	d	d	d	LBMG	France
	Phyllostachys nigra 'henon'	d	d	PCG	USA
	Phyllostachys nigra (Lodd.) Munro f. *boryana* (Mitford) Makino	d	d	RBGE	UK
421.	**Phyllostachys nuda McClure**	Arundinarieae	China	GRIFFIN, MAY	USA
	d	d	d	SCBG, ABG	China
	d	d	d	LBMG	France
	Phyllostachys nuda f. *localis* C.P. Wang & Z.H. Yu	d	d	ABG	China
	Phyllostachys nuda 'Localis'	d	d	LBMG	France
422.	**Phyllostachys parvifolia C.D. Chu & H.Y. Chou**	Arundinarieae	China	WHIOB, SZBG, GXIB, ABG	China
423.	**Phyllostachys platyglossa C.P. Wang & Z.H. Yu**	Arundinarieae	China	WHIOB, SZBG, GXIB, ABG	China
424.	**Phyllostachys prominens W.Y. Hsiung**	Arundinarieae	China	SZBG, GXIB, ABG	China
425.	**Phyllostachys propinqua McClure**	Arundinarieae	China	LBMG	France
	Phyllostachys propinqua McClure f. *lanuginosa* Wen	d	d	GXIB, ABG	China
426.	**Phyllostachys purpurata McClure**	Arundinarieae	China	GRIFFIN, MAY	USA
427.	**Phyllostachys reticulata (Rupr.) K. Koch** *Phyllostachys bambusoides* var. *castillonii* (Lat.-Marl. ex Carrière) Mitf.	Arundinarieae	China	RBGE	UK
	Phyllostachys bambusoides	d	d	TGT	UK
	d	d	d	LBMG	France
	d	d	d	SCBG, WHIOB, LSBG, SZBG, GXIB	China

	d	GRIFFIN, MAY	USA
	d	JBHLCM	Spain
	d	KFRI	India
	d	ABG	China
Phyllostachys bambusoides 'Castillon Inversa'	d	SDBG	USA
Phyllostachys bambusoides var. *castillonis*	d	ABG	China
Phyllostachys bambusoides Siebold et Zucc. f. *castillonis* (Marl. ex Carr.) Makino	d	SZBG	China
Phyllostachys pubescens Mazel ex H. de Leh. f. *luteosulcata* Wen	d	GXIB	China
Phyllostachys pubescens Mazel ex H. de Leh. f. *obtusangula* S.Y. Wang	d	WHIOB	China
Phyllostachys bambusoides f. *lacrima-deae*	d	ABG	China
Phyllostachys bambusoides f. *shouzhu* Yi	d	ABG, WHIOB, SZBG, GXIB	China
Phyllostachys pubescens Mazel ex H. de Leh. f. *gracilis* W.Y. Hsiung		SZBG	China
Phyllostachys bambusoides f. *mixta*	d	ABG	China
Phyllostachys bambusoides f. *nova*	d	ABG	China
Phyllostachys bambusoides 'Violascens'	d	TGT	UK
Phyllostachys bambusoides 'Marliacea'	d	LBMG	France
Phyllostachys bambusoides 'Subvariegata'	d	LBMG	France
Phyllostachys bambusoides 'Tanakae'	d	LBMG	France
Phyllostachys bambusoides 'Castilloni inversa'	d	LBMG	France
Phyllostachys bambusoides 'Castillonis'	d	LBMG	France
Phyllostachys bambusoides f. *lacrima-deae* Keng. f. et. Wen	d	HBG, WHIOB, LSBG, SZBG, GXIB, XMBG	China
Phyllostachys bambusoides 'Holocrysa'	d	LBMG	France

Table 3.2 contd. ....

...*Table 3.2 contd.*

No.	Species	Tribe	Native Range	Repository	Country
	Phyllostachys reticulate	d	d	VVK, KFRI	India
428.	**Phyllostachys rivalis H.R. Zhao & A.T. Liu**	Arundinarieae	China	SZBG, ABG	China
429.	**Phyllostachys robustiramea S.Y. Chen & C.Y. Yao**	Arundinarieae	China	WHIOB, ABG	China
430.	**Phyllostachys rubicunda T.H. Wen** *Phyllostachys concave*	Arundinarieae	China	WHIOB, GXIB, SCBG	China
	d	d	d	ABG	China
431.	**Phyllostachys rubromarginata McClure**	Arundinarieae	China	GRIFFIN	USA
	d	d	d	LBMG	France
	d	d	d	XTBG, SCBG, ABG	China
	Phyllostachys aurita J.L. Lu	d	d	ABG, SZBG, GXIB	China
432.	**Phyllostachys rutila T.H.Wen**	Arundinarieae	China	SZBG, ABG	China
433.	**Phyllostachys sp.**	Arundinarieae	0	SDBG	USA
434.	**Phyllostachys sp.**	Arundinarieae	India	JNTBGRI	India
435.	**Phyllostachys sp.1**	Arundinarieae	0	BBG	Indonesia
436.	**Phyllostachys stimulosa H.R. Zhao & A.T. Liu**	Arundinarieae	China	GXIB	China
437.	**Phyllostachys sulphurea (Carrière) Rivière & C. Rivière**	Arundinarieae	China	JBHLCM	Spain
	d	d	d	KFRI	India
	d	d	d	KIB, XTBG, SZBG, XMBG, SCBG	China
	d	d	d	ABG	China
	Phyllostachys sulphurea (Carrière) Rivière & C. Rivière var. *viridis* R.A. Young	d	d	GRIFFIN	USA
	Phyllostachys sulphurea 'Viridis'	d	d	ABG	China
	Phyllostachys sulphurea 'Robert Young'	d	d	SCBG, ABG	China

No.					
	Phyllostachys sulphurea f. *sulphurea*	d		TGT	UK
	Phyllostachys sulphurea (Carr.) A. et.C. Riv. f. *viridisulcata* (P. X. Zhang) P. X. Zhang			SZBG	China
	Phyllostachys sulphurea 'Houzeau'	d		ABG	China
	Phyllostachys viridis 'Mitis'	d		LBMG	France
	Phyllostachys viridis 'Sulfurea'	d		LBMG	France
438.	***Phyllostachys tianmuensis* Z.P. Wang & N.X. Ma**	Arundinarieae	China	GXIB, ABG	China
439.	***Phyllostachys varioauriculata* S.C. Li & S.H. Wu**	Arundinarieae	China	SZBG, GXIB, ABG	China
440.	***Phyllostachys veitchiana* Rendle** *Phyllostachys rigida*	Arundinarieae	China	ABG	China
441.	***Phyllostachys violascens* Rivière & C. Rivière**	Arundinarieae	China	LBMG	France
	Phyllostachys praecox	d		LBMG	France
	Phyllostachys praecox 'Viridisulcata'	d		LBMG	France
	Phyllostachys praecox C.D. Chu et. C.S. Chao	d		WHIOB, ABG	China
	Phyllostachys praecox f. *notata* S.Y. Chen & C.Y. Yao	d		ABG	China
442.	***Phyllostachys virella* T.H. Wen**	Arundinarieae	China	GXIB, ABG	China
443.	***Phyllostachys viridiglaucescens* (Carrière) Rivière & C. Rivière**	Arundinarieae	China	BGR	Italy
	d	d		JBMPC	Spain
	d	d		SZBG, GXIB, SCBG	China
	d	d		ABG	China
	d	d		LBMG	France
	d	d		RBGE	UK
444.	***Phyllostachys vivax* McClure**	Arundinarieae	China	GRIFFIN	USA
	d	d		SDBG	USA
	d	d		SCBG, WHIOB	China

Table 3.2 contd. ....

...Table 3.2 contd.

No.	Species	Tribe	Native Range	Repository	Country
				LBMG	France
				ABG	China
	Phyllostachys vivax 'Huanvenzhu'			LBMG	France
				ABG	China
	Phyllostachys vivax f. *aureocaulis*			TGT, RBGE	UK
	Phyllostachys vivax 'Aureocaulis'			LBMG	France
				ABG	China
445.	**Phyllostachys yunhoensis S.Y. Chen & C.Y. Yao**	Arundinarieae	China	ABG	China
446.	**Phyllostchys sp.**	Arundinarieae	0	KFRI	India
447.	**Phyllostchys sp.**	Arundinarieae	0	KFRI	India
448.	**Phyllostchys sp.**	Arundinarieae	0	KFRI	India
449.	**Pleioblastus amarus (Keng) Keng f.**	Arundinarieae	China	WHIOB, XTBG, LSBG, GXIB SCBG, ABG	China
	Pleioblastus amarus var. *hangzhouensis*			ABG	China
	Pleioblastus amarus (Keng) Keng var. *pendulifolius* S. Y. Chen			GXIB	China
450.	**Pleioblastus argenteostriatus (Regel) Nakai**	Arundinarieae	Japan	JBMPC	Spain
				JBHLCM	Spain
				SDBG	USA
				SDBG	USA
	Pleioblastus chino (Franch. & Sav.) Makino			RBGE	UK
				XTBG, SCBG	China
				GRIFFIN	USA

Pleioblastus chino f. *angushofolius*		d	ABG	China
Pleioblastus chino 'Elegantissimus'		d	LBMG	France
Pleioblastus chino f. *holocrysa*		d	ABG	China
Pleioblastus chino (Franch. et. Savat.) Makino var. *hisauchii* Makino		d	WHIOB, SZBG, GXIB, XMBG, ABG	China
Pleioblastus pumilus		d	LBMG	France
Pleioblastus humilis (Mitford) Nakai var. *pumilus*		d	RBGE	UK
Sasa argenteostriata (Regel) E. G. Camus			SCBG, CNBG, ABG	China
451.	***Pleioblastus distichus* (Mitford) Nakai**	Arundinarieae	Japan	GRIFFIN	USA
	d		d	LBMG	France
	Pseudosasa disticha (Mitford) Nakai		d	JBHLCM	Spain
452.	***Pleioblastus gramineus* (Bean) Nakai**	Arundinarieae	Japan	LBMG	France
	d		d	WHIOB, XTBG, LSBG, SZBG, GXIB, SCBG, ABG	China
	Pleioblastus gramineus (Bean) Nakai f. *monstrospiralis* Muroi et. H. Hamada		d	SZBG, ABG	China
453.	***Pleioblastus hsienchuensis* T.H.Wen** *Sinobambusa seminuda* Wen	Arundinarieae	China	GXIB, ABG	China
	Pleioblastus juxianensis		d	ABG	China
454.	***Pleioblastus linearis* (Hack.) Nakai**	Arundinarieae	Japan	JBHLCM	Spain
	d		d	LBMG	France
	d		d	TGT	UK
	d		d	WHIOB, SZBG, SCBG	China
	Pleioblastus gozadakensis		d	ABG	China

Table 3.2 contd. ....

...Table 3.2 contd.

No.	Species	Tribe	Native Range	Repository	Country
455.	**Pleioblastus maculatus (McClure) C.D. Chu & C.S. Chao**	Arundinarieae	China	WHIOB, KIB, SZBG, GXIB, ABG	China
	Pleioblastus oleosus Wen	◊	◊	WHIOB, GXIB, XMBG	China
456.	**Pleioblastus ovatoauritus T.H. Wen**	Arundinarieae	0	ABG	China
457.	**Pleioblastus pygmaeus (Miq.) Nakai var. distichus (Mitford) Nakai** =(unresolved)	Arundinarieae	0	RBGE	UK
458.	**Pleioblastus simonii (Carrière) Nakai**	Arundinarieae	Japan	GRIFFIN	USA
	◊	◊	◊	TGT	UK
	◊	◊	◊	GXIB, SCBG	China
	Pleioblastus simonii f. *heterophyllus*	◊	◊	ABG	China
	Pleioblastus simonii f. *albostriatus*	◊	◊	ABG	China
459.	**Pleioblastus solidus S.Y.Chen**	Arundinarieae	China	SZBG, GXIB, ABG	China
460.	**Pleioblastus variegatus (J.Dix) Makino** *Pleioblastus pygmaeus*	Arundinarieae	Japan	SDBG, MBG	USA
	Pleioblastus pygmaeus (Miq.) Nakai	◊	◊	RBGE	UK
	◊	◊	◊	LBMG	France
	◊	◊	◊	RPRC	India
	Bambusa variegata	◊	◊	GBG, RBGPS	Sri Lanka
	Pleioblastus fortunei (Van Houtte) Nakai	◊	◊	GRIFFIN	USA
	◊	◊	◊	JBHLCM	Spain
	◊	◊	◊	TGT	UK
	◊	◊	◊	SDBG	USA
	◊	◊	◊	LBMG	France
	◊	◊	◊	RPRC	India

No.	Species			Gardens	Country
	Pleioblastus shibuyanus 'Tsuboi'	d		LBMG	France
	Sasa fortunei (Van Houtte) Fiori	d		SCBG, WHIOB, KIB, CNBG, SZBG, GXIB, XMBG, HBG	China
	Sasa pygmaea (Miq.) E.G. Camus	d		SCBG, SZBG, XMBG, ABG	China
	Sasa fortunei	d		ABG	China
461.	**Pleioblastus viridistriatus (Regel) Makino**	Arundinarieae	Japan	GRIFFIN, SDBG	USA
	Pleioblastus auricomus (Mitford) D. McClint.	d		RBGE	UK
	Sasa auricoma E.G. Gamus	d		XTBG, CNBG, XMBG, ABG	China
	Pleioblastus viridistriatus	d		SDBG, MBG	USA
	d	d		LBMG	France
	Pleioblastus kongosanensis f. *aureostriatus* Muroi & Yu.Tanaka	d		SZBG, ABG	China
	Pleioblastus viridistriatus 'Chrysophyllus'	d		LBMG	France
	d	d		SDBG	USA
	Pleioblastus viridistriatuss 'Vagans'	d		LBMG	France
	Pleioblastus shibuyanus 'Tsuboi'	d		SDBG	USA
462.	**Pleioblastus yixingensis S.L. Chen & S.Y. Chen**	Arundinarieae	China	SZBG, GXIB, ABG	China
463.	**Pseudosasa amabilis (McClure) Keng. f.**	Arundinarieae	China to Vietnam	GRIFFIN	USA
	d	d		WHIOB, SZBG, GXIB, SCBG	China
	d	d		HBG	China
	d	d		LBMG	France
	d	d		ABG	China
	Pseudosasa amabilis var. *convexa* Z.P. Wang et. G.H. Ye	d		SCBG, SZBG, ABG	China
	Pseudosasa amabilis var. *tenuis*	d		ABG	China
464.	**Pseudosasa cantorii (Munro) Keng f.**	Arundinarieae	China to Hainan	XTBG, SCBG	China

Table 3.2 contd. ...

...Table 3.2 contd.

No.	Species	Tribe	Native Range	Repository	Country
	d	d	d	GRIFFIN	USA
465.	**Pseudosasa hindsii (Munro) S.L. Chen & G.Y. Sheng ex T.G. Liang**	Arundinarieae	China	LBMG	France
	d	d	d	SZBG, SCBG, ABG	China
	Arundinaria hindsii Munro	d	d	GRIFFIN	USA
466.	**Pseudosasa humilis (Mitford) T.Q. Nguyen**	Arundinarieae	Japan	TGT	UK
	Pleioblastus humilis 'albovariegata'	d	d	SDBG	USA
467.	**Pseudosasa japonica (Siebold & Zucc. ex Steud.) Makino ex Nakai**	Arundinarieae	S. Korea, Japan	GRIFFIN	USA
	d	d	d	GRIFFIN, MBG, PCG, SDBG	USA
	d	d	d	WHIOB, ABG, SZBG, GXIB SCBG, HBG	China
	d	d	d	FRI, VVK, RPRC	India
	d	d	d	LBMG	France
	d	d	d	JBMPC, JBHLCM	Spain
	d	d	d	RBGE, TGT	UK
	Pseudosasa japonica f. *akebonosuzi* H. Okamura	d	d	SZBG, ABG	China
	Pseudosasa japonica (Siebold et. Zucc.) Makino var. *Tsutsumiana* Yanagita	d	d	SZBG, ABG	China
	d	d	d	GRIFFIN	USA
	Pseudosasa japonica 'Tsutsumiana'	d	d	LBMG	France
	Pseudosasa japonica f. variata = (unsolved)	d	d	ABG	China
	Pseudosasa japonica 'Variegata'	d	d	LBMG	France
468.	**Pseudosasa longiligula T.H. Wen**	Arundinarieae	China	GXIB	China
469.	**Pseudosasa orthotropa S.L. Chen & T.H. Wen**	Arundinarieae	China	ABG	China

No.		Tribe	Distribution	ABG		Country
470.	*Pseudosasa subsolida* S.L. Chen & G.Y. Sheng	Arundinarieae	China	ABG		China
471.	*Pseudosasa viridula* S.L. Chen & G.Y. Sheng	Arundinarieae	China	XTBG, SCBG		China
472.	*Pseudostachyum polymorphum* Munro	Bambuseae	India to China and Indo-China	VVK		India
	d	d	d	JNTBGRI		India
	d	d	d	SCBG, XTBG, GXIB		China
473.	*Pseudoxytenanthera bourdillonii* (Gamble) H.B.Naithani	Bambuseae	India	JNTBGRI, KFRI		India
474.	*Pseudoxytenanthera monadelpha* (Thwaites) Soderstr. & R. P. Ellis	Bambuseae	India, Sri Lanka, Indo-China	JNTBGRI		India
475.	*Pseudoxytenanthera ritcheyi* (Munro) H.B.Naithani	Bambuseae	India	JNTBGRI, KFRI		India
476.	*Pseudoxytenanthera* sp.	Bambuseae	India (Kerala)	JNTBGRI		India
477.	*Pseudoxytenanthera* sp.	Bambuseae	India (Kerala)	JNTBGRI		India
478.	*Pseudoxytenanthera stocksii* (Munro) T.Q.Nguyen	Bambuseae	India, Vietnam	JNTBGRI, KFRI		India
479.	*Raddia brasiliensis* Bertoloni	Olyreae	Brazil	RBGE		UK
480.	*Rhipidocladum bartlettii* (McClure) McClure	Bambuseae	Mexico to Honduras	FJCBG		Mexico
481.	*Rhipidocladum martinezii* Davidse & R.W.Pohl	Bambuseae	Mexico	FJCBG		Mexico
482.	*Rhipidocladum pittieri* (Hack.) McClure	Bambuseae	Mexico to America	FJCBG		Mexico
483.	*Rhipidocladum racemiflorum* (Steud.) McClure	Bambuseae	Mexico to Surinam and Argentina	FJCBG		Mexico
484.	*Sasa elegantissima* Koidz. *Sasa admirabilis*	Arundinarieae	Japan	LBMG		France
485.	*Sasa glabra* (Nakai) Nakai ex Koidz. f. *alba-striata* Muroi = (unresolved)	Arundinarieae	0	SCBG		China
486.	*Sasa kurilensis* (Rupr.) Makino & Shibata	Arundinarieae	Korea, Sakhalin to Japan	SDBG		USA
487.	*Sasa latifolia* = (unresolved)	Arundinarieae	0	LBMG		France

Table 3.2 contd.…

...Table 3.2 contd.

No.	Species	Tribe	Native Range	Repository	Country
488.	*Sasa oblongula* **C.H.Hu**	Arundinarieae	China	SZBG	China
489.	*Sasa rubrovaginata* **C.H.Hu**	Arundinarieae	China	GXIB	China
490.	*Sasa palmata* **(Burb.) E.G.Camus**	Arundinarieae	Korea to Japan	JNTBGRI	India
	d	d	d	GRIFFIN, SDBG	USA
	Sasa palmata (Burb.) E.G.Camus 'Nebulosa'	d	d	JBHLCM	Spain
	d	d	d	LBMG	France
	Sasa palmata f. *nebulosa*	d	d	TGT	UK
491.	*Sasa senanensis* **(Franch. & Sav.) Rehder**	Arundinarieae	0	GRIFFIN	USA
492.	*Sasa subglabra* **McClure**	Arundinarieae	China	SCBG	China
493.	*Sasa tsuboiana* **Makino**	Arundinarieae	Korea, Japan	SDBG	USA
	d	d	d	LBMG	France
494.	*Sasa veitchii* **(Carrière) Rehder**	Arundinarieae	Sakhalin to Japan	JBHLCM, JBMPC	Spain
	d	d	d	LBMG	France
	d	d	d	RBGE	UK
495.	*Sasaella kogasensis* **(Nakai) Koidz.** *Sasa kogasensis* Nakai	Arundinarieae	Japan	KIB	China
496.	*Sasaella masamuneana* **(Makino) Hatus. & Muroi** *Sasaella glabra* f. *albostriata*	Arundinarieae	Japan	ABG	China
	Sasa masamuneana (Makino) C. S. Chao & Renvoize	d	d	GRIFFIN	USA
	Sasaella masamuneana 'Albostriata'	d	d	SDBG	USA
	d	d	d	LBMG	France
	Sasa masamuneana 'Aureostriata'	d	d	LBMG	France
497.	*Sasaella ramosa* **(Makino) Makino**	Arundinarieae	Japan	TGT	UK

498.	*Sasaella ramosa* **f. albo-striata** = (unresolved)	Arundinarieae	d	ABG	China
499.	*Sasamorpha sinica* **(Keng) Koidz.** *Sasa sinica*	Arundinarieae	China	SCBG, ABG	China
500.	*Schizostachyum auriculatum* **Q.H. Dai & D.Y. Huang**	Bambuseae	China	GXIB	China
501.	*Schizostachyum beddomei* **(C.E.C. Fisch.) R.B. Majumdar**	Bambuseae	India	JNTBGRI	India
502.	*Schizostachyum brachycladum* **(Kurz ex Munro) Kurz**	Bambuseae	Indo-China to Malesia	BBG	Indonesia
	d		d	JNTBGRI, FRI	India
	d		d	SBG	Singapore
	d		d	SCBG, XTBG	China
503.	*Schizostachyum castaneum* **Widjaja**	Bambuseae	Lesser Sunda Islands (Bali)	BBG	Indonesia
504.	*Schizostachyum caudatum* **Backer ex K. Heyne**	Bambuseae	Sumatera to Lesser Sunda Islands (Bali)	BBG	Indonesia
505.	*Schizostachyum cuspidatum* **Widjaja**	Bambuseae	Sumatera, Lesser Sunda Islands (Bali)	BBG	Indonesia
506.	*Schizostachyum dumetorum* **(Hance ex Walp.) Munro**	Bambuseae	China	SCBG	China
507.	*Schizostachyum dullooa* **(Gamble) R.B. Majumdar**	Bambuseae	E. Himalaya to Indo-China	JNTBGRI	India
	d		d	FRI	India
	d		d	BFRI	Bangladesh
	Neohouzeaua dullooa (Gamble) A. Camus		d	AJCBIBG	India
	Tienostachyum dulloa		d	KFRI	India
508.	*Schizostachyum funghomii* **McClure**	Bambuseae	India to China	SCBG, XTBG, SZBG, GXIB	China
	Schyzostachyum subreyorum	Bambuseae	India to China	ABG	China
509.	*Schizostachyum gracile* **(Kurz ex Munro) Holttum**	Bambuseae	Vietnam, Malaya	SBG	Singapore

Table 3.2 contd. ...

...*Table 3.2 contd.*

No.	Species	Tribe	Native Range	Repository	Country
510.	**Schizostachyum hainanense Merr. ex McClure**	Bambuseae	Hainan, Vietnam	SCBG	China
511.	**Schizostachyum iraten Steud.** *Schizostachim irratum*	Bambuseae	Jawa to Lesser Sunda Islands (Bali)	PBG	Indonesia
512.	**Schizostachyum jaculans Holttum** *Schizostachyum jaculans*	Bambuseae	Hainan, Thailand to Malaysia	SCBG	China
	d	d	d	SBG	Singapore
513.	**Schizostachyum latifolium Gamble**	Bambuseae	Malaya to Sumatera, Bali, Bornao, Sulawesi	SBG	Singapore
514.	**Schizostachyum lima (Blanco) Merr.**	Bambuseae	Malesia to Papuasia, Caroline Islands (Palau)	BBG	Indonesia
515.	**Schizostachyum pergracile (Munro) R.B. Majumdar** *Cephalostachyum pergracile*	Bambuseae	India to China	SCBG, XTBG, SZBG, XMBG	China
	Cephalostachyum pergracile Munro	d	d	MAY	USA
	d	d	d	JNTBGRI, KFRI, AJCBIBG	India
	Schizostachyum pergracile	d	d	VVK, FRI	India
	Schizostachyum pergracile ' Khellong'	d	d	VVK	India
516.	**Schizostachyum pseudolima McClure**	Bambuseae	Hainan, Vietnam	SCBG, XTBG, SZBG, GXIB	China
517.	**Schizostachyum silicatum Widjaja**	Bambuseae	Sumatera, Lesser Sunda Islands (Bali)	BBG	Indonesia
518.	**Schizostachyum sp.**	Bambuseae	0	SBG	Singapore
519.	**Schizostachyum sp. 1**	Bambuseae	0	BBG	Indonesia
520.	**Schizostachyum sp.**	Bambuseae	Madagascar, Asia to Pacific	SCBG	China

No.	Species / Synonym	Tribe	Distribution	Gardens	Country
521.	**Schizostachyum virgatum (Munro) H.B.Naithani & Bennet** *Cephalostachyum virgatum*	Bambuseae	Bangladesh to China and Indo-China	SCBG, XTBG, SZBG	China
522.	**Schizostachyum zollingeri Steud.**	Bambuseae	Indo-China to Malesia	SBG	Singapore
	d		d	BBG	Indonesia
523.	**Semiarundinaria densiflora (Rendle) T.H. Wen** *Brachystachyum densiflorum*	Arundinarieae	China	ABG, SCBG	China
524.	**Semiarundinaria fastuosa (Lat.-Marl. ex Mitford) Makino**	Arundinarieae	Japan	GRIFFIN, PCG	USA
	d	d	d	SCBG, SZBG, GXIB, ABG	China
		d	d	LBMG	France
		d	d	JBHLCM	Spain
		d	d	RBGE, TGT	UK
525.	**Semiarundinaria makinoi** =(unresolved)	Arundinarieae	0	LBMG	France
526.	**Semiarundinaria okuboi Makino**	Arundinarieae	Japan	LBMG	France
527.	**Semiarundinaria sinica T.H.Wen**	Arundinarieae	China	GXIB, ABG	China
528.	**Semiarundinaria yashadake (Makino) Makino** *Semiarundinaria yashadake* 'Kimmei'	Arundinarieae	Japan	SDBG	USA
	d	d	d	LBMG	France
529.	**Semiarundinaria sp.** *Brachystachyum* sp.	Arundinarieae	0	SCBG	China
530.	**Semiarundinaria sp.**	Arundinarieae	China to Hainan, Japan	SCBG	China
531.	**Shibataea chiangshanensis T.H.Wen**	Arundinarieae	China	SZBG, ABG	China
532.	**Shibataea chinensis Nakai**	Arundinarieae	China	GRIFFIN	USA
	d	d	d	SCBG, WHIOB, XTBG, LSBG CNBG, ABG, SZBG, GXIB	China
533.	**Shibataea hispida McClure**	Arundinarieae	China	ABG	China

Table 3.2 contd. ...

...*Table 3.2 contd.*

No.	Species	Tribe	Native Range	Repository	Country
534.	**Shibataea kumasasa (Zoll. ex Steud.) Makino**	Arundinarieae	China, Japan	GRIFFIN	USA
	d	d	d	BBG	Indonesia
	d	d	d	AJCBIBG	India
	d	d	d	XMBG, SCBG, ABG	China
	Shibataea kumasaca (Steud.) Makino	d	d	JBHLCM	Spain
	d	d	d	JNTBGRI	India
	d	d	d	LBMG	France
	Shibataea kumasaca arcostriata	d	d	SDBG	USA
535.	**Shibataea lanceifolia C.H. Hu**	Arundinarieae	China	GRIFFIN	USA
	d	d	d	ABG	China
536.	**Shibataea nanpingensis Q.F. Zheng & K.F. Huang**	Arundinarieae	China	ABG	China
537.	**Sinobambusa henryi (McClure) C.D. Chu & C.S. Chao**	Arundinarieae	China	SCBG	China
538.	**Sinobambusa intermedia McClure**	Arundinarieae	China	WHIOB, SCBG, ABG	China
	d	d	d	SDBG	USA
	Pleioblastus longifimbriatus S. Y. Chen		d	WHIOB, ABG	China
539.	**Sinobambusa nephroaurita C.D. Chu & C.S. Chao**		China	SCBG, GXIB	China
540.	**Sinobambusa rubroligula McClure**	Arundinarieae	China to Hainan	LBMG	France
	d	d	d	GXIB	China
541.	**Sinobambusa sp.**	Arundinarieae		ABG	China
542.	**Sinobambusa sp.**	Arundinarieae	0	GRIFFIN	USA
543.	**Sinobambusa tootsik (Makino) Makino ex Nakai**	Arundinarieae	China to Vietnam	SCBG, WHIOB, SZBG	China
	d	d	d	ABG	China

No.	Species	Tribe		Distribution	Germplasm	Country
			d		LBMG	France
	Sinobambusa tootsik var. *laeta* (McClure) Wen		d		SCBG, WHIOB, GXIB, ABG	China
	Sinobambusa tootsik (Makino) Makino ex Nakai var. *tenuifolia* (Koidz.) S. Suzuki		d		GXIB, ABG	China
	Sinobambusa tootsik (Makino) Makino ex Nakai var. *luteoloalbostriata* (S. H. Chen et Z. Z. Wang) T. P. Yi		d		SCBG, XMBG	China
	Sinobambusa tootsik 'Albovariegata'		d		LBMG	France
	Sinobambusa tootsik 'Variegata'		d		SDBG	USA
544.	**Sirochloa parvifolia (Munro) S. Dransf.**	Bambuseae	d	Mayotte, Madagascar	SDBG	USA
545.	**Thamnocalamus spathiflorus (Trin.) Munro**	Arundinarieae	d	Himalaya to Tibet	TGT	UK
	Thamnocalamus spathiflorus var. *crassinodus* (T.P.Yi) C. Stapleton		d		RBGE	UK
546.	**Thyrsostachys oliveri Gamble**	Bambuseae	d	China to Indo-China	FRI, KFRI, VVK, RPRC	India
			d		SCBG, XTBG	China
			d		SDBG	USA
			d		BFRI	Bangladesh
547.	**Thyrsostachys siamensis Gamble**	Bambuseae	d	China to Indo-China	BBG	Indonesia
			d		JNTBGRI, KFRI	India
			d		SCBG, XTBG, SZBG, GXIB XMBG	China
			d		SDBG	USA
			d		BFRI	Bangladesh
			d		GBG, RBGPS	Sri Lanka
			d		SMA	Hong Kong
	Thyrsostachys regia		d		FRI, RPRC, AJCBIBG	India

Table 3.2 contd. ...

...*Table 3.2 contd.*

No.	Species	Tribe	Native Range	Repository	Country
	d	d	d	BFRI	Bangladesh
548.	*Vietnamosasa ciliata* (A.Camus) T.Q. Nguyen	Bambuseae	Indo-China	SBG	Singapore
549.	*Vietnamosasa darlacensis* T.Q. Nguyen	Bambuseae	Vietnam	SBG	Singapore
550.	*Yushania anceps* (Mitford) W.C. Lin	Arundinarieae	W. Himalaya	SDBG	USA
	d	d	d	RBGE, TGT	UK
	Arundinaria anceps	d	d	LBMG	France
551.	*Yushania baishanzuensis* Z.P. Wang & G.H. Ye	Arundinarieae	China	ABG	China
552.	*Yushania boliana* Demoly	Arundinarieae	China	SDBG	USA
553.	*Yushania brevipaniculata* (Hand.-Mazz.) T.P. Yi	Arundinarieae	China	SDBG	USA
554.	*Yushania canoviridis* G.H. Ye & Z.P. Wang	Arundinarieae	China	WHIOB	China
555.	*Yushania confusa* (McClure) Z.P. Wang & G.H. Ye	Arundinarieae	China	LSBG	China
556.	*Yushania maling* (Gamble) R.B. Majumdar & Karthik.	Arundinarieae	Nepal to Assam	TGT	UK
557.	*Yushania menghaiensis* T.P. Yi	Arundinarieae	China	XTBG	China
558.	*Yushania niitakayamensis* (Hayata) Keng f. / *Yushania niitakayamensis*	Arundinarieae	China, Taiwan to Philippines	HBG	China
559.	*Yushania* sp.	Arundinarieae	0	ABG	China
560.	*Yushania uniramosa* Hsueh & T.P. Yi	Arundinarieae	China	ABG	China
561.	Bamboo Hybrid (AKGK-1) - 4 accessions (530-531, 575-1, 739)	Bambuseae	Produced at JTBGRI	JNTBGRI	India
562.	Bamboo Hybrid (AKGK-2) - 3 accessions (556-1, 556-2, 742)	d	d	JNTBGRI	India
563.	Bamboo Hybrid (BG-1) - 2 accessions (990, 991)	d	d	JNTBGRI	India
564.	Bamboo Hybrid (BG-2) - 2 accessions (974, 975)	d	d	JNTBGRI	India
565.	Bamboo Hybrid (BG-3) - 2 accessions (959, 960)	d	d	JNTBGRI	India

566.	Bamboo Hybrid (H-1) - 8 accessions (475-476, 483, 659-660, 745-747)	d	JNTBGRI	India
567.	Bamboo Hybrid (H-2) - 6 accessions (479-480, 484, 736-738)	d	JNTBGRI	India
568.	Bamboo Hybrid (H-3) - 2 accessions (492-493)	d	JNTBGRI	India
569.	Bamboo Hybrid (H-4) -3 accessions (572-1, 572-2, 706)	d	JNTBGRI	India
570.	Bamboo Hybrid (H-5) - 3 accessions (560-561, 740)	d	JNTBGRI	India
571.	Bamboo Hybrid (H-6) - 4 accessions (564-565, 568-569)	d	JNTBGRI	India
572.	Bamboo Hybrid (KCKG-1) - 2 accessions (680-681)	d	JNTBGRI	India
573.	Bamboo Hybrid (KCKG-2) - 2 accessions (682-683)	d	JNTBGRI	India
574.	Bamboo Hybrid (KCKG-3) - 2 accessions (796, 797)	d	JNTBGRI	India

Merostachys (1), *Olmeca* (5), *Otatea* (13), and *Rhipidocladum* (4). The Paleotropical Woody Bamboos are represented by 20 genera and 265 species, viz., *Bambusa* (112), *Bonia* (1), *Cephalostachyum* (2), *Davidsea* (1), *Dendrocalamus* (42), *Dinochloa* (9), *Gigantochloa* (39), *Lingnania* (1), *Melocalamus* (2), *Melocanna* (2), *Neololeba* (1), *Neomicrocalamus* (1), *Ochlandra* (14), *Oxytenanthera* (3), *Pseudostachyum* (1), *Pseudoxytenanthera* (6), *Schizostachyum* (23), *Sirochloa* (1), *Thyrsostachys* (2) and *Vietnamosasa* (2).

Olyreae: *Cryptochloa* (1), *Lithachne* (1), *Olyra* (2) and *Raddia* (1).

Besides, 14 man-made hybrids are also conserved (Koshy, 2010; Koshy and Gopakumar, 2016).

3.7 Dominant Bamboo Genera under Living Collections

Phyllostachys (55 species) represents maximum number of species among the temperate woody bamboos and *Chusquea* (29 species) among the Neotropical Woody Bamboos. *Bambusa* (112 species), *Dendrocalamus* (42 species), and *Gigantochloa* (39 species) top the Paleotropical Woody Bamboos. Most cultivated bamboos are *Bambusa multiplex* (72 entries) and *Bambusa vulgaris* (50 entries) followed by *Phyllostachys reticulata* (42 entries) and *Bambusa textilis* (24 entries). Herbaceous bamboos are poorly represented (Table 3.2).

3.8 What are the Benefits of Bamboo Conservation?

An established living collection of bamboos has many advantages. A well-maintained bambusetum helps: (1). To serve as a permanent display for education: Bamboos being beautiful, they add charm to any landscape. (2). To undertake scientific studies: As semelparous and monocarpic, bamboos flower and fruit at very long intervals, sometimes even up to 120 years (Janzen, 1976). No records are available on the period of flowering and mast seeding for many bamboo species. Reproductive biology of such species remains unknown. When identified species are available in a well-documented bambusetum, it facilitates periodic scientific observations on flowering behaviour, breeding system, pollination mechanism, fruit development and taxonomy. To cite an example, studies conducted on materials from JNTBGRI bambusetum have yielded enormous scientific data on reproductive biology (Koshy and Pushpangadan, 1997; Koshy and Harikumar, 2001; Koshy and Jee, 2001; Koshy et al., 2001; Koshy and Mathew, 2009), propagation (Koshy and Gopakumar, 2005), cytology (Remya Krishnan et al., 2015; Jero et al., 2015), morphology (Koshy et al., 2010), taxonomy (Goh et al., 2013, 2020) and phytochemistry (Baby et al., 2013, Govindan et al., 2016, 2018). (3). To address the needs of the state: If a specific species or its planting stock is required by farmers or other agencies, it can be multiplied and supplied. (4). To open ways for other branches of studies: Bamboo clumps favour growth of a variety of organisms such as fungi, lichens, mushrooms, slugs, snails, ants, birds, and various mammals. A good case in point is that of Giant panda *Ailuropoda melanoleuca* (Zhang et al., 2015). The Logan Botanic Garden is leading an international project to investigate the importance of bamboos in the habitat and diet of giant panda (URL, 33).

3.9 How Should a Modern Bambusetum Function?

All botanic gardens keep record of their living collections. The 19th and early 20th century records were confined to mere listings of species with their countries or regions of origin. The trend changed in the latter half of the 20th century when professionals undertook expeditions giving equal importance to site data, coordinates, and ecological notes. As a result, currently, passport details of all accessions are well documented. Later, most of these gardens started exchange and distribution of their holdings for research and cultivation purposes with this information (Koshy, 2010). However, an excellent model is that of the National Plant Germplasm System (NPGS) of United States Department of Agriculture (USDA) which maintains and distributes the world's largest living collection of plant genetic resources including 64 bamboo species (USDA, 2015; Widrlechner, 1989; URL, 39). They have made available critical information, passport data, and phenotypic characteristics of the plant materials through the web server GRIN - Germplasm Resources Information Network (Volk and Richards, 2008). They distribute seeds, cuttings, dormant scion wood, vegetative propagules, DNA, leaves, fruit, pollen, and in vitro cultures (Byrne et al., 2018). Many bamboo species possess diverse clones for characters like strength, straightness, thickness, and durability which have specific uses in industry and traditional practices. These clones, including "ancient enduring clones", need to be collected and conserved in modern bambuseta. Bambuseta must strive to conserve all native bamboo species, especially the rare ones.

3.10 Seed and Rhizome Banks

Germplasm storage in seed banks is an effective method in conserving plant diversity *ex situ* as it facilitates preservation of abundant genetic material in a small space with minimum risk of genetic damage (Iriondo and Pérez, 1999; Gómez-Campo, 2006). Bamboo seeds of dry and fleshy caryopses have no dormancy period. Mature seeds readily germinate in moist medium. Seed viability in most bamboo species lasts only up to two months. Viability decreases when it is not kept under suitable conditions.

The Millennium Seed Bank Partnership Programme (MSBP) is the largest *ex situ* conservation initiative in the world which focuses on seed collection, long term storage, and germination tests. MSBP shares seeds and germination data for research purposes, species reintroduction, and habitat restoration (Williams, 2007). As for bamboos, the Millennium Seed Bank has collected its billionth seed in 2007 from the African bamboo, *Oxytenanthera abyssinica* (A. Rich.) Munro. The seed was collected in Mali by the MSBP partner institution, the Institut d' Economie Rurale (Williams, 2007). However, a search of MSB Seed List (Millennium Seed Bank— Seed List (kew.org)) shows that bamboo seeds are very poorly represented in seed banks. This appears to be because of the rarity in bamboo seeding since most of the species seed only once, taking a duration of 20–60 or sometimes even 120 years (Janzen, 1976).

Another effective method of bamboo conservation is the establishment of rhizome banks. They function as certified source of bamboo rhizomes/offsets for plantation purposes or further multiplication in nurseries. Though rhizome banks are a potential device for bamboo conservation, they are hardly attempted by established bambuseta. The only known rhizome bank occurs in Uttarakhand state, India, which harbours eight bamboo species, viz., *Bambusa bambos* (L.) Voss, *Dendrocalamus asper* (Schult. f.) Backer, *Dendrocalamus hamiltonii* Nees & Arn. ex Munro, *Dendrocalamus membranaceus* Munro, *Dendrocalamus strictus* (Roxb.) Nees, *Drepanostachyum falcatum* (Nees) Keng f. (=*Arundinaria falcata*), *Melocanna baccifera* (Roxb.) Kurz, and *Thyrsostachys siamensis* Gamble (Chandran, 2008).

3.11 Summary and Conclusion

There are 1,670 species of bamboos in 125 genera in the world. Of them, 560 species and 14 hybrids have been conserved in 139 bambuseta worldwide. The 560 species account for 33.53% of the total species and 60 genera represent 48% of the total genera. Arundinarieae (total 31 genera and 581 species) is represented by 28 genera and 228 species, i.e., 90.32%, and 39.24 %, respectively. Bambuseae (total 73 genera and 966 species) is represented by 28 genera and 327 species, constituting 38.36% and 33.85%, respectively. Olyreae (total 21 genera and 123 species) represents four genera and five species, accounting to 19.04% and 4.07%, respectively. In general, only one third of the total bamboo species is conserved in various bambuseta in the world. There are 122 countries that harbour different species of bamboos but only 37 countries (30.32%) are involved in *ex situ* conservation programmes. Collection and introduction of maximum species with greater emphasis on native and rare taxa must get priority in bamboo conservation. Likewise, conservation of bamboo seeds in seed banks are to be strengthened. The data assembled in this chapter indicate that the woody bamboos, both temperate and tropical, are preferred for introduction due to their high economical and horticultural value. The poor representation of herbaceous bamboos could be ascribed to non-exploitation of their horticulture potential. It is found that 139 botanic gardens function as "Noah's Ark" for bamboos. Seed and rhizome banks provide additional opportunities, though they are yet to gain momentum.

Acknowledgments

Dr. Malay Das, Humboldt Fellow (Germany), NRC Associate (USA) invited me to write this chapter. Dr. K.M. Wong, Singapore Botanic Gardens, has checked the identity of bamboos at Singapore Botanic Gardens and shared the list. Prof. Nian-He Xia, South China Botanical Garden, Chinese Academy of Sciences, sent me a copy of the *Encyclopedia of Chinese Garden Flora, Vol. 6*. Dr. Maria S. Vorontsova, Research Leader (Grasses), Comparative Plant & Fungal Biology, Royal Botanic Gardens, Kew sent me the final dataset on the global distribution of bamboos. Dr. T.S. Nayar, formerly Head, Division of Conservation Biology, Jawaharlal Nehru Tropical Botanic Garden and Research Institute, Thiruvananthapuram and Dr. D. Maity, Department of Botany, University of Calcutta critically reviewed the MS and provided suggestions for improvement. Dr. Anita Balan, Mr. Biju John Koshy, and Mrs. Reeba Jacob extended technical help.

Abbreviations of Repositories

ABG	Anji Bamboo Garden, China
AJCBIBG	Acharya Jagadish Chandra Bose Indian Botanic Garden, Kolkata, India
BBG	Bali Botanic Garden, Indonesia
BFRI	Bangladesh Forest Research Institute, Chittagong, Bangladesh
CNBG	Jiangsu Institute of Botany, China
FJCBG	Francisco Javier Clavijero Botanic Garden, Mexico
FRI	Forest Research Institute, Dehradun, Uttarakhand, India
GBG	Gampaha Botanic Gardens, Henerathgoda, Gampaha, Sri Lanka
GRIFFIN	Plant Genetic Resources Conservation Unit, Griffin, Georgia
GXIB	Guangxi Institute of Botany, China
HBG	Hangzhou Botanical Garden, China
HBGH	Hakgala Botanic Gardens, Hakgala, Sri Lanka
IBCAS	Institute of Botany, CAS, China
JBHLCM	Jardin Botanico-Historico "La Concepcion" de Malaga, Spain
JBMPC	Jardí Botànic Marimurtra Passeig Carles, Spain
JNTBGRI	Jawaharlal Nehru Tropical Botanic Garden and Research Institute, Kerala India
KBG	Kanapaha Botanical Gardens, Florida, USA
KIB	Kunming Institute of Botany, China
KFRI	Kerala Forest Research Institute, Peechi, Kerala, India
LBMG	La Bambouseraie (Maurice Negre Parc Exotique de Prafrance) Generargues, France
LSBG	Lushan Botanical Garden, China
MAY	National Germplasm Repository-Mayaguez, Puerto Rico
MBG	Missouri Botanical Garden, USA
MIA	National Germplasm Repository, Miami, Florida, USA
PBG	Purwodadi Botanic Garden, East Java, Indonesia
PCG	The Palomar College Gardens, California, USA
PZ	Perth Zoo, Australia
RBGE	Royal Botanic Garden Edinburgh, UK
RBGPS	Royal Botanic Gardens, Peradeniya, Sri Lanka
RPRC	Regional Plant Resources Centre, Bhubaneswar, India
SBG	Singapore Botanic Gardens, Singapore
SCBG	South China Botanical Garden, China
SDBG	San Diego Botanic Garden, San Diego, California, USA
SMA	Shing Mun Arboretum, Hong Kong
SZBG	Shenzhen Fairy Lake Botanical Garden, China
TGT	Trebah Garden Trust, Mawnan Smith, Falmouth, UK
VVK	Van Vigyan Kendra, Chessa, India
WHIOB	Wuhan Botanical Garden, China
XMBG	Xiamen Botanical Garden, China
XTBG	Xishuangbanna Tropical Botanical Garden, China

References

Alam, M.K. (2011). Book review, Bamboos at TBGRI. *Bangladesh Journal of Plant Taxonomy*, 18(2): 209–210.

Baby, S., Anil, J.J., Govindan, B., Lukose, S., Gopakumar, B. and Koshy, K.C. (2013). UV induced visual cues in grasses. *Sci. Rep.*, 3: 2738. DOI: 10.1038/srep02738.

Benfield, R.W. (2013). A world garden survey. pp. 8–103. *In*: Benfield, R.W. (ed.). *Garden Tourism*. CABI, Wallingford, UK.

Beniwal, B.S. and Haridasan, K. (1988). Study of bamboos through establishment of Bambusetum. *Indian Forester*, 114(10): 650–655.

BGCIa. GardenSearch. Available online: https://tools.bgci.org/garden_search.php (accessed on 21 February 2022.

BGCIb. PlantSearch. Available online: https://tools.bgci.org/plant_search.php (accessed on 21 February 2022).

Biswas, Sas. (1999). Bamboo diversity and conservation in India. pp. 164–175. *In*: Rao, A.N. and Rao, V.R. (eds.). *Bamboo Conservation, Diversity, Ecogeography, Germplasm, Resource Utilization and Taxonomy*. International Plant Genetic Resources Institute, Rome.

Biswas, Sas. (2008). State-of-the-art Bambusetum of forest research Institute, Dehra Dun, India for its lay-out and design, repository of genetic resource, education, and awareness. *Indian Forester* 134(9): 1261–1263.

[BPG] Bamboo Phylogeny Group. (2012). An updated tribal and subtribal classification of the Bamboos (Poaceae: Bambusoideae). *Bamboo Science and Culture: The Journal of the American Bamboo Society*, 24(1): 1–10.

Byrne, P.E., Volk, G.M., Gardner, C., Gore, M.A., Simon, P.W. and Smith, S. (2018). Sustaining the future of plant breeding: The critical role of the USDA-ARS National Plant Germplasm System. *Crop Science*, 58: 451–468. DOI: 10.2135/cropsci2017.05.0303.

Canavan, S., Richardson, D.M., Visser, V., Le Roux, J.J., Vorontsova, M.S. and Wilson, J.R.U. (2017). The global distribution of bamboos: Assessing correlates of introduction and invasion. *AoB Plants*, 9: plw078. Doi:10.1093/aobpla/plw078.

Chandran, M. (2008). Development of Rhizome banks for bamboo multiplication. *Indian Forester*, 134(3): 445–447.

Christenhusz, M.J.M. (2011). Book review, Bamboos at TBGRI. *Botanical Journal of the Linnean Society*, 167: 131.

Goh, W.L., Chandran, S., Franklin, D.C., Isagi, Y., Koshy, K.C., Sungkaew, S., Yang, H.Q., Xia, N.H. and Wong, K.M. (2013). Multigene region phylogenetic analyses suggest reticulate evolution and a clade of Australian origin among paleotropical woody bamboos (Poaceae: Bambusoideae: Bambuseae). *Plant Systematics and Evolution*, 299: 239–257.

Goh, W.L., Sungkaew, S., Teerawatananon, A., Ohrnberger, D., Widjaja, E.A., Sabu, K.K., Gopakumar, B., Koshy, K.C., Xia, N.H. and Wong, K.M. (2020). The phylogenetic position and taxonomic status of the Southeast and South Asian bamboo genera *Neohouzeaua* and *Ochlandra* (Poaceae: Bambusoideae). *Phytotaxa*, 472(2): 107–122.

Gómez-Campo, C. (2006). Erosion of genetic resources within seed genebanks: The role of seed containers. *Seed Science Research*, 16(4): 291–294. Doi:10.1017/SSR2006260.

Govindan, B., Anil, J.J., Ajikumaran Nair, S.N., Gopakumar, B., Karuna, M.S.L., Venkataraman, R., Koshy, K.C. and Baby, S. (2016). Nutritional properties of the largest bamboo fruit *Melocanna baccifera* and its ecological significance. *Sci. Rep.*, 6: 26135. Doi: 10.1038/srep26135.

Govindan, B., Johnson, A.J., Viswanathan, G., Ramaswamy, V., Koshy, K.C. and Baby, S. (2018). Secondary metabolites from the unique bamboo, *Melocanna baccifera*. *Natural Product Research*, Doi: 10.1080/14786419.2018.1434647.

Henpithaksa, C. (2008). Plants. *In*: *Proceedings of the 46th Kasetsart University Annual Conference, Kasetsart*, 29 January–1 February 2008. Kasetsart University, Bangkok, Thailand. Thai language, pp. 431–440.

Huang, H. (2011). Plant diversity and conservation in China: Planning a strategic bioresource for a sustainable future. *Botanical Journal of the Linnean Society*, 166(3): 282–300.

Huang, H. (ed.). (2018). *Encyclopedia of Chinese Garden Flora.* Vol. 6 (Euphorbiaceae-Gramineae). Science Press. Beijing, pp. 207–379.

Hunter, I. (2003). Bamboo resources, uses and trade: The future? *Journal of Bamboo and Rattan,* 2: 319–326.

[ICFRI] Indian Council of Forestry Research and Education. (2015). *Annual Report,* Dehradun, Uttarakhand, India, 142 pp.

[INBAR] International Network for Bamboo and Rattan. (2014). *Greening Red Earth, Restoring Landscapes, Rebuilding Lives.* INBAR Working Paper No. 76. Beijing, P.R. China, 24 pp.

[IPNI] International Plant Names Index. (2022). The Royal Botanic Gardens, Kew, Harvard University Herbaria & Libraries and Australian National Botanic Gardens. Published on the Internet http://www.ipni.org, Retrieved 15 March 2022.

Iriondo, J.M. and Pérez, C. (1999). Propagation from seeds and seed preservation. pp. 456–57. *In*: Bowes, B.G. (ed.). *A Colour Atlas of Plant Propagation and Conservation.* Manson Publishing, London.

Islam, S.S. (2003). *State of Forest Genetic Resources Conservation and Management in Bangladesh.* Forest Genetic Resources Working Papers, Working Paper FGR/68E. Forest Resources Development Service, Forest Resources Division. FAO, Rome.

Janzen, D.H. (1976). Why bamboos wait so long to flower. *Annual Review of Ecology and Systematics,* 7: 347–391.

Jero Mathu, A., Mathew, P.M., Mathew, P.J., Harikumar, D. and Koshy, K.C. (2015). Cytological study in *Pseudoxytenanthera* (Tribe Bambuseae) occurring in India. *J. Cytol. Genet.,* 16: 61–68.

Kariyawasam, D. (1999). Bamboo resources and utilization in Sri Lanka. pp. 235–247. *In*: Rao, A.N. and Rao, V.R. (eds.). *Bamboo Conservation, Diversity, Ecogeography, Germplasm, Resource Utilization and Taxonomy.* International Plant Genetic Resources Institute, Rome.

[KFRI] Kerala Forest Research Institute. (2005). *Three Decades of Research in KFRI: New Technologies and Information. Peechi, Kerala,* pp. 220–221.

Koshy, K.C. (2010). *Bamboos at TBGRI.* Tropical Botanic Garden and Research Institute, Palode, Thiruvananthapuram, Kerala, India.

Koshy, K.C. (2017). Bambusetum at Jawaharlal Nehru Tropical Botanic Garden and Research Institute: State of the art. pp. 361–378. *In*: Singh, P. and Dash, S.S. (eds.). *Indian Botanic Gardens: Role in Conservation.* Botanical Survey of India, Kolkata.

Koshy, K.C. and Gopakumar, B. (2005). An improvised vegetative propagation technique for self-incompatible bamboos. *Current Science,* 89(9): 1474–1476.

Koshy, K.C. and Gopakumar, B. (2016). Bambusetum.pp. 16–69. *In*: Mohanan, N. and Mathew, P.J. (eds.). *Live Plants of JNTBGRI.* Jawaharlal Nehru Tropical Botanic Garden and Research Institute, Kerala.

Koshy, K.C. and Harikumar, D. (2001). Reproductive biology of *Ochlandra scriptoria,* an endemic reed bamboo of the Western Ghats, India. *Bamboo Science and Culture, J. Amer. Bamboo Soc.,* 15(1): 1–7.

Koshy, K.C. and Jee, G. (2001). Studies on the lack of seed set in *Bambusa vulgaris. Current Science,* 81(4): 375–378.

Koshy, K.C. and Mathew, P.J. (2009). Does *Ochlandra scriptoria* flower annually or once in a lifetime? *Current Science,* 96(6): 769–770.

Koshy, K.C. and Pushpangadan, P. (1997). *Bambusa vulgaris* blooms, a leap towards extinction? *Current Science,* 72(9): 622–624.

Koshy, K.C., Dintu, K.P. and Gopakumar, B. (2010). The enigma of leaf size and plant size in bamboos. *Current Science,* 99(8): 1025–1027.

Koshy, K.C., Harikumar, D. and Narendran, T.C. (2001). Insect visits to some bamboos of the Western Ghats, India. *Current Science,* 81(7): 833–838.

Kumari, P. (2017). Bambusetum in Acharya Jagadish Chandra Bose Indian Botanic Garden. pp. 350–360. *In*: Singh, P. and Dash, S.S. (eds.). *Indian Botanic Gardens: Role in Conservation.* Botanical Survey of India, Kolkata.

Liese, W. and Kohl, M. (2015). *Bamboo: The Plant and Its Uses. Tropical Forestry.* Springer, Basel, Switzerland.

Lucas, S. (2013). *Bamboos.* Reaktion Books Ltd., London, UK.

Maoyi, F. (1999). Bamboo resources and utilization in China. pp. 157–163. *In*: Rao, A.N. and Rao, V.R. (eds.). *Bamboo Conservation, Diversity, Ecogeography, Germplasm, Resource Utilization and Taxonomy*. International Plant Genetic Resources Institute, Rome.

Mejia-Saulés, M.T. and Ordóñez, R.M. 2018. Mexican national living bamboo collection *ex situ* conservation. *Proceedings of the 11th World Bamboo Congress, Mexico*.

Muller, L. (1998). Sundaland transfer. *American Bamboo. Soc. Newsletter*, 19(1 & 2): 10–15.

Offord, C.A. (2017). Germplasm conservation. pp. 281–288. *In*: Thomas, B., Murray, B.G. and Murphy, D.J. (eds.). *Encyclopedia of Applied Plant Science*, Vol. 2 (2nd edn.). Academic Press, Cambridge, MA, USA.

[POWO] Plants of the World Online. (2022). Facilitated by the Royal Botanic Gardens, Kew.http://www.plantsoftheworldonline.org/ Retrieved 02 January 2022.

Remya Krishnan, R.V., Mathew, P.J. and Koshy, K.C. (2015). Polyploidy in Muli Bamboo: *Melocanna baccifera*. *J. Cytol.Genet.*, 16: 1–5.

Roxas, C.A. (1999). Country Reports: Bamboo research in the Philippines. *In*: Rao, A.N. and Rao, V.R. (eds.). *Bamboo Conservation, Diversity, Ecogeography, Germplasm, Resource Utilization, and Taxonomy*. Kunming and Xishuangbanna, Yunnan, China: International Plant Genetic Resources Institute. Rome.

Roxburgh, W. (1814). *Hortus Bengalensis, or a Catalogue of the Plants Growing in the Honourable East India Company's Botanic Grden at Calcutta*. Mission Press, Serampore.

Scurlock, J.M.O., Dayton, D.C. and Hames, B. (2000). Bamboo: An overlooked biomass resource? *Biomass and Bioenergy*, 19: 229–244.

[SFRI] State Forest Research Institute. (2020). *Annual Research Report*. Jabalpur, Madhya Pradesh, India, 99 pp.

Sharma, Y.M.L. (1980). Bamboos in the Asia-Pacific region. pp. 99–120. *In*: Lessard, G. and Chouinard, A. (eds.). *Bamboo Research in Asia. Proceedings of a workshop held in Singapore, 28–30 May*. IDRC. Canada.

Soreng, R.J., Peterson, P.M., Romaschenko, K., Davidse, G., Teisher, J.K., Clark, L.G., Barbera, P., Gillespie, L.J. and Zuloaga, F.O. (2017). A worldwide phylogenetic classification of the Poaceae (Gramineae) II: An update and a comparison of two 2015 classifications. *Journal of Systematics and Evolution*, 55(4): 259–290.

Subramaniam, K.N. (1998). Bamboo genetic resources in India. pp. 32–61. *In*: Vivekanandan, K., Rao, A.N. and Ramanatha Rao, V. (eds.). *Bamboo and Rattan Genetic Resources in Certain Asian Countries*. IPGRI-APO, Serdang, Malaysia.

Sujarwo, W. (2018). Bamboo resources, cultural values, and *ex-situ* conservation in Bali, Indonesia. *Reinwardtia*, 17(1): 67–75.

Tañgau, F.T. and Veracion, V.P. (1989). Philippine Bambusetum: Where bamboos exude their usefulness and charm. *Canopy International*, 15(2): 8–11.

Townsend, R. (2018). Bamboos: *From Victorian curiosity to elephant food*. Royal Botanic Gardens, Kew. https://www.kew.org/read-and-watch/bamboos-victorian-curiosity-elephant-food.

USDA. (2015). Agricultural Research Service. Germplasm Resources Information Network (GRIN). Ag Data Commons. https://doi.org/10.15482/USDA.ADC/1212393. Accessed 2022-01-29.

Volis, S. (2019). Conservation-oriented restoration: A two-for-one method to restore both threatened species and their habitats. *Plant Diversity*, 41(2): 50–58.

Volk, G.M. and Richards, C.M. (2008). Availability of genotypic data for USDA-ARS National Plant Germplasm System Accessions using the Genetic Resources Information Network (GRIN) Database. *Hortscience*, 43(5): 1365–1366.

Vorontosova, M.S. (2011). Book review, Bamboos at TBGRI. *Kew Bulletin*, 66: 195–196.

Wang, P.Y. and Li, D.Z. (2019). *Dendrocalamus menghanensis* (Poaceae, Bambusoideae), a new woody bamboo from Yunnan, China. *PhytoKeys*, 130: 143–150.

Widrlechner, M.P. (1989). Germplasm resources information network and *ex situ* conservation of germplasm. pp. 109–114. *Proceedings of the North American Prairie Conferences. 42*. https://digitalcommons.unl.edu/napcproceedings/42.

Williams, N. (2007). A billion seeds and counting. *Current Biology*, 17(11): 387–388.

Wong, K.M. (2004). *Bamboo, The Amazing Grass*. International Plant Genetic Resources Institute and University of Malaya, Kuala Lumpur.

Wong, K.M. (2011). Book review, Bamboos at TBGRI. *The Gardens' Bulletin Singapore*, 62(2): 337–338.

Wong, K.M. (2022). *List of Bamboos growing in Singapore Botanic Gardens*. Personal Communication.

Wyse Jackson, P.S. (1999). Experimentation on a large scale: An analysis of the holdings and resources of Botanic Gardens. *Botanic Gardens Conservation News*, 3(3): December 1999. UK: Botanic Gardens Conservation International.

Wyse Jackson, P. and Sutherland, L.A. (2013). Role of botanic gardens. pp. 504–521. *In*: Levin, S.A. (ed.). *Encyclopedia of Biodiversity*, Vol. 6 (2nd edn.). Waltham, MA: Academic Press.

Xia, N. (2011). Book review, Bamboos at TBGRI. *Plant Diversity and Resources*, 33(4): 375.

Yang, Q., Duan, Z.B., Wang, Z.L., He, K.H., Sun, Q.X. and Peng, Z.H. (2008). Bamboo resources, utilization, and *ex-situ* conservation in Xishuangbanna, Southeastern China. *Journal of Forestry Research*, 19(1): 79–83.

Zhang, J., Hull, V., Huang, J., Zhou, S., Xu, W., Yang, H., McConnell, W.J., Li, R., Liu, Huang, D., Ouyang, Y.Z., Zhang, H. and Liu, J. (2015). Activity patterns of the giant panda (Ailuropoda melanoleuca). *Journal of Mammalogy*, 96(6): 1116–1127.

URLs Referred

No.	URL

1. Brunei Forestry Centre, Darussalam; http://www.forestry.gov.bn/SitePages/Ex-Situ%20 Conservation.aspx

2. Anji Bamboo Garden: http://www.zj.gov.cn/art/2012/5/30/art_1568751_26245131.html

3. Anji Bamboo Garden: http://www.foreignercn.com/index.php?option=com_content&view=articl e&id=6010:anji-bamboo-expo-garden&catid=68:travel-in-zhejiang&Itemid=140

4. South China Botanical Garden (SCBG): https://www.bamboo-trip.com/scbg

5. South China Botanical Garden (SCBG): http://english.scbg.ac.cn/as/mh/

6. Wuhan Botanical Garden: http://english.whb.cas.cn/rh/rp/200704/t20070401_32818.html

7. Xishuangbanna Tropical Botanical Garden: http://english.xtbg.cas.cn/gh/lc/201001/ t20100122_50301.html

8. Shing Mun Arboretum, Hong Kong: https://www.afcd.gov.hk/english/conservation/con_flo/ con_flo_shing/con_floshing.html

9. Bamboorium, Basar: http://sfri.nic.in/pdf_files/SFRI.pdf

10. Kerala Forest Research Institute, Kerala: https://www.kfri.res.in/bambusetum.asp

11. Botanical Garden of Punjab University, Chandigarh, India: https://botany.puchd.ac.in/

12. cbd doc: https://www.cbd.int/doc/world/in/in-ex-bg-en.pdf

13. Rain Forest Research Institute: https://www.icfre.org/amrit_mahotsav/activity33.30.04.2021-RFRI.pdf

14. Regional Plant Resources Centre Bhubaneswar Botanic Garden: https://www.rprcbbsr.in/View/ bambusetum.aspx

15. Pilikula Development Authority, Karnataka: http://pilikula.com/bamboo_setum.html

16. https://www.bioversityinternational.org/fileadmin/bioversity/publications/Web_version/572/ch26. htm

17. Bambusetum, Rimba Ilmu Botanic Garden, University of Malaya, Kuala Lumpur: https://rimba. um.edu.my/index.php

18. Forest Research Institute Malaysia: https://www.frim.gov.my/attractions/arboreta/

19. Penang Botanic Gardens: https://www.penang-traveltips.com/bambusetum-botanic-garden.htm

20. Forest Research Institute, Yezin; https://www.bioversityinternational.org/fileadmin/bioversity/ publications/Web_version/572/ch28.htm

21. National Kandawgyi Botanical Gardens: http://www.trekthailand.net/myanmar/parks/maymyo/

22. Philippine Bambusetum:https://www.bioversityinternational.org/fileadmin/bioversity/publications/Web_version/572/ch30.htm

23. Singapore Botanic Garden: https://www.nparks.gov.sg/sbg/our-gardens/tyersall-entrance/the-learning-forest

24. Botanic Gardens, Sri Lanka: https://www.botanicgardens.gov.lk/

25. Perth Zoo, Western Australia: http://www.perthzoo.wa.gov.au/animals-plants/plants/

26. Lae Botanic Gardens, Papua New Guinea: http://laebotanicgardens.com/welcome-to-lae-gardens/

27. La Bambouseraie (Maurice Negre Parc Exotique de Prafrance) Generargues, Anduze, France: https://en.wikipedia.org/wiki/Bambouseraie_de_Prafrance

28. The Botanical Garden of Rome, Sapienza University, Rome: https://web.uniroma1.it/ortobotanico/en/bamboo/bamboo

29. Jardí Botànic Marimurtra Passeig Carles Faust, Blanes Girona, Spain: https://marimurtra.cat/wp-content/uploads/2021/10/181021_Marimurtra_TaxonsNumeric.pdf

30. Jardin Botanico-Historico "La Concepcion" de Malaga,Spain: https://laconcepcion.malaga.eu/opencms/export/sites/laconcepcion/.galeria-descargas/2581e918-c71c-11e4-ada8-3d6b25258dd2/Coleccion-de-bambues.pdf

31. Royal Botanic Gardens, Kew, Richmond, London, TW9 3AE: https://www.kew.org/read-and-watch/bamboo-kew-pandas https://www.kew.org/read-and-watch/bamboos-victorian-curiosity-elephant-food

32. Carwinion House and Bamboo Garden: https://www.cornwalls.co.uk/attractions/carwinion-house-and-bamboo-garden.htm

33. Logan Botanic Garden, Scotland: https://www.rbge.org.uk/science-and-conservation/genetics-and-conservation/molecular-ecology/panda-diet/

34. The Lost Gardens of Heligan,UK: https://www.greatgardensofcornwall.co.uk/the-lost-gardens-of-heligan/

35. Royal Botanic Garden Edinburgh: https://data.rbge.org.uk/search/livingcollection/

36. Trebah Garden Trust, UK: https://www.trebahgarden.co.uk/garden/bamboozle

37. Tresco Abbey Gardens: https://www.greatgardensofcornwall.co.uk/tresco-abbey-garden/

38. Lancetilla Botanic Garden & Research Center, Honduras: http://jblancetilla.org/Trifolio%20Sendero%20Hist%20Ingles.pdf

39. US National Plant Germplasm System GRIN-Global: https://npgsweb.ars-grin.gov/gringlobal/cropdetail?type=species&id=251

40. Coastal Georgia Botanical Gardens at the Historic Bamboo Farm: https://coastalbg.uga.edu/the-gardens/barbour-lathrop-bamboo-collection/

41. Kanapaha Botanical Gardens, Florida: https://kanapaha.org/https://kanapaha.org/bamboo-sale

42. Missouri Botanical Garden: https://www.missouribotanicalgarden.org/PlantFinder/PlantFinderSearch.aspx
https://www.missouribotanicalgarden.org/about/additional-information/our-history.aspx

43. The Palomar College Gardens, USA: https://www2.palomar.edu/users/warmstrong/pcarbor1.htm

44. San Diego Botanic Garden (SDBG), San Diego, Encinitas, California, USA: https://www.publicgardens.org/programs/plant-collections-network/collections-showcase/bamboo file:///C:/Users/user/Downloads/Bamboo%20Records%20(Public%20Information).pdf

45. Mexican national living bamboo collection *ex situ* conservation: https://worldbamboo.net/wbcxi/papers/MeijaSaules,%20T.%20and%20R.M.%20Ordonez.pdf

Chapter 4

Genetic Diversity Assessment and Molecular Markers in Bamboos

Eduardo Ruiz-Sanchez,[1,2,*] *Jessica Pérez-Alquicira,*[2,3] *María de la Luz Perez-Garcia*[1,2,4] and *Miguel Angel García-Martínez*[1,2,4]

4.1 Introduction

The bamboos are grass species that are recognized as two kinds: woody bamboos and herbaceous bamboos (Judziewicz et al., 1999). Genetically, the herbaceous bamboos are less studied than the woody bamboo species are. Economically, culturally, and ecologically speaking the woody bamboos are important in many countries in Africa, the Americas, and Asia (Clark et al., 2015). Woody bamboo species have lignified culms with fast growth which are harvested sustainably so that they can be used instead of timber, thus helping to mitigate climate change.

Taxonomically, woody and herbaceous bamboos belong to the subfamily Bambusoideae, one of the 12 subfamilies in the grass family Poaceae (Soreng et al., 2015, 2017). Woody bamboos are classified into two tribes: Arundinarieae and Bambuseae, while herbaceous bamboos belong to the Olyreae tribe (Clark et al., 2015). The woody bamboos belonging to Arundinarieae are known as temperate woody bamboos (TWB), while bamboos in Bambuseae are known as tropical woody bamboos. The tribes are monophyletic lineages, which means that all the species within the tribe share a common ancestor; however, their relationship is controversial depending on which markers are used (chloroplast vs nuclear). Using a few makers or

[1] Departamento de Botánica y Zoología, Centro Universitario de Ciencias Biológicas y Agropecuarias, Universidad de Gudalajara, Mexico.
[2] Laboratorio Nacional de Identificación y Caracterización Vegetal (LaniVeg), Instituto de Botánica, Universidad de Guadalajara, Mexico.
[3] CONACYT, Ciudad de México, Mexico.
[4] Doctorado en Ciencias en Biosistemática, Ecología y Manejo de Recursos Naturales y Agrícolas, Centro Universitario de Ciencias Biológicas y Agropecuarias, Universidad de Guadalajara, Mexico.
* Corresponding author: ruizsanchez.eduardo@gmail.com

the whole chloroplast genome, Bambuseae and Olyreae are sister tribes (Sungkaew et al., 2009; Kelchner and BPG, 2013; Wysocki et al., 2015; Saarela et al., 2018), while with nuclear markers Bambuseae and Arundinarieae are sister tribes (Triplett et al., 2014; Guo et al., 2019; Chalopin et al., 2021).

This last relationship makes the woody character state the result of a single ancestral origin. Geographically, Arundinarieae species are distributed in Africa, Asia, and North America (Wang et al., 2017). Bambuseae are divided into two clades. Paleotropical woody bamboos (PWB) are distributed in Africa, Asia, and northern Australia and Neotropical woody bamboos (NWB) are endemic to the Neotropics in the Americas (Zhou et al., 2017; Tyrrell et al., 2018). Finally, except *Buergersiochloa bambusoides* Pilg. (New Guinea), all other Olyreae species are endemic to the Americas (Judziewicz et al., 1999; Oliveira et al., 2014; Ruiz-Sanchez et al., 2019). In this chapter, we review the molecular markers used to evaluate the genetic diversity and phylogenetic relationships of woody and herbaceous bamboos.

4.2 Molecular Markers Used to Infer Phylogenetic Relationships in Bamboos

4.2.1 Bambusoideae

The grass family Poaceae is a monophyletic lineage. The modern taxonomic classification recognizes Bambusoideae as one of the 12 subfamilies that comprise Poaceae (Soreng et al., 2015, 2017). Phylogenetically, Bambusoideae, Oryzoideae, and Pooideae form the BOP clade, where Bambusoideae is a sister group to the Pooideae subfamily, and Oryzoideae is a sister subfamily to the Bambusoideae and Pooideae clade (Soreng et al., 2017). As noted above, Bambusoideae is classified into three monophyletic tribes: Arundinarieae, Bambuseae, and Olyreae. Nowadays, two competing phylogenetic hypotheses have arisen because of the molecular markers used.

Based on a few chloroplast markers or the whole plastome, Bambuseae was the sister tribe to Olyreae (Bouchenak-Khelladi et al., 2008; Sungkaew et al., 2009; Kelchner and BPG, 2013; Wysocki et al., 2015; Saarela et al., 2018; Chalopin et al., 2021), and this indicates paraphyly among woody bamboos (Fig. 4.1). In contrast, whether using few or many single-copy nuclear markers or the whole nuclear genome, Arundinarieae was found to be the sister tribe to Bambuseae (Triplett et al., 2014; Wysocki et al., 2016; Guo et al., 2019), indicating that woody bamboos are monophyletic (Fig. 4.1). The reticulate origin of woody bamboos with ancient hybridization and allopolyploidy is the explanation for the incongruence between the chloroplast and nuclear phylogenetic hypotheses (Guo et al., 2019; Chalopin et al., 2021).

Here, we give a historical review of the molecular markers (chloroplast and nuclear) used to make phylogenetic inferences in bamboo species. The first molecular marker used to make phylogenetic inferences for the grass family, which includes a bamboo species, was the RNA ribosomal sequence. Hamby and Zimmer (1988), sequenced 18S rRNA and 26S rRNA genes in *Arundinaria gigantea* (Walter) Muhl.,

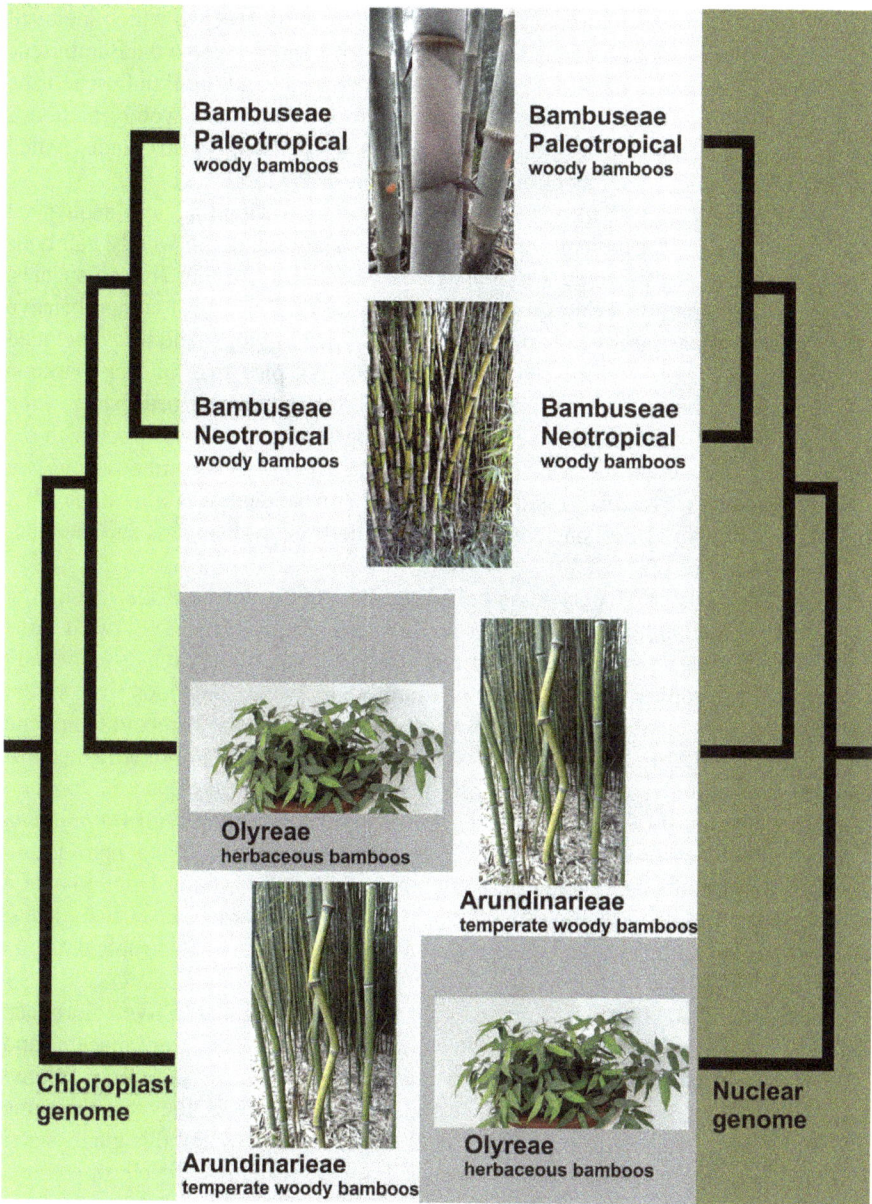

Fig. 4.1. Phylogenetic schematic representation of the three tribes of Bambusoideae based on previous results from whole plastomes (left tree; Wysocki et al., 2015) and nuclear genes from floral transcriptome analysis (right tree; Wysocki et al., 2016).

which was the sister species to the other grasses included. Later, Barker et al. (1995) sequenced the chloroplast gene *rbc*L for a single *Bambusa* species, which turned out to be the sister species to the other grass species sequenced. The same year, Clark et al. (1995) carried out the first comprehensive phylogenetic study of the grass family, sequencing the chloroplast gene *ndh*F for 45 grass species, which included 13 bamboo species. Four belonged to the Olyreae tribe and the rest to the Bambuseae tribe. Bambusoideae s. str. was found to be monophyletic and the Bambuseae tribe was paraphyletic. Olyreae species were sisters to the temperate woody bamboos (now Arundinarieae), and the rest of the species were sisters to this clade, which included Neotropical and Paleotropical woody bamboos (Table 4.1).

Later, Zhang and Clark (2000) increased bamboo sampling and sequenced 24 bamboo species for the chloroplast gene *ndh*F. They found the woody bamboos (temperate + tropical) formed a monophyletic clade, sister to the Olyreae species. The grass phylogeny working group (GPWG) (2001) did the most comprehensive study of the Poaceae family (62 taxa) to that point. Their phylogenetic study included four chloroplast genes (*ndh*F, *phy*B, *rbc*L, and *rpo*B), plus two nuclear markers: GBSSI (granule-bound starch synthase I) and ITS (Internal transcribed spacer of the nuclear ribosomal DNA). However, the Bambusoideae sampling was poor with the inclusion of only seven species: five from Olyreae (herbaceous bamboos) and two woody bamboos (*Pseudosasa* and *Chusquea*). Bambusoideae was a monophyletic subfamily with two clades, one corresponding to herbaceous bamboos and the other to woody bamboos (Table 4.1).

Eight years later, Bouchenak-Khelladi et al. (2008) increased their sampling of Poaceae and Bambusoideae to 395 and 25 taxa, respectively. They used three chloroplast markers (*mat*K, *rbc*L, and *trn*L-F) and found the paraphyly of woody bamboos. Paleotropical and Neotropical (Bambuseae) woody bamboos were sisters to the herbaceous bamboos (Olyereae), and these two clades are sisters to temperate woody bamboos (Arundinarieae). One year later, Sungkaew et al. (2009) did the most comprehensive study of the Bambusoideae subfamily. They included 52 bamboo taxa, four species of Olyeae, nine species of Arundinarieae, and 39 of Bambuseae. They used four chloroplast markers (*atp*B-*rbc*L, *trn*L-F, *rps*16, and *mat*K), and got the same results as Bouchenak-Khelladi et al. (2008): woody bamboos were paraphyletic. Additionally, they found that Bambuseae is composed of two clades, one with the Neotropical woody bamboos and the other with the Paleotropical woody bamboos (Table 4.1).

Kelcher and BPG (2013) increased their molecular sampling to five chloroplast markers (*ndh*F, *rpl*16 intron, *rps*16 intron, *trn*D-*trn*T, and *trn*T-*trn*L spacers) and 33 Bambusoideae taxa. The molecular markers and the taxa sampling differed from the previous study by Sungkaew et al. (2009). However, they got the same results. Olyreae and Bambuseae were found to be sister tribes and this clade, sister to Arundinarieae. In the Bambuseae, two clades were recognized: Neotropical and Paleotropical woody bamboos. To date, Soreng et al. (2015, 2017) have carried out the most comprehensive phylogeny of Poaceae, using 448 taxa and two chloroplast genes (*mat*K and *ndh*F). Their phylogenetic results in the case of Bambusoideae concur with those of previous studies by Bouchenak-Khelladi et al. (2008), Sungkaew et al. (2009), and Kelchner and BPG (2013) (Table 4.1).

Moving from a few chloroplast markers to the whole plastome, Wysocki et al. (2015) did the first phylogenetic analysis of Bambusoideae using the whole chloroplast. Their sampling included 31 Bambusoideae taxa that represented the

Table 4.1. Molecular markers used in phylogenetic studies.

Molecular Marker	Objective	References
18 S rRNA and 26 S rRNA	Infer phylogeny within the grass family (Poaceae) using Ribosomal RNA sequences.	Hamby and Zimmer, 1988
rbcL	Determine phylogenetic relationships among various lineages in the grasses, with particular emphasis on the subfamily Arundinoideae.	Barker et al., 1995
ndhF	Molecular phylogenetic analyses of the Poaceae family.	Clark et al., 1995
rpl16	Phylogenetic analyses of *Chusquea* and the Bambusoideae.	Kelchner and Clark, 1997
ITS and AFLP (*EcoRI* and *MseI*)	A comparison of ITS and AFLP for phylogenetic studies in *Phyllostachys*.	Hodkinson et al., 2000
ITS	Explore the utility of ITS in studies on genetic variation and evolution in the alpine bamboos.	Guo et al., 2001
ITS	Examine the adequacy of ITS for phylogenetic reconstruction in the *Thamnocalamus* group and its allies.	Guo et al., 2002
GBSSI and ITS	Explore phylogenetic relationships of the *Thamnocalamus* group and its allies.	Guo and Li, 2004
rpl16 and 98 morphological characters	Test the monophyly of Bambuseae, Chusqueinae, and Hickeliinae, and examine the conflict between molecular and morphological data sets in the determinate, one-flowered genera of Bambuseae.	Clark et al., 2007
trnL-F and GBSSI	Test the monophyly of *Schizostachyum* and some related genera.	Yang et al., 2007
trnH–psbA, rpl16 and 61 morphological characters	Phylogeny of *Otatea* and recircumscription of Guaduinae.	Ruiz-Sanchez et al., 2008
trnL-F, ITS, and GBSSI	Phylogeny of the major groups of PWB and test generic delimitation.	Yang et al., 2008
ITS and GBSSI	Test the monophyly of various groups within the temperate bamboo clade and evaluate their phylogenetic relationships.	Peng et al., 2008
trnD-trnT, rps16-trnQ, trnC-rpoB, rpl16, and *ndhF*	Test the monophyly of *Neurolepis*.	Fisher et al., 2009
apF-atpH, psbK-psbI, trnL-rpl32, ITS, and 54 morphological characters	Species delimitation in the genus *Otatea*.	Ruiz-Sanchez and Sosa, 2010

Table 4.1 contd. ...

...Table 4.1 contd.

Molecular Marker	Objective	References
Rps16-trnQ, trnC-rpoB, trnH-psbA, trnD-T, and GBSSI	Investigate the phylogenetic relationships among Southeast Asian climbing bamboos and the *Bambusa* complex.	Goh et al., 2010
atpllH, psaA-ORF170, rpl32-trnL, rpoB-trnC, rps16-trnQ, trnD/T, trnS/G, and *trnT/L*	Identify the major clades within tribe Arundinarieae and study molecular variation among different chloroplast regions.	Zeng et al., 2010
trnL–F and ITS	Assess phylogenetic relationships, origins and classification of Arundinarieae.	Hodkinson et al., 2010
atpl–atpH, psaA–ORF170, rpl32–trnL, rps16–trnQ, trnC–rpoB, trnD–trnT, trnH–psbA, trnK–rps16, trnT–trnL, trnV–ndhC, trnG and *ndhF*	Phylogeny of temperate bamboos with an emphasis on *Arundinaria* and allies.	Triplett and Clark, 2010
psbA-trnH, rpl32-trnL, rps16, and GBSSI	Investigate the phylogenetic relationships among *Bambusa* and its allies.	Yang et al., 2010
AFLPs and *trnT-trnL*	Generate a phylogenetic analysis of *Arundinaria* to evaluate patterns of genetic divergence, test the current species-level taxonomy and investigate the potential role of hybridization in the evolution of *Arundinaria.*	Triplett et al., 2010
rbcL, matK, psbK-psbI, rpl16, and *psbA-trnH*	Phylogeny of the genus *Olmeca* and determine whether *Aulonemia clarkiae* and *A. fulgor* should be transferred to *Olmeca.*	Ruiz-Sanchez et al., 2011
ndhF; trnD-trnT; trnC-rpoB, rps16-trnQ, trnT-trnL, and *rpl16*	Test the monophyly of the subtribe Arthrostylidiinae.	Tyrrell et al., 2012
atpl-atpH, psaA-ORF170, rpl32-trnL, rpoB-trnC, rps16-trnQ, trnD-trnT, trnS-trnG, trnT-trnL, and GBSSI	Identify inconsistent relationships between nuclear and plastid trees in Arundinarieae and explore possible causes of the incongruence.	Zeng et al., 2012
ndhF; rpl16, rps16, trnD-trnT, trnT-trnL	Improvement of resolution and establishment of branching order among major clades.	Kelchner et al., 2013
rpl32-trnL, trnT-trnL, rps16-trnQ, trnC-rpoB, GBSSI, and LEAFY	Test the monophyly of *Chimonocalamus* and identify evolutionary processes responsible for the incongruence among gene trees.	Yang et al., 2013
Rps16-trnQ, trnC-rpoB, trnD-T, and GBSSI	Compare nDNA and cpDNA topologies for further insights into lineage evolution among the BDG complex.	Goh et al., 2013
pvcel1, gpa1, and *pabp1*	Investigate the origin of polyploidy in the woody bamboos and examine putative hybrid relationships in the tribe Arundinarieae.	Triplett et al., 2014

Marker / Data	Objective	Reference
Rpl16, ndhA, trnD-trnT, trnT-trnL, ndhF, and *ITS*	Test the monophyly of taxonomic groups within *Euchusquea* to provide a more complete phylogenetic framework for *Chusquea*.	Fisher et al., 2014
Plastome	Explore the use of chloroplast genome sequencing for phylogenetic inference in tribe Arundinarieae.	Ma et al., 2014
atpF–atpH, matK, psbK–psbI, rbcL, rpl32–trnL, ITS, and 47 morphological characters	Species delimitation and estimation of divergence time of the genus *Otatea*.	Ruiz-Sanchez, 2015
Plastome	Perform a full plastome phylogenomic analysis in Bambusoideae.	Wysocki et al., 2015
Plastome	Sequenced the plastome of two Brazilian native species of tribe Bambuseae, performed a phylogeny using 20 species of Bambusoideae with a plastome sequenced and characterized the occurrence, type, and distribution of SRRs in the Bambuseae.	Vieira et al., 2016
SNP data generated from RAD-seq	Test the utility of RAD data in estimating the phylogenetic relationships among temperate bamboos; evaluate and compare mapping and grouping systems for SNP calling and phylogeny inference based on RAD sequencing data.	Wang et al., 2017
ITS	Explore the utility of ITS and its secondary structure to reconstruct the phylogenetic relationships of 21 PWB species.	Ghosh et al., 2017
Rpl32-trnL, trnT-trnL, trnL-trnF; psbA-trnH, rpl16, rps16-trnQ, trnC-rpoB, trnD-trnT, rps16, ndhF [3' end], matK, atpB-rbcL, psbM-petN, trnS-trnfM, ycf4-cemA, trnG-trnT, rps15-ndhF, and *rbcL-psaI*	Provide a phylogeny of PWB, investigate phylogenetic relationships among major clades and investigate the phylogenetic positions of some insufficiently studied genera.	Zhou et al., 2017
Plastome	Identify the rapidly evolving cpDNA loci for subsequent phylogenetic studies on Olyreae.	Wang et al., 2018
ndhF; trnC-rpoB, trnD-trnT, rps16-trnQ, and morphological and anatomical characters	Test whether *Arthrostylidium angustifolium, A. farctum,* and *A. pinifolium* should be classified within the subtribe Guaduinae rather than Arthrostylidiinae.	Tyrrell et al., 2018
ndhF; rps16, rpl16, trnC-rpoB, trnD-trnT, rps16-trnQ, and *trnT-trnL*	Test the monophyly of *Atractantha* and investigate the phylogenetic position of the monotypic *Athroostachys*.	Jesus-Costa et al., 2018
trnD-trnT and ITS	Identify the ancestral area, time of divergence and dispersal events in tribe Olyreae.	Ruiz-Sanchez et al., 2019
74 putative single-copy nuclear genes	Incorporate alleles from multiple nuclear loci to infer the phylogeny of *Phyllostachys*.	Zhang et al., 2019

Table 4.1 contd. ...

...Table 4.1 contd.

Molecular Marker	Objective	References
trnD–trnT, trnS–trnG, rpl32–trnL, ndhF, and ITS	The phylogenetic relationships in Olyrinae.	Oliveira et al., 2020
SNP data generated from ddRAD-seq	Resolve the phylogeny of the *Bambusa-Dendrocalamus-Gigantochloa* complex	Liu et al., 2020
ndhF, rps16-trnQ, trnC–rpoB, trnD–trnT and trnT-trnL, rpl16, and *rps16*	Test the monophyly of the genus *Merostachys* and test previously established informal morphological groupings.	Vinícius et al., 2021
Plastome and mobilome	Clarify the origin(s) of the woody bamboo tribes and resolve the nuclear vs. plastid conflict using genomic tools.	Chalopin et al., 2021
trnD-trnT, trnS-trnG, rpl32-trnL, trnH-psbA and ITS	Re-evaluate the monophyly and phylogenetic position of *Piresia* within the Olyrinae, and clarify the relationships within the genus.	de Carvalho et al., 2021

three tribes (Arundinarieae, Bambuseae, and Olyreae). Their phylogenetic results agreed with those of previous studies (Bouchenak-Khelladi et al., 2008; Sungkaew et al., 2009; Kelchner and BPG, 2013; Soreng et al., 2015). However, their new findings were the long branches found in the *Olyreae* and *Chusquea* species, in contrast to the small branches found in Arundinarieae (temperate woody bamboos). Saarela et al. (2018) did the first Poaceae phylogeny using 250 taxa and the whole plastome. The Bambusoideae sampling included 58 taxa. Their results corroborate the previously reported paraphyly of woody bamboos (Bouchenak-Khelladi et al., 2008; Sungkaew et al., 2009; Kelchner and BPG, 2013; Soreng et al., 2015; Wysocki et al., 2015). They found that some subtribes in Bambuseae and Olyreae were monophyletic. Finally, Chalopin conducted a bigger sampling of Bambusoideae (115 taxa) and the whole plastome, with results that agree with those of previous studies by Wysocki et al. (2015) and Saarela et al. (2018).

As mentioned above, whether a few chloroplast markers or the complete plastid sequences are used, the results are always the same: woody bamboos are paraphyletic in origin, and it is important to mention that the chloroplast is matrilineally inherited. In contrast, nuclear markers are biparentally inherited and ancient hybridization and allopolyploidy have occurred along bamboo lineages. However, low-copy nuclear genes tell a different phylogenetic story. Triplett et al. (2014) did a phylogenetic analysis using three low-copy nuclear genes (cellulase1 (pvcel1), poly-A binding protein1 (pabp1), and G protein a subunit 1 (gpa1)) to elucidate the origin of polyploidy in the woody bamboos and to examine the putative hybrid relationships in Arundinarieae. They sampled 38 Bambusoideae species from 27 genera. They found six ancestral genomes; the allopolyploidy arose independently in temperate and tropical lineages. Hybridization is an important process in the evolution of bamboo species.

Wysocki et al. (2016), using next-generation sequencing, got floral transcriptomes of four species representing the three tribes of Bambusoideae. They used 2,412 nuclear orthologous genes to build a bamboo phylogeny. Their results support the monophyly of woody bamboos. Guo et al. (2019) assembled four bamboo genomes from Arundinarieae, Bambuseae, and Olyreae that represented diploid, tetraploid, and hexaploid lineages. They found that woody bamboos originated from the divergence of herbaceous bamboos, and then experienced complex reticulate evolution through three independent allopolyploid events from four extinct ancestors.

Finally, Chalopin et al. (2021) sampled 51 bamboo species that represented Arundinarieae, Bambuseae, and Olyreae. They then estimated the genome size and sequenced the whole plastome and repeat diversity (interspersed elements, including transposable elements and simple repeats as satellite sequences). They found independent hybridization between the Olyreae ancestor and woody ancestor that gave rise to two Bambuseae lineages. Another independent hybridization occurred between two woody ancestors that gave rise to the Arundinarieae lineage. The retention of the Olyreae plastome associated with differential dominance of nuclear genomes and subsequent diploidization could explain the paraphyly observed in chloroplast phylogeny.

4.2.2 *Arundinarieae*

The temperate woody bamboo tribe Arundinarieae has always been found to come from a monophyletic lineage. This tribe comprises around 600 species, described in 32 genera, with their geographical distribution in Eastern North America, Africa, Eastern Asia, and Madagascar (Clark et al., 2015; Wang et al., 2017). Based on chloroplast markers, 12 lineages are recognized inside of Arundinarieae. However, in recent years, the use of new molecular markers derived from next-generation sequencing is finding more resolutions that could translate into new lineages. TWB are tetraploid species. Base chromosome number x = 12; 2n = 48.

One of the first phylogenetic studies sampling species in this tribe was done by Hodkinson et al. (2000). They compared the utility of the AFLP (amplified fragment length polymorphism) vs ITS in resolving the phylogenetic relationships between species of *Phyllostachys*. They sampled 26 species of temperate woody bamboos, 22 of which were *Phyllostachys* species. They found that the AFLP analysis had a greater resolution capacity to species level than ITS did. One year later, Guo et al. (2001) explored the utility of the nuclear marker ITS on 24 temperate woody bamboo species. They found that *Fargesia* and *Yushania* are not monophyletic lineages. And a year after that, Guo et al. (2002) used ITS to explore the phylogenetic relationship within the *Thamnocalamus* group. They sampled 33 temperate woody bamboo species from nine different genera, and found that *Thamnocalamus*, *Fargesia*, and *Yushania* are polyphyletic (Table 4.1).

Two years later, Guo and Li (2004) explored the utility of the nuclear gene GBSSI in the same *Thamnocalamus* group and compared this new marker with the ITS previously used (Guo et al., 2002). They found poor resolution of the GBSSI tree compared with the ITS tree. Combined marker analysis detected no significant differences from the previous analysis. Peng et al. (2008) did a phylogenetic analysis of 43 species belonging to Arundinarieae (TWB), sampling the two subtribes and genera recognized by Keng and Wang (1996), and sequenced two nuclear genes: GBSSI and ITS. They found that neither subtribe is monophyletic, nor are *Arundinaria* s.l., *Pleioblastus*, *Chimonobambusa*, *Sinobambusa*, or *Phyllostachys* (Table 4.1).

Triplett and Clark (2010) used 12 chloroplast markers (*atp*I-*atp*H, *ndh*F (3' end), *psa*A-ORF170, *rpl*32-trnL, *rps*16-*trn*Q, *trn*C-*rpo*B, *trn*D-*trn*T, *trn*G intron, *trn*H-*psb*A, *trn*K-*rps*16, *trn*T-*trn*L, and *trn*V-*ndh*C) and sampled 88 bamboo species, 82 of which belonged to Arundinarieae. They found six main lineages in Arundinarieae (I. Bergbamboes, II. African alpine bamboos, III. *Chimonocalamus*, IV. *Shibataea* clade, V. *Phyllostachys* clade, and VI. *Arundinaria* clade). Zeng et al. (2010) used eight of the 12 chloroplast markers used by Triplett and Clark (2010) and increased the sample size to 146 species of Arundinarieae. They found the same six lineages established by Triplett and Clark (2010), plus four new lineages (VII. *Thamnocalamus*, VIII. *Indocalamus wilsonii* (Rendle) C.S. Chao & C.D. Chu, IX. *Gaoligongshania*, and X. *Indocalamus sinicus* (Hance) Nakai) (Table 4.1).

Zhang et al. (2012) used the same eight chloroplast markers that Zeng et al. (2010) had worked with, plus the nuclear gene GBSSI and increased the sample size in Arundinarieae. With the chloroplast markers, they found the same 10 lineages previously discovered by Zeng et al. (2010). However, with the GBSSI nuclear gene they found 13 lineages and contrasting phylogenetic topologies. They suggested that their results were a consequence of the different evolutionary trajectories of the chloroplast and the nuclear genome. Hybridization and incomplete lineage sorting could be the causes of the incongruence. Yang et al. (2013) used four chloroplast markers (*rpl32-trn*L, *trn*T-*trn*L, *rps*16-*trn*Q, and *trn*C-*rpo*B), plus two nuclear genes (GBSSI and LEAFY), and sampled 50 Arundinarieae species. With the chloroplast markers, they found a new lineage (XI. *Ampelocalamus calcareus* C.D. Chu & C.S. Chao). The LEAFY gene tree had better resolution than the GBSSI gene tree did. Low informative characters, hybridization, plastid capture, or incomplete lineage sorting were cited as the main processes that caused the incongruence (Table 4.1).

Attigala et al. (2014) used five chloroplast markers (*ndh*F 3' end, *rps*16-*trn*Q, *trn*C-*rpo*B, *trn*D-*trn*T, and *trn*T-*trn*L) and sampled the Sri Lankan *Arundinaria* species, plus the 11 previously identified lineages. Their phylogenetic results led to the addition of the 12th lineage belonging to Arundinarieae (XII. *Kuruna*). Ma et al. (2014) sampled 22 Arundinarieae species representing eight lineages and sequenced the whole chloroplast genome. They achieved a better resolution of the Arundinarieae backbone with strong support. Later, Attigala et al. (2016) explored the relationship between the 12 Arundinarieae lineages using the complete chloroplast genome and 28 temperate woody bamboo species. They found that the 12 lineages were well supported and that the pachymorph rhizome is the ancestral state in Arundinarieae and leptomorph rhizomes likely evolved multiple times. Finally, they reported that pseudospikelets evolved twice in Arundinarieae (Table 4.1).

Wang et al. (2017) changed the traditional Sanger sequencing to the new generation codification using the restriction-site associated DNA sequencing (RAD-seq). They sampled 36 temperate woody bamboo species with an emphasis on the *Phyllostachys* clade. They found a better phylogenetic resolution with strong support and novel phylogenetic relationships compared to previous studies using either chloroplast or nuclear markers. Finally, Zhang et al. (2019) using the next-generation sequencing technology, codified 74 single or low-copy orthologous nuclear genes for 34 temperate woody bamboo species with an emphasis on the *Phyllostachys* genus. They found that *Phyllostachys* is monophyletic, represented by two monophyletic sections. However, as in the previous studies, incomplete lineage sorting and hybridization were responsible for the discordant phylogenetic signals that they found.

4.2.3 *Bambuseae*

The tropical woody bamboo tribe Bambuseae is a monophyletic lineage. This tribe comprises around 875 described species in 71 genera, with their geographical distribution in the Old World and the New World, mainly in tropical areas (Clark et al., 2015; Zhou et al., 2017; Haevermans et al., 2020; Ruiz-Sanchez et al., 2021). Based on chloroplast markers this tribe is divided into two lineages: Paleotropical woody bamboos and Neotropical woody bamboos. PWB is classified into four subtribes: Bambusinae, Hickeliinae, Melocanninae, and Racemobambosinae. PWB are hexaploid species. NWB is classified into three subtribes: Arthrostylidiinae,

Chusqueinae, and Guaduinae. NWB are tetraploid species. Base chromosome numbers x = 10, (11), and 12; 2n = (20) 40, (44), 46, 48, 70, 72.

4.2.4 Paleotropical Woody Bamboos

The first phylogenetic studies on PWB were done by Yang et al. (2007). They explored the utility of the nuclear gene GBSSI vs the chloroplast spacer *trn*L-F in the generic delimitation of *Schizostachyum* and allies (Bambuseae). They sampled 30 woody bamboo species, 25 of them corresponding to the *Schizostachyum* and allies genera. They found better phylogenetic resolution with the GBSSI gene than with the *trn*L-F and proposed some taxonomic changes. Increasing the sampling in Bambuseae genera and species, Yang et al. (2008) carried out a phylogenetic study that included GBSSI, ITS, and *trn*L-F, and sampled 53 Paleotropical woody bamboos. They found contrasting results mainly in the phylogenetic position of the genera *Dinochloa* and *Melocanna*, depending on the nuclear marker used (GBSSI vs ITS). The *trn*L-F tree had low phylogenetic resolution. Yang et al. (2010) did a phylogenetic analysis of the *Bambusa* and allied genera, using GBSSI and three chloroplast markers (*psb*A-*trn*H, *rpl*32-*trn*L, and *rps*16). They sampled 62 NWT species representing eight genera. Their phylogenetic results for the concatenated marker indicated that the Bambusinae subtribe, along with *Melocalamus* and *Thyrsostachys* are monophyletic lineages. *Bambusa* is not monophyletic nor are *Dendrocalamus*, *Gigantochloa*, *Neosinocalmus*, and *Oxytenanthera* (Table 4.1).

Goh et al. (2010) explored the relationships between climbing bamboos and the *Bambusa* complex. They sequenced the GBSSI and four chloroplast markers (*rps*16-*trn*Q, *trn*C-*rpo*B, *trn*H-*psb*A, and *trn*D-T) for 27 species, 24 of them from Bambusinae and three from Melocanninae. They found that the climbing bamboos segregated from *Bambusa* are valid genera. Goh et al. (2013) explored the relationship between and the origin of the Australian PWB species. They sequenced the GBSSI and three chloroplast markers (*rps*16-*trn*Q, *trn*C-*rpo*B, and *trn*D-T) for 51 species, from 17 genera of PWB. They found four lineages: 1. *Bambusa-Dendrocalamus-Gigantochloa* (BDG) complex, 2. *Holttumochloa-Kinabaluchloa* clade, 3. *Dinochloa-Mullerochloa-Neololeba-Sphaerobambos* (DMNS) clade, and 4. *Temburongia simplex* S.Dransf. & K.M. Wong. Also, they found two different phylogenetic hypotheses from contrasting results depending on the marker used (chloroplast or nuclear GBSSI) and stated that hybridization and incomplete lineage sorting could be the causes of the complexity found (Table 4.1).

Ghosh et al. (2017) employed the internal transcribed spacer (ITS1, 5.8S rRNA, and ITS2), and the secondary structure of the ITS, to reconstruct the phylgonetic relationships of 21 Paleotropical woody bamboo species. The sampling included the *Bambusa*, *Dendrocalamus*, *Gigantochloa*, *Melocanna*, and *Pseudobambusa* species. They found that *Bambusa* is not monophyletic. *Dendrocalamus* and *Gigantochloa* are sister to each other, and *Pseudobambusa* is nested in the *B. vulgaris clade*. Meanwhile, *Melocanna* was the sister species to the rest of the genera-species sampled.

The most comprehensive phylogenetic study of the PWB was done by Zhou et al. (2017). They sequenced 18 chloroplast markers (*rpl*32-*trn*L, *trn*T-*trn*L, *trn*L-

*trn*F, *psb*A-*trn*H, *rpl*16 intron, *rps*16-*trn*Q, *trn*C-*rpo*B, *trn*D-*trn*T, *rpl*16 intron, ndhF [3' end], *mat*K, *atp*B-*rbc*L, *psb*M-*pet*N, *trn*S-*trnf*M, *ycf*4-*cem*A, *trn*G-*trn*T, *rps*15-*ndh*F, and *rbc*L-*psa*I) and sequenced 114 species from 40 genera of PWB. They found monophyly for the four recognized subtribes (Bambusinae, Hickeliinae, Melocanninae, and Racemobambosinae). Additionally, they proposed using the name "core Bambusinae" which includes the following genera: *Bonia, Holttumochloa, Kinabaluchloa, Neomicrocalamus, Soejatmia, Temochloa,* and the BDG complex. Liu et al. (2020) moved forward using the next-generation sequencing and the ddRAD-seq method to explore the relationships within the BDG (*Bambusa-Dendrocalamus-Gigantochloa*) complex. They sampled 102 species from this complex and got the SNPs (single nucleotide polymorphisms) to build the phylogeny. They found the monophyly for *Gigantochloa, Melocalamus, Bambusa,* and *Dendrocalamus* to be paraphyletic. The ancient introgression could be responsible for the complex evolutionary history of this complex of genera (Table 4.1).

4.2.5 *Neotropical Woody Bamboos*

In the case of the NWB, the first phylogenetic study to reveal the relationships in this bamboo group was done by Kelchner and Clark (1997). They tested the phylogenetic utility of the chloroplast intron *rpl*16 and sampled 34 Bambusoideae taxa, and 23 taxa belonging to the *Chusquea* genus. They found that Chusqueinae was monophyletic and Guaduinae and Arthrostylidiinae formed a clade sister to Chusqueinae. Later, Clark et al. (2007) increased the Bambusoideae sampling to 46 taxa and sequenced the *rpl*16 intron. The main idea was to evaluate the relationship between one-flowered genera and to test the monophyly of Chusqueinae. They found that Chusqueinae is monophyletic and sister to the clade formed by Arthrostylidiinae and Guaduinae. One year later, Ruiz-Sanchez et al. (2008) tested the monophyly of Guaduinae using the *rpl*16 intron and the *psb*A-*trn*H spacer, both chloroplast markers. Sampling included 26 Bambuseae species. They found the monophyly of Guaduinae if two *Aulonemia* species are included in this subtribe (Table 4.1).

Fisher et al. (2009) tested the monophyly of *Chusquea* and *Neurolepis* (Chusqueinae) by sequencing five chloroplast markers (*ndh*F gene, *rpl*16 intron, and *trn*C-*rpo*B, *trn*D-*trn*T, and *rps*16-*trn*Q spacers) and sampled 27 Bambusoideae taxa, 22 of which were Chusqueinae taxa. They found the *Neurolepis* was paraphyletic, and it was transferred to *Chusquea*. Ruiz-Sanchez et al. (2011) increased the sampling of *Olmeca, Otatea* and included the *Aulonemia* (Arthrostylidiinae) species, and sequenced five chloroplast markers (*rbc*L, *mat*K, *psb*K-*psb*I, *rpl*16, and *psb*A-*trn*H). They transferred the *Aulonemia* species to *Olmeca*, making Guaduinae monophyletic (Table 4.1).

Tyrrell et al. (2012) did the most comprehensive phylogenetic study of Arthrostylidiinae. They sampled 51 taxa, 45 of them representing the Arthrostylidiinae taxa, and sequenced six chloroplast markers (*ndh*F, *trn*D-*trn*T, *trn*C-*rpo*B, *rps*16-*trn*Q, *trn*T-*trn*L, and *rpl*16 intron). They recovered the Arthrostylidiinae subtribe as monophyletic and elevated *Didymogonyx* to a new genus. Fisher et al. (2014) increased the sampling of the *Chusquea* species to 65 and sequenced five chloroplast markers (*ndh*A intron, *ndh*F gene, *rpl*16 intron, *trn*D-*trn*T, and *trn*T-*trn*L spacers) and the nuclear gene ITS. They found support to describe two new subgenera for *Chusquea* (subg. *Platonia* and subg. *Magnifoliae*). Tyrrell et al. (2018) studied the phylogenetic positions of three West Indian *Arthrostylidium* species by sequencing four chloroplast markers (*ndh*F, 3′ *trn*D-*trn*T, *trn*C-*rpo*B, and *rps*16-*trn*Q). They found that these three West Indian *Arthrostylidium* species were nested inside Guaduinae and they were transferred to the new genus, *Tibisia* (Table 4.1).

Jesus-Costa et al. (2018) investigated the phylogenetic position of *Atractantha* in the Arthrostylidiinae subtribe, by sequencing seven chloroplast markers (*ndh*F and *rps*16 genes, *rpl*16 intron, *trn*C-*rpo*B, *trn*D-*trn*T, *rps*16-*trn*Q, and *trn*T-*trn*L) for 44 Bambuseae taxa. They found that *Atractantha* was polyphyletic and one of its species was transferred to *Athroostachys*. Jesus-Costa (2018) did a phylogenetic study of the *Aulonemia* and *Colanthelia* genera. She sampled 73 Bambuseae taxa, 30 of which were *Aulonemia* and *Colanthelia* species, by sequencing seven chloroplast markers (*ndh*F and *rps*16 genes, *rpl*16 intron, *trn*C-*rpo*B, *trn*D-*trn*T, *rps*16-*trn*Q, and *trn*T-*trn*L). She found that neither of the two genera is monophyletic. The most recent phylogenetic study done on NWB was by Vinícius-Silva et al. (2021) who examined the phylogenetic relationships of *Merostachys* by sequencing seven chloroplast markers (*ndh*F and *rps*16 genes, *rpl*16 intron, *trn*C-*rpo*B, *trn*D-*trn*T, *rps*16-*trn*Q, and *trn*T-*trn*L) for 64 Bambuseae taxa, 31 of which were *Merostachys* taxa. They found the monophyly of *Merostachys* divided into two clades (Table 4.1).

4.2.6 Olyreae

The herbaceous bamboo tribe Olyreae is a monophyletic lineage. This tribe comprises 139 described species, in 23 genera, with their geographical distribution in the Neotropics (Tropical America) except for *Buergersiochloa bambusoides*, which is endemic to New Guinea (Zhang and Clark, 2000; Kelchner and BPG, 2013; Oliveira et al., 2014; Clark et al., 2015). Based on chloroplast markers, three subtribes are currently known: Buergersiochloinae, Parianinae, and Olyrinae. The herbaceous bamboos are diploid species. Base chromosome number x = 7, 9, 10, 11, and 12; 2n = 14, 18, 20, 22, and 24.

The first phylogenetic study was conducted by Zhang and Clark (2000). They sequenced the chloroplast gene *ndh*F for seven Olyreae species, five of which are from Olyrinae and the other two from Parianinae. They found that Olyreae was monophyletic and *Buergersiochloa* was the sister taxa to the rest of the species. The Olyrinae s. str. and Parianinae were monophyletic and sister to each other. Kelcher and BPG (2013) increased the molecular sampling to five chloroplast markers (*ndh*F, *rpl*16 intron, *rps*16 intron, *trn*D-*trn*T and *trn*T-*trn*L spacers) and four Olyreae taxa that represent the three subtribes. They found the same result as Zhang and Clark (2002). Oliveira et al. (2014) did the most comprehensive phylogenetic study of Olyreae. They sampled 25 Olyreae species from 13 genera that represent the three subtribes and sequenced the chloroplast spacer *trn*D-*trn*T and the nuclear gene ITS. They found that *Parodiolyra* and *Olyra* are paraphyletic; however, the phylogenetic position of *Buergersiochloa* was uncertain (Table 4.1).

Wysocki et al. (2015) did the first phylogenetic analysis using the whole chloroplast. The sample included eight Olyreae taxa that represent the three subtribes. They corroborated the monophyly of Olyreae as well as the three subtribes, and that *Buergersiochloa* is the sister species to the rest of Olyreae. Saarela et al. (2018) increased the sample size to 10 Olyreae species and included the whole chloroplast in their phylogenetic analysis. They corroborated the results of Wysocki et al. (2015). Wang et al. (2018) sequenced the whole chloroplast of two new Olyreae taxa (*Froesiochloa* and *Rehia*) through the genome-skimming technique. The phylogenetic analysis included 12 Olyreae taxa and their results agree with those previously reported by Saarela et al. (2018). Oliveira et al. (2020) increased the number of taxa sampled to 42 species from 14 genera of Olyreae. They sequenced four chloroplast markers (*trn*D-*trn*T, *trn*S-*trn*G, *rpl*32-*trn*L, and *ndh*F) plus the nuclear gene ITS. They evaluated the generic boundaries within the *Parodiolyra/Raddiella*. They corroborated the paraphyly of *Parodiolyra* and recircumscribed *Parodiolyra*, changing two of its species to a new genus *Taquara*. Most recently, de Carvalho et al. (2021) focused on the genus *Piresia*, included a *Piresiella* taxa from Cuba, and sequenced four chloroplast markers (*trn*D-*trn*T, *trn*S-*trn*G, *rpl*32-*trn*L, and *psb*A-*trn*H) plus the nuclear gene ITS. They found a striking result: *Piresiella* from Cuba is a sister species of *Buergersiochloa* from New Guinea, and they proposed to transfer *Ekmanochloa*, *Mniochloa* and *Piresiella* to Buergersiochloinae. This result has unprecedented biogeographical implications (Table 4.1).

These techniques have been used from 1988—when the first study using the RNA ribosomal sequences was published—to the present, 34 years later. We are still using chloroplast or nuclear markers to understand the phylogenetic relationships among Bambusoideae and the relationships within its tribes and subtribes. However, with the more recent use of next-generation sequencing to get the whole chloroplast and nuclear genomes, multiple low-copy nuclear genes, RNA transcriptomes, or SNPs via RAD-seq or ddRAD-seq, we will see better resolved phylogenies, mainly inside complex, highly diverse bamboo genera. These new phylogenetic results will help us understand the rapid diversification, biogeographic history, and evolution of the bamboos.

4.3 Molecular Markers Used to Analyze Genetic Diversity in Bamboos

We compiled a list of genetic studies on bamboo species that used molecular markers. Most of the studies were done on Arundinarieae and Bambuseae tribes, and the least studied group was Olyreae, with one study using allozymes in the genus *Raddia*. One of the main objectives of most genetic studies was to analyze genetic diversity within and among populations, mainly to provide information to help develop conservation strategies. Other studies used molecular markers to distinguish species complexes, varieties, and hybridization processes. Another common objective was to explore the distribution and structure of clones in populations.

The dominant markers are coded through the presence and absence of bands. The absence of a band usually corresponds to recessive homozygous and its presence to dominant homozygous or heterozygous, though it is not usually possible to distinguish between the latter two genotypes. Among the most used dominant markers are AFLPs (amplified fragment length polymorphism), ISSR (inter simple sequence repeat) and RAPDs (random amplified polymorphic DNA). The major advantage of AFLP is the large number of markers generated and their reproducibility, which is an important tool for population genetics research. The RAPDs and ISSR have been used less in recent times.

The SSR (simple sequence repeat) markers, also known as simple sequence repeats (SSRs), amplify repetitive motifs with variable lengths. They are codominant (the three genotypes can be identified), polymorphic and widely dispersed across the genome. They are useful in plant breeding programs because of their codominant nature. The allozymes are also codominant markers, however we found only two studies using this marker.

Only one study used both RAPD and RFLP. The main objective was to evaluate the potential of combining both markers to detect genetic relationships among bamboo genera and species. This technique consists of digesting RAPD products (RFLP) with three enzymes. The results indicated a high number of polymorphic and reproducible bands, allowing discrimination among genera, species, and varieties of bamboos. This technique is cost-effective and can help in studies of genetic variability for conservation, management, and breeding (Konzen et al., 2017).

Allozymes were the first molecular markers used to analyze genetic diversity. Lee and Chung (1999) inferred relationships between levels of genetic diversity and the reproductive biology of *Sasamorpha borealis* Hack (Arundinarieae). Their main conclusion was that *S. borealis* maintains high levels of genetic diversity (Ho = 0.31 and H_E = 0.21) compared to the mean value of long-lived herbaceous perennial species (N = 4, H_E = 0.20) reported in a review published by Hamrick and Godt (1989). The authors suggested that the combined effects of wind-pollinated and wide geographical distribution are probably the main reason they found a high degree of genetic diversity. The level of genetic differentiation was also high (G_{ST} = 0.310) compared to that of other plants with similar life history traits (Hamrick and Godt, 1989), suggesting that gene flow is low. Another factor might be that there were very few genetically distinct individuals, so the low number of genets in the populations could result in an elevated G_{ST} (Table 4.2).

The second study using allozymes was carried out by Oliveira et al. (2008) with the *Raddia brasiliensis* Bertol. complex (tribe Olyreae) to compare the genetic results with previous morphometric analyses and carry out taxonomic delimitation. The main results indicated that *R. brasiliensis* harbors low levels of genetic diversity (H_E = 0.04–0.17) and high levels of inbreeding compared to the values reported by Hamrick and Godt (1989) in plants with similar life traits (monocots, herbaceous, endemic, and low dispersal capacity). The authors suggest that one of the factors leading to reduced levels of genetic variation is the founder event or genetic bottlenecks within populations. Moreover, the species complex occurs in the Atlantic rainforest or inland forest and has a discontinuous distribution due to intense logging and clearing, both of which have promoted high genetic differentiation among populations and low genetic diversity. The results of this study also indicated that

Table 4.2. Allozymes/RAPDs used on herbaceous and woody bamboos.

Species	Molecular Marker	D	HO	HE	Fst/Gst	Country_Collection	Objective	References
Raddia brasiliensis Bertol.	Allozymes		0.063	0.098	0.0426	Brazil	Analyze allozyme patterns within the same populations	Oliveira et al., 2008
Raddia lancifolia R.P. Oliveira & Longhi-Wagner	Allozymes		0.067	0.101		Brazil	Analyze allozyme patterns within the same populations	Oliveira et al., 2008
Raddia megaphylla R.P. Oliveira & Longhi-Wagner	Allozymes		0.093	0.092	0.177	Brazil	Analyze allozyme patterns within the same populations	Oliveira et al., 2008
Raddia soderstromii R.P. Oliveira, L.G. Clark & Judz.	Allozymes		0.11	0.12	0.187	Brazil	Analyze allozyme patterns within the same populations	Oliveira et al., 2008
Raddia stolonifera R.P. Oliveira & Longhi-Wagner	Allozymes		0.075	0.107		Brazil	Analyze allozyme patterns within the same populations	Oliveira et al., 2008
Sasamorpha borealis (Hack.) Nakai	Allozymes		0.31	0.21	0.31	Korea	Infer relationships between levels of genetic diversity and the reproductive biology of the species	Lee and Chung, 1999
Dendrocalamus giganteus Wall. ex Munro	RAPDs		0.092 ± 0.027			Sri Lanka	Genetic diversity and relationships within populations	Ramanayake et al., 2006
Dendrocalamus giganteus Wall. ex Munro	RAPDs		0.045 ± 0.004			Sri Lanka	Genetic diversity in a population	Ramanayake et al., 2007
Dendrocalamus strictus (Roxb.) Nees	RAPDs			0.13		India	Genetic variability in and among these growth forms	Das et al., 2007
Ochlandra stridula Moon ex Thwaites	RAPDs		0.446 ± 0.210			Sri Lanka	Genetic diversity and relationships within populations	Ramanayake et al., 2006
Phyllostachys vivax McClure	RAPDs					India	Genetic diversity of Indian genotypes of bamboo using RAPD and ISSR markers	Desai et al., 2015
Species identification	RAPDs						Genetic variation among different species, determine the genetic similarities between species	Nayak et al, 2013
Species identification	RAPD-RFLP					Brazil	Validating RAPD-RFLP as a method for estimating the genetic relationships among bamboo taxa	Konzen et al., 2017

there is no congruence between the genetic and morphological results. For example, *R. brasiliensis* and *R. soderstromii* R.P. Oliveira, Clark and Judz are part of the same genetic complex, yet their morphology is very different (Table 4.2).

4.3.1 AFLP Markers

Suyama et al. (2000) were one of the first to conduct studies to use AFLP to analyze the clonal structure at the population level of *Sasa senanensis* (Franch. & Sav.) Rehder on 10 hectares located in Japan. The main result indicated the presence of 22 clones, and the principal clone occurs over 300 m. This study demonstrated the effectiveness of the AFLP markers at distinguishing 22 clones, since only 14 of these clones had been identified by allozyme markers. Furthermore, in the same year as Suyama's study, Loh et al. (2000) examined four genera and 15 species of *Bambusa*, *Dendrocalamus*, *Gigantochloa*, and *Thyrsostachys* using AFLP markers. The six *Bambusa* species were separated into two clusters, while *Thyrsostachys* was well differentiated from the *Bambusa-Giganthocloa-Dendrocalamus* cluster. *B. lako* Widjaja, and *G. atroviolacea* Widjaja exhibited high genetic similarity, suggesting that it is more appropriate to include *B. lako* in the genus *Gigantochloa* than in *Bambusa*. Two *Dendrocalamus* species were very different from *D. brandisii*, clustered within one of the *Bambusa* clusters and *D. giganteus* Munro was a very distant species. These results demonstrate the importance of performing further taxonomic studies of the genus *Dendrocalamus* (Table 4.3).

Marulanda et al. (2002) carried out a study to analyze the genetic distances between the accessions and biotypes of *Guadua angustifolia* Kunth in Colombia and compared them to other *Guadua* species inhabiting Colombia. Three primer combinations were included. The results revealed a high degree of genetic differentiation among the different species analyzed (*G. amplexifolia* J. Presl., *G. macrospiculata* Londoño & L.G. Clark, and *G. superba* Huber). An unexpected result was the high level of genetic diversity found within the accessions of *G. amplexifolia* and low diversity between the accessions. Additionally, two groups were detected, two of them corresponded to *G. angustifolia*, and a third group included *G. amplexifolia*, *G. uncinata*, and *Guadua* sp. Three accessions of *G. amplexifolia*, *G. macrospiculata*, and *G. superba* had the largest genetic distances and were not grouped in the analysis (Table 4.3).

Isagi et al. (2004) analyzed the clonal structure of *Phyllostachys pubescens* (Pradelle) Mazel ex J. Houz. distributed in Japan. This species is one of the most important economic bamboo species in China and grows widely in south China. The population of *P. pubescens* studied was found to arise from at least two genets with different flowering intervals of 67 and 69 years. These differences in flowering correspond to differences between genets, however, some intermediates probably were influenced by environmental conditions. Moreover, Lin et al. (2008) analyzed the genetic similarity among 10 cultivars of *P. pubescens* and two related species, using AFLP and ISSR. The results using both molecular markers were highly correlated, with a high degree of similarity among cultivars and genetic distance ranging from 0.023 to 0.10. In most cases, the genetic data supports the morphological classification of the varieties. The three species were genetically different from each

Table 4.3. AFLP markers used in genetic studies on Arundinarieae and Bambuseae species.

Species	Molecular Marker	D	HO	HE	Fst/Gst	Country Collection	Objective	References
Arundinaria appalachiana Triplett, Weakley & L.G. Clark	AFLP			0.052	0.845	United States	Investigate the potential role of hybridization in the evolution of *Arundinaria*	Triplett et al., 2010
Arundinaria gigantea (Walter) Muhl.	AFLP	0.38–0.49				United States	Clonal identity (sterile vs. flowering)	Mathews et al., 2009
Arundinaria gigantea (Walter) Muhl.	AFLP			0.054	0.845	United States	Investigate the potential role of hybridization in the evolution of *Arundinaria*	Triplett et al., 2010
Arundinaria tecta (Walter) Muhl.	AFLP			0.53	0.845	United States	Investigate the potential role of hybridization in the evolution of *Arundinaria*	Triplett et al., 2010
Bambusa lako Widjaja	AFLP	0.52				Singapore	Genetic variation	Loh et al., 2000
Bambusa longispiculata Gamble ex Brandis	AFLP	0.62				Singapore	Genetic variation	Loh et al., 2000
Bambusa multiplex (Lour.) Raeusch. ex Schult. & Schult. f.	AFLP	0.66				Singapore	Genetic variation	Loh et al., 2000
Bambusa textilis McClure	AFLP	0.6				Singapore	Genetic variation	Loh et al., 2000
Bambusa tulda Roxb.	AFLP	0.51				Singapore	Genetic variation	Loh et al., 2000
Bambusa ventricosa McClure	AFLP	0.58				Singapore	Genetic variation	Loh et al., 2000
Bambusa vulgaris Schrad. ex J.C. Wendl.	AFLP	0.57				Singapore	Genetic variation	Loh et al., 2000
Bashania fangiana (A. Camus) Keng f. & T.H. Wen	AFLP	0.99			0.057	China	Clonal diversity	Ma et al., 2013

Table 4.3 contd. ....

...Table 4.3 contd.

Species	Molecular Marker	D	HO	HE	Fst/Gst	Country_Collection	Objective	References
Dendrocalamus giganteus Wall. ex Munro	AFLP	0.78				Singapore	Genetic variation	Loh et al., 2000
Dendrocalamus hamiltonii Nees & Arn. ex Munro	AFLP	NA				India	Genetic relationships between populations	Waikhom et al., 2012
Gigantochloa atroviolacea Widjaja	AFLP	0.64				Singapore	Genetic variation	Loh et al., 2000
Gigantochloa ridleyi Holttum	AFLP	0.53				Singapore	Genetic variation	Loh et al., 2000
Gigantochloa rostrata K.M. Wong	AFLP	0.57				Singapore	Genetic variation	Loh et al., 2000
Gigantochloa scortechinii Gamble	AFLP	0.54				Singapore	Genetic variation	Loh et al., 2000
Gigantochloa verticillata (Willd.) Munro	AFLP	0.58				Singapore	Genetic variation	Loh et al., 2000
Guadua amplexifolia J. Presl	AFLP	0.31				Colombia	Genetic distances between varieties	Marulanda et al., 2002
Guadua angustifolia Kunth	AFLP	0.29/0.48				Colombia	Genetic distances between varieties	Marulanda et al., 2002
Guadua macrospiculata Londoño & L.G. Clark	AFLP	0.42				Colombia	Genetic distances between varieties	Marulanda et al., 2002
Guadua superba Huber	AFLP	0.48				Colombia	Genetic distances between varieties	Marulanda et al., 2002
Guadua uncinata Londoño & L.G. Clark	AFLP	0.3				Colombia	Genetic distances between varieties	Marulanda et al., 2002
Ochlandra travancorica (Bedd.) Benth. ex Gamble	AFLP	NA			0.456	India	Genetic diversity	Nag et al., 2013
Phyllostachys pubescens (Pradelle) Mazel ex J. Houz.	AFLP	NA				Japan	Clonal structure	Isagi et al., 2004

Species	Marker	Value			Country	Purpose	Reference
Phyllostachys pubescens (Pradelle) Mazel ex J. Houz.	AFLP	NA			China	Genetic diversity and similarity of ten cultivars	Lin et al., 2009
Phyllostachys violascens (Carrière) Rivière & C. Rivière	AFLP	0.87			China	Genetic diversity	Lin et al., 2011
Sasa borealis (Hack.) Makino & Shibata	AFLP		0.477/0.324		South Korea	Genetic diversity	Kim et al., 2015
Sasa pubiculmis Makino	AFLP	NA			Japan	Genetic structure	Miyazaki et al., 2009
Sasa senanensis (Franch. & Sav.) Rehder et Savat.) Rehder	AFLP	0.94			Japan	Clonal structure to population level	Sumaya et al., 2000
Thyrsostachys siamensis Gamble	AFLP	NA			Singapore	Genetic variation	Loh et al., 2000
Trimeresurus stejnegeri (Schmidt)	AFLP			0.040–0.053	Taiwan	Genetic differentiation	Creer et al., 2004

other. The authors emphasize that using AFLP and ISSR were useful to classify cultivars or varieties of a species (Table 4.3).

Miyazaki et al. (2009) studied the genetic structure, flowering patterns, and the rate of seed set for one genet of *Sasa pubiculmis* subsp. *pubiculmis* Makino over four growing seasons (2004–2007). *Sasa* forms extensive, dense populations. Of the main findings was that there is one genotypical genet, which covered an area of 3 ha, and had flowering and non-flowering patches. The authors also found that this species is polycarpic rather than monocarpic. Mathews et al. (2009) studied the clonal diversity of *Arundinaria gigantea*, also known as river cane, one of the three bamboo species native to the southeastern United States. The main goal of this study was to test whether flowering within a stand of river cane is monoclonal or not. The results indicated that most of the fertile culms belonged to the same clone. However, two to three fertile culms belonged to different genotypes. The authors also detected that some ramets were flowering while others were not, as occurs in other bamboos such as *Sasa pubiculmis* subsp. *pubiculmis* (Miyazaki et al., 2009) (Table 4.3).

Triplett et al. (2010) studied the genetic variation of the North American *Arundinaria gigantea* species complex and hybridization among *A. gigantea*, *A. tecta* Muhl., and *A. appalachiana* Triplett, Weakley & L.G. Clark using AFLPs and chloroplast DNA sequences. The main results indicated that the genetic differentiation of the three species concurred with that which had been previously defined based on morphology, anatomy, and ecology. Molecular evidence also indicated that *A. tecta* and *A. appalachiana* are sister species, forming a clade that is divergent from *A. gigantea*, and indicates the presence of hybrids between *A. gigantea* and *A. tecta*. Moreover, low levels of genetic diversity were detected compared with wind-pollinated outcrossing taxa, similar to that of clonal or selfing species (Ellstrand and Roose, 1987) (Table 4.3).

Ma et al. (2013) analyzed the genotypic diversity of the dwarf bamboo *Bashania fangiana* (A. Camus) Keng f. & T.H. Wen in two clonal populations with different genet ages (≤ 30 years versus 70 years) in China. The samples were collected at 30 m intervals from the two populations. Ninety-two genotypes were recorded, indicating populations were multiclonal and highly diverse, unlike the bamboos mentioned above such as *Sasa pubiculmis* subsp. *pubiculmis* and *Arundinaria gigantea*. The largest single clone could occur at 30 m from the closest one. The authors also found that the genotypic diversity and genet density did not differ between genets of different ages (Table 4.3).

Lin et al. (2011) studied the genetic diversity of different cultivars of the species *Phyllostachys violascens* (Carrière) Rivière & C. Rivière using three different molecular markers: AFLP, SRAP (sequence-related amplified polymorphism), and AFLP. The results indicated the cultivars could be divided into four groups, which concurs with the morphological information. The authors suggest that the three types of molecular markers are useful for identifying the cultivars, but that AFLP was the most efficient method for assessing genetic diversity based on two indexes (RP—resolving power, which describes relative band informativeness, and MI—marker index, which is equivalent to the polymorphism information content) (Table 4.3).

Waikhom et al. (2012) characterized landraces of *Dendrocalamus hamiltonii* Nees & Arn. ex Munro based on morphology and genetic diversity using AFLP. Fermented shoots of this bamboo are used as a traditional food by different ethnic groups in northeastern India. Overall, the genetic and morphological results were congruent though there were some differences. Particularly, morphological characteristics could be influenced by changes in environmental conditions such as soil type, light, temperature, and moisture regime. The landraces were genetically grouped according to their geographic origin. Moreover, *D. hamiltonii* was found to harbor noteworthy genetic variation. The authors also associated the levels of biochemical traits to genotypes that can be used in breeding programs (Table 4.3).

Nag et al. (2013) analyzed the genetic diversity of *Ochlandra travancorica* (Bedd.) Benth. ex Gamble, an industrially important reed bamboo, using AFLPs and RAPDs. The results indicated high levels of genetic diversity and more of the variation occurred within populations (54%) than among populations (46%). Three different groups were detected. The F_{ST} value was very high, i.e., populations are not panmictic and are isolated. These markers were useful to detect high degrees of polymorphism to identify germplasm resources for industrial use. The genetic diversity of *Sasa borealis* (Hack.) Makino & Shibata distributed in South Korea was analyzed (Kim et al., 2015). The level of genetic diversity was relatively high (h = 0.14) and genetic differentiation among populations was high (G_{ST} = 0.32) (Table 4.3).

4.3.2 ISSR Markers

Desai et al. (2015) assessed the genetic diversity of 13 Indian bamboo genotypes using RAPD and ISSR markers. Five *Bambusa* species (nine samples) and *Phyllostachys vivax* McClure (four samples) were analyzed. Many markers were used to characterize the genotypes. For the RAPD markers 30 primers were used yielding a total of 645 amplified fragments, and 12 ISSR primers were analyzed, resulting in 241 polymorphic bands. The final database included 846 polymorphic markers. The 13 genotypes were differentiated and were grouped into three clusters. The genetic inferences from both markers were highly correlated, however some samples were grouped into different clusters depending on the marker used. The *P. vivax* samples were grouped into the same cluster using both RAPD and ISSR, while some *Bambusa* samples were grouped with *Phyllosatchys* (Table 4.4).

Melocanna baccifera (Roxb.) Kurz is an economically important, dominant bamboo species in Manipur (Northeast India) that benefits the local ecosystem by improving soil conservation and biodiversity. However, the overexploitation of this species by locals has led to a reduction in population size that will cause environmental degradation. Nilkanta et al. (2017) assessed the genetic diversity of seven populations from five Manipur districts. There were significant levels of genetic variation within populations (Nei's genetic diversity, H = 0.163; Shannon index, *I* = 0.256). *M. baccifera* is known to be self-incompatible, outcrosses, and has a diverse distribution throughout Manipur. Most probably, these factors are contributing to the high genetic variability of this species. However, high genetic

Table 4.4. ISSR/SSR markers used in genetic studies on Arundinarieae and Bambuseae species.

Species	Molecular Marker	D	HO	HE	Fst/Gst	Country_Collection	Objective	References
Bambusa vulgaris Schrad. ex J.C. Wendl.	ISSR					India	Genetic diversity of Indian genotypes of bamboo using RAPD and ISSR markers	Desai et al., 2015
Bambusa balcooa Roxb.	ISSR					India	Genetic diversity of Indian genotypes of bamboo using RAPD and ISSR markers	Desai et al., 2015
Bambusa bambos (L.) Voss	ISSR					India	Genetic diversity of Indian genotypes of bamboo using RAPD and ISSR markers	Desai et al., 2015
Bambusa multiplex (Lour.) Raeusch. ex Schult. & Schult. f.	ISSR					India	Genetic diversity of Indian genotypes of bamboo using RAPD and ISSR markers	Desai et al., 2015
Bambusa tulda Roxb.	ISSR					India	Genetic diversity of Indian genotypes of bamboo using RAPD and ISSR markers	Desai et al., 2015
Dendrocalamopsis beecheyana (Munro) Keng f.	ISSR			0.134	0.53	China	Genetic diversity within and among the populations	He et al., 2020
Dendrocalamus giganteus Wall. ex Munro	ISSR			0.04	0.84	China	Genetic variation and differentiation among seven populations	Tian et al., 2012
Dendrocalamus membranaceus Munro	ISSR			0.16	0.25	China	Genetic variation and differentiation among 12 populations	Yang et al., 2012
Guadua weberbaueri Pilg.	ISSR			0.3		Brazil	Diversity and genetic structure with and without anthropic interference in an area of mixed forest with palm and bamboo species in the Western Brazilian Amazon	da Silva et al., 2020
Melocanna baccifera (Roxb.) Kurz	ISSR			0.163	0.19	India	Genetic diversity and population genetic structure of *M. baccifera* in 5 districts of Manipur using ISSR markers	Nilkanta et al., 2017
Oxytenanthera abyssinica (A. Rich.) Munro	ISSR			0.2		Ethiopia	Genetic diversity, population structure, and gene flow	Oumer et al., 2020

Species	Marker				Objective	Country	Reference
Phyllostachys pubescens (Pradelle) Mazel ex J. Houz.	ISSR	NA	NA	NA	Genetic diversity and similarity of ten cultivars	China	Lin et al., 2009
Phyllostachys violascens (Carrière) Rivière & C. Rivière	ISSR	0.87			Genetic diversity	China	Lin et al., 2011
Phyllostachys vivax McClure	ISSR				Genetic diversity of Indian genotypes of bamboo using RAPD and ISSR markers	India	Desai et al., 2015
Dendrocalamus hamiltonii Nees & Arn. ex Munro	SSR			0.165/--	Genetic diversity and structure	India	Meena et al., 2019
Dendrocalamus sinicus L.C. Chia & J.L. Sun	SSR	0.448	0.541	0.306/--	Genetic structure and differentiation	China	Yang et al., 2018
Fargesia spathacea Franch.	SSR	0.461	0.366	0.952/--	Genetic structure	China	Huang et al., 2020
Guadua aff. *chaparensis*	SSR	0.43	0.8		Genetic structure and diversity	Brazil	Silva et al., 2020
Guadua aff. *lynnclarkiae*	SSR	0.39	0.46		Genetic structure and diversity	Brazil	Silva et al., 2020
Guadua amplexifolia J. Presl	SSR	0.21	0.38	--/0.10	Genetic diversity	Mexico	Pérez-Alquicira et al., 2021
Guadua angustifolia Kunth	SSR	0.42	0.56	0.098/--	Genetic structure and diversity	Colombia	Posso, 2011
Guadua chacoensis (Rojas Acosta) Londoño & P.M. Peterson		0.004	0.017		Develop simple sequence repeat (SSR) markers for genetic studies	Brazil	Rossarolla et al., 2020
Guadua inermis Rupr. ex E. Fourn.	SSR	0.2	0.3	0.47/0.44	Genetic diversity	Mexico	Pérez-Alquicira et al., 2021
Guadua tuxtlensis Londoño & Ruiz-Sanchez	SSR	0.33	0.38	--/0.04	Genetic diversity	Mexico	Pérez-Alquicira et al., 2021
Kuruna debilis (Thwaites) Attigala, Kathriar. & L.G. Clark	SSR	0.75	0.7	0.113/--	Genetic diversity and population structure	India and Sri Lanka	Attigala et al., 2017
Phyllostachys edulis (Carrière) J. Houz.	SSR	0.32	0.45	0.175/--	Genetic diversity and population differentiation	China	Jiang et al., 2017
Sasa cernua Makino	SSR				Clonal identification	Japan	Kitamura and Kawahara, 2009

diversity within small populations is expected if the reduction in population size has occurred in a few generations. This might be the reason for observing large variability within populations. The coefficient of genetic differentiation among populations was moderate (G_{ST} = 0.194) and according to the authors, high gene flow was detected (Nm 2.54) despite this species having a poor seed dispersal mechanism. Humans may also have mediated the movement of genotypes (Table 4.4).

Oxytenanthera abyssinica (A. Rich.) Munro is an economically important, Ethiopian lowland bamboo. This species has experienced population decline because of anthropogenic factors. Oumer et al. (2020) analyzed the genetic diversity, population structure and gene flow of *O. abyssinica* populations. A total of 140 individuals were analyzed from 13 populations. Genetic diversity was high compared to that of other bamboos species (gene diversity, H ± SD 0.2702 ± 0.1945 and Shannon index, I ± SD 0.4061 ± 0.2595). The levels of genetic differentiation were moderate to high (G_{ST} = 0.244) and the estimated number of migrants was Nm = 1.54. Additionally, the authors indicated that one of the populations (Gambella Region) has notable genetic differences from the other populations and needs to be evaluated with more genetic markers before it can be considered a new species (Table 4.4).

Dendrocalamus giganteus Munro, one of the largest woody bamboos, is an economically important species used as raw material for industrial pulp, construction, and furniture. It is widely grown in Southeast Asia and the Yunnan Province of China. The natural distribution of this species is thought to be in southern Myanmar and northwestern Thailand as well as China's Yunnan Province; and it usually grows along river valleys and hilly, forested areas. Tian et al. (2012) examined the genetic variation and differentiation among seven populations of *D. giganteus* in Yunnan Province, China. Genetic diversity within populations was low (H_E = 0.041) compared to that of other monocotyledons (H_E = 0.144) (Hamrick and Godt, 1989) and there was a high degree of genetic differentiation among populations (G_{ST} = 0.847). The authors suggested that the long vegetative phase and sporadic flowering of this species have caused the reduction of genetic diversity. Another possible explanation is that the populations studied were cultured rather than naturally occurring. Thus, the samples analyzed may be ancient clones selected for less flowering, to prevent the plants from dying after flowering. The Mantel test indicated that there was no significant correlation between genetic and geographic distances among populations. According to the low levels of genetic diversity and high genetic differentiation, the authors suggest *in situ* conservation strategies be implemented for all populations and sampling for *ex situ* conservation collections (Table 4.4).

Yang et al. (2012) studied the degree of genetic variation and differentiation among 12 populations of *Dendrocalamus membranaceus* Munro. This species is highly valuable in economic and ecological terms and is one of the most frequently occurring bamboos in Southeast Asia. *D. membranaceus* is used as a vegetable crop and as a raw material for furniture, construction, and industrial pulp. Populations have decreased because of overexploitation in Yunnan. Expected diversity was estimated to be 0.164 and Shannon's index was 0.249 at the population level. These values are higher than those of other bamboo species, probably due to the high ploidy level (hexaploid) of *D. membranaceus* (Table 4.4).

He et al. (2020) studied the genetic diversity of 17 accessions from six populations of *Dendrocalamopsis beecheyana* var. *pubescens* in two provinces of China (Fujian and Guangxi). This species is an ornamental bamboo that is widely cultivated in the southern region of Guangdong Province, Hong Kong, and Taiwan. In China, this species is used as construction material and the bamboo shoots are used in traditional Chinese cuisine. The main results indicated that the 17 accessions formed six major groups and the six populations were grouped into three groups. Genetic differentiation among populations was high ($G_{ST} = 0.53$) and thus low levels of gene flow were observed. Moreover, levels of diversity within the six populations were high, even though this species flowers every 10–20 years in an unpredictable mode and propagation is mainly by clones. Thus, explanations for this diversity are habitat heterogeneity of long-lived plants and wide geographical distribution (Table 4.4).

Silva et al. (2020) characterized the structure and genetic diversity of *Guadua weberbaueri* Pilg. in a native and anthropized areas using ISSR, in the state of Acre, Brazil. This region has high species richness and endemism, and more than half of its territory is dedicated to the protection of this biodiversity. The forest has several species with economic potential; for example, bamboo is a non-timber resource with multiple uses. Acre harbors the largest native bamboo reserves. The native population had higher levels of genetic diversity than anthropized environments did, indicating a significant risk for species conservation if management policies are not implemented (Table 4.4).

4.3.3 RAPD Markers

Nayak et al. (2003) identified and examined the genetic relationship of 12 bamboo species. The analyses included 10 pairs of primers to identify the 12 species. The most related species were *Bambusa vulgaris* Schrader ex Wendl and *Bambusa vulgaris* var. *striata* ex Wendl, along with *Bambusa ventricosa* Maclure. *Bambusa* multiplex var. *Silver* stripe and *Bambusa multiplex* (Lour.) Raeushel ex. Schult & Sehult.f were very closely related and there was no variation with *Bambusa ventricosa* McClure. Another cluster included *Bambusa arundinacea* Retz., *Cephalostachyum pergracile* Munro and *Bambusa balcooa* Roxb. (Table 4.2).

Das et al. (2005) developed SCAR (sequence-characterized amplified region) primers from two species-specific RAPD markers from *Bambusa balcooa* Roxb. and *B. tulda* Roxb. to allow the molecular identification of these two species. They found that the two pairs of primers amplified bands only from *B. balcooa* and *B. tulda* and did not amplify in other bamboo species tested. One year later, Bhattacharya et al. (2006) tested the species-identity of flowering populations of *B. tulda* by amplifying a *B. tulda*-specific SCAR marker, Tuldo$_{609}$. Then, they did a fingerprinting through RAPD analysis. They did not find any polymorphism from the different population sampled of *B. tulda*.

Dendrocalamus strictus (Roxb.) Nees is widely distributed in India, particularly in semi-dry and dry areas. *D. strictus* has different growth forms based on edaphic factors and climate conditions. Das et al. (2017) studied the genetic variability within and among individuals with different growth forms in three different locations of India, using 10 RAPD primers. The results indicated the growth forms are genetically different. Geographic separation, as well as physiological/flowering barriers have decreased this genetic divergence and the different growth forms correspond to ecotypes (Table 4.2).

Ramanayake et al. (2007) analyzed the phenotypic variations and genetic diversity of *Dendrocalamus giganteus* Munro (giant bamboo) introduced in Sri Lanka. Six primers and 24 polymorphic bands were analyzed. The genetic distance was from 0 to 0.115 (mean = 0.045). The phenotypic similarities and low genetic diversity suggest that the population studied mainly propagates vegetatively (Table 4.2).

Das et al. (2007) evaluated the phylogenetic relationships of 15 bamboo species (*Bambusa affinis* Munro., *B. arundinacea* Retz., *B. atra* Lindl., *B. auriculata* Kurz., *B. balcooa*, *B. multiplex* 'Riviereorum' R. Maire, *B. oliveriana* Gamble, *B. striata* Lodd. ex Wendl., *B. tulda*, *B. vulgaris* Schrad. ex Wendl., *B. wamin*, Camus., *Dendrocalamus giganteus* Munro., *D. strictus* (Roxb.) Nees., *Gigantochloa atroviolacea* Widjaja and *Pseudobambusa kurzii* (Munro) Ohrnberger) using 120 polymorphic RAPDs. Their results indicated that the dendrogram obtained is congruent with previous taxonomic classification.

Bhattacharya et al. (2009) analyzed the genetic diversity of nine populations of *Thamnocalamus spathiflorus* (Trin.) Munro subsp. *spatiflorus* by amplifying 22 fingerprinting random primers. They did not find any polymorphism from the different population sampled of *T. spathiflorus* subsp. *spathiflorus.*

4.3.4 SSR Markers

Kitamura and Kawahara (2009) identified clonal genotypes in the dwarf bamboo *Sasa cernua* Makino. Their main objective was to analyze the sporadic flowering of *S. cernua* to understand whether the flowering unit is genetically controlled. The distribution of clones was analyzed in one patch in Central Hokkaido, Japan. In 2006, flowering occurred on 60.5% of living culms in a 1600 m^2 patch. Their results indicate that all flowering clumps belonged to the same clone and had probably originated from a single clone of a sporadically flowering patch. The non-flowering clump was found to belong to a single clone. The results further indicated that only a portion of a clone flowers and dies (Table 4.4).

Phyllostachys edulis (Carrière) J. Houz is one of the most economically important bamboos, and the third largest source of timber. It is distributed in southern China and accounts for 70% of bamboo plantations. Human activity has resulted in habitat deterioration and loss of germplasm. Using 20 SSR markers Jiang et al. (2017) evaluated the genetic structure of *P. edulis*, including 34 populations and 803 individuals across its entire distribution in China. Genetic diversity was moderate (H = 0.376), as was genetic differentiation (G_{ST} = 0.162). This genetic differentiation was lower than the values reported for other woody bamboos. Bayesian analysis and the sNMF/ANLS-AS method indicated the presence of two and three clusters, respectively. Long flowering intervals (67–120 years) and low pollen dispersal caused by infrequent, sparse flowering and other factors such as human moving genotypes could explain the high genetic diversity observed in several populations. Due to habitat deterioration and the low capacity for regeneration through seedlings, the authors stated there is an urgent need to preserve germplasm, particularly for those populations harboring high levels of genetic diversity (Table 4.4).

Attigala et al. (2017) studied the genetic diversity and population structure of the temperate woody bamboo *Kuruna debilis* (Thwaites) Attigala, Kathriar. & L.G. Clark distributed in Sri Lanka and southern India. This species is threatened due to deforestation and habitat fragmentation. The study was carried out in six known Sri Lanka populations, and only 28 individuals were analyzed because of the rarity of this species. Allelic diversity was high and genetic differentiation was moderate ($F_{ST} = 0.113$). The six populations were grouped into three genetic clusters consistent with the spatial proximity of the populations. The authors proposed that populations with high genetic diversity be targeted for conservation (Table 4.4).

The *Fargesia spathacea* Franch. complex comprises 15 closely related species with a sympatric distribution in China. Their classification is based on morphological characteristics, which has caused controversy. Huang et al. (2021) studied the genetic diversity, population structure and introgression processes, as well as the phenotypic variation and ecological factors underlying this species' genetic structure. The levels of genetic diversity covered a wide range ($H_E = 0.07$–0.81) and the inbreeding coefficient values were negative, indicating an excess of heterozygotes in the populations. Phylogenetic analyses revealed two major groups with most of the species representing different lineages. The differences between the two groups were influenced by variations in elevation. Based on genetic and morphological information the identity of three species was confirmed (*F. decurvata* J.L. Lu, *F. spathacea*, and *F. murielae* (Gamble) T.P. Yi) and the invalidation of four other species was suggested (*F. scabrida* T.P. Yi, *F. robusta* T.P. Yi, *F. denudata* T.P. Yi, *F. murielae*, and *F. nitida* (Mitford) Keng f. ex T.P. Yi) (Table 4.4).

Dendrocalamus sinicus Chia & J.L. Sun is among the largest and strongest bamboos. It is distributed in the south and southwestern regions of the Yunnan Province in southwestern China. Yang et al. (2018) analyzed the dispersal or vicariance speciation in woody bamboos using *D. sinicus* as a model. This study included three chloroplast DNA fragments (cpDNA) and eight SSR markers, and 232 individuals from 18 populations across the species entire geographic range. There was a high level of genetic differentiation. The two markers revealed two main genetic groups that corresponded to morphological variation (straight culm and sinuous culm). Eleven haplotypes and a strong phylogeographic structure were detected. The demographic analyses indicated that the straight culm group had stayed stable, and the Sinuous group had possibly undergone a recent population expansion. Moreover, the cpDNA markers indicated that *D. sinicus* had experienced dispersal and isolation, however SSR data revealed that contemporary gene flow had occurred, with gene flow direction mainly from the intermediate region to the northern or southern areas (Table 4.4).

Meena et al. (2019) studied the genetic diversity and population structure of 19 natural stands of *D. hamiltonii* distributed across the northeast Himalayas. Sixty-eight SSR primers were tested, with 17 primers exhibiting positive and polymorphic results. A total of 130 alleles were detected in 535 individuals. *D. hamiltonii* had low levels of genetic diversity (h = 0.175, I = 0.291) and moderate levels of genetic differentiation (F_{ST} = 0.16). The clustering patterns followed geographic distribution. Two populations exhibited genetic admixture and were recommended for *in situ* conservation, as did six populations with high genetic diversity located in different geographical regions (Table 4.4).

Rossarolla et al. (2020) identified and characterized SSR markers for *Guadua chacoensis* (Rojas) Londoño & P.M. Peterson and evaluated their transferability to other bamboo species. In total, 35 SSR markers were tested, after PCR optimization, 10 exhibited high polymorphism in *G. chacoensis* and amplified successfully in other bamboo species, including the species of *Guadua*, *Dendrocalamus*, *Farguesia* and *Bambusa* (Table 4.4).

Silva et al. (2020) analyzed the genetic diversity and structure of *Guadua aff. chaparensis* Londoño and Zurita and *G. aff. lynnclarkiae* species in the southwestern region of the Brazilian Amazon. The expected heterozygosity for *G. aff. chaparensis* was high (H_E = 0.50), but the observed heterozygosity was lower (H_O = 0.43). Similar values were found for *G. aff. lynnclarkiae* (H_E = 0.46; H_O = 0.39). The populations did not exhibit inbreeding, and a high genetic structure was evident (G_{ST} = 0.46). A lower spatial genetic correlation was detected in individuals of *G. aff. chaparensis* because they were geographically distant. In the *G. aff. lynnclarkiae* population, individuals were much closer geographically, leading to greater kinship and a higher co-ancestry coefficient. The assignment analyses returned five groups of populations, with low genetic similarity. The authors concluded that the populations have unique characteristics that should be taken into consideration for conservation and population management (Table 4.4).

Pérez-Alquicira et al. (2021) evaluated the genetic diversity of three *Guadua* Knuth species (*G. amplexifolia* J. Presl, *G. inermis* Rupr. ex Fourn and *G. tuxtlensis* Londoño & L.G. Clark) and focused on the analysis of *G. inermis'* genetic structure. Three main questions were addressed: (1) Are the three *Guadua* species genetically differentiated? (2) Does the vulnerable species *G. inermis* have low levels of genetic diversity? (3) What is the relative contribution of geographic and environmental factors to the genetic structure of *G. inermis*? The authors found that the three *Guadua* species were genetically differentiated. For *G. inermis*, higher levels of genetic diversity were found compared with monocot species and genetic differentiation among populations was high. There was a significant association between genetic distance and the maximum temperature of the warmest month, but geographic distance did not influence the genetic distance (Table 4.3).

4.3.5 SNP (Single Nucleotide Polymorphism) Markers

Zhou et al. (2019) analyzed the genetic diversity of Moso bamboo, *Phyllostachys edulis*. Three samples were included, wild type (WT) with normally colored internodes in green and two variants. These samples were re-sequenced with an average coverage of 30 x. A total of 4,700,803 unique single nucleotide polymorphisms (Uni-SNPs) and 268,150 unique InDels (Uni-Indels) were identified. There were 215,297 unique regions with SV and 65,935 unique regions with copy-number variations (CNVs). These genetic differences explain the divergence among the three accessions. Genetic variation was linked to genes in pathways such as ribosome biogenesis, caffeine, among others. Another study examined the SNP distribution in a partial sequence of six genes involved in nitrogen use efficiency (Liu et al., 2021). Thirty-two bamboos were analyzed. The nucleotide sequences of these six genes were relatively conserved, and haplotype diversity was relatively high (nucleotide diversity θw = 0.5137 and total nucleotide polymorphisms πT = 0.03332). The results showed that the six genes are evolving under neutrality, and that the NRT2.1 and AMT2.1 gene sequences may have experienced negative selection. The 32 bamboo species were divided into five categories.

4.4 Transcriptomes

The study of RNAseq has helped to detect the genes that participate in flowering, vegetative growth, and drought resistance, among other traits. Gao et al. (2014) investigated the transcriptome at four important stages of flower development in *Phyllostachys edulis*. The plant material included leaf samples from non-flowering plants and mixed flower samples. More than 67 million reads were obtained from these two samples. Based on sequence homology, a total of 18,309 genes were divided into the three main gene ontology (GO) categories: biological process, cellular component, and molecular function. More than 714 floral development-related bamboo genes were highly expressed in the panicle tissues. Transcriptome analysis was also run on *D. latiflorus* Munro (Liu et al., 2012). The results showed 105 unigenes encoding key enzymes related to the lignin biosynthesis, and 621 SSR markers were detected.

Molecular markers have helped elucidate the genetic diversity and structure of populations of bamboo species. Overall, the levels of genetic diversity were moderate to high in bamboos, even though clonal reproduction predominates. Polyploidy has probably played an important role, leading to high levels of diversity. Most of the studies we reviewed that used dominant markers identified genetic lineages among cultivated and native populations and revealed the genetic relationships among bamboos. Classical taxonomy studies are based on morphology and growth habit; however, phenomena such as phenotypic plasticity and convergence can lead to errors in species identification. Therefore, genetic analyses serve as a complementary source of information. Only a few studies have included massive sequencing; SNPs markers have helped us understand linkage disequilibrium and nucleotide diversity. Transcriptomes have revealed the genes that participate in flower development. Further massive sequencing studies are necessary to characterize differences and similarities among DNA regions, and to identify the genes related to flowering, clonality, and growth to improve breeding programs. The genetic information provided by bamboo species can indeed increase the effectiveness of conservation and biodiversity management.

References

Attigala, L., Gallaher, T., Nason, J. and Clark, L. (2017). Genetic diversity and population structure of the threatened temperate woody bamboo *Kuruna debilis* (Poaceae: Bambusoideae: Arundinarieae) from Sri Lanka based on microsatellite analysis. *J. Natl. Sci. Found.*, 45: 53–65.

Attigala, L., Kathriarachchi, H.S. and Clark, L.G. (2016). Taxonomic revision of the temperate woody bamboo genus *Kuruna* (Poaceae: Bambusoideae: Arundinarieae). *Syst. Bot.*, 41: 174–196.

Attigala, L., Triplett, J.K., Kathriarachchi, H. and Clark, L.G. (2014). A new genus and a major temperate bamboo lineage of the Arundinarieae (Poaceae: Bambusoideae) from Sri Lanka based on a multi-locus plastid phylogeny. *Phytotaxa*, 174: 187–205.

Barker, N.P., Linder, H.P. and Harley, E.H. (1995). Polyphyly of Arundinoideae (Poaceae): Evidence from rbcL sequence data. *Syst. Bot.*, 20: 423–435.

Bhattacharya, S., Das, M., Bar, R. and Pal, A. (2006). Morphological and molecular characterization of *Bambusa tulda* with a note on flowering. *Ann. Bot.*, 98: 529–535.

Bhattacharya, S., Ghosh, J.S., Das, M. and Pal, A. (2009). Morphological and molecular characterization of *Thamnocalamus spathiflorus* subsp. *spathiflorus* at population level. *Plant Syst. Evol.*, 282: 13–20.

Bouchenak-Khelladi, Y., Salamin, N., Savolainen, V., Forest, F., van der Bank, M., Chase, M.W. and Hodkinson, T.R. (2008). Large multigene phylogenetic trees of the grasses (Poaceae): Progress toward complete tribal and generic level sampling. *Mol. Phylogenet Evol.*, 47: 488–505.

Clark, L.G., Dransfield, S., Triplett, J. and Sánchez-Ken, J.G. (2007). Phylogenetic relationships among the one-flowered, determinate genera of Bambuseae (Poaceae: Bambusoideae). *Aliso*, 23: 315–332.

Clark, L.G., Londoño, X. and Ruiz-Sanchez, E. (2015). Bamboo taxonomy and habitat. pp. 1–30. *In*: Liese, W. and Köhl, M. (eds.). *Bamboo: The Plant and Its Uses*. Springer International Publishing Switzerland, New York, USA.

Clark, L.G., Zhang, W. and Wendel, J.F. (1995). A phylogeny of the grass family (Poaceae) based on ndhF sequence data. *Syst. Bot.*, 20: 436–460.

Chalopin, D., Clark, L.G., Wysocki, W.P., Park, M., Duvall, M.R. and Bennetzen, J.L. (2021). Integrated genomic analyses from low-depth sequencing help resolve phylogenetic incongruence in the Bamboos (Poaceae: Bambusoideae). *Front. Plant Sci.*, 12: 725728.

Creer, S., Thorpe, R.S., Malhotra, A., Chou, W.H. and Stenson, A.G. (2004). The utility of AFLPs for supporting mitochondrial DNA phylogeographical analyses in the Taiwanese bamboo viper, Trimeresurus stejnegeri. *J. of Evolution Biol.*, 17: 100–107.

Das, M., Bhattacharya, S. and Pal, A. (2005). Generation and characterization of SCARs by cloning and sequencing of RAPD products: A strategy for species-specific marker development in bamboo. *Ann. Bot.*, 95: 835–841.

Das, M., Bhattacharya, S., Basak, J. and Pal, A. (2007). Phylogenetic relationships among the bamboo species as revealed by morphological characters and polymorphism analyses. *Biol. Plantarum*, 51: 667–672.

Das, S., Singh, Y.P., Negi, Y.K. and Shrivastav, P.C. (2017). Genetic variability in different growth forms of *Dendrocalamus strictus*: Deogun revisited. *N.Z.J. For. Sci.*, 47: 23.

Desai, P., Gajera, B., Mankad, M., Shah, S., Patel, A., Patil, G., Subhash, N. and Kumar, N. (2015). Comparative assessment of genetic diversity among Indian bamboo genotypes using RAPD and ISSR markers. *Mol. Biol. Reports*, 42: 1265–1273.

de Carvalho, M.L.S., de Jesus, I.S., Bezerra, H.B., Oliveira, I.L.C., van den Berg, C., Schnadelbach, A.S., Clark, L.G. and Oliveira, R.P. (2021). Phylogenetics of *Piresia* (Poaceae: Bambusoideae) reveals unexpected generic relationships within Olyreae with taxonomic and biogeographic implications. *Taxon.*, 70: 492–514.

Ellstrand, N.C. and Roose, M.L. (1987). Patterns of genotypic diversity in clonal plant species. *Am. J. Bot.*, 74: 123–131.

Fisher, A.E., Clark, L.G. and Kelchner, S.A. (2014). Molecular phylogeny estimation of the bamboo genus *Chusquea* (Poaceae: Bambusoideae: Bambuseae) and description of two new subgenera. *Syst. Bot.*, 39: 829–844.

Fisher, A.E., Triplett, J.K., Ho, C., Schiller, A.D., Oltrogge, K., Schorder, E.S., Kelchner, S.A. and Clark, L.G. (2009). Paraphyly in the bamboo subtribe Chusqueinae (Poaceae: Bambusoideae), and a revised infrageneric classification for *Chusquea*. *Syst. Bot.*, 34: 673–683.

Gao, J., Zhang, Y., Zhang, C., Qi, F., Li, X., Mu, S. and Peng, Z. (2014). Characterization of the floral transcriptome of Moso bamboo (*Phyllostachys edulis*) at different flowering developmental stages by transcriptome sequencing and RNA-seq analysis. *Plos One*, 9: e98910.

Ghosh, J.S., Bhattacharya, S. and Pal, A. (2017). Molecular phylogeny of 21 tropical bamboo species reconstructed by integrating non-coding internal transcribed spacer (ITS1 and 2) sequences and their consensus secondary structure. *Genetica*, 145: 319–333.

Goh, W.L., Chandran, S., Franklin, D.C., Isagi, Y., Koshy, K.C., Sungkaew, S., Yang, H.Q., Xia, N.H. and Wong, K.M. (2013). Multigene region phylogenetic analyses suggest reticulate evolution and a clade of Australian origin among paleotropical woody bamboos (Poaceae: Bambusoideae: Bambuseae). *Plant Syst. Evol.*, 299: 239–257.

Goh, W.L., Chandran, S., Lin, R.S., Xia, N.H. and Wong, K.M. (2010). Phylogenetic relationships among Southeast Asian climbing bamboos (Poaceae: Bambusoideae) and the *Bambusa* complex. Biochem. *Syst. Ecol.*, 38: 764–773.

GPWG (Grass Phylogeny Working Group). (2001). Phylogeny and subfamilial classification of the grasses (Poaceae). *Ann. Missouri Bot.*, 88: 373–457.

Guo, Z.H., Chen, Y.Y. and Li, D.Z. (2001). Phylogenetic studies on *Thamnocalamus* group and its allies (Bambusoideae: Poaceae) based on ITS sequence data. *Mol. Phylogenet. Evol.*, 22: 20–30.

Guo, Z.H., Chen, Y.Y. Li, D.Z. and Yang, J.B. (2001). Genetic variation and evolution of the alpine bamboos (Poaceae: Bambusoideae) using DNA sequence data. *J. Plant Res.*, 114: 315–322.

Guo, Z.H. and Li, D.Z. (2004). Phylogenetics of the *Thamnocalamus* group and its allies (Gramineae: Bambusoideae): Inference from the sequences of GBSSI gene and ITS spacer. *Mol. Phylogenet. Evol.*, 30: 1–12.

Guo, Z.H., Ma, P.M., Yang, G.Q., Hu, J.Y., Liu, Y.L., Xia, E.H., Zhong, M.C., Zhao, L., Sun, G.L., Xu, Y.X., Zhao, Y.J., Zhang, Y.C., Zhang, Y.X., Zhang, X.M., Zhou, M.Y., Guo, Y., Guo, C., Liu, J.X. and Li, D.Z. (2019). Genome sequences provide insights into the reticulate origin and unique traits of woody bamboos. *Mol. Plant*, 12: 1353–1365.

Haevermans, T., Mantuano, D., Zhou, M.Y., Lamxay, V., Haevermans, A., Blanc, P. and Li, D.Z. (2020). Discovery of the first succulent bamboo (Poaceae: Bambusoideae) in a new genus from Laos' karst areas, with a unique adaptation to seasonal drought. *PhytoKeys*, 156: 125–137.

Hamby, R.K. and Zimmer, E.A. (1988). Ribosomal RNA sequences for inferring phylogeny within the grass family (Poaceae). *Plant Syst. Evol.*, 160(1): 29–37.

Hamrick, J.L. and Godt, M.J.W. (1989). Allozyme diversity in plant species. pp. 43–63. *In*: Brown, A.H.D., Clegg, M.T., Kahler, A.L. and Weir, B.S. (eds.). *Plant Population Genetics, Breeding Germplasm Resources*. Sinauer Associates, Massachusetts, Sunderland Massachusetts, USA.

He, T.Y., Qu, Y.Q., Chen, L.Y., Xu, W., Rong, J.D., Chen, L.G., Fan, L.L., Tarin, M.W.K. and Zheng, Y.S. (2020). Genetic diversity analysis of *Dendrocalamopsis Beecheyana* var. *pubescens* based on ISSR markers. *Appl. Ecol. Environ. Res.*, 17: 12507–12519.

Hodkinson, T.R., Renvoize, S.A., Chonghaile, G.N., Stapleton, C. and Chase, M.W. (2000). A comparison of ITS nuclear rDNA sequence data and AFLP markers for phylogenetic studies in *Phyllostachys* (Bambusoideae, Poaceae). *J. Plant Res.*, 113: 259–269.

Huang, L., Xing, X.C., Li, W.W., Zhou, Y., Zhang, Y.Q., Xue, C., Ren, Y. and Kang, J.Q. (2021). Population genetic structure of the giant panda staple food bamboo (*Fargesia spathacea* complex) and its taxonomic implications. *J. Syst. Evol.*, 59: 1051–1064.

Isagi, Y., Shimada, K., Kushima, H., Tanaka, N., Nagao, A., Ishikawa, T., OnoDera, H. and Watanabe, S. (2004). Clonal structure and flowering traits of a bamboo [*Phyllostachys pubescens* (Mazel) Ohwi] stand grown from a simultaneous flowering as revealed by AFLP analysis. *Mol. Ecol.*, 13: 2017–2021.

Jesus-Costa, C. (2018). *Estudos filogenéticos moleculares e taxonômicos na subtribo Arthrostylidiinae* (Poaceae: Bambusoideae: Bambuseae). PhD thesis, Universidade Federal de Viçosa.

Jesus-Costa, C., Clark, L.G. and Santos-Gonçalves, A.P. (2018). Molecular phylogeny of *Atractantha*, and the phylogenetic position and circumscription of *Athroostachys* (Poaceae: Bambusoideae: Bambuseae: Arthrostylidiinae). *Syst. Bot.*, 43: 656–663.

Jiang, W., Bai, T., Dai, H., Wei, Q., Zhan, W. and Ding, Y. (2017). Microsatellite markers revealed moderate genetic diversity and population differentiation of Moso bamboo (*Phyllostachys edulis*)—a primarily asexual reproduction species in China. *Tree Genet. Genomes*, 13: 130.

Judziewicz, E.J., Clark, L.G., Londoño, X. and Stern, M.J. (1999). *American Bamboos*. Smithsonian Institution Press, Washington.

Kelchner, S.A. and Clark, L.G. (1997). Molecular evolution and phylogenetic utility of the chloroplast rpl16 intron in Chusquea and the Bambusoideae (Poaceae). *Mol. Phylogenet. Evol.*, 8: 385–397.

Kelchner, S.A. and BPG [Bamboo Phylogeny Group]. (2013). Higher level phylogenetic relationships within the bamboos (Poaceae: Bambusoideae) based on five plastid markers. *Mol. Phylogenet. Evol.*, 67: 404–413.

Keng, P.C. and Wang, Z.P. (1996). *Flora Reipublicae Popularis Sinicae, delectis Florae Reipublicae Popularis Sinicae agendae Academiae Sinicae edita*; Vol. 9, pt. 1 (Gramineae 1: Bambusoideae). Science Press, Beijing.

Kim, I.R., Yu, D. and Choi, H.K. (2015). A phytogeographical study of *Sasa borealis* populations based on AFLP analysis. *Korean J. Plant Taxo.*, 45: 29–35.

Kitamura, K. and Kawahara, T. (2009). Clonal identification by microsatellite loci in sporadic flowering of a dwarf bamboo species, *Sasa cernua*. *J. Plant Res.*, 122: 299–304.

Konzen, E.R., Peron, R., Ito, M.A., Brondani, G.E. and Tsai, S.M. (2017). Molecular identification of bamboo genera and species based on RAPD-RFLP markers. *Silva Fennica*, 51: 1691.

Lee, N.C. and Chung, M.G. (1999). High levels of genetic variation in Korean populations of *Sasamorpha borealis* (Poaceae). *Bot. Bull. Acad. Sin.*, 40: 311–317.

Lin, X.C., Ruan, X., Lou, Y.F., Guo, X. and Fang, W. (2008). Genetic similarity among cultivars of *Phyllostachys pubescens*. *Plant Syst. Evol.*, 277: 67–73.

Lin, X., Lou, Y., Zhang, Y., Yuan, X., He, J. and Fang, W. (2011). Identification of genetic diversity among cultivars of *Phyllostachys violascens* using ISSR, SRAP, and AFLP markers. *Bot. Rev.*, 77: 223–232.

Liu, M., Qiao, G., Jiang, J., Yang, H., Xie, L., Xie, J. and Zhuo, R. (2012). Transcriptome sequencing and *de novo* analysis for ma bamboo (*Dendrocalamus latiflorus* Munro) using the Illumina platform. *Plos One*, 7: e46766.

Liu, X., Luo, M., Chen, X. and Ding, C. (2021). Identification of single nucleotide polymorphisms and analysis of linkage disequilibrium in different bamboo species using the candidate gene approach. *Phyton*, 90: 1697–1709.

Liu, J.X., Zhou, M.Y., Yang, G.Q., Zhang, Y.X., Ma, P.F., Guo, C., Vorontsova, M.S. and Li, D.Z. (2020). ddRAD analyses reveal a credible phylogenetic relationship of the four main genera of *Bambusa-Dendrocalamus-Gigantochloa* complex (Poaceae: Bambusoideae). *Mol. Phylogenet. Evol.*, 146: 106758.

Loh, J., Kiew, R., Set, O., Gan, L. and Gan, Y. (2000). A study of genetic variation and relationships within the bamboo subtribe Bambusinae using AFLP. *Ann. Bot.*, 85: 607–612.

Ma, Q.Q., Song, H.X., Zhou, S.Q., Yang, W.Q., Li, D.S. and Chen, J.S. (2013). Genetic structure in dwarf bamboo (*Bashania fangiana*) clonal populations with different genet ages. *Plos One*, 8(11): e78784.

Ma, P.F., Zhang, Y.X., Zeng, C.X., Guo, Z.H. and Li, D.Z. (2014). Chloroplast phylogenomic analyses resolve deep-level relationships of an intractable bamboo tribe Arundinarieae (Poaceae). *Syst. Biol.*, 63: 933–950.

Marulanda, M.L., Márquez, P. and Londoño, X. (2002). AFLP analysis of *Guadua angustifolia* (Poaceae: Bambusoideae) in Columbia with emphasis on the coffee region. *J. Am. Bamboo Soc.*, 16: 32–42.

Mathews, K.G., Huguelet, J., Lanning, M., Wilson, T. and Young, R.S. (2009). Clonal diversity of *Arundinaria gigantea* (Poaceae; Bambusoideae) in western North Carolina and its relationship to sexual reproduction: An assessment using AFLP fingerprints. *Castanea*, 74: 213–223.

Meena, R.K., Bhandhari, M.S., Barhwal, S. and Ginwal, H.S. (2019). Genetic diversity and structure of *Dendrocalamus hamiltonii* Nees & Arn. ex Munro natural metapopulation: A commercially important bamboo species of northeast Himalayas. *3 Biotech.*, 9: 60.

Miyazaki, Y., Ohnishi, N., Takafumi, H. and Hiura, T. (2009). Genets of dwarf bamboo do not die after one flowering event: Evidence from genetic structure and flowering pattern. *J. Plant Res.*, 122: 523–528.

Nag, A., Gupta, P., Sharma, V., Sood, A., Ahuja, P.S. and Sharma, R.K. (2013). AFLP and RAPD based genetic diversity assessment of industrially important reed bamboo (*Ochlandra travancorica* Benth). *J. Plant Biochem. Biotechnol.*, 22: 144–149.

Nayak, S., Rout, G.R. and Das, P. (2003). Evaluation of genetic variability in bamboo using RAPD Markers. *Plant, Soil Environ.*, 49: 24–28.

Nilkanta, H., Amom, T., Tikendra, L., Rahaman, H. and Nongdam, P. (2017). ISSR marker-based population genetic study of *Melocanna baccifera* (Roxb.) Kurz: A commercially important bamboo of Manipur, Northeast India. *Scientifica*, 2017: 3757238.

Oliveira, I.L., Matos, A.O., Silva, C., de Carvalho, M.L.S., Tyrrell, C.D., Clark, L.G. and Oliveira, R.P. (2020). Delving deeper into the phylogenetics of the herbaceous bamboos (Poaceae: Bambusoideae, Olyreae): Evaluation of generic boundaries within the *Parodiolyra/Raddiella* clade uncovers a new genus. *Bot. J. Linn. Soc.*, 192: 61–81.

Oliveira, R.P., Borba, E.L., Longhi-Wagner, H.M., Pereira, A.C.S. and Lambert, S.M. (2008). Genetic and morphological variability in the *Raddia brasiliensis* complex (Poaceae: Bambusoideae). *Plant Syst. Evol.*, 274: 25–35.

Oliveira, R.P., Clark, L.G., Schnadelbach, A.S., Monteiro, S.H.N., Longhi-Wagner, H.M. and van den Berg, C. (2014). A molecular phylogeny of *Raddia* (Poaceae, Olyreae) and its allies based on noncoding plastid and nuclear spacers. *Mol. Phylogenet. Evol.*, 78: 105–117.

Oumer, O.A., Dagne, K., Feyissa, T., Tesfaye, K., Durai, J. and Hyder, M.Z. (2020). Genetic diversity, population structure, and gene flow analysis of lowland bamboo [*Oxytenanthera abyssinica* (A. Rich.) Munro] in Ethiopia. *Ecol. Evol.*, 10: 11217–11236.

Peng, S., Yang, H.Q. and Li, D.Z. (2008). Highly heterogeneous generic delimitation within the temperate bamboo clade (Poaceae: Bambusoideae): Evidence from GBSSI and ITS sequences. *Taxon.*, 57: 799–810.

Pérez-Alquicira, J., Aguilera-Lopez, S., Rico, Y. and Ruiz-Sanchez, E. (2021). A population genetics study of three native Mexican woody bamboo species of *Guadua* (Poaceae: Bambusoideae: Bambuseae: Guaduinae) using nuclear microsatellite markers. *Bot. Sci.*, 99: 542–559.

Posso, T.A.M. (2011). Diversidad genética y estructura poblacional de Guadua angustifolia Kunth en el eje cafetero colombiano. Tesis de Magister en Ciencias Agrarias con énfasis en fitomejoramiento. Universidad Nacional de Colombia.

Ramanayake, S.M.S.D., Meemaduma, V.N. and Weerawardene, T.E. (2010). Genetic diversity in a population of *Dendrocalamus giganteus* Wall, ex Munro (giant bamboo) in the Royal Botanic Gardens in Peradeniya. *J. Natl. Sci. Found. Sri Lanka*, 35: 207–210.

Rossarolla, M.D., Tomazetti, T.C., Vieira, L.N., Guerra, M.P., Klabunde, G.H.F., Scherer, R.F., Pescador, R. and Nodari, R.O. (2020). Identification and characterization of SSR markers of *Guadua chacoensis* (Rojas) Londoño & P.M. Peterson and transferability to other bamboo species. *3 Biotech.*, 10: 273.

Ruiz-Sanchez, E., Sosa, V. and Mejia-Saules, M.T. (2008). Phylogenetics of *Otatea* inferred from morphology and chloroplast DNA sequence data and recircumscription of Guaduinae (Poaceae: Bambusoideae). *Syst. Bot.*, 33: 277–283.

Ruiz-Sanchez, E., Sosa, V. and Mejía-Saules, M.T. (2011). Molecular phylogenetics of the Mesoamerican bamboo *Olmeca* (Poaceae:, Bambuseae): Implications for taxonomy. *Taxon.*, 60: 89–98.

Ruiz-Sanchez, E., Sosa, V., Ortiz-Rodriguez, A.E. and Davidse, G. (2019). Historical biogeography of the herbaceous bamboo tribe Olyreae (Bambusoideae: Poaceae). *Folia Geobot.*, 54: 177–189. Doi: 10.1007/s12224-019-09342-7.

Ruiz-Sanchez, E., Tyrrell, C.D., Londoño, X., Oliveira, R.P. and Clark, L.G. (2021). Diversity, distribution, and classification of Neotropical woody bamboos (Poaceae: Bambusoideae) in the 21st Century. *Bot. Sci.*, 99: 198–228.

Saarela, J.M., Burke, S.V., Wysocki, W.P., Barrett, M.D., Clark, L.G., Craine, J.M., Peterson, P.M., Soreng, R.J., Vorontsova, M.S. and Duvall, M.R. (2018). A 250 plastome phylogeny of the grass family (Poaceae): Topological support under different data partitions. *Peer J.*, 6: e4299.

Silva, S.M.M., Martins, K., Costa, F.H.S., De Campos, T. and Schewinski-Pereira, J. (2020). Genetic structure and diversity of native *Guadua* species (Poaceae: Bambusoideae) in natural populations of the Brazilian Amazon rainforest. *An. Acad. Bras. Cienc.*, 92: e20190083.

Soreng, R.J., Peterson, P.M., Romaschenko, K., Davidse, G., Teisher, J.K., Clark, L.G., Barberá, P., Gillespie, L.J. and Zuloaga, F.O. (2017). A worldwide phylogenetic classification of the Poaceae (Gramineae) II: An update and a comparison of two 2015 classifications. *J. Syst. Evol.*, 55: 259–290.

Soreng, R.J., Peterson, P.M., Romaschenko, K., Davidse, G., Zuloaga, F.O., Judziewicz, E.J., Filgueiras, T.S., Davis, J.I. and Morrone, O. (2015). A worldwide phylogenetic classification of the Poaceae (Gramineae). *J. Syst. Evol.*, 53: 117–137.

Sungkaew, S., Stapleton, C.M., Salamin, N. and Hodkinson, T.R. (2009). Non-monophyly of the woody bamboos (Bambuseae; Poaceae): A multigene region phylogenetic analysis of Bambusoideae ss. *J. Plant Res.*, 122: 95–108.

Suyama, Y., Obayashi, K. and Hayashi, I. (2000). Clonal structure in a dwarf bamboo (*Sasa senanensis*) population inferred from amplified fragment length polymorphism (AFLP) fingerprints. *Mol. Ecol.*, 9: 901–906.

Tian, B., Yang, H.Q., Wong, K.M., Liu, A. and Ruan, Z.Y. (2012). ISSR analysis shows low genetic diversity versus high genetic differentiation for giant bamboo, *Dendrocalamus giganteus* (Poaceae: Bambusoideae), in China populations. *Genet. Resour. Crop Evolut.*, 59: 901–908.

Triplett, J.K. and Clark, L.G. (2010). Phylogeny of the temperate bamboos (Poaceae: Bambusoideae: Bambuseae) with an emphasis on *Arundinaria* and allies. *Syst. Bot.*, 35: 102–120.

Triplett, J.K., Clark, L.G., Fisher, A.E. and Wen, J. (2014). Independent allopolyploidization events preceded speciation in the temperate and tropical woody bamboos. *New Phytol.*, 204: 66–73.

Triplett, J.K., Oltrogge, K.A. and Clark, L.G. (2010). Phylogenetic relationships and natural hybridization among the North American woody bamboos (Poaceae: Bambusoideae: *Arundinaria*). *Am. J. Bot.*, 97: 471–92.

Tyrrell, C.D., Londoño, X., Prieto, R.O., Attigala, L., McDonald, K. and Clark, L.G. (2018). Molecular phylogeny and cryptic morphology reveal a new genus of West Indian woody bamboo (Poaceae: Bambusoideae: Bambuseae) hidden by convergent character evolution. *Taxon.*, 67: 916–930.

Tyrrell, C.D., Santos-Gonçalves, A.P., Londoño, X. and Clark, L.G. (2012). Molecular phylogeny of the arthrostylidioid bamboos (Poaceae: Bambusoideae: Bambuseae: Arthrostylidiinae) and new genus *Didymogonyx. Mol. Phylogenet. Evol.*, 65: 136–148.

Vieira, L.D.N., Dos Anjos, K.G., Faoro, H., Fraga, H.P.D.F., Greco, T.M., Pedrosa, F.D.O., de Sousa, E.M., Rogalski, M., de Souza, R.F. and Guerra, M.P. (2016). Phylogenetic inference and SSR characterization of tropical woody bamboos tribe Bambuseae (Poaceae: Bambusoideae) based on complete plastid genome sequences. *Curr. Genet.*, 62: 443–453.

Vinícius-Silva, R., Fregonezi, J.N., Clark, L.G. and Santos-Gonçalves, A.P. (2021). Morphological evolution and molecular phylogenetics of the *Merostachys* clade (Poaceae: Bambusoideae: Bambuseae: Arthrostylidiinae) based on multi-locus plastid sequences. *Bot. J. Linn. Soc.*, 195: 53–76.

Waikhom, S.D., Ghosh, S., Talukdar, N.C. and Mandi, S.S. (2012). Assessment of genetic diversity of landraces of *Dendrocalamus hamiltonii* using AFLP markers and association with biochemical traits. *Genet. Mol. Res.*, 11: 2107–2121.

Wang, W., Chen, S. and Zhang, X. (2018). Whole-genome comparison reveals divergent IR borders and mutation hotspots in chloroplast genomes of herbaceous bamboos (Bambusoideae: Olyreae). *Molecules*, 23: 1537.

Wang, X., Ye, X., Zhao, L., Li, D., Guo, Z. and Zhuang, H. (2017). Genome-wide RAD sequencing data provide unprecedented resolution of the phylogeny of temperate bamboos (Poaceae: Bambusoideae). *Sci. Rep.*, 7: 11546.

Wysocki, W.P., Clark, L.G., Attigala, L., Ruiz-Sanchez, E. and Duvall, M.R. (2015). Evolution of the bamboos (Bambusoideae; Poaceae): A full plastome phylogenomic analysis. *BMC Evol. Biol.*, 15: 50.

Wysocki, W.P., Ruiz-Sanchez, E., Yin, Y. and Duvall, M.R. (2016). The floral transcriptomes of four bamboo species (Bambusoideae; Poaceae): Support for common ancestry among woody bamboos. *BMC Genomics*, 17: 384.

Yang, H.Q., An, M.Y., Gu, Z.J. and Tian, B. (2012). Genetic diversity and differentiation of *Dendrocalamus membranaceus* (Poaceae: Bambusoideae), a declining bamboo species in Yunnan, China, as based on Inter-Simple Sequence Repeat (ISSR) Analysis. *Int. J. Mol. Sci.*, 13: 4446–4457.

Yang, H.Q., Peng, S. and Li, D.Z. (20070. Generic delimitations of *Schizostachyum* and its allies (Gramineae: Bambusoideae) inferred from GBSSI and *trnL-F* sequence phylogenies. *Taxon.*, 56: 45–54.

Yang, H.Q., Yang, J.B., Peng, Z.H., Gao, J., Yang, Y.M., Peng, S. and Li, D.Z. (2008). A molecular phylogenetic and fruit evolutionary analysis of the major groups of the paleotropical woody bamboos (Gramineae: Bambusoideae) based on nuclear ITS, GBSSI gene and plastid *trnL-F* DNA sequences. *Mol. Phylogenet. Evol.*, 48: 809–824.

Yang, J.B., Dong, Y.R., Wong, K.M., Gu, Z.J., Yang, H.Q. and Li, D.Z. (2018). Genetic structure and differentiation in *Dendrocalamus sinicus* (Poaceae: Bambusoideae) populations provide insight into evolutionary history and speciation of woody bamboos. *Sci. Rep.*, 8: 16933.

Yang, J.B., Yang, H.Q., Li, D.Z., Wong, K.M. and Yang, Y.M. (2010). Phylogeny of *Bambusa* and its allies (Poaceae: Bambusoideae) inferred from nuclear GBSSI gene and plastid *psbA-trnH, rpl32-trnL and rps16* intron DNA sequences. *Taxon.*, 59: 1102–1110.

Yang, H.M., Zhang, Y.X., Yang, J.B. and Li, D.Z. (2013). The monophyly of *Chimonocalamus* and conflicting gene trees in Arundinarieae (Poaceae: Bambusoideae) inferred from four plastid and two nuclear markers. *Mol. Phylogenet. Evol.*, 68: 340–356.

Zeng, C.X., Zhang, Y.X., Triplett, J.K., Yang, J.B. and Li, D.Z. (2010). Large multi-locus plastid phylogeny of the tribe Arundinarieae (Poaceae: Bambusoideae) reveals ten major lineages and low rate of molecular divergence. *Mol. Phylogenet. Evol.*, 56: 821–839.

Zhang, W. and Clark, L.G. (2000). Phylogeny and classification of the Bambusoideae (Poaceae). pp. 35–42. *In:* Jacobs, S.W.L. and Everett, J.E. (eds.). *Grasses: Systematics and Evolution*. CSIRO Publishing, Victoria, Australia.

Zhang, L.N., Ma, P.F., Zhang, Y.X., Zeng, C.X., Zhao, L. and Li, D.Z. (2019). Using nuclear loci and allelic variation to disentangle the phylogeny of *Phyllostachys* (Poaceae, Bambusoideae). *Mol. Phylogenet. Evol.*, 137: 222–235.

Zhang, Y.X., Zeng, C.X. and Li, D.Z. (2012). Complex evolution in Arundinarieae (Poaceae: Bambusoideae): Incongruence between plastid and nuclear GBSSI gene phylogenies. *Mol. Phylogenet. Evol.*, 63: 777–797.

Zhou, M.B., Wu, J.J., Ramakrishnan, M., Meng, X.W. and Vinod, K.K. (2019). Prospects for the study of genetic variation among Moso bamboo wild-type and variants through genome resequencing. *Trees*, 33: 371–381.

Zhou, M.Y., Zhang, Y.X., Haevermans, T. and Li, D.Z. (2017). Towards a complete generic-level plastid phylogeny of the paleotropical woody bamboos (Poaceae: Bambusoideae). *Taxon.*, 66: 539–553.

Chapter 5

Genetic Diversity Assessment and Molecular Markers in African Bamboos

Oumer Abdie[1,]* and *Muhamed Adem*[2]

5.1 Introduction

Bamboos, one of the most important non-timber forest resources or a potential alternative to wood and wood products (Ekhuemelo et al., 2018) and fastest-growing plant in the world (100 cm per day) (Zhang et al., 2011; Li et al., 2021). Taxonomically with 1,700 species and 127 generas belonging to the family Poaceae, constitute a single subfamily Bambusoideae. Molecular phylogenetic analysis suggested that Bambusoideae falls into the Bambusoideae-Oryzoideae-Pooideae (BOP) clade, which is a phylogenetic sister Pooideae (Saarela et al., 2018).

Bamboo plant grows naturally in areas receiving annual rainfall ranging from 1,200 to 4,000 mm, with an average annual temperature of 8–36°C. They also can grow well in different soil types, ranging rich alluvium to hard lateritic, sandy, and loamy soils. The multi-functional use of bamboo grass considering its rapid growth rates, resilience, and the possibility of multiple harvesting in a few years' time are significant advantages over the other forest species (Diver, 2006; Akinlabi et al., 2017). Bamboo matures in 3–5 years and thereafter can be harvested annually for about 20 years (Tariyal, 2016; Dalagnol et al., 2018) or longer, depending on the gregarious flowering period, after which bamboo dies. Its gregarious flowering interval can be between 20 –120 years, depending on the species (Huy and Trinh, 2019).

Bamboo serves as an important part of nature-based development. Known in some parts of the world as "green gold", "poor man's timber", this fast-growing grass

---

[1] Department of Biotechnology, College of Natural and Computational Science, Wolkite University, Wolkite, Ethiopia.

[2] Department of Forestry, College of Agriculture and Natural Resources, Madda Walabu University, Bale Robe, Ethiopia.

Email: muhammed.adem@aau.edu.et

* Corresponding author: oumer.abdie@aau.edu.et, oumar.abdie@gmail.com

plant covers over 31.5 million hectares (ha) of land across the tropics and subtropics. It is a multipurpose plant with over 10, 000 documented uses and applications, with rapid regeneration capacity and the possibility of annual harvesting within a few years of planting has significant advantages over the other forest species (Diver, 2006; Akinlabi et al., 2017). It has been proven to help combat several global challenges, including rural poverty, land degradation, deforestation, urban development, unsustainable resource use (or reducing pressure on forestry resources) and climate change (Yuen et al., 2017; Ekhuemelo et al., 2018; Kaushal et al., 2018; Huy and Trinh, 2019). Two and half billion people are estimated to be directly involved in the production and consumption of bamboos (Scurlock et al., 2000). It has been also proven to address many global challenges and contributes to the United Nations Sustainable Development Goals: SDG 1 (no poverty), SDG 7 (affordable and clean energy), SDG 11 (sustainable and resilient housing), SDG 12 (efficient use of resources), SDG 13 (address climate change), and SDG 15 (life on land) (Yuen et al., 2017; Ekhuemelo et al., 2018; Kaushal et al., 2018; Huy and Trinh, 2019). Due to its potential for soil erosion and water recharge, bamboo also provides an opportunity for restoration of degraded areas and watershed development (Kaushal et al., 2020). Furthermore, bamboo is a herbal medicine for a mild case patient for the currently occurred outbreak corona virus disease (COVID-19) (Xu and Zhang, 2020; Shahrajabian et al., 2020), a food for humans in bamboo growing African countries particularly in Benishangul-Gumuz Regional state of Ethiopia (Oumer et al., 2020) and feed and fodder for livestock, and it contributes to ensuring food security (Halvorson et al., 2011; Choudhury et al., 2013; Nongdam and Tikendra, 2014; Mulatu et al., 2019; Andriarimalala et al., 2019). Bamboo has huge economic potential; the global production and local consumption are estimated at the USD 60 billion, and the international export is valued at USD 2 billion per annum (International Network for Bamboo and Rattan (INBAR), 2019).

The most common and useful bamboos are those with 'woody' (lignified) stems belonging to the tribes Bambuseae and Arundinarieae that show a wide geographical and altitudinal distribution. The roughly 1,300 woody bamboo species often play critical roles in maintaining the ecology of forest habitats and have great economic importance to humans (Kelchner and Bamboo Phylogenetic Group (BPG), 2013). Herbaceous bamboos (tribe Olyreae) differ from woody bamboos; they do not have or possess only weakly lignified culms, usually no culm leaves, and no outer ligules. In addition, they are less popularly known, and they are found in tropical forests, predominantly in the New World (BPG, 2012; Kelchner and BPG, 2013).

Based on the molecular and morphological characteristics, true bamboos (Bambusoideae) are greatly supported as a monophyletic lineage and categorized into three tribes (Fig. 5.1): Bambuseae (tropical woody bamboos), Arundinarieae (temperate woody bamboos), and Olyreae (herbaceous bamboos) (Sungkaew et al., 2009; BPG, 2012; Kelchner and BPG, 2013; Saarela et al., 2018). The Bambuseae tribe differs from the Olyreae based on the presence of abaxial ligules (Zhang, 2000; Grass Phylogeny Working Group (GPWG), 2001). Within the tribe Bambuseae, there are two clades comprising neotropical woody bamboos (NWB) and paleotropical woody bamboos (PWB). The NWB clade is composed of three

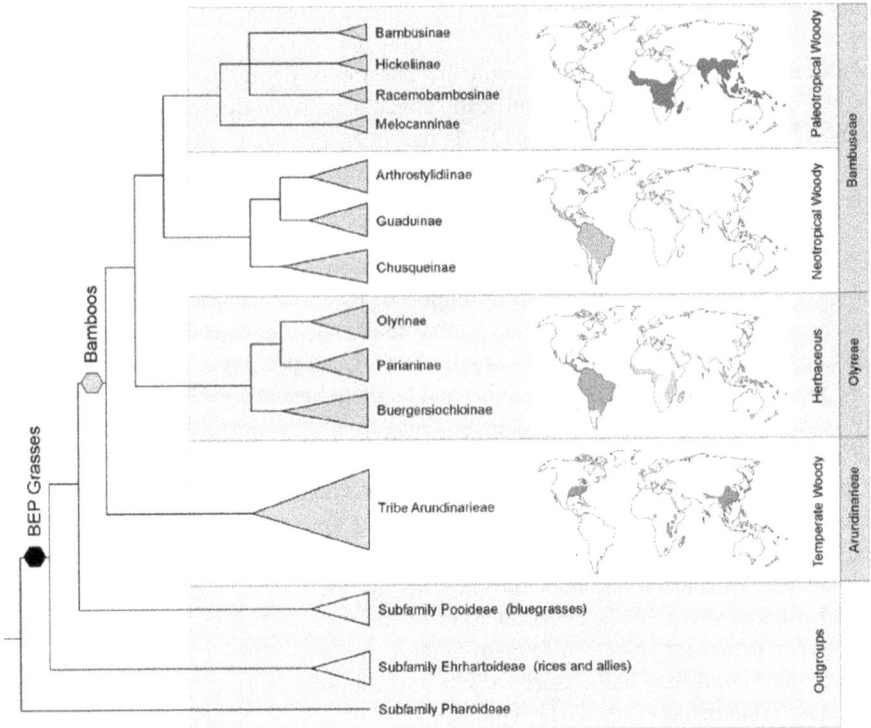

Fig. 5.1. Subtribe relationships and geographic distributions of the four major lineages of bamboos based on plastid (chloroplast) DNA sequences from five loci: *ndhF*, *rpl16* intron, *rps16* intron, *trnD–trnT* intergenic spacer, and *trnL–trnF* intergenic spacer (Kelchner and BPG, 2013).

subtribes: Guaduinae, Arthrostylidiinae, and Chusqueinae whereas clade PWB has Bambusinae, Racemobambosinae, Melocanninae, and Hickeliinae subtribes. The Olyreae contains three subtribes: Buergersiochloinae and Olyrinae Parianinae (Clark et al., 2007; Sungkaew et al., 2009; BPG, 2012; Kelchner and BPG, 2013).

5.2 Bamboo Cytogenetics, Ploidy Level, and Genome Size

Chromosome records of bamboos show that seemingly consistent ploidy levels define each bamboo clade. The PWB are mostly hexaploid and the NWB and the Arundinarieae (temperate woody bamboos) are tetraploid. Woody bamboo genera have a typical basic chromosome number of $x = 12$, except for Chusquea ($x = 10$) (Pohl and Clark, 1992; Chen et al., 2003; Clark et al., 2010). The previously reported higher number chromosome in bamboos was $2n = 108$ recorded in *Bambusa schizostachyoides* (Kurz) Gamble (Devi and Sharma, 1993) followed by $2n = 104$ in six Chinese species (Chen et al., 2003). But $2n = 192$ is the highest chromosome number in the entire bamboo family reported for the first time in Pseudoxytenanthera (Mathu et al., 2015).

Genome size assessments are useful in studying evolutionary and adaptation mechanisms. Further, this information is a prerequisite for genome sequencing and analysis projects. However, only 1% of angiosperms have been investigated for DNA content estimations (Pellicer et al., 2018). Moreover, such information is limited to only few species of bamboo. Flow cytometric studies explored genome size variation among temperate and tropical bamboo species that ranged from 2.04 Gb to 2.6 Gb in temperate and 1.14 Gb to 1.6 Gb in tropical bamboo species (Gui et al., 2007). These inferences also suggest that polyploidy is the imperative powerhouse in the evolution of woody bamboos. Recently, two independent flow cytometric studies on 37 bamboo species (Kumar et al., 2011) and a tetraploid *Phyllostachys pubescens* (Gui et al., 2007) showed that genome size in different bamboo species ranges from 1.2 Gb to 2.9 Gb, which is slightly higher in range as compared to prior studies by Gielis (1997). Further, these estimates revealed that the genome sizes in bamboo species are more than three to seven folds larger than the genome sizes of Nipponbare (Japonica rice) and 10–24 times larger than *Arabidopsis thaliana* genome size and much smaller than bread wheat (allohexaploid 16 Gb) and durum wheat (allotetraploid 11 Gb) species, respectively (Ganal and Röder, 2007; Mayer et al., 2014).

5.3 Geographical Distribution and Area Coverage of Bamboo

Bamboo is widely distributed in the tropical, subtropical, and temperate countries in Asia, Latin America, and Africa from sea level to highlands (Fig. 5.2) (BPG, 2012; Kelchner and BPG, 2013). About 31.5 million hectares of the earth's surface is covered by bamboos.

After Asia and South America, Africa is the third richest continent in terms of bamboo species. Bamboo is common in most of sub-Saharan Africa, from Ethiopia all the way down to South Africa and Madagascar. The continent with 31 countries has 43 species (Table 5.1) occurring on around 1.5 million hectares. Of these, 40 (33 of them are naturally occurring) are mainly found in Madagascar while the remaining three are in mainland Africa (Ohrnberger, 1999; Embaye, 2000; Akinlabi et al., 2017).

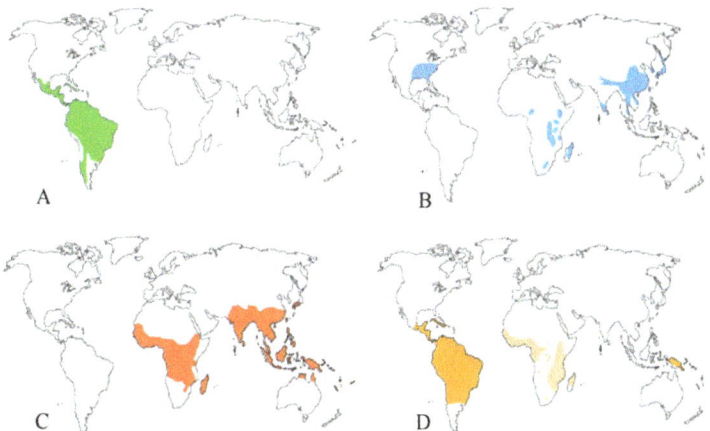

Fig. 5.2. Worldwide distribution of bamboo. (A) Neotropical woody bamboos, (B) Temperate woody bamboos, (C) Paleotropical woody bamboos, and (D) Herbaceous bamboos (Chokthaweepanich, 2014).

Table 5.1. Numbers of species of Bambuseae occurring in the countries of Africa.

S. No.	A	B	S. No.	A	B
1.	Madagascar	33 (32)	17.	Cote d'Ivoire	1 (0)
2.	Tanzania	4 (1)	18.	Eritrea	1 (0)
3.	Malawi	3 (0)	19.	Gambia	1 (0)
4.	Uganda	3 (0)	20.	Ghana	1 (0)
5	Zambia	3 (0)	21.	Guinea	1 (0)
6.	Cameroon	2 (0)	22.	Guinea-Bissau	1 (0)
7.	Congo	2 (0)	23.	Kenya	1 (0)
8.	Democratic Republic of Congo	2 (0)	24.	Lesotho	1 (0)
9.	Ethiopia	2 (0)	25.	Mozambique	1 (0)
10.	Sudan	2 (0)	26.	Nigeria	1 (0)
11.	Zimbabwe	2 (0)	27.	Rwanda	1 (0)
12.	Angola	1 (0)	28.	Senegal	1 (0)
13.	Benin	1 (0)	29.	Sierra Leone	1 (0)
14.	Burundi	1 (0)	30.	South Africa	1 (0)
15.	The Central Africa Republic	1 (0)	31.	Togo	1 (0)
16.	Comoro Island	1 (0)			

Note: A - Country, B - Number of naturally occurring species (species without documented distributions in other nations).
Source: (Ohrnberger, 1999; Akinlabi et al., 2017).

According to the world bamboo resources assessment report, Ethiopia, Kenya, and Uganda possess most of the bamboo resources in Africa (Lobovikov et al., 2007; Zhao et al., 2018). Ethiopia contributes the lion part with two woody bamboo species: the African Alpine Bamboo or highland bamboo (*Yushanealpina* K. Shumann Lin; synonym: *Arundinaria alpina* K. Schumann) and the monotypic genus lowland bamboo (*Oxytenanthera abyssinica* A. Richard) Munro constituting more than 1.44 million hectares (Zhao et al., 2018). This constitutes about 67% of the total area of bamboo on the continent and 7% of the world (Embaye, 2000; Kelbesa et al., 2000). The most predominant bamboo species in the continent; the lowland bamboo (*O. abyssinica*) is also predominant in Ethiopia accounts for 85% of the total national coverage and the rest 15% is covered by highland bamboo (*A. alpina*) (Embaye, 2000; Embaye et al., 2003). The lowland bamboo covers a range of elevation between 540–1,750 m and highland bamboo at a higher elevation above 2,480 m (Zhao et al., 2018; Oumer et al., 2020).

5.4 Importance of Genetic Diversity and Population Structure Study

Assessment of genetic diversity and population structure study is a prerequisite to promote genetic, evolutionary, and functional studies of plants to have a comprehensive understanding of the biology and other key characteristics that enable to properly exploit plant resources in general and bamboo, in particular (Das et al., 2008; Sui et al., 2009; Peng et al., 2013). Genetic diversity and population structure within different populations of a particular species is the main building block for understanding evolutionary and speciation aspects of that species. In addition, genetic diversity is the basis of an organism's ability to adapt to changes in its environment, crucial for the effective conservation, management, and efficient utilization of plant genetic resources and can be affected by many factors (Amos and Harwood, 1998). Geographical (e.g., landscape, latitude, longitude, and altitude) and environmental (e.g., temperature and precipitation) factors affect the genetic diversity and population structure of a species and the individuals among populations (Wellenreuther et al., 2011; Pauls et al., 2013).

5.5 Molecular Markers Used for the Genetic Diversity and Population Structure Study of Bamboos Found in Africa

Genetic markers are important developments in the field of plant breeding and conservation (Kebriyaee et al., 2012). They can be broadly grouped into two categories: classical markers and DNA/molecular markers. Morphological, cytological, and biochemical markers are types of classical markers whereas the molecular markers are short or long DNA fragments or sequences that act as sign or flags (Collard et al., 2005). For the advancement and widely used molecular marker, one of the breakthroughs in biological sciences in general and particularly molecular research, , is the invention of Polymerase Chain Reaction (PCR) that amplifies DNA *in vitro* (Mullis and Faloona, 1987). Since then, molecular marker techniques were grouped into non-PCR based and PCR-based, depending on the requirement of PCR or not to produce different DNA fragments *in vitro*.

DNA based molecular markers are way better than morphological markers as the former are cost-effective, highly informative, not tissue or age-specific, neutral and are not influenced by environmental factors (Kumar et al., 2011; Uchoi et al., 2017). The use of those DNA based markers has given rise to a significant improvement in understanding phylogenetic relationships within the bamboo species with the help of genetic diversity studies, which was previously constricted with limited phenotypic based characteristics (Das et al., 2008; Bhandari et al., 2021). The most practical and most important application of the use of molecular markers in bamboo is doubtlessly the precise identification of bamboo species and genetic variation assessment within species within a short period of time. Taxonomists and others working with bamboo are very much aware with the difficulties associated with identification and genetic variation assessment of bamboo genotypes, which itself is important for effectively selecting superior genotypes.

An ideal DNA marker should be co-dominant, evenly distributed throughout genome, highly reproducible and having ability to detect higher level of polymorphism (Mondini et al., 2009). Some examples of DNA markers that used for genetic diversity study are restriction fragment length polymorphism (RFLP), amplified fragment length polymorphism (AFLP), simple sequence repeats (SSRs), inter simple sequence repeats (ISSRs), and single-nucleotide polymorphisms (SNPs) (Jiang, 2013).

Although many molecular techniques are used for the genetic diversity study of bamboos globally, only few techniques are tested for the genetic diversity, population structure, and gene flow study of naturally occurring African bamboos. Among molecular techniques used for genetic diversity study of bamboos in the African region, SSR markers (Muhamed, 2019), ISSR markers (Oumer et al., 2020), and cpDNA genes (Oumer et al., 2021) could be listed.

5.6 Simple Sequence Repeat (SSR) Markers for the Genetic Diversity and Population Structure Study of *Arundinaria alpina* (K. Schum)

Simple sequence repeat (SSR) markers which are one of the powerful microsatellite markers in plant biology refer to DNA sequences of 1–6 bp in length that are in tandem repeated a variable number of times (Jurka and Pethiyagoda, 1995). When compared with other markers, SSR markers provides many desirable features including, PCR screening, hyper-variability, multi-allelic nature, co-dominant inheritance, reproducibility, relative abundance, and extensive genome coverage, followed by denaturing gel electrophoresis for allele size determination (Kumar et al., 2009; Manoj et al., 2013). Microsatellites are also fluorescent-labelled and detected (Schuelke, 2000). As a result, SSR were considered as reliable for resolving the morphological issue and for construction of QTL mapping (Jeennor and Volaert, 2014), high-density linkage maps (Sugita et al., 2013), cultivar identification (Ercisli et al., 2011), genetic diversity analysis (Moretzsohn et al., 2004), marker-assisted selection (Sui et al., 2009). Muhamed (2019) is the first to use SSR markers (Table 5.2) for the genetic diversity and population structure study of Ethiopian highland bamboo (*Arundinaria alpina* (K. Schum)) in Ethiopia.

Areas of Ethiopia with abundant highland bamboo resources include Jima (Agaro, Gera), Gore, Bore, Gujji, Hagere Selam, Western Shewa (Ambo, Tikure-Inchini, Shenen, Jibat Mountain), Bale (Harena forest and Shedem Kebele), Western Aris (Degaga, Munesa, Shashemene) in Oromiya regional state, Awi-zone/Inijibara, Gojam/Choke mountains, South Wello (Denkoro forest) and South Gonder (Debere Tabor), Debresina/Wofwasha, and Ankober in Amhara regional state and Sidama, Bonga (Ameya, Wushwush), Bench-Maji, Kefecho Shekech (Ameya, Baha-Chapa, Gada, Gecha-Masha, Andracha, Chencha/Arbaminch, Mizan Teferi/Kulish, Dawaro) and Gurage/Indibir-Jembero in Southern Nations, Nationalities and Peoples Regional State (Desalegn and Tadesse, 2014).

According to the genetic diversity and population structure study using SSR markers by Muhamed (2019) from the 150 Ethiopian highland bamboo individuals representing 15 populations, a total of 49 alleles, ranging from two to four with average of 3.06 alleles per locus were observed and the actual number of alleles for individual populations varied from 2.063 (Agaro) to 2.563 (Debresina and Rira). The highest genetic identity (0.980) was observed between Kosober and Konta and between Agaro and Konta. Observation of a high similarity rate suggested a close relationship among populations. Furthermore, paired comparisons among populations based on locus-by-locus analysis for the 16 microsatellite loci revealed

Table 5.2. SSR primers used for the genetic diversity study of *A. alpina* (Muhamed, 2019).

Repeat Motif	(5'-3')	Tm (°C)	Size (bp)	Product Size (bp)
$(CT)_9$	F: GAGGCTCCTTGGACATCACC	60.11	20	202
	R: CCGACGAATAAGGCAGGCTT	60.46	20	
$(TCC)_5$	F: CGAAGTGGAAGAGGTCGTAG	57.45	20	237
	R: CTTTCCCTTGCCTCCCTTC	57.73	19	
$(AGAGG)_6$	F: CATTCTGTTGATGGCGCTAT	56.29	20	201
	R: CAGGGAGAACGACCAACTAC	57.64	20	
$(GA)_{21}$	F: GGAGCGCAAATCCCATAAAG	57.22	20	243
	R: CTTGCTTCGTTGCCAGTATC	57.19	20	
$(GA)_{14}$	F: TTTGGGAGAGGGATTTTGCT	57.01	20	205
	R: AACTCAGTGCATCAGATCGT	56.95	20	
$(TC)_{10}$	F: ACTGATATTCACCCTGCAGT	56.22	20	212
	R: ACAGTGGCGAAGATGAAGAT	56.93	20	
$(AG)_9$	F: GGCTGCATAGTTCAAAGGAC	56.78	20	218
	R: TGAGCGAACTACCCAAACTT	57.08	20	
$(TC)_6$	F: TTACAATTTGAGGGCCCTGT	57.01	20	206
	R: CATCTGGGTGCTGTTCTAGT	56.93	20	
$(CT)_8$	F: CGAGGAGCCACTGATCTCAC	59.89	20	200
	R: TCTCTTCCATCCCGAACCCT	59.96	20	
$(CCT)_5$	F: GCCTCCTTGAGATCCTCCTT	59.78	20	202
	R: TGCAGCAGCAAGAACAGC	60.01	18	
$(GCG)_6$	F: TTTGGCGTAATCCCTGTAGC	60.01	20	213
	R: TACCACAGCAGCAGCAACAC	61.09	20	
$(GCC)_6$	F: GACAGAAGCGGAAGTTGGAC	59.85	20	190
	R: GAGAAGCAGCAGTGGAGAGC	60.44	20	
$(TCG)_6$	F: GAGGGACTGGATGATGGTGT	59.77	20	202
	R: GAACGAGCCGTTCCAAATAG	59.71	20	
$(GT)_6$	F: CCCTGATGGTAGAAACACCG	60.37	20	214
	R: AAAAATCCAAATGAGCATCAA	57.24	21	
$(GGC)_6$	F: TCGTTCTACATCCCGAGGTC	60.07	20	206
	R: ACTGACCTGAACCGAACACC	60.01	20	
$(CTC)_6$	F: CTGTGACTGTGGATTGGTGG	59.99	20	200
	R: CCTCTGACGCTGGAGCTG	60.88	18	

Note: Tm = melting temperature.

that Fst ranged from 0.000 to 0.265. The value of F-statistical analysis showed that there are no loci that showed complete differentiation as there is no loci with Fst value one. According to F-statistics majority of the loci showed weak genetic differentiation.

On the other hand, the study by Muhamed (2019), SSR based genetic diversity analysis in Ethiopian highland bamboo (*A. alpina*) populations, Debresina population showed the highest genetic diversity while Agaro had the lowest genetic diversity. In terms of population genetic differentiation, Kosober population has the highest genetic differentiation within itself, while Masha population has the lowest genetic differentiation. Both population structure and individual phylogeny merged the 15 highland bamboo populations into nine groups, but individuals of the same population did not cluster together, rather dispersed, implying a higher level of genetic admixture. The authors concluded from genetic diversity analysis that it's understood that the genetic base of the Ethiopian highland bamboo populations is neither wide nor narrow, rather it is moderate. As Kosober population has the highest genetic differentiation, for this population deserves more conservation attention. In general, the study had attempted to investigate the genetic diversity and structure of highland bamboo populations which could be used as a basis for conservation interventions, improvement programs, and/or for further investigations.

5.7 Inter Simple Sequence Repeat (ISSR) Markers for the Genetic Diversity and Population Structure Study of *Oxytenanthera abyssinica* (A. Rich.)

The application of ISSR markers in plants' genetic diversity study and in designing a conservation strategy is well recognized with better performance than morphological, cytological, and biochemical methods as well as better reproducibility and polymorphism than RAPD and RFLP techniques (Tesfaye et al., 2014). The ISSR markers are used to study the genetic diversity of different species of plants either in combination with other markers or solely. But the use of ISSR markers in bamboo in general and particularly in *O. abyssinica* is limited or none. These markers have been in used to study various crops and cereals of African origin but there have been few numbers of research on trees and forest plants (Oumer et al., 2020). Oumer et al. (2020) is the first to use 19 ISSR markers (Table 5.3) after the initial screening of 108 primers for the genetic diversity and population structure study of 130 individuals representing 13 populations (Table 5.4) of lowland bamboo (*Oxytenanthera abyssinica* (A. Rich.) Munro) from natural bamboo growing areas in Ethiopia.

There was no research practice made on bamboos using molecular markers in general and particularly ISSR markers in the African continent until Oumer et al. (2020) broke the ice. Oumer et al. (2020) to study the genetic diversity, population structure, and gene flow analysis of *O. abyssinica* on the naturally grown 130 individuals representing 13 populations (Table 5.4) in Ethiopia using ISSR primers performed the heterozygosity, level of polymorphism, marker efficiency, Nei's gene diversity (H), and Shannon's information index (I) analysis, analysis of molecular variance (AMOVA), analysis for cluster, principal coordinates (PCoA), and admixture analyses.

The bands generated by the ISSR primers were clear and polymorphic. Representative gel image of UBC-834, UBC-840, and UBC-888 are described in Fig. 5.3a–c.

Table 5.3. ISSR primers used for the genetic diversity study of *O. abyssinica* (Oumer et al., 2020).

No.	Primer Name	Nucleotide Sequence	Repeat Motifs	Anchorage Property	T_a (°C)
1.	UBC810	$(GA)_8T$	Dinucleotide	3'-anchored	42
2.	UBC812	$(GA)_8A$	,,	,,	42
3.	UBC815	$(CT)_8G$	,,	,,	42
4.	UBC824	$(TC)_8G$	,,	,,	43
5.	UBC834	$(AG)_8YT$	,,	,,	43
6.	UBC835	$(AG)_8YC$	,,	,,	45
7.	UBC840	$(GA)_8YT$	,,	,,	42
8.	UBC841	$(GA)_8YC$	,,	,,	43
9.	UBC844	$(CT)_8RC$	,,	,,	43
10.	UBC848	$(CA)_8RG$	,,	,,	47
11.	UBC861	$(ACC)_6$	Tri	Unanchored	55
12.	UBC864	$(ATG)_6$	,,	,,	39
13.	UBC873	$(GACA)_4$	Tetra	,,	42
14.	UBC876	$(GATA)_2(GACA)_2$	,,	,,	39
15.	UBC880	$(GGAGA)_3$	Penta	,,	45
16.	UBC881	$(GGGTG)_3$	,,	,,	47
17.	UBC888	$BDB(CA)_7$	5'-anchored	5'-anchored	47
18.	UBC889	$DBD(AC)_7$	,,	,,	45
19.	ISSR_2	$(GACAC)_4$	Penta	,,	55

Note: Y = C or T → (PYramidine), R = A or G → (PuRine), B = C, G, or T → (not A), D = A, G, or T → (not C), T_a = annealing temperature.

Table 5.4. Samples of ISSR marker investigated *O. abyssinica* population collection sites along with their GPS location (Oumer et al., 2020).

Region	Zone	District	Specific Collection Site	GPS Reading		Altitude m.a.s.l
				X	Y	
Beneshangul-Gumuz	Metekel	Guba	Yarenja	11° 16' 13.1"	035° 22' 15.4"	824
		Dangur	Misreta	11° 18' 50.3"	036° 14' 10.6"	1240
		Mandura	Etsitsa	11° 09' 14.5"	036° 19' 50.3"	1039
		Pawe/Almu	Mender 30	11° 18' 32.5"	036° 24' 40.2"	1118
	Assosa	Bambasi	Ambesa Chaka	09° 53' 55.0"	034° 40' 01.8"	1518
		Assosa	Tsetse Adurnunu	10° 09' 29.9"	034° 31' 37.1"	1507
		Kurmuk	HorAzab	10° 32' 33.7"	034° 28' 57.9"	1275
	Kemash	Kemash	Kemash	09° 29' 31.4"	035° 52' 35.2"	1234
		Yasso	Dangacho	09° 52' 27.5"	036° 05' 32.6"	1176
Oromia	West Wollega	Gimbi	Aba Sena Forest	09° 01' 32.2"	035° 59' 54.1"	1407
	BunoBedele	DabuHena	Didhessa Valley	08° 40' 21.1"	036° 23' 32.9"	1399
Gambella	Gambella	Abol	Penkwe	08° 14' 13.1"	034° 31' 06.2"	435
SNNPs	Konta	Konta	Koyshe	06° 43' 35.6"	036° 34' 26.8"	958

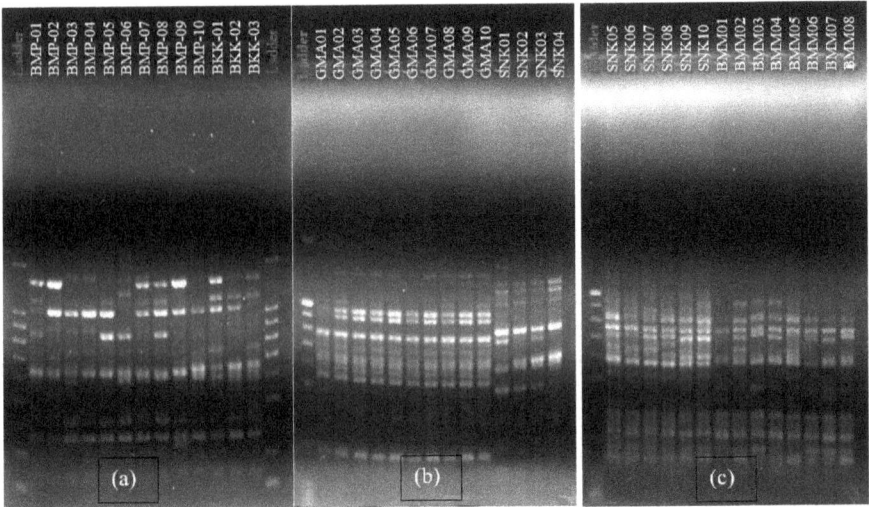

Fig. 5.3. A representative of ISSR-PCR electrophoresis profile of O. abyssinica using (a) UBC-834, (b) UBC-845, and (c) UBC-888 (Oumer et al., 2020).

The results indicated high genetic variation (84.48%) at species level. The H, I, observed and effective number of alleles at the species level were 0.2702, 0.4061, 1.8448, and 1.4744, respectively, suggesting a relatively high level of genetic diversity. However, genetic differentiation at the population level was relatively low (Table 5.6). On the other hand, AMOVA, using grouped populations, revealed that most (61.05%) of the diversity was distributed within the populations with $F_{ST} = 0.38949$, $F_{SC} = 0.10486$, and $F_{CT} = 0.31797$ (Table 5.5).

Furthermore, cluster analysis grouped the populations into markedly distinct clusters, suggesting confined propagation in distinct geographic regions. STRUCTURE analyses showed K = 2 for all populations and K = 11 excluding

Table 5.5. Analysis of molecular variance (AMOVA) for the 130 individuals of the 13 populations of with seven administrative Zonal groups and without grouping *O. abyssinica* (Oumer et al., 2020).

	Source of Variation	d.f.	Sum of Squares	Variance Components	Percentage of Variation
With seven Zonal grouping	Among Groups	6	2124.524	16.38900 Va	31.80
	Among Populations within Groups	6	409.983	3.68630 Vb	7.15
	Within Populations	117	3681.700	31.46752 Vc	61.05
	Total	129	6216.208	51.54282	
Fixation Indices: $F_{SC} = 0.10486$ $F_{ST} = 0.38949$ $F_{CT} = 0.31797$					
Without grouping	Among Populations	12	2534.508	17.97415 Va	36.35
	Within Populations	117	3681.700	31.46752 Vb	63.65
	Total	129	6216.208	49.44167	
Fixation Index $F_{ST} = 0.36354$					

Table 5.6. Genetic diversity within populations and genetic differentiation parameters of 13 populations of *O. abyssinica* (Oumer et al., 2020).

Region-Zone	Code-Population	NPL	PPL (%)	(H ± SD)	(I ± SD)	Na	Ne
Gambella-Gambella	GGAM-Abol	143	41.09	0.1978 ± 0.2390	0.2767 ± 0.3336	1.4109 ± 0.4927	1.3857 ± 0.4686
SNNPs-Konta Special Wereda	SNNPs-Koyshe	153	43.97	0.2102 ± 0.2408	0.2942 ± 0.3359	1.4397 ± 0.4971	1.4097 ± 0.4730
B-Gumuz-Metekel Zone	BGM-Mandura	141	40.52	0.1842 ± 0.2285	0.2611 ± 0.3214	1.4052 ± 0.4916	1.3484 ± 0.4408
	BGM-Dangur	155	44.54	0.2112 ± 0.2400	0.2962 ± 0.3349	1.4454 ± 0.4977	1.4102 ± 0.3349
	BGM-Guba	164	47.13	0.2193 ± 0.2383	0.3088 ± 0.3330	1.4713 ± 0.4999	1.4224 ± 0.4661
	BGM-Pawe	155	44.54	0.2078 ± 0.2366	0.2927 ± 0.3310	1.4454 ± 0.4977	1.3996 ± 0.4622
	Zone Mean	**153.75**	**44.1825**	**0.2056 ± 0.2358**	**0.2897 ± 0.3301**	**1.4418 ± 0.4967**	**1.3952 ± 0.4260**
B-Gumuz-Kemash Zone	BGK-Kemash	155	44.54	0.2129 ± 0.2411	0.2981 ± 0.3363	1.4454 ± 0.4977	1.4150 ± 0.4735
	BGK-Yasso	143	41.09	0.1922 ± 0.2342	0.2705 ± 0.3281	1.4109 ± 0.4927	1.3690 ± 0.4543
	Zone Mean	**149**	**42.815**	**0.2025 ± 0.2376**	**0.2843 ± 0.3322**	**1.4282 ± 0.4952**	**1.3920 ± 0.4639**
B-Gumuz-Assosa Zone	BGA-Assosa	147	42.24	0.2040 ± 0.2407	0.2851 ± 0.3356	1.4224 ± 0.4947	1.3987 ± 0.4730
	BGA-Bambasi	156	44.83	0.2056 ± 0.2353	0.2902 ± 0.3292	1.4483 ± 0.4980	1.3937 ± 0.4583
	BGA-Kurmuk	146	41.95	0.1962 ± 0.2353	0.2761 ± 0.3293	1.4195 ± 0.4942	1.3773 ± 0.4571
	Zone Mean	**149.67**	**43.01**	**0.2019 ± 0.2371**	**0.2838 ± 0.3314**	**1.4301 ± 0.4956**	**1.3899 ± 0.4628**
Oromia-West Wellega	ORWW-Gimbi	148	42.53	0.2023 ± 0.2399	0.2842 ± 0.3344	1.4253 ± 0.4951	1.3961 ± 0.4717
Oromia-Buno-Bedelle	ORBB-Dabu Hena	158	45.40	0.2165 ± 0.2407	0.3035 ± 0.3360	1.4540 ± 0.4986	1.4207 ± 0.4721
	Mean of Populations	**151.077**	**43.4131**	**0.2046 ± 0.2377**	**0.2875 ± 0.3322**	**1.4341 ± 0.4960**	**1.3959 ± 0.4543**
	Overall for Species	**294**	**84.48**	**0.2702 ± 0.1945**	**0.4061 ± 0.2595**	**1.8448 ± 0.3626**	**1.4744 ± 0.3960**
Summary of Genic Variation Statistics for All Loci Total and (Mean ± SD)	Ht	0.2708 ± 0.0379					
	Hs	0.2047 ± 0.0327					
	Gst	0.2442					
	Nm	1.5474					

Abbreviations: NPL: Number of polymorphic loci, PPL: percentage of polymorphic loci, H: Nei's gene diversity, I: Shannon's information index, na: observed number of alleles, ne: effective number of alleles, Ht: total genetic diversity, Hs: genetic diversity within populations, Gst: the coefficient of gene differentiation and Nm: estimate of gene flow among populations.

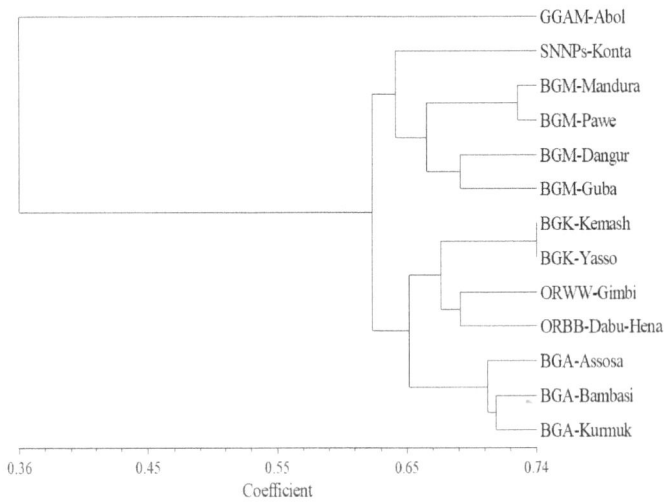

Fig. 5.4. UPGMA based dendrogram for 13 *O. abyssinica* populations based on Jaccard's similarity coefficient using 19 ISSR primers (Oumer et al., 2020).

Gambella population. Using these markers dendrogram construction methods of UPGMA (Fig. 5.4) and NJ (Fig. 5.5), researchers found strong evidence that the genetic diversity of the lowland bamboo is associated with distinct geographic regions and that isolates of the Gambella Region, with their unique genetic origin, are quite different from other bamboos found in the country.

5.8　Chloroplast (cpDNA) Genes for the Population Genetic Diversity and Structure Study of *Oxytenanthera abyssinica* (A. Rich.)

Most molecular phylogenetics and genetic diversity studies on bamboos have utilized nuclear and/or plastid DNA sequences. Chloroplast or plastid is a cellular organelle resulting from endosymbiosis between independent living cyanobacteria and a non-photosynthetic host ~ 1.5 billion years ago (Dyall et al., 2004). Each has its own genome that is usually non-recombinant and uniparentally inherited (Birky, 1995). Chloroplast genome of highest plant have conserved quadripartite structure, consisting of two copies of a large inverted repeat (IR), and two sections of unique DNA, which are referred to as the "large single copy regions (LSC)" and "small single copy regions (SSC)" (Jansen et al., 2005).

　　Chloroplast DNA sequence and its characteristic for population genetics study is well known and strongly preferable due to its relatively small size (compared to mitochondria and nuclear genome (Wagner, 1992)) and conserved region property (Kim et al., 1999; Chokthaweepanich, 2014; Attigala, 2017). In its evolution, it is assumed to be conserved in terms of nucleotide substitution with very few rearrangements that allow the molecule to be used to solve phylogenetic relations and taxonomic anomalies, especially at deep levels of evolution (Kim et al., 1999).

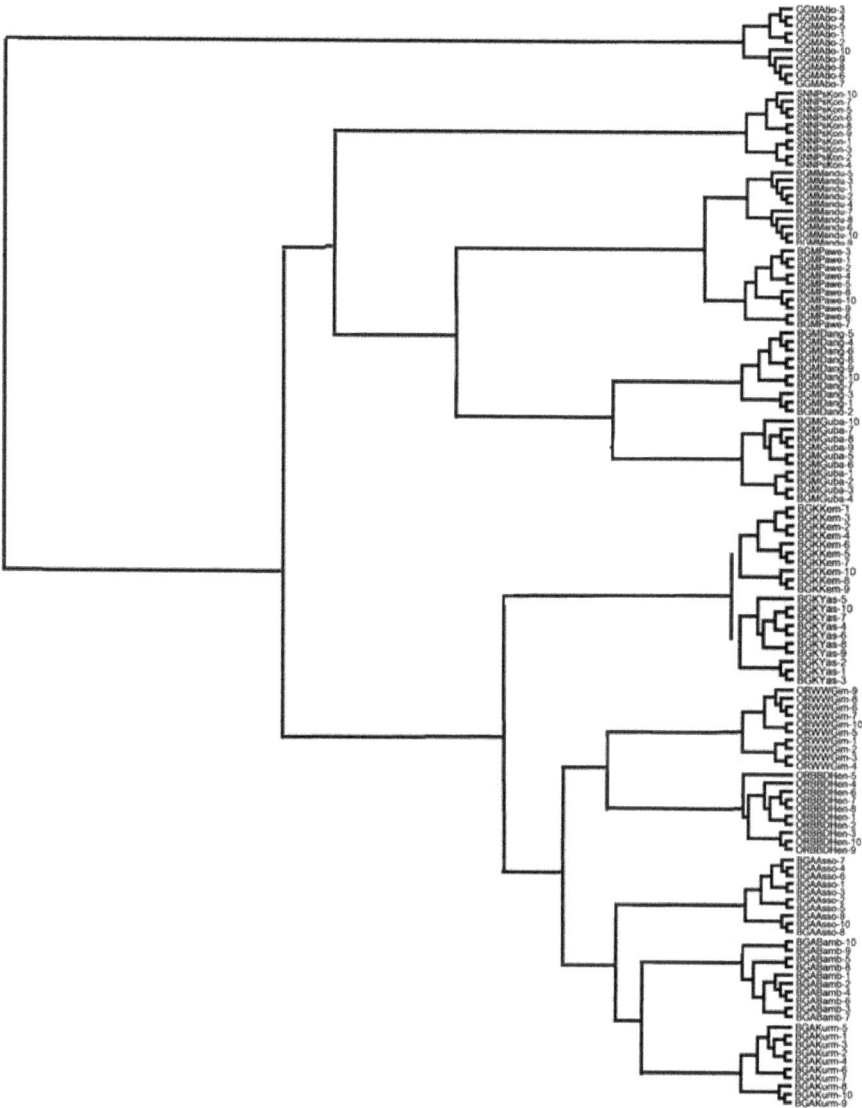

Fig. 5.5. The neighbour joining (NJ) analysis of 130 individuals of *O. abyssinica* based on Jaccard's similarity coefficient (Oumer et al., 2020).

The focus of plant molecular systematics has shifted towards to more rapid development of cpDNA loci. Simple genetics property, high copy within typical genomic DNA, and structural stability makes cpDNA preferable for evolutionary and population genetics studies in many plants since it has protein-coding genes (*matK, ndhF*, etc.), introns (*trnL, rpll6, rps16*, etc.) and intergenic spacer (intergenic spacers between *rbcL* and *atpB, trnT* and *trnL, trnL* and *trnF*) regions (Scarcelli et al., 2011; Daniell et al., 2016). As the evolutionary rates of cpDNA with nucleotide substitutions and structural changes are highly conserved (Soltis et al., 2004), both coding and non-coding cpDNA regions are used to elucidate genetic diversity and population structure, the evolutionary relationships at the generic and higher taxonomic levels (Shaw et al., 2007; Shaw et al., 2014). Some of these cpDNA markers have been used on crops and trees of African origin. Oumer et al. (2021) is the first to use cpDNA markers for the genetic diversity and population structure study of Ethiopian lowland bamboo [*Oxytenanthera abyssinica* (A. Rich.) Munro] from 13 natural bamboo growing areas in Ethiopia.

One of the greatly available bamboo species in Africa in general and particularly in Ethiopia is *Oxytenanthera abyssinica* (A. Rich.). The plant is widely distributed in the continent and Ethiopia could be described as the depot of the species. It accounts 85% of bamboo in the country and 67% of the continent. According to the study by Oumer et al. (2021) using the pair-end sequenced PCR product of selected cpDNA genes (Fig. 5.6) on *O. abyssinica* population in Ethiopia, the GC and AT content analysis showed that the highest AT content (66.6%) was observed in

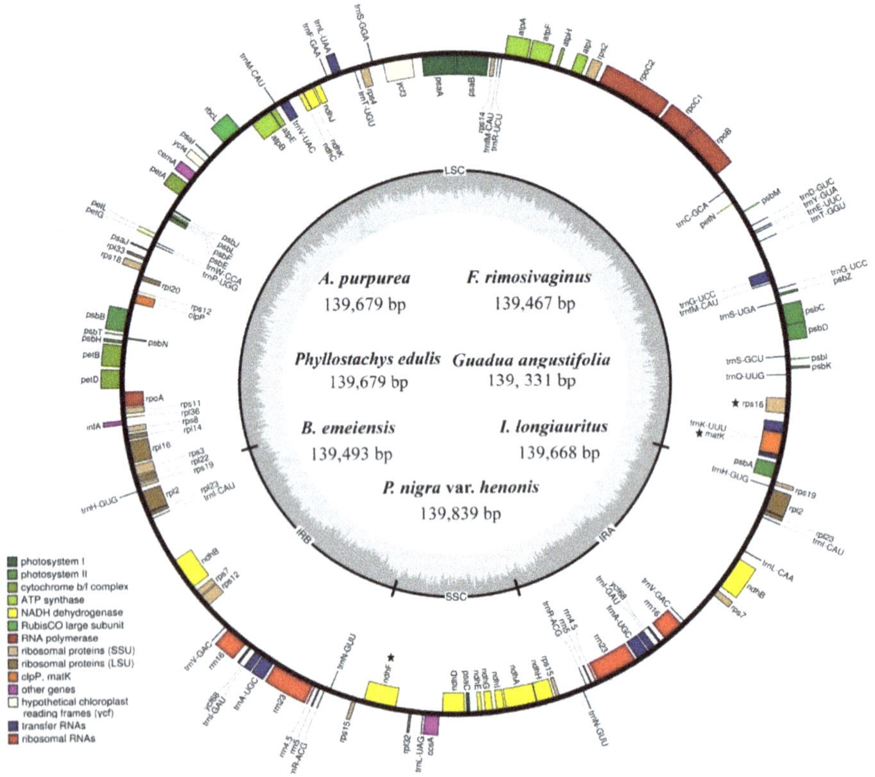

Fig. 5.6. Schematic gene map representation of some woody bamboo chloroplast genomes (Zhang et al., 2011; Wu et al., 2015). Inner thick arcs represent the inverted repeat regions (IRA; IRB), the large and small single copy (LSC; SSC) region. Genes inside and outside of the circle are transcribed clockwise and counterclockwise and colored according to their functional groups. *: genes used for the genetic diversity and population structure study of *O. abyssinica* by Oumer et al. (2021).

Bambasi, Kurmuk, and Pawe populations and average GC and AT content and total sequence were 66.5%, 33.5%, and 3,658.6 in the aggregate cpDNA genes. Since the comparison was between different populations of a single species, there was no significant difference on GC and AT content observed and the result of each sample was close to the average.

On the other hand, the nucleotide diversity analysis showed that Metekel Zone found extremely higher average number of nucleotide differences, nucleotide diversity, and population mutation rates per 100 sites. Furthermore, higher frequency of InDel makes Metekel, the distant population (samples from Gambella and South nations and nationalities and peoples Region), and Assosa Zone populations more diverse than other populations. This confirmed the significant role of InDel in genetic differentiation and population structure of *O. abyssinica* populations (Oumer et al., 2021).

Oumer et al. (2021) also found the Metekel Zone (Fig. 5.7) as the most diverse population, and authors recommended to conservation of lowland bamboo of the Metekel Zone. The Assosa Zone of Bambasi and Kurmuk populations has found the source of evolution of *O. abyssinica*, and the Gambella population shows a difference from other *O. abyssinica* populations found in Ethiopia. Furthermore, the authors indicate the possible existence of extra species of bamboo in the Gambella region. Additionally, they suggest additional work using multiple and advanced techniques on the species since their study was performed only by few chloroplast coding and non-coding genes.

Oumer et al. (2021) concluded that populations of *O. abyssinica* collected in Ethiopia show clear diversity based on their geographic location (Fig. 5.8). Sample collection sites that have relatively closer distance such as Assosa, Kemash, and Oromia, showed smaller nucleotide diversity, InDel, genetic differentiation, and DNA divergence but higher gene flow between populations.

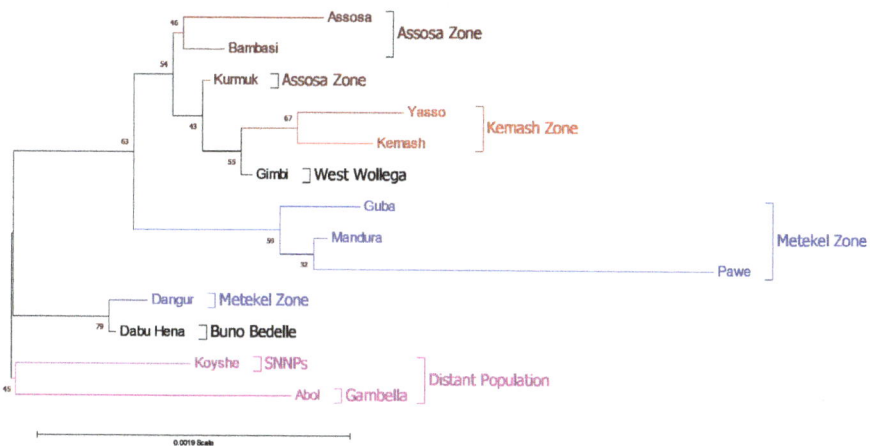

Fig. 5.7. The NJ tree of *O. abyssinica* obtained based on pair-wise distance from the combined data of matK, ndhF, and rps16 (Oumer et al., 2021).

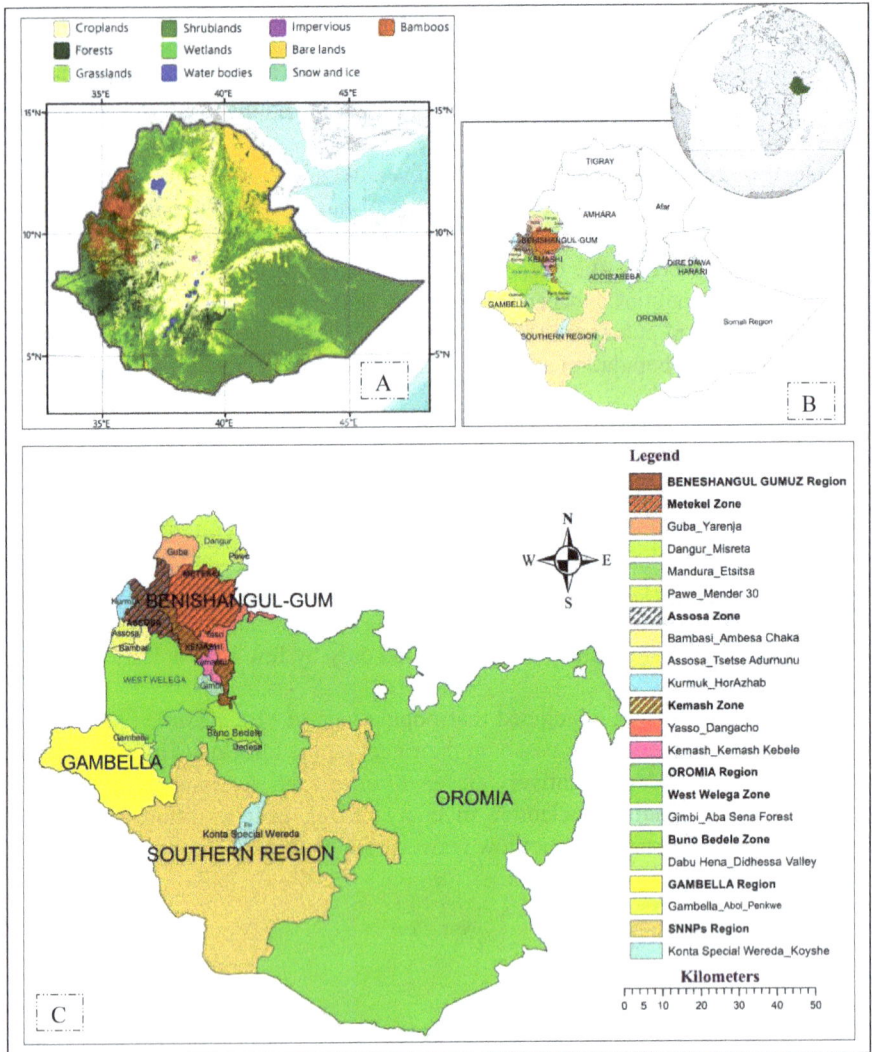

Fig. 5.8. Maps showing bamboo distribution and sample collection area (A) Ethiopia's bamboo cover map in the Finer Resolution Observation and Monitoring - Global Land Cover (FROM-GLC) classification scheme with other land cover classes, (B) Map of Ethiopia showing sample collection area, and (C) Clipped map showing sample collection area (Oumer et al., 2021).

5.9 Conclusions

Africa has naturally a great bamboo potential and resource. Recently, the study of these resources has started studying by different researchers with different disciplines. Additionally, many species from other continents are also being introduced and tested and multiplied in multiple geographical conditions. Many African countries in general and Ethiopia, Ghana, Cameroon, Kenya, Uganda, etc., apart from having ample resource of naturally grown bamboo plants, are introducing, testing, and multiplying multiple bamboo species in different geographical conditions. As of the trials from Ethiopia, Ghana, and Kenya; many of the species have adopted African soil and environment (Mulatu et al., 2016).

Despite the availability of wide variety of bamboo in, very little is known about the population genetic diversity, structure, and gene flow analysis of bamboo', especially in growing in natural forests. Although nowadays many young African researchers are trying to investigate the issues of bamboo in different disciplines, but limited expertise, lack of facilities and research funds pulls back the molecular-based studies of bamboo in the continent.

Although many molecular marker techniques are used for the genetic diversity study of bamboos globally, only a few techniques are tested for the genetic diversity, population structure, and gene flow study of naturally occurring African bamboos. Molecular markers tested for the genetic diversity, population structure, and gene flow analysis from bamboos of African origin shows clear geographic distance-based genetic diversity and structure. Despite Oumer et al. (2020) by using ISSR markers and Oumer et al. (2021) by using cpDNA genes described the possible existence of additional bamboo species in Ethiopia, it has not been proved whether the species exists in any other country or there is an additional species resource to the continent. Additionally, Muhamed (2019) also found different morphological varieties of highland bamboo that have not been properly investigated for their molecular characteristics. Thus, critical investigations and studies using advanced molecular markers are mandatory for bamboos found in the continent for conservation of the natural population as well as to fill the gaps in the area.

References

Akinlabi, E.T., Anane-Fenin, K. and Akwada, D.R. (2017). *Bamboo: The Multipurpose Plant*. Springer, Switzerland.

Amos, W. and Harwood, J. (1998). Factors affecting levels of genetic diversity in natural populations. *Philosophical Transactions of the Royal Society of London. Series B: Biological Sciences*, 353(1366): 177–186.

Andriarimalala, J.H., Kpomasse, C.C., Salgado, P., Ralisoa, N. and Durai, J. (2019). Nutritional potential of bamboo leaves for feeding dairy cattle. *Pesquisa Agropecuária Tropical*, 49.

Attigala, L., Gallaher, T., Nason, J. and Clark, L.G. (2017). Genetic diversity and population structure of the threatened temperate woody bamboo Kuruna debilis (Poaceae: Bambusoideae: Arundinarieae) from Sri Lanka based on microsatellite analysis. *Journal of the National Science Foundation of Sri Lanka*, 45(1): 53.

Bamboo Phylogeny Group (BPG). (2012). An updated tribal and subtribal classification of the bamboos (Poaceae: Bambusoideae). *The Journal of the American Bamboo Society*, 24(1): 1–10.

Bhandari, S., Tyagi, K., Singh, B. and Goutam, U. (2021). Role of molecular markers to study genetic diversity in Bamboo: A Review. *Plant Cell Biotechnol. Mol. Biol.*, 22: 86–97.

Birky, C.W. (1995). Uniparental inheritance of mitochondrial and chloroplast genes: Mechanisms and evolution. *Proceedings of the National Academy of Sciences*, 92(25): 11331–11338.

Boban, S., Maurya, S. and Jha, Z. (2022). DNA fingerprinting: An overview on genetic diversity studies in the botanical taxa of Indian Bamboo. *Genetic Resources and Crop Evolution*, 1–30.

Bystriakova, N., Kapos, V. and Lysenko, I. (2004). *Bamboo Biodiversity*. UNEP-WCMC/INBAR.

Chen, R.Y., Li, X.L., Song, W.Q., Liang, G.L., Zhang, P.X., Lin, R.S., Zong, W.X., Chen, C.B. and Feng, X.L. (2003). Chromosome atlas of various bamboo species. *In: Chromosome atlas of major economic plants genome in China IV*.

Chokthaweepanich, H. (2014). *Phylogenetics and Evolution of the Paleotropical Woody Bamboos (Poaceae: Bambusoideae: Bambuseae*. Iowa State University.

Choudhary, A.K., Kumar, S., Patil, B.S., Sharma, M., Kemal, S., Ontagodi, T.P., Datta, S., Patil, P., Chaturvedi, S.K., Sultana, R. and Hegde, V.S. (2013). Narrowing yield gaps through genetic improvement for Fusarium wilt resistance in three pulse crops of the semi-arid tropics. *SABRAO Journal of Breeding and Genetics*, 45(03): 341–370.

Clark, L.G., Davidse, G. and Ellis, R.P. (2010). Natural hybridization in bamboos: Evidence from Chusquea Sect. Swallenochloa (Poaceae: Bambusoideae). Hibridación natural en bambú: Evidencia de Chusquea Sección Swallenochloa (Poaceae: Bambusoideae). *National Geographical Research*, 5(4): 459–476.

Clark, L.G., Dransfield, S., Triplett, J. and Sánchez-Ken, J.G. (2007). Phylogenetic relationships among the one-flowered, determinate genera of Bambuseae (Poaceae: Bambusoideae). *Aliso: A Journal of Systematic and Floristic Botany*, 23(1): 315–332.

Collard, B.C., Jahufer, M.Z.Z., Brouwer, J.B. and Pang, E.C.K. (2005). An introduction to markers, quantitative trait loci (QTL) mapping and marker-assisted selection for crop improvement: The basic concepts. *Euphytica*, 142(1): 169–196.

Dalagnol, R., Wagner, F.H., Galvão, L.S. and Nelson, B.W. (2018). Life cycle of bamboo in the southwestern Amazon and its relation to fire events. *Biogeosciences*, 15(20): 6087–6104.

Daniell, H., Lin, C.S., Yu, M. and Chang, W.J. (2016). Chloroplast genomes: Diversity, evolution, and applications in genetic engineering. *Genome Biology*, 17(1): 1–29.

Das, M., Bhattacharya, S., Singh, P., Filgueiras, T.S. and Pal, A. (2008). Bamboo taxonomy and diversity in the era of molecular markers. *Advances in Botanical Research*, 47: 225–268.

Dereso, Y. (2019). Regeneration study of lowland bamboo (*Oxytenanthera abyssinica* AR Munro) in Mandura district, Northwest Ethiopia. *Biodiversity International Journal*, 3(1): 18–26.

Desalegn, G. and Tadesse, W. (2014). Resource potential of bamboo, challenges, and future directions towards sustainable management and utilization in Ethiopia. *Forest Systems*, 23(2): 294–299.

Devi, S.T. and Sharma, G.S. (1993). Chromosome numbers in some bamboo species of Manipur. *BIC India Bulletin*, 3(1): 16–21.

Diver, S. (2006). Bamboo. *In: A Multipurpose Agroforestry Crop*. ATTRA, Appropriate Technology Transfer for Rural Areas.

Dyall, S.D., Brown, M.T. and Johnson, P.J. (2004). Ancient invasions: From endosymbionts to organelles. *Science*, 304(5668): 253–257.

Ekhuemelo, D.O., Tembe, E.T. and Ugwueze, F.A. (2018). Bamboo: A potential alternative to wood and wood products. *South Asian Journal of Biological Research*, 1(1): 9–24.

Embaye, K. (2000). The indigenous bamboo forests of Ethiopia: An overview. *AMBIO: A Journal of the Human Environment*, 29(8): 518–521.

Embaye, K., Christersson, L., Ledin, S. and Weih, M. (2003). Bamboo as bioresource in Ethiopia: Management strategy to improve seedling performance (Oxytenanthera abyssinica). *Bioresource Technology*, 88(1): 33–39.

Ercisli, S., Ipek, A. and Barut, E. (2011). SSR marker-based DNA fingerprinting and cultivar identification of olives (Olea europaea). *Biochemical Genetics*, 49(9): 555–561.

Ganal, M.W. and Röder, M.S. (2007). Microsatellite and SNP markers in wheat breeding. *In: Genomics-assisted Crop Improvement*. Springer, Dordrecht, pp. 1–24.

Gielis, J. (1997). Genetic variability and relationships in Phyllostachys using random amplified polymorphic DNA. *In: The Bamboos: Linn. Soc. Symp. Ser.* Vol. 19, pp. 107–124.

Grass Phylogeny Working Group, Barker, N.P., Clark, L.G., Davis, J.I., Duvall, M.R., Guala, G.F., Hsiao, C., Kellogg, E.A., Linder, H.P., Mason-Gamer, R.J. and Mathews, S.Y. (2001). Phylogeny and subfamilial classification of the grasses (Poaceae). *Annals of the Missouri Botanical Garden*, 373–457.

Gui, Y., Wang, S., Quan, L., Zhou, C., Long, S., Zheng, H., Jin, L., Zhang, X., Ma, N. and Fan, L. (2007). Genome size and sequence composition of Moso bamboo: A comparative study. *Science in China Series C: Life Sciences*, 50(5): 700–705.

Halvorson, J.J., Cassida, K.A., Turner, K.E. and Belesky, D.P. (2011). Nutritive value of bamboo as browse for livestock. *Renewable Agriculture and Food Systems*, 26(2): 161–170.

Huy, B. and Trinh, T.L. (2019). A manual for bamboo forest biomass and carbon assessment. *International Bamboo and Rattan Organization, Beijing, China.*

International Network for Bamboo and Rattan (INBAR). (2019). *Bamboo and Rattan Commodities in the International Market.*

Jansen, R.K., Raubeson, L.A., Boore, J.L., Depamphilis, C.W., Chumley, T.W., Haberle, R.C., Wyman, S.K., Alverson, A.J., Peery, R., Herman, S.J. and Fourcade, H.M. (2005). Methods for obtaining and analyzing whole chloroplast genome sequences. *In: Methods in Enzymology.* Vol. 395. Academic Press, U.S.A., pp. 348–384.

Jeennor, S. and Volkaert, H. (2014). Mapping of quantitative trait loci (QTLs) for oil yield using SSRs and gene-based markers in African oil palm (Elaeis guineensis Jacq.). *Tree Genetics & Genomes,* 10(1): 1–14.

Jiang, G.L. (2013). Molecular markers and marker-assisted breeding in plants. *Plant Breeding from Laboratories to Fields,* 3: 45–83.

Jurka, J. and Pethiyagoda, C. (1995). Simple repetitive DNA sequences from primates: Compilation and analysis. *Journal of Molecular Evolution,* 40(2): 120–126.

Kassahun, T. (2014). Review of bamboo value chain in Ethiopia. *International Journal of African Society Culture and Traditions,* 2(3): 52–67.

Kaushal, R., Kumar, A., Jayaraman, D., Mandal, D., Rao, I.V.R., Dogra, P. and Mishra, P.K. (2018). Research methodologies for field monitoring, analysis, and evaluation of resource conservation aspects of bamboos. *INBAR,* 9(4): 124.

Kaushal, R., Tewari, S., Banik, R.L., Thapliyal, S.D., Singh, I., Reza, S. and Durai, J. (2020). Root distribution and soil properties under 12-year-old sympodial bamboo plantation in Central Himalayan Tarai Region, India. *Agroforestry Systems,* 94(3): 917–932.

Kebriyaee, D., Kordrostami, M., Rezadoost, M.H. and Lahiji, H.S. (2012). QTL analysis of agronomic traits in rice using SSR and AFLP markers. *Notulae Scientia Biologicae,* 4(2): 116–123.

Kelbessa, E., Bekele, T., Gebrehiwot, A., Hadera, G. and Ababa, A. (2000). A socioeconomic case study of the Bamboo sector in Ethiopia. *An Analysis of the Production-to-Consumption System.* Addis Ababa, Ethiopia.

Kelchner, S.A. and Bamboo Phylogenetic Group. (2013). Higher level phylogenetic relationships within the bamboos (Poaceae: Bambusoideae) based on five plastid markers. *Molecular Phylogenetics and Evolution,* 67(2): 404–413.

Kigomo, B.N. and Kamiri, J.F. (1985). Observations on the growth and yield of Oxytenanthera abyssinica (A. Rich) Munro in plantation. *East African Agricultural and Forestry Journal,* 51(1): 22–29.

Kim, S.C., Crawford, D.J., Jansen, R.K. and Santos-Guerra, A. (1999). The use of a non-coding region of chloroplast DNA in phylogenetic studies of the subtribeSonchinae (Asteraceae: Lactuceae). *Plant Systematics and Evolution,* 215(1): 85–99.

Kindu, M. and Mulatu, Y. (2010). Status of bamboo resource development, utilization, and research in Ethiopia: A review. *Ethiopian Journal of Natural Resources,* 1: 79–98.

Kumar, P.P., Turner, I.M., Nagaraja Rao, A. and Arumuganathan, K. (2011). Estimation of nuclear DNA content of various bamboo and rattan species. *Plant Biotechnology Reports,* 5(4): 317–322.

Kumar, P., Gupta, V.K., Misra, A.K., Modi, D.R. and Pandey, B.K. (2009). Potential of molecular markers in plant biotechnology. *Plant Omics,* 2(4): 141–162.

Li, W., Shi, C., Li, K., Zhang, Q.J., Tong, Y., Zhang, Y., Wang, J., Clark, L. and Gao, L.Z. (2021). Draft genome of the herbaceous bamboo Raddia distichophylla. G3, 11(2): jkaa049.

Liese, W. (2008). Bamboo plantations. *The Two Bamboos of Ethiopia.* University of Mamburg, Germany.

Lobovikov, M., Paudel, S., Ball, L., Piazza, M., Guardia, M., Wu, J. and Ren, H. (2007). *World Bamboo Resources: A thematic Study Prepared in the Framework of the Global Forest Resources Assessment 2005* (No. 18). Food & Agriculture Org.

Mathu, A.J., Mathew, P., Mathew, P., Harikumar, D. and Koshy, K. (2015). Cytological study in Pseudoxytenanthera (Tribe Bambuseae) occurring in INDIA. *Journal of Cytological Genetics,* 16(NS): 61–68.

Mayer, K.F., Rogers, J., Doležel, J., Pozniak, C., Eversole, K., Feuillet, C., Gill, B., Friebe, B., Lukaszewski, A.J. and Sourdille, P. (2014). A chromosome-based draft sequence of the hexaploid bread wheat (Triticum aestivum) genome. *Science,* 345(6194): 1251788.

Mondini, L., Noorani, A. and Pagnotta, M.A. (2009). Assessing plant genetic diversity by molecular tools. *Diversity*, 1(1): 19–35.

Moretzsohn, M.D.C., Hopkins, M.S., Mitchell, S.E., Kresovich, S., Valls, J.F.M. and Ferreira, M.E. (2004). Genetic diversity of peanut (Arachis hypogaea L.) and its wild relatives based on the analysis of hypervariable regions of the genome. *BMC Plant Biology*, 4(1): 1–10.

Muhamed, A. (2019). *RNA-Seq Transcriptome Profiling of Ethiopian Lowland Bamboo (Oxytenanthera Abyssinica (A.Rich) Munro Under Drought and Salt Stresses and SSR Based Genetic Diversity Analysis of Ethiopian Highland Bamboo (Arundinaria Alpina K. Schum).* PhD thesis, Addis Ababa University, Addis Ababa, Ethiopia.

Mulatu, Y., Alemayehu, A. and Tadesse, Z. (2016). *Bamboo Species Introduced in Ethiopia: Biological, ecological and management Aspects.* Ethiopian Environment and Forest Research Institute (EEFRI).

Mulatu, Y., Bahiru, T., Kidane, B., Getahun, A. and Belay, A. (2019). Proximate and mineral composition of indigenous bamboo shoots of Ethiopia. *Greener Journal of Agricultural Sciences*, 9(2): 215–221.

Mullis, K.B. and Faloona, F.A. (1987). Specific synthesis of DNA *in vitro* via a polymerase-catalyzed chain reaction. *In: Methods in Enzymology*, Vol. 155. Academic Press, pp. 335–350.

Nongdam, P. and Tikendra, L. (2014). The nutritional facts of bamboo shoots and their usage as important traditional foods of Northeast India. *International Scholarly Research Notices, 2014*.

Ohrnberger, D. (1999). *The Bamboos of the World Annotated Nomenclature and Literature of the Species and the Higher and Lower Taxa.* Elsevier, Amsterdam, p. 585.

Oumer, O.A., Dagne, K., Feyissa, T., Tesfaye, K., Durai, J. and Hyder, M.Z. (2020). Genetic diversity, population structure, and gene flow analysis of lowland bamboo [*Oxytenanthera abyssinica* (A. Rich.) Munro] in Ethiopia. *Ecology and Evolution*, 10(20): 11217–11236.

Oumer, O.A., Tesfaye, K., Feyissa, T., Yibeyen, D., Durai, J. and Hyder, M.Z. (2021). cpDNA-Gene-sequence-based genetic diversity, population structure, and gene flow analysis of Ethiopian Lowland Bamboo (Bambusinea: Oxytenanthera abyssinica (A. Rich.) Munro). *International Journal of Forestry Research, 2021*.

Pauls, S.U., Nowak, C., Bálint, M. and Pfenninger, M. (2013). The impact of global climate change on genetic diversity within populations and species. *Molecular Ecology*, 22(4): 925–946.

Pellicer, J., Hidalgo, O., Dodsworth, S. and Leitch, I.J. (2018). Genome size diversity and its impact on the evolution of land plants. *Genes*, 9(2): 88.

Peng, Z., Lu, Y., Li, L., Zhao, Q., Feng, Q.I., Gao, Z., Lu, H., Hu, T., Yao, N., Liu, K. and Li, Y. (2013). The draft genome of the fast-growing non-timber forest species Moso bamboo (Phyllostachys heterocycla). *Nature Genetics*, 45(4): 456–461.

Pohl, R.W. and Clark, L.G. (1992). New chromosome counts for Chusquea and Aulonemia (Poaceae: Bambusoideae). *American Journal of Botany*, 79(4): 478–480.

Saarela, J.M., Burke, S.V., Wysocki, W.P., Barrett, M.D., Clark, L.G., Craine, J.M., Peterson, P.M., Soreng, R.J., Vorontsova, M.S. and Duvall, M.R. (2018). A 250 plastome phylogeny of the grass family (Poaceae): Topological support under different data partitions. *Peer J*, 6: e4299.

Scarcelli, N., Barnaud, A., Eiserhardt, W., Treier, U.A., Seveno, M., d'Anfray, A., Vigouroux, Y. and Pintaud, J.C. (2011). A set of 100 chloroplast DNA primer pairs to study population genetics and phylogeny in monocotyledons. *PLoS One*, 6(5): e19954.

Schuelke, M. (2000). An economic method for the fluorescent labeling of PCR fragments. *Nature Biotechnology*, 18(2): 233–234.

Scurlock, J.M.O., Dayton, D.C. and Hames, B. (2000). Bamboo: An overlooked biomass resource? *Biomass and Boienergy*, 19: 229–244.

Shahrajabian, M.H., Sun, W., Shen, H. and Cheng, Q. (2020). Chinese herbal medicine for SARS and SARS-CoV-2 treatment and prevention, encouraging using herbal medicine for COVID-19 outbreak. *Acta Agriculturae Scandinavica, Section B—Soil & Plant Science*, 70(5): 437–443.

Shaw, J., Lickey, E.B., Schilling, E.E. and Small, R.L. (2007). Comparison of whole chloroplast genome sequences to choose noncoding regions for phylogenetic studies in angiosperms: The tortoise and the hare III. *American Journal of Botany*, 94(3): 275–288.

Shaw, J., Shafer, H.L., Leonard, O.R., Kovach, M.J., Schorr, M. and Morris, A.B. (2014). Chloroplast DNA sequence utility for the lowest phylogenetic and phylogeographic inferences in angiosperms: The tortoise and the hare IV. *American Journal of Botany*, 101(11): 1987–2004.

Soltis, P.S. and Soltis, D.E. (2004). The origin and diversification of angiosperms. *American Journal of Botany*, 91(10): 1614–1626.

Sugita, T., Semi, Y., Sawada, H., Utoyama, Y., Hosomi, Y., Yoshimoto, E., Maehata, Y., Fukuoka, H., Nagata, R. and Ohyama, A. (2013). Development of simple sequence repeat markers and construction of a high-density linkage map of Capsicum annuum. *Molecular Breeding*, 31(4): 909–920.

Sui, X.X., Wang, M.N. and Chen, X.M. (2009). Molecular mapping of a stripe rust resistance gene in spring wheat cultivar Zak. *Phytopathology*, 99(10): 1209–1215.

Sungkaew, S., Stapleton, C., Salamin, N. and Hodkinson, T.R. (2009). Non-monophyly of the woody bamboos (Bambuseae; Poaceae): A multi-gene region phylogenetic analysis of Bambusoideae ss. *Journal of Plant Research*, 122(1): 95–108.

Tariyal, K. (2016). Bamboo as a successful carbon sequestration substrate in Uttarakhand: A brief analysis. *Int. J. Curr. Adv. Res*, 5: 736–738.

Tesfaye, K., Govers, K., Bekele, E. and Borsch, T. (2014). ISSR fingerprinting of Coffea arabica throughout Ethiopia reveals high variability in wild populations and distinguishes them from landraces. *Plant Systematics and Evolution*, 300(5): 881–897.

Uchoi, A., Malik, S.K., Choudhary, R., Kumar, S., Pal, D., Rohini, M.R. and Chaudhury, R. (2017). Molecular markers in assessing genetic variation of Indian citron (Citrus medica L.) cultivars collected from different parts of India. *Indian J. of Biotechnology*, 16(3): 346–356.

Wagner, D.B. (1992). Nuclear, chloroplast, and mitochondrial DNA polymorphisms as biochemical markers in population genetic analyses of forest trees. *New Forests*, 6(1): 373–390.

Wellenreuther, M., Sanchez-Guillen, R.A., Cordero-Rivera, A., Svensson, E.I. and Hansson, B. (2011). Environmental and climatic determinants of molecular diversity and genetic population structure in a coenagrionid damselfly. *PLoS One*, 6(6): e20440.

Wu, M., Lan, S., Cai, B., Chen, S., Chen, H. and Zhou, S. (2015). The complete chloroplast genome of Guadua angustifolia and comparative analyses of neotropical-paleotropical bamboos. *PloS One*, 10(12): e0143792.

Xu, J. and Zhang, Y. (2020). Traditional Chinese medicine treatment of COVID-19. *Complementary Therapies in Clinical Practice*, 39: 101165.

Yuen, J.Q., Fung, T. and Ziegler, A.D. (2017). Carbon stocks in bamboo ecosystems worldwide: Estimates and uncertainties. *Forest Ecology and Management*, 393: 113–138.

Zhang, Y.J., Ma, P.F. and Li, D.Z. (2011). High-throughput sequencing of six bamboo chloroplast genomes: Phylogenetic implications for temperate woody bamboos (Poaceae: Bambusoideae). *PloS One*, 6(5): e20596.

Zhang, W. (2000). Phylogeny and classification of the Bambusoideae (Poaceae). *In*: Jacobs, S.W.L. and Everett, J. (eds.). *Grasses: Systematics and Evolution*. CSIRO Publishing.

Zhao, Y., Feng, D., Jayaraman, D., Belay, D., Sebrala, H., Ngugi, J., Maina, E., Akombo, R., Otuoma, J., Mutyaba, J., Kissa, S., Qi, S., Assefa, F., Oduor, N.M., Ndawula, A.K., Li, Y. and Gong, P. (2018). Bamboo mapping of Ethiopia, Kenya, and Uganda for the year 2016 using multi-temporal Landsat imagery. *International Journal of Applied Earth Observation and Geoinformation*, 66: 116–125.

Chapter 6

Breeding System, Molecular Genetic Map, and Artificial Hybridization in Woody Bamboos

PeiTong Dou, TiZe Xia and *HanQi Yang**

6.1 Introduction

Woody bamboos (hereinafter referred to as 'bamboos') are one of the most unique taxa with sexual reproductive characteristics among seed plants. The flowering cycle of bamboos is often decades or even 120 to 150 years, and most bamboos die after flowering (Sun et al., 2015; Zheng et al., 2020). Furthermore, the flowering asynchrony usually results in infertility or low seed setting rates under natural conditions (Du et al., 2000; Gu et al., 2012; Xie et al., 2019). The above characteristics not only severely hinder our understanding of the sexual reproductive characteristics and genetic law of bamboo, but also pose a major threat or obstacle to the cultivation, breeding, genetic improvement, and management of bamboo forests (Jiang, 2002). In this chapter, we have summarized the main achievements related to the bamboo breeding system, genetic map, and artificial hybridization in the last two decades. It could not only improve our understanding on the reproductive biological characteristics of bamboos, but also promote the future research on breeding of new varieties, forest management, and sustainable use of bamboo resources.

6.2 The Breeding System of Woody Bamboo

Breeding system is the basis of survival and evolution of seed plants. Meanwhile, bamboo flowering probably gives rise to a mass death, and brings about huge

Institute of Highland Forest Science, Chinese Academy of Forestry, Kunming, Yunnan, China.
Emails: lkyzksdpt@163.com; 1462027635@qq.com
* Corresponding author: yanghanqikm@aliyun.com

economic losses and environmental damage (Zheng et al., 2020). Therefore, the research on the breeding systems and the associated genetic improvement and generation of artificial hybrids have become hot spots in bamboo germplasm conservation and utilization (Jiang, 2002). Botanists are also interested in the flowering and fruiting events of bamboo all over the world. However, because of the long flowering cycle and randomness, the current research progress is far from a comprehensive understanding of the bamboo breeding system. The achievements have mainly focused on reproductive biology, especially the morphology and structure of flower organs, pollen morphology and viability, embryology, gene flow, etc. Here, we review the research progress on the breeding system in bamboos.

6.2.1 Flowering and Fruiting Phenomena of Woody Bamboos

6.2.1.1 Flowering Cycle

Flowering is the transformation signal from vegetative growth to reproductive growth in seed plants. Woody bamboos are the typical perennial plants and are only an arbor-like group in Poaceae occurring in a forest environment (Sun et al., 2015). The existing data suggest that the flowering cycle of bamboos ranges from 1 to more than 100 years depending on the species, and in many instances it ranges from 20–60 years (Table 6.1). Meanwhile, many species, such as the *Bambusa bambos* and *B. polymorpha*, had several records of flowering intervals, indicating the complexity and unpredictability of the flowering cycle within woody bamboos (Table 6.1).

6.2.1.2 Flowering Types

At the early study stage, Brandis (1899) divided the bamboo flowering into three types according to habits observed, namely annual, periodic, and uncertain. However, the applicability of this classification is limited because only the rare bamboo species have been observed and counted. According to the number of flowering clumps in a bamboo forest, Janzen (1976) classified flowering status of bamboos into mass flowering and sporadic flowering. Du et al. (2000) suggested that a bamboo forest was called mass flowering if more than 50% flowering occurs in clumps. Mass flowering is also known as massive synchronized, or population flowering and the proportion of mass flowering is higher in the wild bamboos than that of the cultivated bamboos (Table 6.2). For example, *Melocalamus*, *Cephalostachyum*, *Gigantochloa*, *Indosasa*, and *Ampelocalamus* are characterized by mass flowering. Among the cultivated bamboos, the proportion of bamboos with sporadic flowering is higher, such as *Bambusa*, *Dendrocalamus*, and *Phyllostachys* (Du et al., 2000). In addition, two flowering types seem to have no relevance with the rhizome types. Two typical monopodium (or scattered) bamboos, *Phyllostachys edulis* and *Ph. reticulata*, were observed as mass flowering (Du et al., 2000). Sporadic flowering also occurred in sympodium bamboos, such as *Bambusa tulda* (Bhattacharya et al., 2006), *B. emeiensis* (Zheng et al., 2020), *B. multiplex* (Lin et al., 2015) and *Dendrocalamus sinicus* (Xie et al., 2019), amphipodium bamboos (e.g., *Shibataea chinensis*, Lin and Ding, 2013) as well as monopodium bamboos (e.g., *Phyllostachys vivax*, Zheng et al., 2016; *Pseudosasa amabilis* var. *convexa*, Zhang et al., 2003).

Table 6.1. Flowering interval record of woody bamboos (Data are mainly adopted from Du et al., 2000; Zheng et al., 2020).

Species	Flowering Cycle (years)	References
Ampelocalamus scandens	29	Zhang et al., 1992a
Acidosasa notata	> 50	Xu et al., 2012
Arundinaria fargesii	50–60	Liu and Fu, 2007
A. racemosa	30–31	Jiao, 1956; Janzen, 1976
Bambusa arnhemica	41–51	Franklin, 2010
B. balcooa	32–34	Fan and Qiu, 1987
B. bambos	30–40, 47–52	Kurz, 1876; Bourdillon, 1895; Nicholls, 1895; Brandis, 1899; Brandis, 1906; Troup, 1921; Blatter and Parker, 1929; Blatter, 1930; Dutra, 1938; Ueda, 1960; Mitra and Nayak, 1972; Janzen, 1976; Fu, 1985; Bennet et al., 1990; Tewari, 1993; Seethalakshmi and Muktesh, 1998; Poudyal, 2009
B. copelandii	47–48	Raizada, 1948; McClure, 1966
B. teres	35, 35–60	Fu, 1985; Fan and Qiu, 1987; Poudyal, 2009
B. polymorpha	54–60, > 50, > 68, 80	Kwe, 1904; Bradley, 1914; Seifriz, 1923; Jiao, 1956; Ueda, 1960; Janzen, 1976
B. tuda	48	Ram and Gopal, 1981; Bhattacharya et al., 2006
B. vulgaris	150+	Janzen, 1976
Cephalostachyum pingbianense	46	Du et al., 2000
Chimonobambusa quadrangularis	100+	McClure, 1966
C. utilis	~ 60	Zhang et al., 1994
Chusquea abietifolia	30–34	Seifriz, 1920; Seifriz, 1950; Janzen, 1976
Ch. culeou	12, 14–20, 61–62	Janzen, 1976; Pearson et al., 1994; González and Donoso, 1999; Jaksic and Lima, 2010; Tagle et al., 2013; Guerreiro, 2014
Ch. lorentziana	32	Guerreiro, 2014
Ch. montana	41	Guerreiro, 2014
Ch. quila	15–20, 45	Janzen, 1976; Guerreiro, 2014
Ch. ramosissima	23, 29	Dutra, 1938; Guerreiro, 2014
Ch. tenella	15–16	Dutra, 1938; Guerreiro, 2014
Ch. valdiviensis	50–70	González and Donoso, 1999; Tagle et al., 2013
Dendrocalamus faleatun	28–30	Janzen, 1976; Zhou, 1984; Lin et al., 2009
D. giganteus	40, ~ 76	Janzen, 1976; Lin, 2009
D. hamiltonii	25, 30, 44	Gupta, 1972; Janzen, 1976; Fan and Qiu, 1987
D. hookeri	117	Janzen, 1976

Table 6.1 contd. ...

...Table 6.1 contd.

Species	Flowering Cycle (years)	References
D. strictus	8–9, 12–15, 20–70, 7–70	Troup, 1921; Kadambi, 1949; Jiao, 1956; Ueda, 1960; McClure, 1966; Shah, 1968; Wang and Chen, 1971; Gupta, 1972; Khan, 1972; Janzen, 1976; Fu, 1985; Fan and Qiu, 1987; Wu, 1988
Drepanostachyum falcatum	20–30, 35	Lowndes, 1947; Jiao, 1956; Fu, 1985; Fan and Qiu, 1987; Lin, 2009
D. intermedium	10	Brandis, 1899
Fargesia denudata	50–60, 63	Shao, 1986; Yu et al., 1987; Liu and Fu, 2007
F. murielae	35, 80–100, 110	Tredici, 1998; Li and Denich, 2004
F. nitida	50–60	Qin, 1995
F. robusta	50–60	Liu and Fu, 2007
F. scabrida	50–60	Qin, 1995; Liu and Fu, 2007
F. spathacea	35, 110	Li and Denich, 2004
Guadua chacoensis	31	Guerreiro, 2014
G. paraguayana	38	Guerreiro, 2014
G. sarcocarpa	26–29	Nelson, 1994; de Carvalho et al., 2013
G. trinii	30–32	Dutra, 1938; Janzen, 1976; Guerreiro, 2014
G. weberbaueri	27–28	Guerreiro, 2014
Himalayacalamus falconeri	20–38	Tingle, 1904; Bean, 1907; Holttum, 1956; Janzen, 1976; Fan and Qiu, 1987
Indocalamus latifolius	~ 100	Anonymous, 1961
I. tessellatus	60, > 115	McClure, 1966; Yuan et al., 2008
Kuruna wightiana	1	Janzen, 1976; Fan and Qiu, 1987
Merostachys anomala	30	Dutra, 1938
M. burchellii	30	Dutra, 1938
M. clausenii	32	Guerreiro, 2014
M. fistulosa	30–34	Janzen, 1976; Jaksic and Lima, 2010
Melocanna humilis	7–10, 26–50	Ueda, 1960; Janzen, 1976; Ram and Gopal, 1981; Fu, 1985; Fan and Qiu, 1987; Lin et al., 2009; Singleton et al., 2010; Govindan et al., 2016
M. multiramea	31–33	Budke et al., 2010; Guerreiro, 2014
M. skvortzovii	30–34	Guerreiro, 2014
M. sp.	11	Janzen, 1976
Nastus elegantissimus	3	Kurz, 1876
Neehouzeaua dullooa	14–17	Gupta, 1972
Neololeba amahussana	1	Fan and Qiu, 1987
Ochlandra scriptoria	1	McClure, 1966

Table 6.1 contd. ...

...Table 6.1 contd.

Species	Flowering Cycle (years)	References
O. stridula	1	Janzen, 1976, Zhou, 1984, Lin et al., 2009
O. travancorica	7, 28–30	Broun, 1887; Jiao, 1956; Janzen, 1976; Fu, 1985; Fan and Qiu, 1987
Oldeania alpina	~ 40	Janzen, 1976
Otatea acuminata subsp. *aztecorum*	30–35	Anonymous, 2004
O. ramirezii	8–30	Ruiz-Sanchez, 2013
Oxytenanthera abyssinica	7–21	Fanshawe, 1972; Janzen, 1976
Phyllostachys aurea	13–19	Janzen, 1976
Ph. dulcis	42–43	Adamson, 1978
Ph. edulis	> 48, 67	Janzen, 1976; Watanabe et al., 1980
Ph. fimbriligula	> 60	Chen et al., 1995
Ph. glauca	~ 45, 50–60 or 120	Zhang, 1977; Yue et al., 2018
Ph. heteroclada	50–60, ~ 80	Li and Denich, 2004; Wang and Wu, 2009
Ph. nigra	40–50	Lin et al., 2010a
Ph. nigra var. henonis	40–50, 58–63	Kawamura, 1927; Wu, 1988; Li and Denich, 2004; Lin, 2009
Ph. reticulata	40–50, 60 or 100, 115, 120, > 100	Kawamura, 1927; Ueda, 1960; Numata, 1970; Chen, 1973; Janzen, 1976; Lu, 1980; Zhou, 1984; Li and Denich, 2004
Pleioblastus argenteastriatus	> 26	Lin et al., 2017
Pl. fortunei	> 26	Lin et al., 2017
Pl. fortunei	~ 45	Yue et al., 2018
Pl. simonii var. variegata	30	McClure, 1966
Rhipidocladum neumannii	21	Guerreiro, 2014
Sarocalamus faberi	44–55	Yu et al., 1987; Qin et al., 1989
Sasa kurilensis	> 100	Inoue et al., 2014
Sasaella kogasensis 'Aureostriatus'	> 31	Lin et al., 2017
Schizostachyum dullooa	37–48	González et al., 2002; Giordano et al., 2009; Marchesini et al., 2010; Nath and Das, 2010
Thamnocalamus spathiflorus	10–11, 16–17	Brandis, 1899; Janzen, 1976
Thyrsostachys oliver	48	McClure, 1966
Yushania confusa	88	Li and Denich, 2004
Yushania maling	50+	Ray, 1952

Note: The Latin names of all bamboo species adopt the latest names, so the names of some bamboo species are different from those in the original article, and the same below.

Table 6.2. Summary of the flowering and fruiting phenomena for bamboo species (mainly based on Du et al., 2000; Zheng et al. 2020).

Species	Flowering Time	Flowering Place	Living States	Flowering Type	Dead or Deadless	Fruiting State	References
Ampelocalamus patellaris	1996	Ruili, Longchuan, YN	wild	whol.	dead	seed	Du et al., 2000
A. scandens	1990–1991	Chishui Guizhou Prov.	wild	whol.	dead	seed	Zhang et al., 1992a
Arundinaria fargesii	1978–1983	Zhenba County, Shanxi Prov.	wild	whol.	dead	seed	Zhou, 1984
Bambusa blumeana	1991–2000	Mengla County, YN	cult.	frag.	deadl.	seedl.	Du et al., 2000
B. burmanica	1995–1998	Ruili, Yingjiang, YN	wild	frag.	deadl.	seed	Du et al., 2000
B. distegia	1990	Tengchong County, YN	cult.	frag.	dead	seedl.	Du et al., 2000
B. emeiensis	1996	Midu County, YN	cult.	frag.	dead	seed	Du et al., 2000
B. papillata	1999–2000	Guangzhou, Kunming	cult.	frag.	dead	seed	Du et al., 2000
B. rigida	1998–1999	Kunming	cult.	frag.	deadl.	seedl.	Du et al., 2000
B. sinospinosa	1998–1999	Mengzi, Kunming	cult.	frag.	deadl.	seedl.	Du et al., 2000
B. sp.	1990	Kunming	cult.	frag.	deadl.	seedl.	Du et al., 2000
B. ventricosa	1999	Kunming	cult.	frag.	deadl.	seedl.	Du et al., 2000
Cephalostachyum chinense	1994–1998	Yuanyang County, YN	wild	whol.	dead	seed	Du et al., 2000
C. latifolium	1996	Pingbian County, YN	wild	whol.	dead	seed	Du et al., 2000
C. pergracile	1986–1988	Xishuangbanna, YN	wild	whol.	dead	seed	Du et al., 2000
C. scandens	1992–1997	Yingjiang, Tengchong county, YN	wild	whol.	dead	seed	Du et al., 2000
Chimonobambusa utilis	1990	Tongzi, Guizhou Provin	wild	whol.	dead	seed	Zhang et al., 1994
Chimonocalamus pallens	1996–1997	Jinping County, YN	wild	frag.	dead	seed	Du et al., 2000
Ch. dumosus	1995	Pingbian County, YN	wild	whol.	dead	seed	Du et al., 2000
Dendrocalamus barbatus	1987	Luchun County, YN	cult.	frag.	dead	seed	Du et al., 2000

Table 6.2 contd. ...

...Table 6.2 contd.

Species	Flowering Time	Flowering Place	Living States	Flowering Type	Dead or Deadless	Fruiting State	References
D. barbatus var. internodiradicatus	1995	Menglun, Kunming	cult.	frag.	dead	seedl.	Du et al., 2000
D. birmanicus	1998	Mangshi, YN	cult.	frag.	dead	seedl.	Du et al., 2000
D. brandisii	1989	Mojiang County, YN	cult.	frag.	dead	seedl.	Du et al., 2000
D. fugongensis	1993	Fugong County, YN	cult.	frag.	dead	seedl.	Du et al., 2000
D. giganteus	1994	Pu'er County, YN	cult.	frag.	dead	seedl.	Du et al., 2000
D. hamiltonii	Often	Xishuangbanna, YN	cult.	frag.	dead	seedl.	Du et al., 2000
D. latiflorus	1990	Mile County, YN	cult.	frag.	dead	seed	Du et al., 2000
D. semiscandens	1994	Mengla County, YN	wild	frag.	dead	seed	Du et al., 2000
D. pachystayum	1997	Shiping County, YN	cult.	frag.	dead	seedl.	Du et al., 2000
D. peculiaris	1998	Jinping County, YN	cult.	frag.	dead	seed	Du et al., 2000
D. sinicus	1987	Menghai County, YN	cult.	frag.	dead	seedl.	Du et al., 2000
D. yunnanicus	1987–1990	Hekou, Jinping County, YN	cult.	frag.	dead	seedl.	Du et al., 2000
Fargesia edulis	1997–1998	Gaoligong Mt., YN	wild	frag.	both	seed	Du et al., 2000
F. fungosa	1992	Zhaotong County, YN	wild	whol.	dead	seed	Du et al., 2000
F. hygrophila	1994, 1997	Cangshan Mt., YN	wild	frag.	deadl.	seedl.	Du et al., 2000
F. melanostachys	1994	Gaoligong Mt., YN	wild	frag.	dead	seed	Du et al., 2000
F. papyrifera	1993	Yunlong County, YN	wild	frag.	dead	seed	Du et al., 2000
F. tomentosa	1997	Cangshan Mt., YN	wild	frag.	deadl.	seed	Du et al., 2000
Gaoligongshania megalothyrsa	1994–1997	Gongshan County, YN	wild	whol.	dead	seed	Du et al., 2000
Gigan nigrociliata	1991–1993	Mengla County, YN	wild	whol.	dead	seed	Du et al., 2000
Indocalamus sp.	1994	Mengla County, YN	wild	whol.	deadl.	seedl.	Du et al., 2000
Indosasa austro-yunnanensis	1985	Simao, Kunming	wild	whol.	deadl.	seedl.	Du et al., 2000

Lingnania intermedia	1998–2000	Kunming	cult.	frag.	deadl.	seedl.	Du et al., 2000
Melocalamus rugosa	1995–1999	West part, YN	wild	whol.	dead	seed	Du et al., 2000
M. scandens	1998	Kunming	wild	whol.	dead	seed	Du et al., 2000
M. compactiflorus	1985–2000	Cangyuan, Menghai County, YN	wild	whol.	dead	seed	Du et al., 2000
Neomicrocalamus prainii	1998–1999	Kunming	wild	frag.	deadl.	seed	Du et al., 2000
Phyllostachys nidularia	1999	Zhejiang, Kunming	cult.	frag.	deadl.	seedl.	Du et al., 2000
Ph. reticulata	1988–1990	Zhejiang Prov.	cult.	frag.	dead	sterility	Zhang et al., 1992b
Ph. heteroclada	1995	Yiliang County, YN	cult.	frag.	dead	seed	Du et al., 2000
Ph. edulis	Often	Many place	cult.	whol.	dead	seed	Du et al., 2000
Ph. glauca f. *yunzhu*	1998	Gejiu, Kunming	cult.	whol.	dead	seedl.	Du et al., 2000
Pleioblastus gramineus	1999	Kunming	cult.	frag.	deadl.	seed	Du et al., 2000
Pseudostachyum polymorphum	1993–1994	Cangyuan County, YN	wild	frag.	dead	seed	Du et al., 2000
Schizostachyum dumetorum	1999	Yangchun Guangdong Prov.	wild	whol.	dead	seed	Du et al., 2000
S. pingbianensis	1998–1999	Pingbian County, YN	wild	whol.	both	seed	Du et al., 2000
Chimonobambusa sichuanensis	2008	Nanjing	cult.	frag.	dead	seed	Lin et al., 2009
Ch. utilis	1986–1989	Nanjing	cult.	whol.	dead	seed	Zhou, 1984
Phyllostachys arcana f. *luteosulcata*	2007	Nanjing	cult.	frag.	dead	seed	Lin and Ding, 2007
Ph. glauca	2014	Jinan, Linqing Shandong Prov.	cult.	whol.	dead	seed	Yue et al., 2018
Ph. heteroclada	2018	Nanjing	cult.	whol.	both	seedl.	Zheng et al., 2020
Ph. heteroclada	1987–1989	Nanjing	cult.	part.	dead	seedl.	Zhou, 1984
Ph. nidularia	1984	Nanjing	cult.	frag.	deadl.	seedl.	Zhou, 1984
Ph. viridiglaucescens	2007	Nanjing	cult.	frag.	dead	seed	Lin and Ding, 2007
Ph. vivax	2015	Yangzhou Jiangsu Prov.	cult.	frag.	dead	seed	Zheng et al., 2016

Table 6.2 contd. ...

...Table 6.2 contd.

Species	Flowering Time	Flowering Place	Living States	Flowering Type	Dead or Deadless	Fruiting State	References
Pleioblastus argenteostriatus f. *angustifolius*	2015	Huangshan Anhui Prov.	cult.	frag.	dead	seed	Zheng et al., 2016
Pl. fortunei	2015–2017	Nanjing, Yancheng Jiangsu Prov.	cult.	whol.	dead	seed	Zheng et al., 2016
Pl. simonii f. *heterophyllus*	2007–2011	Nanjing	cult.	frag.	dead	seed	Lin and Ding, 2013
Pl. yixingensis	2017	Nanjing	cult.	whol.	dead	seed	Zhang et al., 2018
Pseudosasa amabilis var. *convexa*	Around 2002	Nanjing	cult.	frag.	dead	seed	Zhang et al., 2003
Sasaella kogasensis 'Aureostriatus'	2015–2017	Nanjing	cult.	whol.	dead	seed	Zheng et al., 2016
Bambusa multiplex	2011–2012	Nanjing	cult.	frag.	deadl.	seed	Lin et al., 2015
Dendrocalamus latiflorus	2001	Nanjing County, Fujian Prov.	cult.	frag.	dead	seed	Xing et al., 2005
D. membranaceus	2013–2015	Jinghong County YN	wild	frag.	dead	seed	Xie et al., 2016
D. membranaceus	2013–2015	Jinghong County YN	wild	whol.	both	seed	Xie et al., 2016
D. sinicus	2012, 2015	Ximeng County YN	cult.	frag.	dead	seed	Xie et al., 2019
Phyllostachys heteroclada	2004–2005	Ya'an, Sichuan Prov.	wild	both	deadl.	seed	Wang and Wu, 2009
Ph. violascens	2004–2009	Lin'an, Zhejiang Prov.	cult.	frag.	dead	seed	Lin et al., 2010b
Pleioblastus fortunei	2016–2018	Nanjing	cult.	whol.	both	seed	Fu et al., 2020
Pseudosasa viridula	2019	Bamboo germplasm garden of Jiangxi Agricultural University	cult.	whol.	deadl.	seed	Zhao et al., 2020
Oligostachyum spongiosum	2019	Bamboo germplasm garden of Jiangxi Agricultural University	cult.	whol.	dead	seed	Liu et al., 2021

Notes: 1. The records of different flowering characteristics of the same species were sorted out.

2. The abbreviation 'cult.' means cultivated; 'whol.' means massive-synchronized flowering; 'frag.' means sporadically flowering; 'Part.' means partial flowering; 'deadl.' means survived after flowering; 'seedl.' means no seed after flowering.

Recently, Zheng et al. (2020) summarized four flowering habits in bamboos. Besides mass and sporadic flowering habits included also were the partial flowering and the combined massive synchronized and sporadic flowering (Table 6.3). The combined massive synchronized and sporadic flowering meant that bamboos that were sporadic or small-area flowering before or after mass flowering, and genera such as *Chusquea*, *Dendrocalamus*, and *Phyllostachys* also exhibit this flowering habit. For the partial flowering bamboos, such as *Phyllosasa* and *Pleioblastus*, the flowering area in the bamboo forest was patchy, and between mass flowering and sporadic flowering. Some genera, e.g., *Bambusa*, *Chimonobambusa*, *Dendrocalamus*, *Phyllostachys* and *Schizostachyum*, simultaneously exhibit the flowering characteristics of mass, sporadic, and combined flowering. This phenomenon may be related to the distribution area, and the species with a large distribution range tend to have multiplex types of flowering patterns.

Based on the consequences of flowering, bamboos' flowering types can be divided into death post-flowering, immortality post-flowering, and coexistence of death and immortality post-flowering (Du et al., 2000). According to the published records (Table 6.2), species such as the *Ampelocalamus*, *Cephalostachyum*, *Dendrocalamus*, *Melocalamus*, and S*chizostachyum* die after flowering. *Neomicrocalamus* and *Indosasa* can survive after flowering. Post-flowering death and rejuvenation occur in both *Fargesia* and *Phyllostachys*. In addition, the scattered bamboos will usually die within 1–2 years after flowering, and the sympodium bamboo will often regain their growth after several years of rejuvenation and renewal (Lin et al., 2015).

In summary, the flowering types of bamboos are largely determined by their biological characteristics. There is no clear relationship between flowering habit and taxonomic status at the genus level. On the other hand, the flowering types of bamboo forests are closely related to the origin of bamboo forests, namely wild or cultivated types. Usually, wild bamboo species tend to mass flowering, while the cultivated species tend to sporadic flowering (Du et al., 2000; Chen et al. 2017).

6.2.2 Breeding System of Woody Bamboos

The breeding system refers to the integration of all sexual characteristics that affect the genetic characteristics of offspring, mainly including flower morphological characteristics, life span of flower organs, type and frequency of pollinators, degree of self-compatibility, and mating system. Among them, the mating system is the core. At present, the propagation of bamboo plants mainly depends on seeds, bamboo rhizomes, and bamboo culms. Since bamboo takes a long time to bloom and seeds or fruits are usually unavailable, asexual reproduction is the main method of bamboo afforestation in China and in Southeast Asia (Tan and Qi, 1994; Chen and Ma, 2005; Sun et al., 2015).

6.2.2.1 Inflorescence

The inflorescence of bamboos is compound inflorescence, and the basic unit of inflorescence is spikelet. According to developmental characteristics and the arrangement differences of spikelets on the inflorescence axis, inflorescence is

Table 6.3. Flowering types of woody bamboos (mainly based on Zheng et al., 2020; Du et al., 2000).

Flowering Type	Bamboo Species
Sporadically flowering	*Bambusa blumeana, B. burmanica, B. distegia, B. emeiensis, B. multiplex, B. oldhamii, B. papillata, B. remotiflora, B. rigida, B. sinospinosa, B.* sp., *B. ventricosa, B. vulgaris, B. xiashanensis, B. xueana, Chimonobambusa ningnanica, Ch. pachystachys, Ch. tumidissinoda, Chimonocalamus pallens, Dendrocalamus barbatus, D. barbatus var. internodiradicatus, D. birmanicus, D. brandisii, D. fugongensis, D. giganteus, D. hamiltonii, D. latiflorus, D. manipureanus, D. membranaceus, D. minor, D. pachystachys, D. peculiaris, D. semiscandens, D. sinicus, D.* sp., *D. yunnanicus, Drepanostachyum falcatum* var. *glomerata, Fargesia dracocephala, F. edulis, F. frigidis, F. hygrophila, F. melanostachys, F. papyrifera, F. tomentosa Indosasa patens, I. sinica, Lingnania intermedia, Neomicrocalamus prainii, Phyllostachys angusta, Ph. arcana, Ph. bissetii, Ph. heteroclada, Ph. nidularia, Ph. praecox 'Viridisulcata', Ph. reticulata, Ph. violascens, Ph. violascens* f. *preveynalis, Pleioblastus gramineus, Pl. simonii* f. *heterophyllus, Pseudosasa amabilis* var. *convexa, Pseudostachyum polymorphum, Schizostachyum dumetorum, Sch. funghomii, Sch. pingbianensis, Semiarundinaria densiflora* var. *villosum, Yushania* sp.
Massive-synchronized flowering	*Arundinaria fargesii, Acidosasa purpurea, Ampelocalamus patellaris, A. scandens, A. stoloniformis, Bambusa arnhemica, B. bambos, Cephalostachyum chinense, Ce. latifolium, Ce. pergracile, Ce. pingbianense, Ce. scandens, Chimonobambusa pachystachys, Ch. quadrangularis, Ch. rigidula, Ch. szechuanensis, Ch. tumidissinoda, Ch. utilis, Chimonocalamus dumosus, Chusquea abietifolia, Ch. quila, Ch. ramosissima, Dendrocalamus longispathus, Fargesia denudata, F. fungosa, F. murielae, F. nitida, F. obliqua, F. qinlingensis, F. robusta, F. scabrida, F. spathacea, Gaoligongshania megalothyrsa, Gigantochloa albociliata, G. nigrociliata, Guadua trinii, Indocalamus* sp., *Indocalamus tessellatus, I. wilsonii, I. angustata, Indosasa angustata, I. austro-yunnanensis, Melocalamus compactiflorus, M. rugosa, M. scandens, Otatea ramirezii, Phyllostachys atrovaginata, Ph. edulis, Ph. fimbriligula, Ph. glabrata, Ph. glauca* f. *Yunzhu, Ph. glauca, Ph. heteroclada* f. *solide, Ph. meyeri, Ph. nigra* var. *henonis, Ph. propinqua, Ph. rutila, Ph. sulphurea* var. *viridis, Ph. vivax, Pleioblastus amarus, Pl. fortunei, Pl. argenteostriatus, Pl. linearis, Pl. maculatus, Pseudosasa japonica, P. viridula, Oligostachyum spongiosum, Sarocalamus faberi, Sasa kurilensis, S. palmata, S. senanensis, S. sinica, S. veitchii* var. *hirsuta, Sasaella kogasensis 'Aureostriatus', Schizostachyum dumetorum, Shibataea chinensis, Sinobambusa tootsik, Yushania confusa*
Combined massive-synchronized and sporadic flowering	*Acidosasa notata, Arundinaria fargesii, Bambusa tulda, Chimonobambusa pachystachys, Chusquea ramosissima, Ch. culeou, Ch. macrostachya, Ch. montana, Ch. uliginosa, Ch. valdiviensis, Dendrocalamus hamiltonii, D. hamiltonii* var. *hamiltonii, D. membranaceus, D. strictus, Indocalamus latifolius, Melocanna humilis, Phyllostachys heteroclada, Ph. reticulata, Ph. reticulata* f. *shouzhu, Ph. reticulata* f. *tanakae, Ph. rubromarginata, Sasa cernua, Schizostachyum dullooa*
Partial flowering	*Acidosasa notata, Indocalamus* sp., *Phyllosasa tranquillans* f. *shiroshima, Ph. iridescens, Ph. reticulata* f, *lacrima-deae, Pleioblastus amarus* var. *pendulifolius, Pl.* simonii

divided into semelauctant and interauctant inflorescence (Yi et al., 2008). Like common grasses, semelauctant inflorescence consists of a variety of the flowering branches (or spikelets), and their development and maturity occurs simultaneously.

Semelauctant inflorescence is also called determinate inflorescence or genuine inflorescence. The attachment site of semelauctant inflorescence is at the top of the uppermost vegetative leaf (flag leaf or uppermost leaf) of the plant vegetative body. The inflorescence axis and its branches (including spikelet stalk) are mostly solid, there are no obvious nodes at the branches (including spikelet stalk), occasionally with small scaly bracts, without buds in axils. Semelauctant inflorescence usually occurs in subtropical or temperate genera, such as *Fargesia, Sasa, Thamnocalamus* (Bhattacharya et al., 2006; Yi et al., 2008).

Another type of inflorescence is interauctant inflorescence, and its basic structure is pseudospikelet, which is formed by a spikelet on the top of a shortened twig. There is a first outer leave (or prophyll) on the inner side of the branchlet base, and the leaf organs above it are bracts like glume or lemma shape. A bud-subtending bract often has twigs in its axils that develop into secondary spikelets, which have the potential to develop into another spikelet and eventually form a cluster of spikelets. Therefore, such pseudospikelets clusters are formed by the successive development of pseudospikelets at all levels. Interauctant inflorescences are also called indeterminate inflorescence, and pseudospikelets can be born in each node of the vegetative branches to form spikes, conical, and other inflorescence patterns. Interauctant inflorescence usually occurs in tropical or subtropical genera, such as *Bambusa, Dendrocalamus, Phyllostachys* (Yi et al., 2008).

6.2.2.2 Floret

The spikelets of bamboo contain one to more florets, and the florets have three or six stamens. The filaments of stamens are slender, most of them are separated or partially commissure, and anthers are two, locular longitudinal. The pistil is located at the upper part of the ovary, with one chamber, one inverted ovule, and one or two to three styles. Its ovary is ovoid or nearly spherical. Depending on whether the stigma extends out of the lemma or not, the florets can also be divided into short style and long style (Zhang and Ma, 1990). The stigma of short style florets is slightly exposed during flowering, but it does not extend beyond the lemma, and the stigma retracts when the lemma closes over time. The florets of *Chimonobambusa sichuanensis, Pseudosasa amabilis* var. *convexa, Semiarundinaria densiflora, Pseudosasa viridula, B. multiplex*, and *Phyllostachys reticulata* belong to this type (Zhang and Ma, 1990, Zhao et al., 2020). The stigma of long style florets is slender, and the lemma is not fully open at flowering. The stigmas can extend out of the lemma to receive pollen, then become withered after pollination. The florets of *Ph. Violascens* f. *notata, Ph. Viridiglaucescens, Ph. Glauca, Ph. Arcana* '*Luteosulcata*', and *Shibataea chinensis* belong to the long style type (Zhang and Ma, 1990) (see Table 6.3).

6.2.2.3 Pollen

The morphology of bamboo pollen is generally similar, spherical, or nearly spherical, with a single germination hole, concave and nearly circular, and the diameter of a single pollen grain is 20 ~ 100 μm (Yao et al., 2020). The pollen grain size is positively correlated with the length of the anthers, that is, the longer anthers bear larger pollen grains, and the outer wall decoration of pollen grains is mostly fine granular (Yao et al., 2020; Zhang et al., 1990). Pollen viability refers to the ability of pollen to survive, develop, and germinate. It is determined by genetic material and is influenced by environmental factors, such as temperature, light, and humidity. Under natural conditions, the pollen germination rate of bamboos is generally low and is accompanied by male sterility and abortion caused by poor pollination (Chen and Ma, 2005; Chen et al., 2017). For example, *Phyllostachys nidularia* (adhesion between anthers) and *Pleioblastus maculatus* (no anthers) exhibit male infertility. Pollen abortion is influenced by genetic factors and environmental conditions, which may be prevalent in bamboos. The genetic factors may cause abnormal shrink and cavity pollen grains in some anthers during pollen development (Liu et al., 2021). Meanwhile, in the process of flower development and blooming, if the adverse factors such as low temperature, high temperature, diseases, and insect pests occur, the flower organs will be damaged and lead to pollen abortion (Zhang and Ma, 1990, 1992). The types and degree of pollen abortion are varied according to bamboo species. The pollen abortion rate can reach more than 50% at the blooming peak of *Pl. intermedius* and *Ph. Heteroclada* (Chen and Ma, 2005).

6.2.2.4 Pollination

Effective pollination of seed plants is a key step to successful sexual reproduction. When the pistil is inadequately pollinated, namely limited in pollen quantity or quality, the seed yield will decrease. So far, most of the literature has recorded that woody bamboos were dichogamy and protogyny (Nadgauda et al., 1993; Chen et al., 2017). Wind pollination is generally considered to be the main pollination mechanism of bamboos. Most documented bamboos, such as *Bambus tulda*, *Ochlandra travancorica*, *Phyllostachys edulis*, *Dendrocalamus strictus*, *D. sinicus*, and *Arundinaria gigantea* (Venkatesh, 1984; Nadgauda et al., 1993; Gagnon and Platt, 2008; Chen et al., 2017; Xie et al., 2019; Chakraborty et al., 2021), were wind pollinated. Therefore, the spatial distribution of flowering bamboos will affect the transmission of pollen in a bamboo forest. When flowering bamboos are dispersed from each other and exceed the effective pollination distance, it will reduce the pollination effectiveness (Xie et al., 2019). This situation reduces the out-crossing rate and increases the possibility of inbreeding, which will lead to an increased homozygosity of genes, harmful allele expression, and inbreeding decline in offspring. Such inbreeding depression can be expressed post-zygotically by reduced seed germination in bamboo.

In addition to wind pollination, the insect pollination was reported in some bamboo species. During the flowering period of *D. membranaceus*, *D. sinicus*, and *Phyllostachys niduraria*, a bee (*Apis cerana*) is a common flower-visiting insect (Xie et al., 2019). Bees usually visited flowers from 10:00 a.m. to 12:00 a.m., but rainy days would affect insect flight. Hence, the number of visitors would decrease accordingly, which affected pollen transmission. To sum up, the development of florets, spatial distribution of flowering individuals and pollinators are important factors affecting sexual reproduction (Table 6.4). Therefore, under the condition

Table 6.4. Seed setting rate of woody bamboos.

Bamboo Species	Seed Setting Rate %	Factors Affecting Seed Setting Rate	Flowering Period	References
Arundinaria simonii f. *heterophylla*	31.2	Sporadic flowering and pollen tube growth were inhibited	March–mid and late May	Lin and Ding, 2013; Lin, 2009
Bambusa multiplex	2.56–4.69	Flowering year after year, bamboo age structure shows aging trend	October–late May of the next year	Lin et al., 2015
B. pervariabilis × (*Phyllostachys edulis* + *Dendrocalamus latiflorus*)	8	–	–	Zhang and Chen, 1986
B. sinospinosa × (*Ph. edulis* + *D. minor*)	3.7	–	–	Zhang and Chen, 1986
B. textilis × *B. pervariabilis*	13.6	–	–	Zhang and Chen, 1986
B. textilis × *B. sinospinosa*	10.8	–	–	Zhang and Chen, 1986
D. latiflorus	18.85–21.28	Nutritional status of parents, pollen viability (pollen quality and climatic factors at flowering)	November–February of the next year	Xing et al., 2003
D. latiflorus × (*B. pervariabilis* + *B. textilis*)	8.1–14.5	–	–	Zhang and Chen, 1986
D. latiflorus × *B. sinospinosa*	0.6–1.6	–	–	Zhang and Chen, 1986
D. minor × *D. latiflorus*	22	–	–	Zhang and Chen, 1986
D. membranaceus	1.76–7.49	Area of flowering bamboo clumps, pollen vigor, pollinators	November–May of the next year	Xie et al., 2016
D. sinicus	0.34–0.64	Sporadic flowering, hermaphroditism of florets and pollen vigor	October–May of the next year	Gu et al., 2012
Oligostachyum spongiosum	8.1	Effects of pests and precipitation	Late April–late May	Liu et al., 2021
Pseudosasa viridula	0.12	Pests	January–May	Zhao et al., 2020

Table 6.4 contd. ...

...Table 6.4 contd.

Bamboo Species	Seed Setting Rate %	Factors Affecting Seed Setting Rate	Flowering Period	References
Ph. glauca	0.24	Sporadic flowering Climatic conditions at flowering (temperature, light)	April–September	Yue et al., 2018
Ph. reticulata	0	Abortive pollen, poor pollination, dysplasia of style and stigma	February–May	Zhang et al., 1992b
Ph. iridescens	6.1	Ovaries or anthers withered and Pests	Full flowering period: April–May, September–November	Zhang and Ma, 1989
Ph. Edulis × B. pervariabilis	1.3–3.8	–	–	Zhang and Chen, 1986
Ph. Edulis × B. textilis	1.0–2.0	–	–	Zhang and Chen, 1986
Ph. Edulis × B. sinospinosa	0.47–1.56	–	–	Zhang and Chen, 1986
Shibataea chinensis	0	Pollen abortion, protogynous dichogamy, short stigma life and basically no receptivity	October–March of the next year	Lin and Ding, 2013; Lin, 2009

Note: "—" represents an unrecorded phenomenon.

of normal development of florets, we can improve the pollination conditions of bamboos, through artificial assisted pollination, to increase the seed setting rate.

6.2.2.5 Mating System

The study on plant mating system can be divided into two levels: the first level is the population selfing rate and outcrossing rate, and the second one is to study the process and pattern of gene flow (He and Ge, 2001). The existing research results show that bamboo mating system is of the following main traits: (1) usually self-compatible; (2) no agamospermy; and (3) mixed mating system.

The plant mating system is affected by various factors such as flowering population density, floral synchronization, and post-polling mechanisms (Table 6.4). Generally, self-pollination is dominant in sporadic flowering populations, while outcrossing is dominant in mass flowering populations (e.g., Matsuo et al., 2014; Zhong et al., 2017). The mating systems of two wild temperate bamboo species, *Sasa cernua* and *Ph. edulis* was found self-compatible, and self-pollination was dominant in sporadic flowering while outcrossing was dominant in mass flowering (Kitamura and Takayuki, 2011). Furthermore, based on morphological observation and artificial pollination experiments, the pollination and breeding system of two tropical bamboos, namely, wild *D. membranaceus* and the cultivated *D. sinicus* were studied. The results also suggested that they belonged to a mixed mating system, self-compatible but outcrossing was dominant. There was no agamospermy in either species (Fig. 6.1) (Chen et al., 2017; Xie et al., 2019). Recently, mixed inflorescence types, including early solitary spikelet and late pseudospikelet in the sporadic flowering populations of *Bambusa tulda*, were reported by Chakraborty et al. (2021). Among the flowering events of *B. tulda*, the percentage of seed settings

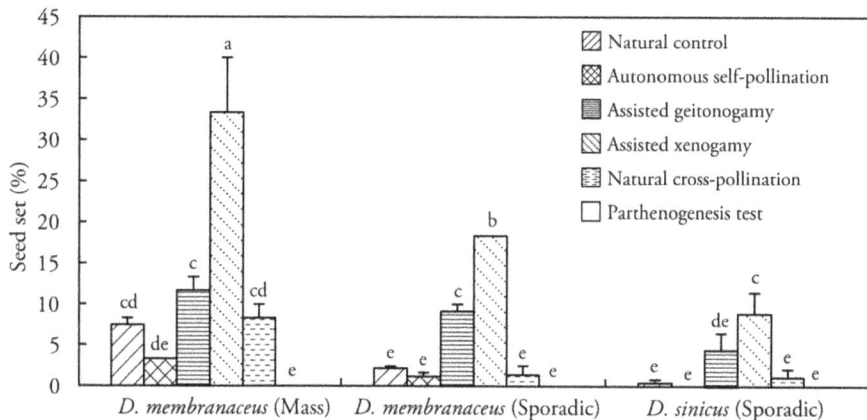

Fig. 6.1. Difference of seed setting rate between *Dendrocalamus membranaceus* and *D. sinicus* under different pollination conditions (Chen et al., 2017).

Note: Average seed set of *D. membranaceus* and *D. sinicus* from different flowering populations subjected to six pollination treatments. The results of the same treatments were compared across populations and means with the same letter are not significantly different at P < 0.05.

in pseudospikelet (17.3 ~ 25.7%) was significantly higher than solitary spikelets (3.2 ~ 9.6%), while the pollen germination rate of self-pollination (4.2 ~ 8.6%) was significantly lower than that of the cross pollination (32.9 ~ 41.3%) (Chakraborty et al., 2021).

The self-compatibility may be the result of evolution driven by a lack of pollinators or insufficient pollination. Due to inbreeding depression, most of the inbred offspring will be eliminated before maturity, and their mating pattern will change with pollination conditions (Xie et al., 2019). To avoid population decline caused by self-crossing, the maturation time of pistils and stamens of flowering bamboo clumps on the same inflorescence should not be overlapped, which is specifically manifested as the first maturation of pistils to prevent self-pollination. But it is still unable to prevent pollination between ramets of the same clone (Nadgauda et al., 1993; Chen et al., 2017).

6.2.2.6 Fruits or Seeds

Fruit characters, especially fruit types, have played an important role in phylogenetics, breeding improvement, and forest cultivation of woody bamboos. According to Tzvelev (1976), Kaden proposed that caryopsis was the only fruit type found in Poaceae. Furthermore, Sendulsky et al. (1987) subdivided caryopsis in Poaceae into five types: (1) basic or typical caryopsis with a thin dry pericarp fully adnating to the seed, (2) follicoid, caryopsis with a free pericarp adjoining the seed, (3) bacoid or berry-like, (4) nucoid or nut-like, and (5) cistoid type with pericarp, when moistened, separating entirely from the seed. According to the caryopsis classification of Sendulsky et al. (1987), three caryopsis types usually occur in the paleotropical woody bamboos, i.e., basic, bacoid, and nucoid caryopsis (Yang et al., 2008). The basic caryopsis occurs in species such as *Phyllostachys edulis* and *Dendrocalaums membranaceus*, and the berry-like caryopsis is found in *Melocanna humilis* and *Melocalamus arrectus*. Moreover, *D. sinicus, Chimonobambusa utilis,* and *Qiongzhuea tumidinoda* bear nut-like caryopsis (Yang et al., 2008).

6.2.2.7 Seeds Setting Rate

It is common in bamboos to bear fruit after flowering (Table 6.2). Among 68 flowering bamboos recorded in China, 38 species (56%) produced fruits or seeds, indicating that most bamboos could produce fruits after flowering like other seed plants (Du et al., 2000). On the other hand, the seed-setting rate of bamboos is considerably low under natural pollination conditions (Table 6.4). Empirically a higher rate of seed setting was correlated with flowering area of bamboo forests. Three *Sasa* (namely *Sasa senanensis, S. kurilensis,* and *S. palmata*) exhibited the higher seed-setting rate with larger flowering area (Mizuki et al., 2014). A larger flowering area of bamboo forests implied a higher capacity for pollen dispersal (Chakraborty et al., 2021) and higher availability of pollen grains, then pollinators also could improve seed-setting rates (Xie et al., 2019). On the other hand, Chen et al. (2017) found that the seed-setting rate of *Dendrocalamus sinicus* was 0.42% under natural conditions, and after artificial xenogamy, it increased about 20 times to 8.89%. Lin and Ding (2013) also found that artificially assisted pollination could increase the seed-setting rate of *Arundinaria simonii* f. *heterophylla* from 31.2% to 67.5%. These studies indicated that artificial xenogamy could significantly improve the seed-setting rate and promote sexual reproduction. Therefore, artificial pollination can be used as an important technical means to produce new bamboo varieties.

The reasons for the low seed-setting rate include: (1) genetic factors, such as male sterility, pollen abortion, low pollen viability, truncated embryo sac development and affinity of fertilization. Within bamboos with a low seed-setting rate, especially some cultivated bamboos, there are obvious dysplasia of style and anther, abortive pollen, which lead to the early abortion of seeds (Chen and Ma, 2005); (2) insect pests that eat ovary, anther, and stigma which seriously damage the whole process of sexual reproduction; (3) floral non-synchronization; (4) adverse environmental factors, like rainy weather during the flowering period; and (5) wind pollination. Compared with insect pollination, wind pollination has lower pollination effectiveness, especially for the sporadical flowering bamboos and the cultivated bamboos. Overall, no fruiting or low fruiting rate of bamboos may be related to its biological characteristics, the morphological development of flower organs, pest interference, and pollination effectiveness during the flowering period (Table 6.4).

6.3 Genetic Map of Woody Bamboo

Compared with crops and other economic tree species, woody bamboo has a longer generation cycle, unpredictable flowering time, inbreeding depression, mass death after flowering and difficulty to backcross making it more difficult to establish a genetic map in woody bamboo. Despite all these limitations, scientists are trying to establish a genetic map of important economic bamboos. Yuan et al. (2005) obtained the F1 population of *Dendrocalamus latiflorus* by artificial pollination and constructed the linkage map using the RAPD marker according to the pseudo-testcrossing mapping strategy. The findings suggest that the genetic loci of *D. latiflorus* were highly heterozygous and had several markers independent of their parents. A total of seven linkage groups was obtained by multipoint linkage analysis, covering a total map distance of 305.7 centiMorgan (cM). Among these, the longest linkage group was 100.0 cM, while the shortest one was 12.4 cM and the average map distance between markers was around 12.74 cM. This work provided a basis for establishing a high-density genetic map of *D. latiflorus*. Recently, Guo et al. (2019) genotyped the inbred progeny population of *D. latiflorus* through ddRAD sequencing and constructed the first high-density genetic linkage map in Bambusoideae. These ddRAD markers detected 36 linkage groups, corresponding to 36 chromosomes of *D. latiflorus*. The map covered a total of 3,113 cM (93.29% of the genome) on chromosomes ranging from 4.91 to 131.69 cM in size. Among them, 19.9% (720) markers on *D. latiflorus* were homologous with rice genome, and 681 markers were in the highly collinear region. Moreover, the ratio between rice chromosome and *D. latiflorus* linkage group was 1:3, which was consistent with their chromosome pairs (12, 36), and the results also suggested that *D. latiflorus* was of a hexaploid origin.

The genetic map is an important tool for QTL mapping, molecular marker assisted breeding, and gene cloning. Its quality depends on the universality and saturation of the map. Generally, it is required that an average interval of 20 cM or less between the markers to map the linkage group. For the genetic linkage of QTL mapping, the average interval between markers should be less than 10 cM. Due to the unique reproductive biological characteristics, it is difficult to establish a high-density genetic linkage map for most bamboos. Fortunately, with the development of sequencing technology and functional genomics, and with the reduction of cost, many markers, e.g., express sequence tags (ESTs) have been deposited in GenBank and other databases. They have become important resources for the marker development, which may provide an important way to establish a high-density genetic map for woody bamboos. What is more important, we should pay more attention to mapping populations and mapping strategies for bamboos, which is the only way to fundamentally speed up the establishment of the genetic linkage map of bamboos.

6.4 Artificial Hybridization and Application to Woody Bamboo

Bamboo reforestation and forest regeneration may employ seed or seedlings from sexual reproduction, and vegetative seedlings from asexual reproduction with the rhizome or culm. Due to many factors, such as long generation cycle, low pollination rate, poor pollen fertility, mass death after flowering, insect damage to flower organs, etc., the seed setting rate of bamboos is generally low under natural conditions, and the study on hybrid breeding of bamboos is seriously hindered. Nevertheless, as an important and conventional breeding method, natural or artificial crossbreeding is conducive to cultivating new varieties with strong adaptability and excellent timber properties. Chinese scientists have tried to carry out crossbreeding of bamboos since the 1970s. Flowering bamboo clumps or culms are collected at the same location, when blooming is noticed as parents for intergeneric and interspecific crossbreeding. Excellent individuals are screened out from the hybrid offspring for further asexual propagation and then subjected to afforestation and forest regeneration.

Zhang and Chen (1986) used *Bambusa pervariabilis* × *Dendrocalamus latiflorus* as the parents, and selected an excellent variety of edible shoots, 'Chengma 7', from the hybrid offspring. Through the crossing of *B. pervariabilis* × (*D. latiflorus* + *B. textilis*), the 'Chengmaqing 1', a variety with rapid growth, good timber, stress resistance, and beautiful culm shape was bred (Yi et al., 2008). Ning and Dai (1995) selected *B. pervariabilis* as the female and *B. grandis* as the male to crossbreed into 'Chenglv' 3, 6, 8, and 30. Those varieties were of the traits of large diameter grades at breast high, strong asexual fecundity, and good paper-making quality. In the past 20 years, the hybrids 'Chenglv' have been widely cultivated in tropical and subtropical areas of southwestern and southern China (Table 6.5).

Table 6.5. New hybrid varieties of the woody bamboos (based on Xu, 2019).

Female	Male	Offspring	Fine Traits of Offspring
B. pervariabilis	*B. grandis*	Chenglv 3, 6, 8, 30	Large culm diameter (DBH = 10.1cm), strong asexual reproduction ability and good paper-making performance.
B. pervariabilis	*D. latiflorus*	Chengma 1, 7, 25	Fast growth, strong stress resistance, can be used as bamboo shoots and timber.
B. pervariabilis	*D. latiflorus*, *B. textilis*	Chengmaqing 1	Rapid growth, good material, strong stress resistance and beautiful appearance.
B. textilis	*D. latiflorus*	Qingma 11	Beautiful appearance and good papermaking performance of bamboo.
D. hamiltonii	*D. latiflorus*	Banma 1	The bamboo shoots taste delicious and produce more bamboo shoots.

6.5 Research Prospects

Woody bamboo is an important type of forestry germplasm resources in the subtropical and tropical regions of Asia, Africa, Central and South America. Bamboo forests are valuable economic forests with the most perfect combination of economic, ecological, and cultural values. Meanwhile, the bamboo industry is representative of traditional, green, and sustainable forestry industries in China and in many other countries. Therefore, bamboos have a great potential to improve rural and agricultural development in the world, especially in "Rural Revitalization of China".

Considering the importance of bamboo resources, bamboos will remain as one of the main research areas of plant science in the future. Some hot spots may include: (i) conservation of bamboo germplasm resources. Excavation of excellent germplasm resources and the improvement of bamboo forest management technologies are urgently needed for the upgrading of China's bamboo industry; (ii) to understand regulation mechanism of bamboo flowering. The death of bamboo forests after mass flowering will result in tremendous ecological and economic disasters. To date, we have not fully revealed the mechanism underlying bamboo flowering; (iii) breeding of excellent bamboo varieties. Designing molecular breeding is an up-to-date approach in seed industry, and very likely it would also be applied to bamboo breeding before long. At present, it is still an important, feasible, and effective method to create and select new varieties with commercially important traits using the traditional crossbreeding technology.

Author Contribution

PeiTong Dou wrote and revised this chapter.
TiZe Xia wrote part of the content.
HanQi Yang formulated the content and objectives of this chapter and wrote and revised the content.

Acknowledgements

Authors thank all the researchers who supported the preparation of the book and the authors of the references cited.

Funding

This work was supported by the Fundamental Research Funds of the Chinese Academy of Forestry (CAFYBB2021SZ001), the National Natural Science Foundation of China (31870574), and the Department of Sciences and Technology of Xizang Autonomous Region (XZ201801-GA-11). The funders had no role in study design, data collection and analysis, decision to publish, or preparation of the manuscript.

Conflict of Interest

The authors declare that they have no competing interests.

References

Adamson, W.C. (1978). Flowering interval of sweetshoot bamboo. *Econ. Bot.*, 32(4): 360–362.

Anonymous. (1961). A preliminary study on the cause of the flowering and fruiting of *Indocalamus latifolius*. *J. Zhejiang Agric. Sci.*, (7): 352–353.

Anonymous. (2004). Otatea in western Mexico is in bloom. *World Bamboo Rattan*, 2(1): 1.

Bean, W.J. (1907). The flowering of cultivated bamboos. *Bulletin of Miscellaneous Information* (*Royal Botanic Gardens, Kew*), 6: 228–233.

Bennet, S.S.R., Gaur, R.C. and Sharma, P.N. (1990). *Thirty-seven Bamboos Growing in India New Delhi*. Controller of Publications, Government of India, India.

Bhattacharya, S., Das, M., Bar, R. et al. (2006). Morphological and molecular characterization of *Bambusa tulda* with a note on flowering. *Ann. Bot.*, 98(3): 529–535.

Blatter, E. (1930). The flowering of bamboos. *Part II. J. Bombay Nat. Hist. Soc.*, 33: 135–141.

Blatter, E.B. and Parker, R.N. (1929). The Indian bamboos brought up to date. *Indian Forester*, 55: 541–562.

Bourdillon, T.F. (1895). Seeding of the thorny bamboo. *Indian Forester*, 21: 228–229.

Bradley, J.W. (1914). Flowering of Kya-thaung bamboo (*Bambusa polymorpha*) in the Prome Division, Burma. *Indian Forester*, 40: 526–529.

Brandis, D. (1899). Biological notes on Indian bamboos. *Indian Forester*, 25: 1–25.

Brandis, D. (1906). On some bamboos in Martaban south of Toungoo between the Salwin and Sitang Rivers-II. *Indian Forester*, 32: 179–295.

Broun, A.F. (1887). Seeding of bamboos. *Indian Forester*, 579.

Budke, J.C., Alberti, M.S., Zanardi, C. et al. (2010). Bamboo dieback and tree regeneration responses in a subtropical forest of south America. *Forest Ecol. Manage.*, 260(8): 1345–1349.

Chakraborty, S., Biswas, P., Dutta S. et al. (2021). Studies on reproductive development and breeding habit of the commercially important bamboo *Bambusa tulda* Roxb. *Plants* (*Basel*), 10(11): 2375.

Chen, G.C. and Ma, N.X. (2005). Advances in studies on genetics and breeding of bamboos. *Fores. Res.*, 18(6): 749–754.

Chen, L.N., Cui, Y.Z., Wong, K.M. et al. (2017). Breeding system and pollination of two closely related bamboo species. *AoB Plants*, 9(3): plx021.

Chen, M.Y. (1973). Giant timber bamboo in Alabama. *J. Forestr.*, 71: 777.

Chen, Y.L., Ren, D.T. and Zhu, B.Y. (1995). Observations on flowering habit of *Phyllostachys fimbriligura* and its rejuvenation measures. *J. Zhejiang Forestr. Sci. Techn.*, 15: 50–56.

de Carvalho, A.L., Nelson, B.W., Bianchini, M.C. et al. (2013). Bamboo-dominated forests of the southwest Amazon: Detection, spatial extent, life cycle length and flowering waves. *PLoS One*, 8: e54852.

Du, F., Xue, J.R., Yang, Y.M. et. al. (2000). Study on flowering phenomenon and its type of bamboo in Yunnan in past fifteen years. *Sci. Sil. Sin.*, 36(6): 57–68.

Dutra, J. (1938). Bambusees de Rio Grande du sud Revista. *Sudamerica Bot.*, 5: 145–152.

Fan, F.S. and Qiu, F.G. (1987). The bamboo production and scientific research in India. *J. Bamboo Res.*, 6: 50–68.

Fanshawe, D.B. (1972). The bamboo, *Oxytenanthera abyssinica*: Its ecology, silviculture, and utilization. *Kirkia*, 8: 157–166.

Franklin, D.C. (2010). Synchrony and asynchrony: Observations and hypotheses for the flowering wave in a long-lived semelparous bamboo. *J. Biogeogr.*, 31: 773–786.

Fu, H.J., Fang, T.T., Yang, M. et. al. (2020). Studies on flowering biological characteristics of *Pleioblastus pygmaeus*. *Fores. Res.*, 33(2): 54–60.

Fu, Q.Y. (1985). *Bamboo is a Potential Raw Material for Papermaking in Tropical Countries*. Sichuan Papermaking, pp. 34–41.

Gagnon, P.R. and Platt, W.J. (2008). Reproductive and seedling ecology of a semelparous native bamboo (*Arundinaria gigantea*, Poaceae). *J. Torrey Bot. Soc.*, 135: 309–316.

Giordano, C.V., Rodolfo, A.S. and Austin, A.T. (2009). Gregarious bamboo flowering opens a window of opportunity for regeneration in a temperate forest of Patagonia. *New Phytol.*, 181(4): 880–889.

González, M.E. and Donoso, Y.C. (1999). Seed and litter fall in *Chusquea quila* (Poaceae: Bambusoideae), after synchronous flowering in south-central Chile. *Rev. Chil Hist. Nat.*, 72: 169–180.

González, M.E., Veblen, T.T., Donoso, C. et al. (2002). Tree regeneration responses in a lowland Nothofagus-dominated forest after bamboo dieback in south-central Chile. *Plant Ecol.*, 161(1): 59–73.

Govindan, B., Johnson, A.J., Nair, S.N.A. et al. (2016). Nutritional properties of the largest bamboo fruit *Melocanna baccifera* and its ecological significance. *Sci. Rep.*, 6: 26135.

Gu, Z.J., Yang, H.Q., Sun, M.S. et al. (2012). Distribution characteristics flowering, and seeding of *Dendrocalamus sinicus* in Yunnan, China. *Fores. Res.*, 25(1): 1–5.

Guerreiro, C. (2014). Flowering cycles of woody bamboos native to southern south America. *J. Plant Res.*, 127(2): 307–313.

Guo, Z.H., Ma, P.F., Yang, G.Q. et al. (2019). Genome sequences provide insights into the reticulate origin and unique traits of woody bamboos. *Mol. Plant.*, 12(10): 1353–1365.

Gupta, K.K. (1972). Flowering in different species of bamboos in Cachar district of Assam in recent times. *Indian Forest*, 98: 83–85.

He, T.H. and Ge, S. (2001). Mating system, paternity analysis and gene flow in plant populations. *Acta Phytoecol. Sin.*, 25(2): 144–154.

Holttum, R.E. (1956). The typification of the generic name *Bambusa* and the status of the name *Arundo bambos* L. *Taxon.*, 5: 26–28.

Inoue, M., Ayaka, S., Ayumi, M. et al. (2014). Clonal structure, seed set, and self-pollination rate in mass-flowering bamboo species during off-year flowering events. *PLoS One*, 9: e105051.

Jaksic, F.M. and Lima, M. (2010). Myths and facts on ratadas: Bamboo blooms, rainfall peaks, and rodent outbreaks in south America. *Austr. Ecol.*, 28(3): 237–251.

Janzen, D.H. (1976). Why bamboos wait so long to flower? *Ann. Rev. Ecol. Syst.*, 7: 347–391.

Jiang, Z.H. (2002). *World Bamboo and Rattan* (in Chinese). Liaoning Sci. Tech. Press, Shenyang, China.

Jiao, Q.Y. (1956). Discussion on the flowering habit and stage development of perennial bamboo and sisal hemp. *Plant Physiol. J.*, (2): 13–19.

Kadambi, K. (1949). On the ecology and silviculture of *Dendrocalamus strictus* in the bamboo forests of Bhadravati division, Mysore State, and comparative notes on the species *Bambusa arundinacea*, *Oxytenanthera monostigma* and *Oxytenanthera stocksu*. *Indian Forester*, 75: 289–299.

Kawamura, S. (1927). On the periodical flowering of the bamboo. *Jpn. J. Bot.*, 3: 335–349.

Khan, M.A.W. (1972). Propagation of *Bambusa vulgaris*: Its scope in forestry. *Indian Forester*, 359–362.

Kitamura, K. and Takayuki, K. (2011). Estimation of outcrossing rates at small-scale flowering sites of the dwarf bamboo species, *Sasa cernua*. *J. Plant Res.*, 124(6): 683–688.

Kurz, S. (1876). Bamboo and its use. *Indian Forester*, 1: 219–269.

Kwe, T. (1904). The flowering of *Bambusa polymorpha*. *Indian Forester*, 30: 244–245.

Li, Z. and Denich, M. (2004). Is Shennongjia a suitable site for reintroducing giant panda? An appraisal on food supply. *Environmentalist*, 24(3): 165–170.

Lin, E.P. (2009). *Functions of AP1/SQUA-? REV-?TB1-like Genes and Isolation and Expression of microRNAs in Phyllostachys Praecox.* Dissertation's thesis. Zhejiang University, Hangzhou, China.

Lin, S.Y. and Ding, Y.L. (2007). Studies on the floral biological characteristics of three bamboo species of *Phyllostachys*. *J. Forestr. Eng.*, 21: 52–55.

Lin, S.Y. and Ding, Y.L. (2013). Studies on the breeding system in Shibataea chinensis and *Arundinaria simonii* f. *heterophylla*. *J. Nanjing Forest. Univ.*, 37(3): 1–5.

Lin, S.Y., Fan, T.T., Jiang, M.Y. et al. (2017). The revision of scientific names for three dwarf bamboo species (cultivar) based on the floral morphology. *J. Nanjing Forestr. Univ.*, 41(1): 189–193.

Lin, S.Y., Hao, J.J., Hua, X. et al. (2009). The megasporogenesis, microsporogenesis, and the development of their female and male gametophyte in *Menstruocalamus sichuanensis*. *J. Nanjing Forest. Univ.*, 33(3): 9–12.

Lin, S.Y., Shi, W.W., Miu, B.B. et al. (2010a). Research advances in reproduction biology of bamboos. *World Bamboo Rattan*, 8(2): 1–6.

Lin, X.C., Yuan, X.L., Lin, R. et. al. (2010b). Studies on floral biology of *Phyllostachys violascens*. *J. Fores. Environ.*, 30(4): 333–337.

Lin, S.Y., Li, J., Zhao R. et al. (2015). Research on the flowering biological characteristics of *Bambusa multiplex* in Nanjing City. *J. Nanjing Forest. Univ.*, 39(2): 52–56.

Liu, H.W., Zhao, W.Q., Xiao, J. et al. (2021). Flowering biological characteristics of *Oligostachyum spongiosum*. *Acta Bat. Boreal.*, 41(11): 1853–1862.

Liu, Y.Y. and Fu, J.H. (2007). Bamboo in habitat of giant panda and its flowering phenomenon. *World Bamboo Rattan*, 5: 1–4.

Lowndes, D.G. (1947). Flowering of bamboos. *J. Bombay Nat. Hist. Soc.*, 47: 180.

Lu, J.L. (1980). Studies on flowering and regeneration of *Phyllostachys reticulata*. *J. Henan Agric. Univ.*, 11–20.

Marchesini, V.A., Sala, O.E. and Austin, A.T. (2010). Ecological consequences of a massive flowering event of bamboo (*Chusquea culeou*) in a temperate forest of Patagonia, Argentina. *J. Veget. Sci.*, 20(3): 424–432.

Matsuo, A., Tomimatsu, H., Suzuki, J.I. et al. (2014). Female and male fitness consequences of clonal growth in a dwarf bamboo population with a high degree of clonal intermingling. *Ann. Bot.*, 114(5): 1035–1041.

McClure, F.A. (1966). *The Bamboos*. Harvard University Press, Cambridge, MA, USA.

Mitra, G.N. and Nayak, Y. (1972). Chemical composition of bamboo seeds (*Bambusa arundinacea* Wild). *Indian Forest.*, 98: 479–481.

Mizuki, I., Sato, A., Matsuo, A. et al. (2014). Clonal structure, seed set, and self-pollination rate in mass-flowering bamboo species during off-year flowering events. *PLoS One*, 9: e105051.

Nadgauda, R.S., John, C.K. and Mascarenhas, A.F. (1993). Floral biology and breeding behavior in the bamboo *Dendrocalamus strictus* Nees. *Tree Phys.*, 13: 401–408.

Nath, A.J. and Das, A.K. (2010). Gregarious flowering of a long-lived tropical semelparous bamboo *Schizostachyum dullooa* in Assam. *Curr. Sci.*, 99(2): 154–155.

Nelson, B.W. (1994). Natural forest disturbance and change in the Brazilian Amazon. *Remote Sens. Rev.*, 10: 105–125.

Nicholls, J. (1895). The flowering of the thorny bamboo. *Indian Forester*, 21: 90–95.

Ning, C.Q. and Dai, Q.H. (1995). Study on hybrid breeding of *Bambusa pervariabilis × Dendrocalamopsis daii*. *Guangxi Forest. Sci.*, 24(4): 167–168.

Numata, M. (1970). Conservation implications of bamboo flowering and death in Japan. *Biol. Conserv.*, 2(3): 227–229.

Pearson, A.K., Pearson, O.P. and Gomez, I.A. (1994). Biology of the bamboo *Chusquea culeou* (Poaceae: Bambusoideae) in southern Argentina. *Vegetatio.*, 111: 93–126.

Poudyal, P.P. (2009). Bamboo flowering in Sikkim and elsewhere in India. *Mag. Am. Bamboo Soc.*, 30: 9–10.

Qin, Z.S. (1995). Study on reproductive characteristic of *Bashania fangenia*. *Acta Bot. Boreali Occidentalia Sinica*, 15: 229–233.

Qin, Z.S., Cai, X.S. and Huang, J.Y. (1989). Seed characteristics and natural regeneration of arrow bamboo (*Bashania fangenia*). *J. Bamboo Res.*, 8: 1–12.

Raizada, M.B. (1948). A litter-known Burmese bamboo (*Sibocalamus copelandi*). *Indian Forester*, 74: 7–10.

Ram, H.Y.M. and Gopal, B.H. (1981). Some observations on the flowering of bamboos in Mizoram. *Curr. Sci.*, 50: 708–710.

Ray, P.K. (1952). Gregarious flowering of a common hill bamboo *Arundinaria maling* Gamble. *Indian Forester*, 78: 89–91.

Ruiz-Sanchez, E. (2013). *Otatea ramirezii* (Poaceae: Bambusoideae: Bambuseae) flower description and the importance of the Mexican national living bamboo collection. *Phytotaxa*, 150(1): 54–60.

Seethalakshmi, K.K. and Muktesh, K.M.S. (1998). *Bamboos of India: A Compendium*. International Network for Bamboo and Rattan, Beijing, China.

Seifriz, W. (1920). The length of the life cycle of a climbing bamboo. A striking case of sexual periodicity in *Chusquea abietifolia* Griseb. *Am. J. Bot.*, 7: 83–94.

Seifriz, W. (1923). Observations on the causes of gregarious flowering in plants. *Am. J. Bot.*, 10: 93–112.

Seifriz, W. (1950). Gregarious flowering of Chusquea. *Nature*, 165: 635–636.

Sendulsky, T., Filgueiras, T.S. and Burman, A.G., (1987). Fruits, embryos, and seedlings. *In*: Soderstrom, T.R., Hilu, K.W., Campbell, C.S. and Barkworth, M.E. (eds.). *Grass Systematics and Evolution*. Smithsonian Institution Press, Washington, DC, USA.

Shah, N.C. (1968). Flowering of the bamboo, *Dendrocalamus hookerii* and *Dendrocalamus strictus* in Assam and Bihar states. *Indian Forester*, 94: 717.

Shao, J.X. (1986). Preliminary survey on the ecological characteristics of *Fargesia denudate*. *Chin. J. Ecol.*, 5: 41–44.

Singleton, G.R., Belmain, S.R., Brown, P.R. et al. (2010). *Rodent Outbreaks: Ecology and Impacts Los Banos*. Int. Rice Res. Inst., PA, Philippines.

Sun, M.S., Yan, B., Xu, T. et al. (2015). *Resources and Utilization of Bamboo Plants*. Sci. Press, Beijing, China.

Tagle, L., Roberto, M., Mireya, B. et al. (2013). Determination of minimal age of five species of Chusquea bamboos through rhizome analysis as a tool to predict the flowering in southern Chile. *Rev. Chil. Hist. Nat.*, 86: 423–432.

Tan, H.C. and Qi, L.P. (1994). Study on seedling propagation by clones of clump bamboo. *J. Bamboo Res.*, (1): 62–73.

Tewari, D.N. (1993). *A Monograph on Bamboo*. Dehra Dun: International Book Distributors.

Tingle, A. (1904). The flowering of the bamboo. *Nature*, 70: 342.

Tredici, P.D. (1998). The first and final flowering of Muriel's bamboo. *Arnoldia*, 58: 11–17.

Troup, R.S. (1921). *The Silviculture of Indian Trees*. Oxford University Press, New York, USA.

Tzvelev, N.N. (1976). *Poaceae USSR (Zlaki SSSR)*. Nauka, Leningrad.

Ueda, K. (1960). Studies on the physiology of bamboo: With reference to practical application. *Bull. Tokyo Univ. Forests*, 30: 1–167.

Venkatesh, C.S. (1984). Dichogamy and breeding system in a tropical bamboo *Ochlandra travancorica*. *Biotropica*, 16: 309–312.

Wang, T.T. and Chen, M.Y. (1971). Studies on bamboo flowering in Taiwan. *Taipei Nat. Taiwan Univ. Forest Exp. Sta Tech. Bull.*, 87: 27.

Wang, X.H. and Wu, H.M. (2009). Biological characteristics study of *Phyllostachys Heteroclada*'s flowering. *J. Chengdu Univ.*, 28(3): 195–198.

Watanabe, M., Ueda, K., Manabe, I. et al. (1980). Flowering, seeding, germination, and flowering periodicity of *Phyllostachys pubescens*. *Jpn. Forestr. Soc.*, 64(3): 107–111.

Wu, G.M. (1988). Reproduction of bamboo species of *Phyllos, tachys*: I. Flowering of bamboo species of *Phyllostachys*. *J. Nanjing Forestry Univ.*, (1): 60–67.

Xie, N., Chen, L.N., Dong, Y.R. et al. (2019). Mixed mating system and variable mating patterns in tropical woody bamboos. *BMC Plant Biol.*, 19: 418.

Xie, N., Chen, L.N., Wong, K.M. et al. (2016). Seed set and natural regeneration of *Dendrocalamus membranaceus* Munro after mass and sporadic flowering in Yunnan, China. *PLoS One*, 11(4): e0153845.

Xing, X.T. (2003). *Study on the Genetic Variation of Populations and Improved Seeds Breeding of Dendrocalamus latiflorus* Munro. Chinese Academy Forestry, Beijing.

Xing, X.T., Fu, M.Y. and Xiao, X.T. (2005). Biological characteristics of flowering and controlled pollination of *Dendrocalamus latiflorus* Munro. *J. Beijing Forest. Univ.*, 27(6): 103–107.

Xu, P.F. (2019). *Hybridization between Phyllostachys edulis and P. violascens, and Identification and Early Growth Characteristics of the Hybrids*. M.D. Dissertation, Zhejiang Agriculture and Forestry University, Hangzhou.

Xu, Y.K., Li, Q., Zhang, S.H. et al. (2012). Textual research on the flowering history of *Pleioblastus intermedius* and analysis of death. *For. By-Product Speciality China*, (1): 86–87.

Yang, H.Q., Yang, J.B., Peng, Z.H. et al. (2008). A molecular phylogenetic and fruit evolutionary analysis of the major groups of the paleotropical woody bamboos (Gramineae: Bambusoideae) based on nuclear ITS, GBSSI gene and plastid trnL-F DNA sequences. *Mol. Phylogen. Evol.*, 48(3): 809–824.

Yao, W.J., Jiang, M.Y., Wang, X. et al. (2020). Biological analysis of flowering and pollen germination in *Sasaella kongosanensis* 'Aureostriatus'. *J. Northeast Forestr. Univ.*, 48(3): 13–18.

Yi, T.P. (2008). *Iconographia Bambusoidearum Sinicarum*. Sci. Press, Beijing, China.

Yu, Q.Z., Wu, M., Zhao, B.H. et al. (1987). A preliminary study on flowering habits of staple food bamboo of giant pandas. *Sichuan Forestr. Sci. Technol.*, 8: 49–54.

Yuan, J.L., Fu, M.Y. and Jiang, J.M. (2005). The genetic linkage map construction of *Dendrocalamus latiflorus* Munro. *J. Forest. Engineer*, 19(4): 21–23.

Yuan, X.L., Huang, Q.C. and Peng, H.Z. (2008). Correct understanding of bamboo flowering. *Zhejiang Forestr.*, (1): 32–33.

Yue, X.H., Zhao, R. and Lin, S.Y. (2018). Flowering biological characteristics of *Phyllostachys glauca* McClure. *Jiangsu Agric. Sci.*, 46(10): 117–122.

Zhang, C.X., Xie, Y.F. and Ding, Y.L. (2003). The studies of leaf senescence of *Pseudosasa amabilis* var. *convexa* during flowering and seeding stage. *J. Nanjing Forestr. Univ.*, 27: 59–61.

Zhang, G.C. and Chen, F.S. (1986). Studies on crossbreeding of Bamboo. *Guangdong Forest Sci. Technol.*, (3): 1–5.

Zhang, H., Zou, J.Y. and Wang, Y.Y. (1994). Inquiry to flowering and seed bearing of *Chimonobambusa quadrangularis* Makino in Jinfoshan of Tongzi county. *J. Bamboo Res.*, 13(2): 66–69.

Zhang, J.X., Luo, W. and Ming, Y. (1992a). A study on flowering and fruitage of *Ampelocalamus scandens*. *J. Bamboo Res.*, 11(3): 97–99.

Zhang, S.S. (1977). Methods of flowering withered and restored in bamboo forest. *Sichuan Forestr. Sci. Technol.*, 19–21.

Zhang, W.Y. and Ma, N.X. (1989). Biological characteristics of bamboo plants in flowering stage. *Forest Res.*, 2(6): 596–600.

Zhang, W.Y. and Ma, N.X. (1990). Vitality of bamboo pollens and natural pollination in bamboo plants. *Forest Res.*, 3(3): 250–255.

Zhang, W.Y. and Ma, N.X. (1992). A study on flowering and fruiting of *Phyllostachys bambusoides*. *J. Bamboo Res.*, 11(2): 11–25.

Zhang, W.Y., Ma, N.X., Wu, L.L. et al. (1992b). A study on flowering and fruiting of *Phyllostachys bambusoides*. *J. Bamboo Res.*, 11(2): 15–25.

Zhang, Y., Zhang, L., Lan, F.R. et al. (2018). The first record of flowering and bearing about *Pleioblastus yixingensis* (Bambusoideae: Poaceae). *J. Trop. Subtrop. Bot.*, 26: 171–177.

Zhao, W.Q., Wu, Z.C., Xiao, J. et al. (2020). Flowering biological characteristics of *Pseudosasa viridula*. *Forest Res.*, 33(3): 31–38.

Zheng, X., Jiang, M.Y., Zhang, L. et al. (2016). Fruit morphological characteristics of thirteen bamboo species. *J. Plant Resour. Environ.*, 25: 96–103.

Zheng, X., Lin, S., Fu, H. et al. (2020). The bamboo flowering cycle sheds light on flowering diversity. *Front. Plant Sci.*, 11: 381.

Zhong, Y., Yue, J., Lou, C. et al. (2017). Floral organ and breeding system of *Dendrocalamus latiflorus*. *Sci. Silv. Sin.*, 53(1): 1–10.

Zhou, F.C. (1984). *Silviculture of Bamboo Forests*. China Forestry Publishing House, Beijing, China.

Chapter 7
Circadian Clock Genes and their Role in Bamboo Flowering

Smritikana Dutta,[1,2] *Sukanya Chakraborty*[1] and *Malay Das*[1,*]

7.1 Introduction

Among different exogenous cues, light is one of the most important signals that govern the timing of flower induction in an angiosperm. For determining flowering time, both duration of the light versus the dark phase (photoperiod) and spectrum of light (e.g., red light or blue light) should be considered. To sense the minute differences in the range of light that the plants are exposed to, they have adopted different photoreceptors, which are enabled to accept lights of specific wavelengths. Although the duration of day and night are perfectly balanced in the environment, plants can measure either the length of light or dark to respond photoperiodically (Hamer, 2009). This internal timekeeping mechanism of the plants, which helps them to sense the deviation of photoperiod, is known as the circadian clock.

The term 'circadian', which means "about a day" in Latin was first introduced by Franz Halberg in 1959. The specific rhythmic behavior of an organism with respect to the circadian clock is known as circadian rhythm. Despite a sudden deprivation of exogenous signals up to a limited time, the internal diurnal/circadian rhythm is not perturbed, suggesting the existence of an endogenous biological clock (free-running period), which primarily governs the plant response towards photoperiodism. The first witness of the presence of such an endogenous clock response was noticed in *Tamarindus indicus*. The leaves moved in a cyclic manner when the plant was placed in a constantly dark place. A similar trend was also noticed in the case of the opening and closing of its flowers and was called the "flower clock" (Linnaeus, 1751). About a century later, it was experimentally validated that the endogenous clock response or free-running period literally exists in plants, and it was 22–23 hours in *Mimosa*

[1] Department of Life Sciences, Presidency University, Kolkata, India.
[2] Special Centre for Molecular Medicine, Jawaharlal Nehru University, New Delhi, India.
* Corresponding author: malay.dbs@presiuniv.ac.in

pudica (de Candolle, 1832) and 24 hours in *Neurospora crassa* (Sulzman et al., 1984). Over the past few decades, extensive research has suggested that the circadian rhythm and its endogenous regulation can be inherited.

7.2 Bamboo Flowering: Types and their Ecological Impact

Over decades, extensive research works suggest that the circadian rhythm and its endogenous regulation of the behavior of bamboo remain elusive due to two major reasons. Firstly, flowering happens in stands or culms originating from the same genetic stock (seed or rhizome) after a prolonged, predetermined period of vegetative growth, extending up to 120 years (Janzen, 1976; Zheng et al., 2020). Secondly, this unusual flowering is followed by mass death of the flowering culms, known as monocarpy or semelparity. Therefore, such large-scale plant death occurring over a wide geographical area causes ecological as well as socioeconomic imbalance (John and Nadgauda, 2002). Thus, the flowering of bamboo remains an enigma for decades (Ramanayake, 2006).

The duration of the vegetative growth phase varies across species (Guangchu, 2002). Based on the duration of the vegetative phase and the nature of flowering, bamboo flowering is of three major types (Brandis, 1906). In the annual flowering bamboos, flowering happens in every year, but seeds are either not produced or produced with very low viability, and the plants generally remain alive. This phenomenon has been observed in *Bambusa atra, Schizostachyum brachycladum,* and *Ochlandra stridula* (Gamble, 1896; Rhind, 1945; Sharma, 1991; Ramanayake and Yakandawala, 1998; Koshy, 1997).

The flowering of some woody bamboo species occurs simultaneously over a wide geographical area, or even across continents, a phenomenon known as gregarious flowering (Brandis, 1906). The endogenous time clock of a particular species leads to this type of synchronous flowering (Thapliyal et al., 2015). Gregarious flowering generally persists for two to three years until the bamboos die and spreading of flowering takes place over the entire area. This type of flowering has been noticed in *Fargesia murielae, Arundinaria falcate, B. bambos,* and in *Thamnocalamus spathiflorus* subsp. *Spathiflorus* (Bhattacharya et al., 2009). Another kind of bamboo flowering is sporadic flowering, where a few culms undergo flowering, e.g., in *Chusquea* sp. (Seifriz, 1950), and *B. tulda* (Bhattacharya et al., 2006). Sporadic flowering can be triggered by adverse environmental conditions such as drought. However, irregular flowering populations that were initially sporadic flowering may over time be transformed into gregarious populations.

To address the ecological importance of late-flowering behavior in bamboo, three major hypotheses were proposed: (i) habitat modification hypothesis, (ii) seed predator hypothesis, and (iii) resource allocation hypothesis (Wang et al., 2016). According to the habitat modification hypothesis, late-flowering prepares the ground to increase the chance of successful growth and germination of the seedlings. The adult culms grow vigorously by vegetative propagation, which in turn changes the micro climatic condition as well as stores massive bioresources in advance for the future generation (Stearns, 1980). The second concept, i.e., the seed predator hypothesis, suggests that true delayed flowering produces many seeds, which are too numerous for the predators to consume (Kakishima et al., 2011). Therefore, the remaining seeds get a chance to germinate and thereby initiate the regeneration of forests. The third hypothesis could explain the utility of the long vegetative phase (Gadgil and Bossert, 1970). According to this theory, the extended vegetative phase is dedicated to the acquisition of sufficient resources for flowering. On one hand, the requirement for energy gradually increases as the plants move towards the reproductive phase. On the other, the total vegetative growth terminates during this phase (Harper and Ogden, 1970). Therefore, culms sacrifice their lives for the sake of successful reproduction (Abrahamson and Caswell, 1982).

Bamboo serves as one of the most predominant species in most of the Asian forests. The synchronous mass flowering in bamboo is followed by the massive production of seeds and subsequently the death of culms. This causes a gross decline in the forest population as well as drastic changes in forest dynamics (Marchesini et al., 2009; Sertse et al., 2011; Austin et al., 2012). Other ecological consequences of mass flowering include its impact on the availability of light intensity, nutrient cycling, decomposition of organic matters, interactions among species, and survival of seedlings in the forest. For example, drastic changes in light intensity after the mass death of bamboo culm results into quick growth of many tall, tree species, which in turn influence the dynamics of the forest population (Taylor et al., 2004; Marchesini et al., 2009). As a result, this altered population dynamics may impact the interaction among the organisms present at diverse trophic levels (Raffaele et al., 2007). Bamboo flowering has devastating effects on the giant panda. Pandas usually feed on young, bamboo shoots. Therefore, flowering, and sudden, mass death results into severe scarcity of food for pandas. In addition, growth of bamboo shoots helps in proper nutrient cycling and litter production to maintain the fertility of soil (Christanty et al., 1997; Singh and Singh, 1999). Therefore, large scale flowering causes depletion in important soil mineral contents, especially nitrogen (Takahashi et al., 2007). After the occurrence of flowering and mass extinction of the bamboo vegetation, restoration of vegetation may take up to 15 years to develop (Janzen, 1976).

At the time of gregarious flowering, large number of seeds are produced which are subsequently consumed by rodents. These seeds are highly enriched in terms of their nutrient contents and result into a sudden increase in the reproduction rate of rodents. However, with the arrival of monsoon seasons, germination of majority seeds takes place, which may cause food scarcity for the rodents. This in turn forces the rodents to migrate to nearby agricultural field causing massive crop loss and ultimately famine. Therefore, bamboo flowering is believed to be a bad omen worldwide (Mohan Ram and Hari Gopal, 1981; John and Nadgauda, 2002).

7.3 Photoperiodism in Flowering: What is Known about Bamboo?

One of the most important environmental stimuli that can regulate flowering time is light (Levy and Dean, 1998). The amount of light available in a daily 24-hour cycle is known as photoperiod. The duration of day length changes in a cyclic manner throughout the year. Both plants and animals can sense changes in photoperiod and consequently respond towards it (Levy and Dean, 1998). Role of photoperiodism in the determination of flowering time was first discovered by Garner and Allard (1920). Their findings revealed that exposure to even a very short light period can successfully induce flowering in soybean and tobacco. Plants can be classified into three major groups based on their requirement for day length to induce flowering. There are short-day plants (SDP), e.g., *O. sativa*, long-day plants (LDP), e.g., *A. thaliana*, and day-neutral plants (DNP), e.g., *Lycopersicum esculentum*. In SDP, induction of flowering requires photoperiod, which is shorter than the critical day length (CDL), while in case of LDP it is longer than CDL. In case of DNPs, flowering time is not affected by photoperiod. Thus, CDL acts as the regulatory switch that controls the transition from non-inductive to inductive phase via photoperiod. However, the length of CDL may vary from plant to plant.

Extensive studies have been conducted over decades to understand the critical role of day length on seasonal regulation of flowering (Klebs, 1913). However, not many studies could have been conducted to assess the impact of photoperiod in bamboo flowering due to many reasons such as presence of prolonged vegetative phase (Janzen, 1976; Zheng et al., 2020) and the actual season of induction of flowering varying across species (Clerget, 2021). Despite such limitations, the photosensitivity of bamboo has already been studied in *Pleioblastus variegatus* based on multiple vegetative characters (Tanaka and Takimoto, 1972). The flowering time of *P. variegatus* was in spring and the seedlings were found to be critically sensitive to CDL. When grown under SD and LD, the plants revealed noticeable phenotypic differences with respect to chlorophyll content, internode elongation, and shoot branching. Interestingly, under the SD condition, when the dark phase was interrupted with a short exposure of light, the phenotypes resemble that of the LD condition. A recent finding suggests that majority of the bamboo species either flowers in April–July or October–December suggesting existence of day length sensitivity in the plant group with respect to flowering (Clerget, 2021). In addition, it has been identified that bamboos maintain a lunar clock (exact sun-moon phasing of the year), which was retained in their cellular memory from the day of seedling emergence. The induction of flowering happens only after occurrence of the species-specific lunar phasing (Clerget, 2021). However, the impact of light duration on gregarious or sporadic flowering is yet to be discovered.

7.4 Early Background on Important Photoreceptors Involved in Flowering

The duration (photoperiod) and spectrum of light (e.g., red, or blue light) play important roles in floral transition (Lin, 2000). In the photoperiodic pathway, a group of photoreceptors first receive a light signal and eventually transmits it to downstream circadian clock genes to relay the floral inductive signal (Thomas and Vince-Prue, 1997). Plants harbor a range of photoreceptors to detect light having differing spectra. For example, phytochromes (PHY) can absorb red (655 nm) and far-red light (850 nm), while blue light (580 nm) and UV-A (400 nm) are absorbed by cryptochromes (CRY) and zeitlupe (ZTL)/lov kelch protein 2 (LKP2)/flavin binding kelch repeat F-box 1 (FKF1) (Yanovsky and Kay, 2003).

7.4.1 Red Light Photoreceptors

Phytochromes are the group of chromoproteins which have two inter-convertibles, photo sensory isomeric forms. They are Pr (red-light-absorbing) and Pfr (far-red-light absorbing) (Kendrick and Kronenberg, 1994; Hughes, 1999). In *A. thaliana* PHY family consists of five members, which are PHYA, B, C, D, and E (Quail, 2002). The quantity, quality, and timing of light exposure can affect the regulation of PHY signaling in different manners (Quail et al., 1995). PHYB inhibits floral induction by degrading CO proteins (Endo et al., 2013). In *A. thaliana*, mutant PHYB plants revealed early flowering in both SD and LD in comparison to wild type (Mockler et al., 1999). This finding was further supported by PHYB mutant pea (Iv-1) (Weller and Reid, 1993), and sorghum plants (Ma3R) (Pao and Morgan, 1986). On the other hand, double mutant plants harboring PHYA mutation in combination either with PHYB or PHYC revealed early flowering in rice. This confirms their positive effect on floral induction (Takano et al., 2005). Moreover, both PHYD and PHYE can play inhibitory effects on floral induction (Lin, 2000). In majority plants, PHYB/D/E and PHYA interact antagonistically with respect to flowering (Figs. 7.1A, 7.1B).

7.4.2 Blue Light Photoreceptors

Majority of the blue light photoreceptors have an inductive effect on flowering, which is in contrary to the effects observed for red light photoreceptors. Two important blue light photoreceptors, i.e., cryptochromes (*CRY1* and *CRY2*) are genetically redundant and promote flowering by entertaining the clock components of *A. thaliana* (Somers, 2005). Among three *CRY* homologs (*OsCRY1a*, *OsCRY1b*, and *OsCRY2*), only *OsCRY2* performs flower promoting activity (Hirose et al., 2006; Fig. 7.1A). Over expression of *CRY2* in tomato resulted in altered rate of leaf production and delayed flowering (Giliberto et al., 2005). The *CRYs* induce flowering through an interconnected loop with the *PHYS* (Fig. 7.1A). On the other hand, another group of blue light photoreceptors *ZTL/LKP2/FKF1* promotes flowering by directly regulating the circadian clock genes. The *ZTL* mutants possessed altered circadian rhythms of the clock components (Kevei et al., 2006). However, in case of *ZTL* the circadian rhythm exists at the translational level, which is crucial for the regulation of the downstream clock gene *TOC1* (Kim et al., 2003). The circadian rhythm of *FKF1* exists in the mRNA level (Imaizumi et al., 2003).

7.5 Early Background on Molecular Mechanisms Controlling the Circadian Clock Regulation of Flowering

In plants, the circadian clock is composed of a series of multiple transcriptional and translational feedback loops (TTFLs) to maintain the rhythmic pattern in expression (Zhang and Kay, 2010). The TTFLs are interconnected through complex positive and negative interactions among clock genes (Fig. 7.1A; Fogelmark and Troein, 2014; Hsu et al., 2014; McClung, 2014). A mathematical study predicted that presence of a greater number of interlocked interactions enable the flexibility of the clock action (Rand et al., 2006). Detailed transcriptome profiling studies on *A. thaliana* suggests that 30% of the total transcripts of the cell are controlled by circadian clock (Michael

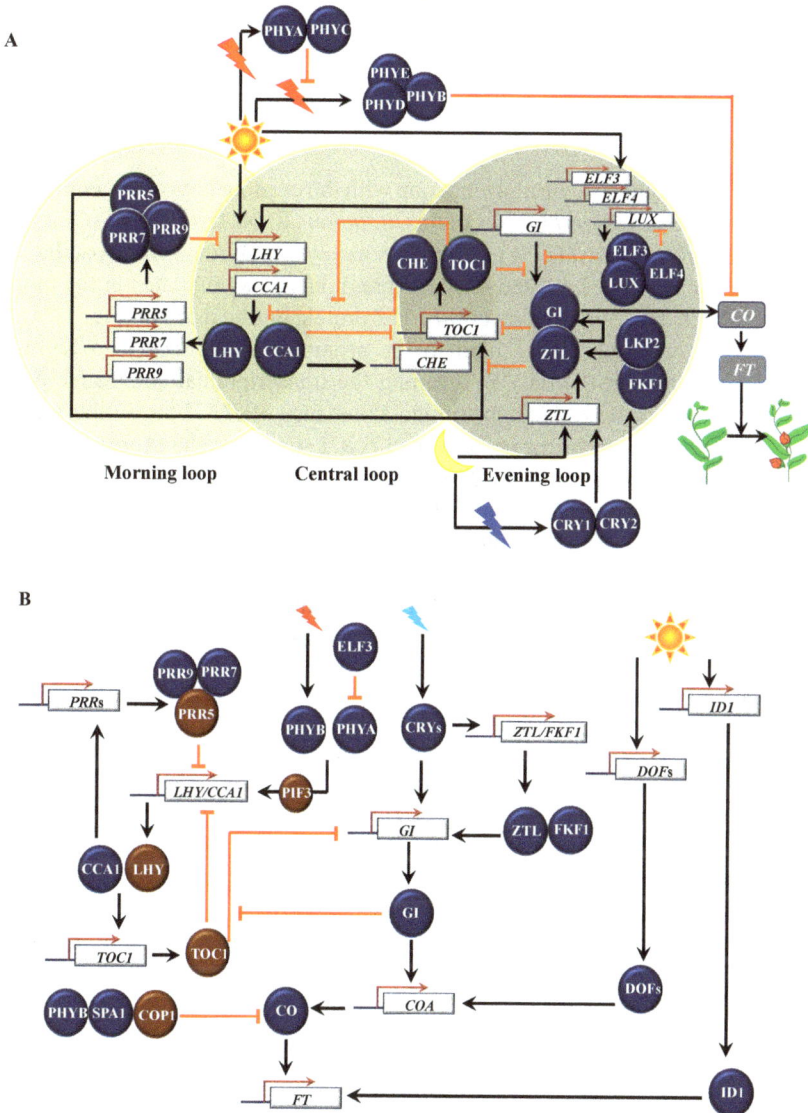

Fig. 7.1. Schematic representation of transcriptional and translational feedback loop of the genes involved in photoperiodic pathway.

Note: (A) Arabidopsis (B) Bamboo. Arrows indicate activation and T-bars indicate repression of gene activity. Blue circles denote upregulated proteins and brown denotes downregulated.

Source: This figure was prepared based on the information obtained from published reports of Zhang et al., 2012; Peng et al., 2013; Gao et al., 2014; Biswas et al., 2016; Guo et al., 2016; Dutta et al., 2018; Wang et al., 2020; Basak et al., 2021.

et al., 2008). Similar observations were also recorded in *O. sativa*, *Carica papaya*, *Zea mays*, *Glycin max* and *Populus trichocarpa* (Zdepski et al., 2008; Hayes et al., 2010; Hoffman et al., 2010; Khan et al., 2010; Filichkin et al., 2011; Marcolino-Gomes et al., 2014). In *A. thaliana*, the circadian clock is composed of three defined

TTFLs, which are a central loop, a morning loop, and an evening loop (Fig. 7.1A; McClung, 2008; Harmer, 2009; McClung and Gutiérrez, 2010; Pruneda-Paz and Kay, 2010).

7.5.1 The Central Loop

The central loop connects the morning loop with the evening loop and is composed of *Timing Of Cab Expression 1* (*TOC1*), and *Circadian Clock-Associated 1* (*CCA1*) and/ *Late Elongated Hypocotyl* (*LHY*) (Alabadí et al., 2001). During central oscillation of the circadian clock, the *LHY* and/*CCA1* expressions are upregulated in the morning which results in the accumulation of the LHY protein during midday (Fig. 7.1A; Kim et al., 2003). *LHY* and *CCA1* bind to the promoter of the *TOC1* to repress its expression throughout the day. By evening the transcriptional abundance of *LHY* drops, which in turn results into maximum abundance of *TOC1* transcripts (Alabadí et al., 2001). Consequently, the up-regulated *TOC1* stimulates the expression of LHY. The loss of function mutation of *TOC1* in *A. thaliana* resulted in lower transcript accumulation for *CCA1/LHY*, which further confirms the positive regulatory role of *TOC1* on *CCA1/LHY* (Ito et al., 2008). The *TOC1* mediated activation of *LHY* involves an additional candidate *CCA1 Hiking Expedition* (*CHE*), which represses *LHY* in the absence of *TOC1*. The *TOC1* represses the expression of *CHE* to activate the transcriptional level of *LHY* (Fig. 7.1A; Pruneda-Paz et al., 2009).

7.5.2 The Morning Loop

The second loop, i.e., the morning loop consisted of *Pseudo-Response Regulators* (*PRR* 5, 7, and 9) and *LHY* (Fig. 7.1A; Nakamichi et al., 2005; Salome and McClung, 2005). The *PRRs* are close homologs of *TOC1* and in morning they suppress *LHY* transcription by binding to its promoter (Nakamichi et al., 2010). The *PRR5* also regulates the *LHY* expression in an indirect way by enhancing the phosphorylation of *TOC1* protein (Wang et al., 2010). Interestingly, *PRRs* can also repress their own transcription by binding to their promoters (Fig. 7.1A; Nakamichi et al., 2010).

7.5.3 The Evening Loop

The third one is the evening loop, which mainly demonstrates the regulation pattern between *TOC1* and *Gigantea* (*GI*). The GI passes circadian output to the master clock integrator constans (*CO*; Herrero et al., 2012). Thus, maintaining appropriate *GI* transcript level is essential for the photoperiodic induction of flowering. The regulation of *TOC1* and *GI* is governed by evening complex (EC), which is composed of ELF3-ELF4-LUX proteins (Herrero et al., 2012). Both EC and *TOC1* can suppress *GI* through-out the day, while by evening *GI* level attains stability by blue light mediated stabilization of *Zeitlupe* (*ZTL*) (Kim et al., 2007). In presence of blue light, *ZTL* with the help of two additional candidates (*FKF1* and *LKP2*) mediate *TOC1* degradation and thereby reaffirm stabilization of *GI*. This is quite evident in the loss of function mutant lines *ZTL FKF1 LKP2* of *A. thaliana*. The higher accumulation of *TOC1* was noticed in the triple mutant line of *ZTL FKF1 LKP2*, but not in the single mutant *ZTL* (Baudry et al., 2010). By the end of night, *EC* was repressed by its autoregulatory mechanism to get the circadian clock ready for its next cycle (Helfer et al., 2011).

It is quite evident from the discussions above that the key components of the circadian clock of flowering are the *LHY*, *TOC1*, *ZTL*, and *GI* and have therefore been reviewed elaborately in the following sections.

7.6 *LHY/CCA1*, *TOC1/PRR*, *ZTL*, *GI* Gene Homologs Identified from Diverse Angiosperm Plants

In *A. thaliana*, *LHY* and *CCA1* genes belong to the MYB domain transcription factor family. The MYB proteins are usually composed of an N terminal MYB DNA binding domain along with a C terminal modular region (Ambawat et al., 2013). However, based on the number of adjacent MYB repeats these proteins can be grouped as 1R, R2R3, 3R, and 4R-MYB (Dubos et al., 2010). Among them the R2R3-MYB proteins are most abundantly found in the plant kingdom (Jiang et al., 2004; Wilkins et al., 2009). For instance, the *LHY/CCA1* gene clustered with R2R3-MYB (Lu et al., 2009). The *LHY/CCA1* gene of dicotyledonous and monocotyledonous plants had evolved from its common eudicot ancestor (Fig. 7.2A; Takata et al., 2008). However, after the divergence of monocotyledonous lineage successive duplication events (α, β) resulted into the emergence of multiple homologs in *P. vulgaris*, *P. trichocarpa* and *A. thaliana* (Fig. 7.2A; Takata et al., 2008). The *LHY/CCA1* genes have been identified from several dicotyledonous and many monocotyledonous plant species apart from the reference genomes, *A. thaliana* and *O sativa*.

TOC1 is a member of PRR gene family and an important regulator of the plant circadian clock (Farre and Liu, 2013). The *A. thaliana* PRR gene family is composed of five genes (*TOC1*, *PRR3*, *PRR5*, *PRR7*, and *PRR9*), which enormously contribute towards the clock-mediated functions that include flower induction (Nakamichi et al., 2007; Nakamichi et al., 2012). All PRRs including *TOC1* contain a N terminal pseudo-response regulator domain (PRR), a C terminal CONSTANS, CONSTANS LIKE, and *TOC1* (CCT) domain (Fig. 7.2B; Putterill et al., 1995; Mizuno and Nakamichi, 2005). The *PRR* gene family had been evolved from a common ancestor, true Response Regulator (Fig. 7.2B). During evolution, *TOC1* diverged first, which was subsequently followed by the divergence of next two clades PRR5-PRR9 and PRR3-PRR7 (Fig. 7.2B; Satbhai et al., 2011). However, gene duplication events in monocotyledons resulted in emergence of PRR37 and PRR73, which were homologous to PRR7 and emergence of PRR59 and PRR95, which were homologous to PRR5 (Fig. 7.2B; Takata et al., 2010). Recent findings suggest that members of TOC1/PRR gene family are well conserved diverse monocotyledonous plants and are usually present in multiple copies. For example, five *TOC1/PRR* genes had been found in *O sativa* (Poaceae; *OsTOC1*, *OsPRR37*, *OsPRR73*, *OsPRR59*, *OsPRR95*; Murakami et al., 2003). Other than the model monocot *O. sativa* the *TOC1/PRR* gene family had been identified from many other Poaceae family members.

ZTL is an important blue light photoreceptor, which belongs to the F-box family (Zoltowski and Imaizumi, 2014). In *A. thaliana*, ZTL gene family is composed of three members, which are ZTL, flavin-binding kelch repeat, Fbox1 (FKF1) and LOV kelch protein 2 (LKP2, Imaizumi et al., 2003). It is now known that ZTL proteins

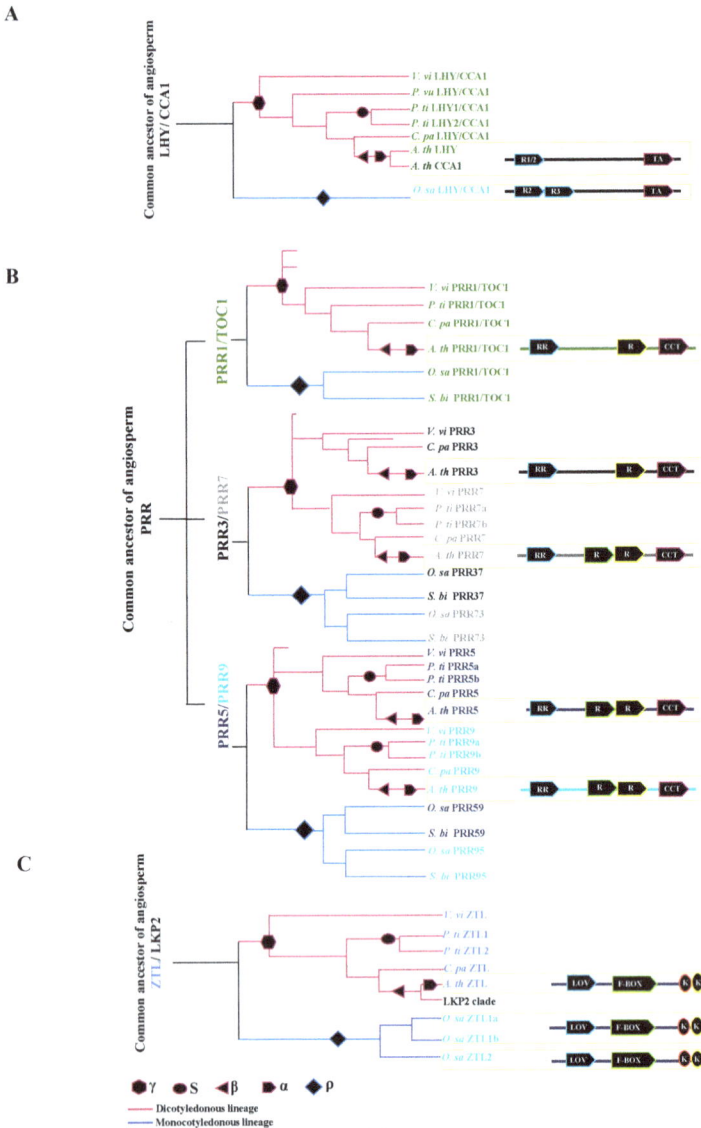

Fig. 7.2. Phylogenetic relationship and diversification of functional domains of clock gene families in angiosperms.

Note: (A) LHY/CCA1 (B) PRR (C) ZTL. The pink lines indicate the dicotyledonous subgroups, and the blue lines indicate the monocotyledonous groups. Occurrences of different duplication events are marked by different symbols.

Abbreviations: V. vi: *Vitis vinifera*, P. al: *Pheseolus vulgaris*, P. ti: *Populus tricocarpa*, C. pa: *Carica papaya*, A. th: *Arabidopsis thaliana*, O. sa: *Oryza sativa*, S. bi: *Sorghum bicolor*, R1: mybR1, R2: mybR2, R3: mybR3, TA: transcriptional activation domain, RR: response receiver domai, R: repressor domain, CCT: *CONSTANS, CONSTANS*-like, TOC1 domain, LOV: light-oxygen-voltage domain, K: kelch repeat.

Source: This figure was prepared based on the information obtained from published reports of Jin and Martin, 1999; Lariguet and Dunand, 2004; Takata et al., 2008, 2010; Holm et al., 2010; Lou et al., 2012; Toda et al., 2019.

possess an N terminal ight, oxygen or voltage (LOV) domain, a F-box domain in the middle followed by six Kelch-repeats (Somers et al., 2000). The LOV domain is implicated for the absorption of blue light, while the F-box is responsible for recognition of substrate for degradation (Fig. 7.2C; Somers et al., 2000; Briggs et al., 2002). In angiosperms, the ZTL homologs evolved from a common ancestor which underwent a single round of duplication prior to the divergence of monocot and dicots. Another round of duplication occurred in both monocotyledonous and dicotyledonous lineages, which resulted into emergence of ZTL1 and ZTL2 in monocots and ZTL and LKP2 in dicot plants (Fig. 7.2C; Lou et al., 2012). Unlike the reference dicotyledonous species, multiple ZTL homologs had been identified from different dicots. Interestingly, no homolog of ZTL could be identified in the tetraploid genome *B. rapa* (Lou et al., 2012). Like the dicotyledonous lineage multiple copies of ZTL were found in many Poaceae species including *O. sativa* (OsZT1a, OsZTL1b, OsZTL2) (Lariguet and Dunand, 2005; Murakami et al., 2007).

GI is an important plant specific nuclear protein, which transmits circadian clock signal to the clock integrator gene CO (Koornneef et al., 1998). GI proteins possess variable numbers of transmembrane domains, which are found scattered throughout the protein. For instance, in *A. thaliana* eleven transmembrane domains have been identified, while only five are present in *O. sativa*, *T aestivum*, *B. distachyon*, and *G. max* (Fowler et al., 1999; Li et al., 2013). Apart from transmembrane domains the GI proteins also possess four C terminal nuclear localization domains (NLS) (Li et al., 2013). Both in *A. thaliana* and *O. sativa*, the transmembrane domains are responsible for the localization of the GI protein in the plasma membrane, whereas the NLS domains are responsible for the localization of the protein in the nucleus (Fowler et al., 1999; Li et al., 2013). The evolutionary origin of *GI* gene is more recent in comparison to other circadian clock genes, and it is believed that GI originated after the divergence of land plants (Mishra and Panigrahi, 2015). Therefore, no GI homolog could be identified in any prokaryotes and lower plant group such as liverworts till to date (Corellou et al., 2009; Holm et al., 2010). However, the GI homologs are ubiquitously present in the angiosperms and are associated with multiple biological functions such as carbohydrate metabolism (Dalchau et al., 2011), starch accumulation (Eimert et al., 1995), phytohormone signaling (Tseng et al., 2004), fruit setting (Brock et al., 2007), movement of cotyledon (Tseng et al., 2004), transpiration (Sothern et al., 2002) and stress tolerance such as cold (Cao et al., 2005), herbicide (Qian et al., 2014), drought (Riboni et al., 2013) and salt (Park et al., 2013). A single or multiple homologs of GI had been found in many dicotyledonous species including *A. thaliana*. Like dicotyledonous plants, GI homologs are conserved in Poaceae species, which includes *O. sativa* (OsGI) (Hayama et al., 2002).

7.7 Circadian Clock in Bamboo Flowering

In bamboo, large numbers of studies have been conducted in ecological aspect due to their diverse flowering habit. However, the reason behind extreme delayed flowering could not be unraveled due to lack of molecular studies such as identification and molecular characterization of conserved flowering pathways as well as the novel molecular players over decades. Therefore, molecular studies on bamboo flowering have started flourishing recently. Circadian control serves as a major regulating factor in flower induction. However, bamboos can grow all over the world from 51° N in Japan to 47° S in South Argentina except in European countries. Therefore, bamboos are facing a large amount of deviation not only for duration of light but also for wavelengths. Therefore, circadian response of bamboo towards flowering may vary among temperate and tropical habit and could be an interesting area to discover.

Firstly, *in silico* identification of multiple homologs of photoperiodic pathway genes starting from the photoreceptors (*PHYA, PHYB, CRY1, CRY2*) to clock components (*CCA1, ELF3, ELF4, TOC1, COP1, FKF1, ZTL, GI*) in the sequenced temperate bamboo genome *P. edulis* suggested the existence of conserved photoperiodic pathway genes (Biswas et al., 2016; Basak et al., 2021) see Table 7.1. In the early days of molecular studies, non-targeted approaches have been put forward to identify their transcriptional activity during flower induction in bamboos such as *D. latiforus* (Zhang et al., 2012; Wang et al., 2020), *P. edulis* (Peng et al., 2013; Gao et al., 2014), *Phyllostachys violascens* (Jiao et al., 2019), *Guadua inermis, Otatea acuminata, Phyllostachys aurea, Lithachne paucifora* (Wysocki et al., 2016), and *Fargesia macclureana* (Li et al., 2019). For example, in *D. latiforus* several transcription factors related to circadian clock have been found upregulated such as zinc finger homeodomain (zf-HD), R2R3 type of MYB, PRRs. The key clock components such as LHY, PRR1, PRR95, ELF3, ELF4, and COP1 were also found upregulated during floral induction during flower initiation of *D. latiforus* (Zhang et al., 2012). In contrast, CCA1, TOC1, and GI could not be identified in this study. However, multiple copies of *Indeterminate 1* (*ID1*) are abundantly expressed in flowering tissue (Zhang et al., 2012). ID1 is found primarily a key photoperiod-independent flowering regulator in rice. However, in *Bambusa multiplex* ID1 is highly responsive to photoperiod and diurnally regulated due to the presence of a cis-responsive element (CAANNNNATC) which is absent in rice (Guo et al., 2016; Fig. 7.1B). This finding suggests that the circadian clock in bamboo may be regulated in a novel way compared to the model plant systems. Floral transcriptomic and proteomic analysis of *D. latiforus* identified several photoreceptors and central clock components involved in flowering (Wang et al., 2020). Both the red (*PHYA, PHYB*) and blue (*CRYs*) light receptors were up regulated in flowering associated leaves, but not in vegetative ones. In addition, detailed comparison between comparable leaves obtained from flowering and non-flowering plants suggested that accumulation of *PHYs* resulted into down-regulation of *LHY* and upregulation of *CCA1*. The elevated level of *CCA1* can upregulate *TOC1*. However, the expression of *TOC1* was found extremely suppressed in the flowering tissue. This may be due to CRY mediated upregulation of *GI*. However, the elevated level of *GI* failed to activate *CO*. To complete the feedback loop, the *LHY/CCA1* upregulates *PRR9* and *PRR5*, which in turn activate *CO*. The evening loop component ELF3 was also found upregulated in the flowering stage. Thus, this study provides a nice summary of the circadian clock response that has been identified in a bamboo species based on the observation made over three years starting from July to October (Wang et al., 2020; Fig. 7.1B). In floral transcriptomes of *P. edulis*, photoreceptors (*CRY1* and Phytochrome-

Table 7.1. Important photoperiod responsive genes that have been identified so far in bamboo and their putative role in flowering.

Circadian Clock Components	Homologs	Bamboo Species	Experimental Method	Nature of Deregulation in Flowering tissues	Putative Role in Flowering	References
Photoreceptors	PHYA	*D. latiforus*	Floral transcriptome	Upregulated in mature leaves of flowering culm	Upregulation of LHY/CCA1	Wang et al., 2020
	PHYB	*D. latiforus*	Floral transcriptome	Upregulated in mature leaves of flowering culm	Upregulation of LHY/CCA1, degradation of CO in dark	Wang et al., 2020
	CRY1	*D. latiforus*	Floral transcriptome	Upregulated in mature leaves of flowering culm	Stabilization of GI, upregulation of ZTL/FKF1	Wang et al., 2020
		P. edulis	Floral transcriptome	Down regulated in floral bud	Stabilization of GI	Gao et al., 2014
	CRY2	*D. latiforus*	Floral transcriptome	Upregulated in mature leaves of flowering culm	Stabilization of GI, upregulation of ZTL/FKF1	Wang et al., 2020
	PIF3	*D. latiforus*	Floral transcriptome	Upregulated in young leaves of flowering culm	Upregulation of LHY/CCA1	Wang et al., 2020
		P. edulis	Floral transcriptome	Upregulated in floral bud	Upregulation of LHY/CCA1, downregulation of CO	Gao et al., 2014
Morning loop component	PRR1	*D. latiforus*	Floral transcriptome	Upregulated in flower bud, mature leaves of flowering culm	Upregulation of LHY/CCA1, downregulation of GI	Zhang et al., 2012
	PRR3	*D. latiforus*	Floral transcriptome	Downregulated in mature leaves of flowering culm	Inhibition of GI mediated TOC1 degradation	Wang et al., 2020
	PRR5	*D. latiforus*	Floral transcriptome	Upregulated in mature leaves of flowering culm	Upregulation of CDF1, downregulation of LHY/CCA1	Wang et al., 2020
	PRR7	*D. latiforus*	Floral transcriptome	Downregulated in mature leaves of flowering culm	Upregulation of CDF1, downregulation of LHY/CCA1	Wang et al., 2020

Table 7.1 contd. ...

...Table 7.1 contd.

Circadian Clock Components	Homologs	Bamboo Species	Experimental Method	Nature of Deregulation in Flowering tissues	Putative Role in Flowering	References
	PRR9	D. latiforus	Floral transcriptome	Upregulated in mature leaves of flowering culm	Upregulation of CDF1, downregulation of LHY/CCA1	Wang et al., 2020
	PRR95	D. latiforus	Floral transcriptome	Upregulated in flower bud	Downregulation of LHY/CCA1	Zhang et al., 2012
Central loop component	TOC1	D. latiforus	Floral transcriptome	Upregulated in young leaves of flowering culm	Upregulation of LHY/CCA1, downregulation of GI	Wang et al., 2020
		B. tulda	Targeted identification and characterization	Upregulated in young leaves of flowering culm and internode	Daylength mediated circadian response towards flowering	Dutta et al., 2018
	LHY	D. latiforus	Floral transcriptome	Upregulated in flower bud	Upregulation of PRRs, downregulation of TOC1	Zhang et al., 2012
		B. tulda	Targeted identification and characterization	Upregulated in young leaves of flowering culm and immature inflorescence bud	Red light mediated circadian response towards flowering	Dutta et al., 2018
	CCA1	D. latiforus	Floral transcriptome	Upregulated in mature leaves of flowering culm	Upregulation of PRRs, downregulation of TOC1	Wang et al., 2020
	GI	D. latiforus	Floral transcriptome	Upregulated in mature leaves of flowering culm	Upregulation of CO, degradation of TOC1	Wang et al., 2020
		B. tulda	Targeted identification and characterization	Upregulated in young leaves and internode	Light mediated circadian response to upregulate CO	Dutta et al., 2018
		P. edulis	Floral transcriptome	Down regulated in floral bud	Upregulation of CO	Gao et al., 2014
	ZTL	B. tulda	Targeted identification and characterization	Upregulated in young leaves of flowering culm	Light mediated circadian response for floral induction	Dutta et al., 2018

Category	Gene	Species	Method	Expression	Function	Reference
Evening loop component	ELF3	D. latiforus	Floral transcriptome	Upregulated in flower bud	Downregulation of PHYB	Zhang et al., 2012; Wang et al., 2020
	ELF4	D. latiforus	Floral transcriptome	Upregulated in flower bud	Downregulation of GI	Zhang et al., 2012; Wang et al., 2020
	COP1	D. latiforus	Floral transcriptome	Upregulated in young leaves of flowering culm	Degradation of CO in dark	Wang et al., 2020
	SPA1	D. latiforus	Floral transcriptome	Upregulated in mature leaves of flowering culm	Degradation of CO in dark	Wang et al., 2020
Clock Integrator	COLs	D. latiforus	Floral transcriptome	Downregulated in flower bud, mature leaves of flowering culm	Upregulation of FT and flower induction	Zhang et al., 2012; Wang et al., 2020
	COLs	P. edulis	Floral transcriptome	Downregulated in flowering tissue	Upregulation of FT and flower induction	Peng et al., 2013
	COI	P. violascens	Targeted identification and characterization	Upregulated in immature leaves	Inhibition of flowering through interaction with 14-3-3	Xiao et al., 2018
	COL1, COL4, COL11	P. edulis	Targeted identification and characterization	Upregulated in leaves	Maximum expression peak at the end of the day revealed flower inductive role	Liu et al., 2019
	COA	B. tulda	Targeted identification and characterization	Upregulated in young leaves of flowering culm	Maximum expression peak at the end of the day revealed flower inductive role	Dutta et al., 2018
Other light responsive	ID1	D. latiforus	Floral transcriptome	Upregulated in flower bud	Photoperiod induced flowering	Zhang et al., 2012
	ID1	B. multiplex	Targeted identification and characterization	Upregulated in immature leaves of flowering culm	Photoperiod induced flowering due to presence of a cis-responsive element CAANNNATC	Guo et al., 2016

interacting factor 3) and clock genes such as *LHY* and *GI* are found upregulated in flower associated tissues but not *PHYA, PHYB, ELF3, ELF4, TOC1, CCA1* (Gao et al., 2014). However, all these findings demonstrate the circadian clock regulation in temperate species (Table 7.1).

A recent study in tropical species, *Bambusa tulda* also revealed that the key clock component such as *LHY, ZTL, TOC1,* and *GI* are conserved and highly expressed in leaves associated with flowering supporting their involvement in photoperiodic pathway (Dutta et al., 2018). In addition, *BtLHY, BtTOC1,* and *BtGI* were phylogenetically closer to that of temperate bamboo *P. heterocycla* but not the *BtZTL*. Apart from tissue specific expression, appropriate circadian oscillations of the clock genes are essential for photoperiod induced flowering. The diurnal expression patterns of *BtLHY, BtZTL* genes attained a transcriptional peak in the morning and *BtTOC1, BtGI* in the afternoon, respectively, in bamboo which were comparable to rice (Fig. 7.1B). In contrast, the *BtZTL* transcript demonstrated bimodal peak in SD but unimodal peak in LD unlike rice. The expression divergence of *BtZTL* may lead to a new avenue of photoperiod regulated flower induction in bamboo (Dutta et al., 2018) (Table 7.1).

7.8　Circadian Clock Integrator Gene *CONSTANS (CO)* in Bamboo and other Plants

The circadian clock integrator gene *CO* accepts light signal that comes through the clock genes and upregulates *FT* to induce flowering. This gene was named as '*CONSTANS*' due to its 'constant' nature of flower induction (Redei, 1962). It is ubiquitously present in all flowering plants and the gene family is known as 'constans like' genes (*COLs*) (Griffiths et al., 2003). The amino terminal end of *CO* protein generally consists of a single or tandemly duplicated B-box motifs and retains important signatures to render clues on the evolution of this gene in the plant kingdom. Recent studies based on the number and structural conservation of B-boxes revealed that the angiosperm *COLs* were evolved from members of the green algae Chlorophyta (Serrano et al., 2009).

CO like genes (*COLs*) are a multi-gene, large family and are ubiquitously present across angiosperms (Griffiths et al., 2003). For instance, there are 32 and 17 *COL* genes present in *A. thaliana* and *O. sativa*, respectively, and these are possibly results of recent tandem duplication events (Putterill et al., 1995). Apart from these two reference plants, *COLs* have also been identified from other plant species such as *Glycine max* (Fabaceae; *GmCOL1a* to *GmCOL13b*; Wu et al., 2014), *Lotus japonicus* (Fabaceae; *LjCOLa*; Yamashino et al., 2013), *Medicago truncatula* (Fabaceae; *MtCOLa*; Hecht et al., 2005, 2007), *Pisum sativum* (Fabaceae; *PsCOL1a*; Hecht et al., 2005), *Tarenaya hassleriana* (Cleomaceae; *ThCOL*; Schranz and Mitchell-Olds, 2006), *Solanum tuberosum* (Solanaceae; *StCO1, StCO2*; González-Schain et al., 2012; Kloosterman et al., 2013; Navarro et al., 2011), *Vitis vinifera* (Vitaceae; *VtCO*; Almada et al., 2009), *Pharbitis nil* (Convolvulaceae; *PnCO*; Liu et al., 2001), *Raphanus sativus* (Brassicaceae; *RsaCOL1* to *RsaCOL20*; Hu et al., 2018) *Populus trichocarpa* (Salicaceae; *PtCO1, PtCO2*; Bohlenius et al., 2006; Hsu et al., 2012), *Hordeum vulgare* (Poaceae; *HvCO1, HvCO2, HvCO3*; Campoli et al.,

2012; Griffiths et al., 2003), *Triticum aestivum* (Poaceae; *TaHd1, TaHd1, TaHd1*; Nemoto et al., 2003), *Sorghum bicolor* (Poaceae; *SbCO*; Yang et al., 2014), *Zea mays* (Poaceae; *ZmCONZ1*; Miller et al., 2008), *Lolium perenne* (Poaceae; *LpCO*), and *Festuca pratensis* (Poaceae; *FpCO*; Armstead et al., 2005).

In addition, 41 *COLs* were found associated with flowering in temperate bamboo *P. edulis* (Gao et al., 2014). However, detailed sequence characterization and functional analysis were performed for only 14 *COLs* in *P. edulis* (Liu et al., 2019). In *P. violascens*, two functionally diverged *COLs* have been isolated (Xiao et al., 2018). In addition to *COLs* (CO5, CO6, CO7, and CO8), 27 putative CCAAT box binding zf-HD genes have also been found upregulated during flowering in *D. latiflorus* (Zhang et al., 2012). In tropical species *B. tulda*, two diurnally responsive *CO* homologs has also been reported, among which *BtCOA* was associated with flowering, while *BtCOB* was not (Dutta et al., 2018; Table 7.1).

7.9 Transcriptional Regulation of *CO*: Model Plants vs. Bamboo

A specific diurnal pattern of *CO* expression is essential for the induction of *FT* (Sawa et al., 2007). In *A. thaliana*, the transcriptional upregulation of *CO* occurs in the dark and down-regulation happens throughout the day under SD condition (8 hours light and 16 hours darkness). In contrast, under LD condition (16 hours light and 8 hours darkness), *CO* mRNA level needs to be enhanced at the end of the day, preceding activation of *FT* (Suarez-Lopez et al., 2001; Yanovsky and Kay, 2002).

Floral transcriptomic studies reported down-regulation of *CO* in the flowering associated tissues in *P. edulis* (Peng et al., 2013), *D. latiflorus* (Wang et al., 2020). One possible reason may be the presence of repeat insertions within the genic and regulatory regions of respective homologs (Peng et al., 2013). In addition, *CO* transcripts attain its maximum expression peak at the end of the day and hence it is important to monitor their diurnal pattern of expression. The *COLs* are highly duplicated in bamboos and are involved in diverse developmental process in addition to flowering. A study on detailed diurnal expression analysis of 14 *COLs* in *P. edulis* had identified four diverse diurnal expression patterns (Liu et al., 2019) (Table 7.1.) Among them, only *COL1, COL4*, and *COL11* revealed elevated expression pattern at the end of the day under both SD and LD, which is important for floral induction suggesting their putative involvement in flowering. On the other hand, diurnal regulation of both the *PvCOs* confirmed that they regulate flowering negatively in *P. violascens* (Xiao et al., 2018). In the tropical species *B. tulda* among two *COLs*, *BtCOA* reaches its maximum expression peak at the end of the day which was just opposite in *BtCOB* suggesting existence of functional divergence among the two copies with respect to flowering (Dutta et al., 2018) (Table 7.1).

The transcriptional activity of *CO* is regulated by many transcription factors such as flowering bHLHs (FBHs), cycling DOF factor (CDF) and by the circadian clock outputs (Toledo-Ortiz, 2003; Imaizumi et al., 2005; Ito et al., 2012). The FBHs (FBH1, FBH2, FBH3, and FBH4) upregulate the transcriptional expression of *CO* by binding with the E-box element (CANNTG) located in the promoter region of *CO*. However, the over-expressed *A. thaliana* lines of FBH suggested that FBHs are responsible for upregulation of *CO* but were unable to regulate the diurnal expression of *CO* mRNA. The CDFs (CDF1, CDF2, CDF3, CDF5), belong to the family of DOF domain transcription factors and repress the transcriptional expression of *CO* (Imaizumi et al., 2005). In addition, the expression of *CDFs* was found to be circadian clock regulated (Fornara et al., 2009). For instance, the expression level of *CDFs* got elevated in the morning and started to decline from late afternoon. The *CDF1* binds to the DOF consensus binding site (AAAG) TGT, present at the proximal *CO* promoter (Yanagisawa and Schmidt, 1999; Imaizumi et al., 2005).

Approximately, 28 DOF transcription factors are predominantly expressed in panicles of *P. edulis* and were homologous to DOF3 of *Jatropha curcas* and DOF12 of rice. Both are responsive to the circadian clock and are involved in photoperiod induced flowering process in *J. curcas* and rice, respectively (Gao et al., 2014). Among 28 Dofs, *PeDOF4* and *PeDOF5* are highly expressed in leaf tissue and sequentially similar with rice DOF112. In addition, majority of the bamboo *DOFs* are enriched with light responsive cis elements at their promoter region revealing their possible involvement in photoperiod induced flowering process.

7.10 Post Translational Regulation of *CO*: What is Happening in Bamboo?

The transcriptional regulatory mechanism elucidates how *CO* expresses at the end of the day but could not conclude about the stability of the *CO* protein. Post-translational regulation of *CO* protein is important for simultaneous maintenance of its protein level along with the transcript level to activate the pathway integrator *FT* (Suarez-Lopez et al., 2001). The stability of *CO* was influenced by two major kinds of external cues such as light and temperature.

Both duration and quality of light affect the stability of *CO* proteins (Valverde et al., 2004). For instance, *CO* is generally degraded under red light and stabilized under blue light (Valverde et al., 2004). In the morning, *CO* protein gets degraded through *PHYB* in the presence of red light (Endo et al., 2013). However, in the afternoon far-red light mediated upregulation of *PHYA* stabilizes the *CO* protein (Valverde et al., 2004). Therefore, the antagonistic effect of *PHYA* and *PHYB* confer great impact on the stabilization of *CO*. Another *CO* stabilization protein, phytochrome dependent late flowering (*PHL*) also inhibits *PHYB* mediated degradation of *CO* in the late afternoon (Endo et al., 2013). In temperate bamboo *D. latiflorus*, accumulation of *PHYB* was reported in the flowering associated leaves, while PHL could not be identified. This may be linked to the low expression of *CO* in respective tissues (Wang et al., 2020; Fig. 7.3).

At night, *CO* protein eventually gets degraded via constitutive photomorphogenesis 1 (*COP1*) and Suppressor of *PHYA* (*SPA1*) complex by ubiquitin mediated proteasomal degradation (Jang et al., 2008). Therefore, COP1-SPA complex mediated degradation of *CO* at night resulted into *CO* mediated *FT* expression only in the late afternoon (Laubinger et al., 2006; Jang et al., 2008). However, the inhibitory effect of the *COP1-SPA1* complex on *CO* protein stability throughout the day was limited by cryptochrome 1 (*CRY1*) and *CRY2* in a blue light dependent manner. The *CRY1* and *CRY2* directly interact with the *COP-SPA1* complex to stabilize *CO* (Liu et

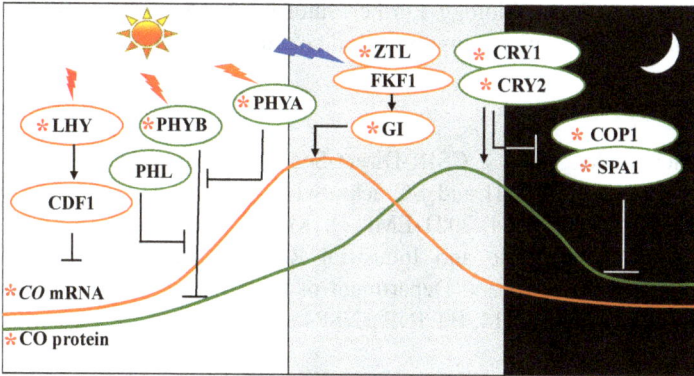

Fig. 7.3. Model depicting CO regulation at transcriptional, translational, and post-translational level in *A. thaliana*.

Note: Asterisks indicate that homologous genes have been in bamboo. Arrows indicate positive effects and line bars denote inhibitory effects.

Abbreviations: LHY: late elongated hypocotyl, CDF: cycling DOF factor, CO: Constans, PHY: phytochrome, ZTL: zeitlupe, FKF1: flavin binding, kelch repeat, F: Box 1, GI: Gigantea, CRY: cryptochrome, COP1:- constitutive photomorphogenesis 1, SPA: suppressor of PHYA.

Source: Figure prepared is based on the information available in Locke et al., 2006; McClung, 2006; Zeilinger et al., 2006.

al., 2011). In the presence of blue light, *CRY1* interacts with *SPA1* and *CRY2* interacts with both *SPA1* and *COP1* to interrupt the formation of *COP1-SPA1* complex (Zuo et al., 2011). Therefore, *PHYA-PHYB-PHL* complex and *COP1-SPA1-CRY1-CRY2* complex play a pivotal role in CO stabilization (Fig. 7.3).

Although, *CRYs* and *SPA1* proteins are found accumulated in the flowering tissue, the *COP1* level was noticeably low, which may interrupt the formation of CRY mediated *COP1-SPA1* complex to maintain *CO* protein level in dark in *D. latiflorus* (Wang et al., 2020; Fig. 7.3). Therefore, upregulation, stabilization, and maintenance of *CO* protein throughout the day may involve additional molecular players in bamboo.

7.11 Conclusions and Future Perspective

Flowering behavior of the woody bamboos and their dependence on diverse environmental cues remained unexplored over decades. Bamboos can grow across four subcontinents in the world and hence experience massive environmental variations. It is known for decades that various environmental factors, particularly light, immensely affect flowering time. However, not many studies have been conducted to assess impact of light intensity and duration on induction of bamboo flowering. Long generation time and inaccessibility of flowering population that are predominantly growing in forests pose real challenge to track incidents of flowering and collection of tissues for experimental purposes. Nevertheless, analyzing available floral transcriptome, proteome, and a few targeted studies confirmed presence of key clock genes in bamboo. Many of them are present in multiple copies and rudimentary evidence suggest that functional divergence among them may play crucial role in conferring unique flower biology. Further, studies are required to characterize more genes to understand relationships among all circadian clock genes in bamboo.

Acknowledgements

SD acknowledges CSIR-Direct-Senior Research Fellowship [08/155(0055)/2019-EMR-I] and SC acknowledges CSIR-Direct-Senior Research Fellowship [08/0155(11864)/2021-EMR-I]. MD acknowledges financial support from Council of Scientific and Industrial Research, India [38(1386)/14/EMR-II], [38(1493)/19/EMR-II] and Department of Biotechnology, Govt. of India (BT/PR10778/PBD/16/1070/2014, BT/INF/22/SP45088/2022).

References

Abrahamson, W.G. and Caswell, H. (1982). On the comparative allocation of biomass energy, and nutrients in plants. *Ecology*, 63: 982–991.

Alabadí, D., Oyama, T., Yanovsky., M.J., Harmon, F.J., Más, P. and Kay, S.A. (2001). Reciprocal regulation between *TOC1* and *LHY/CCA1* within the *Arabidopsis* circadian clock. *Science*, 293: 880–883.

Almada, R., Cabrer, N., Casaretto, J.A., Ruiz-Lara, S.V.E. and González, V.E. (2009). *VvCO* and *VvCOL1*, two *CONSTANS* homologous genes, are regulated during flower induction and dormancy in grapevine buds. *Plant Cell Rep.*, 28: 11931203.

Ambawat, S., Sharma, P., Yadav, N.R. and Yadav, R.C. (2013). MYB transcription factor genes as regulators for plant responses: An overview. *Physiology and Molecular Biology of Plants*, 19: 307–321.

Armstead, I.P., Skot, L., Turner, L.B., Skot, K., Donnison, I.S., Humphreys, M.O. and King, I.P. (2005). Identification of perennial ryegrass (*Lolium perenne* L.) and meadow fescue (Festuca pratensis Huds.) candidate orthologous sequences to the rice Hd1(Se1) and barley HvCO1 *CONSTANS*-like genes through comparative mapping and microsynteny. *New Phytologist*, 167: 239–247.

Austin, A.T. and Marchesini, V.A. (2012). Gregarious flowering and death of understory bamboo slow litter decomposition and nitrogen turnover in a southern temperate forest in Patagonia, Argentina. *Funct. Ecol.*, 26(1): 265–273.

Basak, M., Dutta, S., Biswas, S., Chakraborty, S., Rahaman, T., Sarkar, A., Dey, S., Biswas, P. and Das, M. (2021). Genomic insights into growth and development of bamboos: What have we learnt and what more to discover? *Trees*, 35: 1771–1791.

Baudry, A., Ito, S., Song, Y.H., Strait, A.A., Kiba, T., Lu, S., Henriques, R., Pruneda-Paz, J.L., Chua, N.H., Tobin, E.M., Kay, S.A. and Imaizumi, T. (2010). F-box proteins *FKF1* and *LKP2* act in concert with *Zeitlupe* to control *Arabidopsis* clock progression. *Plant Cell*, 22: 606–622.

Bhattacharya S., Das, M., Bar, R. and Pal, A. (2006). Morphological and molecular characterization of *Bambusa tulda* Roxb. with a note on flowering. *Annals of Botany*, 98(3): 529–535.

Bhattacharya S., Das, M., Ghosh, J.S. and Pal, A. (2009). Morphological and molecular characterization of *Thamnocalamus spathiflorus* subsp. *Spathiflorus* at population level. *Plant Systematics and Evolution*, 282: 13–20.

Biswas, P., Chakraborty, S., Dutta, S., Pal, A. and Das, M. (2016). Bamboo flowering from the perspective of comparative genomics and transcriptomics. *Front. Plant Sci.*, 7: 1900.

Böhlenius, H., Huang, T., Charbonnel-Campaa, L., Brunner, A.M., Jansson, S., Strauss, S.H. and Nilsson, O. (2006). CO/FT regulatory module controls timing of flowering and seasonal growth cessation in trees. *Science*, 312: 1040–1043.

Brandis, D. (1906). *Indian Trees*. Book Agency, Delhi, India.

Briggs, W.R. and Christie, J.M. (2002). *Phototropins 1 and 2*: Versatile plant blue-light receptors. *Trends in Plant Science*, 7: 204–210.

Brock, M.T., Tiffin, P. and Weinig, C. (2007). Sequence diversity and haplotype associations with phenotypic responses to crowding: *GIGANTEA* affects fruit set in *Arabidopsis thaliana*. *Molecular Ecology*, 16: 3050–3062.

Campoli, C., Drosse, B., Searle, I., Coupland, G. and von Korff, M. (2012). Functional characterization of *HvCO1*, the barley (*Hordeum vulgare*) flowering time ortholog of *CONSTANS*. *Plant J.*, 69: 868–880.

Cao, S., Ye, M. and Jiang, S. (2005). Involvement of *Gigantea* gene in the regulation of the cold stress response in *Arabidopsis*. *Plant Cell Rep.*, 24: 683–690.

Christanty, L., Mailly, D. and Kimmins, J.P. (1997). Without bamboo, the land dies: A conceptual model of the biogeochemical role of bamboo in an Indonesian agroforestry system. *For. Ecol. Manag.*, 91: 83–91.

Clerget, B. (2021). Bamboos flower after the return of almost the same sun-moon phasing as at seedling emergence. *BioRxiv*. Doi.org/10.1101/2021.06.11.448081.

Corellou, F., Schwartz, C., Motta, J., Djouani-Tahri, E.B., Sanchez, F. and Bouget, F. (2009). Clocks in the green lineage: Comparative functional analysis of the circadian architecture of the picoeukaryote *Ostreococcus*. *Plant Cell*, 21: 3436–3449.

Dalchau, N., Baek, S.J., Briggs, H.M., Robertson, F.C., Dodd, A.N. and Gardner, M.J. (2011). The circadian oscillator gene *Gigantea* mediates a long-term response of the *Arabidopsis thaliana* circadian clock to sucrose. *Proceedings of the National Academy of Sciences USA*, 108: 5104–5109.

de Candolle, A.P. (1832). *Physiologie Végétale*. Bechet Jeune, Paris.

Dubos, C., Stracke, R., Grotewold, E., Weisshaar, B., Martin, C. and Lepiniec, L. (2010). MYB transcription factors in *Arabidopsis*. *Trends Plant Sci.*, 15: 573–581.

Dutta, S., Biswas, P., Chakraborty, S., Mitra, D., Pal, A. and Das, M. (2018). Identification, characterization, and gene expression analyses of important flowering genes related to photoperiodic pathway in bamboo. *BMC Genom.*, 19: 190.

Eimert, K., Wang, S., Lue, W. and Chen, J. (1995). Monogenic recessive mutations causing both late floral initiation and excess starch accumulation in *Arabidopsis*. *Plant Cell*, 7: 1703–1712.

Endo, M., Tanigawa, Y., Murakami, T., Araki, T. and Nagatani, A. (2013). *Phytochrome*-dependent late-flowering accelerates flowering through physical interactions with *Phytochrome B* and *CONSTANS*. *Proceedings of the National Academy of Sciences, USA*, 110: 18017–18022.

Farre, E.M. and Liu, T. (2013). The *PRR* family of transcriptional regulators reflects the complexity and evolution of plant circadian clocks. *Curr. Opin. Plant Biol.*, 16: 621–629.

Filichkin, S.A., Priest, H.D., Givan, S.A., Shen, R., Bryant, D.W., Fox, S.E., Wong, W.K. and Mockler, T.C. (2011). Genome-wide mapping of alternative splicing in *Arabidopsis thaliana*. *Genome Res.*, 20: 45–58.

Fogelmark, K. and Troein, C. (2014). Rethinking transcriptional activation in the Arabidopsis circadian clock. *PLoS Comput. Biol.*, 10: e1003705.

Fornara, F., Panigrahi, K.C.S., Gissot, L., Sauerbrunn, N., Rühl, M., Jarillo, J.A. and Coupland, G. (2009). *Arabidopsis* DOF transcription factors act redundantly to reduce *CONSTANS* expression and are essential for a photoperiodic flowering response. *Dev. Cell.* 17: 75–86.

Fowler, S., Lee, K., Onouchi, H., Samach, A., Richardson, K., Morris, B., Coupland, G. and Putterill, J. (1999). Arabidopsis and encodes a protein with several possible membrane spanning domains. *EMBO Journal*, 18: 4679–4688.

Gadgil, M. and Bossert, W.H. (1970). Life historical consequences of natural selection. *Am. Nat.*, 104: 1–24.

Gamble, J.S. (1896). *The Bambuseae of British India*. Bengal Secretariat Press, Calcutta, 133 pp.

Gao, J., Zhang, Y., Zhang, C., Qi, F., Li, X., Mu, S. and Peng, Z. (2014). Characterization of the floral transcriptome of Moso bamboo (*Phyllostachys edulis*) at different flowering developmental stages by transcriptome sequencing and RNA-seq analysis. *PLoS One*, 9: e98910.

Giliberto, L., Perrotta, G., Pallara, P., Weller, J.L., Fraser, P.D., Bramley, P.M., Fiore, A., Tavazza, M. and Giuliano, G. (2005). Manipulation of the blue light photoreceptor *cryptochrome 2* in tomato affects vegetative development, flowering time, and fruit antioxidant content. *Plant Physiol.*, 137: 199–208.

González-Schain., N.D., Díaz-Mendoza, M., Żurczak, M. and Suárez-López, P. (2012). Potato *CONSTANS* is involved in photoperiodic tuberization in a graft-transmissible manner: Regulation of tuberization by potato *CONSTANS*. *The Plant Journal*, 70: 678–690.

Griffiths, S., Dunford, R.P., Coupland, G. and Laurie, D.A. (2003). The evolution of *CONSTANS*-like gene families in barley, rice, and *Arabidopsis*. *Plant Physiology*, 131: 1855–1867.

Guangchu, Z. (2002). *A Manual of Bamboo Hybridization*. INBAR Technical Report No. 21.

Guo, X., Guan, Y., Xiao, G., Zai-en, X., Haiyun, Y. and Fang, W. (2016). Isolation and characterization of an Indeterminate1 gene, BmID1, from bamboo (*Bambusa multiplex*). *J. Plant. Biochem. Biotechnol.*, 25: 30–39.

Harmer, S.L. (2009). The circadian system in higher plants. *Annu. Rev. Plant Biol.*, 60: 357–377.

Harper, J.L. and Ogden, J. (1970). The reproductive strategy of higher plants: I. The concept of strategy with special reference to *Senecio vulgaris* L. *J. Ecol.*, 58: 681–698.

Hayama, R., Izawa, T. and Shimamoto, K. (2002). Isolation of rice genes possibly involved in the photoperiodic control of flowering by a fluorescent differential display method. *Plant Cell Physiol.*, 43: 494–504.

Hayes, K.R., Beatty, M., Meng, X., Simmons, C.R., Habben, J.E. and Danilevskaya, O.N. (2010). Maize global transcriptomics reveals pervasive leaf diurnal rhythms but rhythms in developing ears are largely limited to the core oscillator. *PLoS One*, 5: e12887.

Hecht, V., Foucher, F., Ferrandiz, C., Macknight, R., Navarro, C., Morin, J., Vardy, M.E., Ellis, N., Beltran, J.P. and Rameau, C. (2005). Conservation of *Arabidopsis* flowering genes in model legumes. *Plant Physiology*, 137: 1420–1434.

Hecht, V., Knowles, C.L., Vander Schoor, J.K., Liew, L.C., Jones, S.E., Lambert, M.J.M. and Weller, J.L. (2007). Pea *Late Bloomer1* is a *Gigantea* ortholog with roles in photoperiodic flowering, de-etiolation, and transcriptional regulation of circadian clock gene homologs. *Plant Physiology*, 144: 648–661.

Helfer, A., Nusinow, D.A., Chow, B.Y., Gehrke, A.R., Bulyk, M.L. and Kay, S.A. (2011). Lux Arrhythmo encodes a nighttime repressor of circadian gene expression in the Arabidopsis core clock. *Curr. Biol.*, 21: 126–133.

Herrero, E., Kolmos, E., Bujdoso, N., Yuan, Y., Wang, M., Berns, M.C., Uhlworm, H., Coupland, G., Saini, R., Jaskolski, M., Webb, A., Goncalves, J. and Davis, S.J. (2012). *Early Flowering4* recruitment of *Early Flowering3* in the nucleus sustains the Arabidopsis circadian clock. *Plant Cell*, 24: 428–443.

Hirose, F., Shinomura, T., Tanabata, T., Shimada, H. and Takano, M. (2006). Involvement of rice cryptochromes in de-etiolation responses and flowering. *Plant Cell Physiology*, 47: 915–925.

Hoffman, D.E., Jonsson, P., Bylesjo, M., Trygg, J., Antti, H., Eriksson, M.E. and Moritz, T. (2010). Changes in diurnal patterns within the *Populus* transcriptome and metabolome in response to photoperiod variation. *Plant Cell Environ.*, 33: 1298–1313.

Holm, K., Källman, T., Gyllenstrand, N., Hedman, H. and Lagercrantz, U. (2010). Does the core circadian clock in the moss Physcomitrella patens (Bryophyta) comprise a single loop? *BMC Plant Biol.*, 10: 109.

Hsu, C.Y., Adams, J.P., No, K., Liang, H., Meilan, R., Pechanova, O., Barakat, A., Carlson, J.E., Page, G.P. and Yuceer, C. (2012). Overexpression of CONSTANS homologs CO1 and CO2 fails to alter normal reproductive onset and fall bud set in woody perennial poplar. *PLoS One*, 7: e45448.

Hsu, P.Y., Devisetty, U.K. and Harmer, S.L. (2014). Accurate timekeeping is controlled by a cycling activator in Arabidopsis. *eLife 2*: e00473.

Hu, T., Wei, Q., Wang, W., Hu, H., Mao, W., Zhu, Q. and Bao, C. (2018). Genome-wide identification and characterization of *CONSTANS*-like gene family in radish (*Raphanus sativus*). *PLoS One*, 13(9): e0204137.

Hughes, J. (1999). Prokaryotes and phytochrome. *Plant Physiology*, 121: 1059–1068.

Imaizumi, T., Schultz, T.F., Harmon, F.G., Ho, L.A. and Kay, S.A. (2005). FKF1 F-box protein mediates cyclic degradation of a repressor of *CONSTANS* in *Arabidopsis*. *Science*, 309: 293–297.

Imaizumi, T., Tran, H.G., Swartz, T.E., Briggs, W.R. and Kay, S.A. (2003). FKF1 is essential for photoperiodic specific light signalling in *Arabidopsis*. *Nature*, 426: 302–306.

Ito, S., Niwa, Y., Nakamichi, N., Kawamura, H., Yamashino, T. and Mizuno, T. (2008). Insight into missing genetic links between two evening-expressed pseudo-response regulator genes *TOC1* and *PRR5* in the circadian clock-controlled circuitry in *Arabidopsis thaliana*. *Plant Cell Physiology*, 49: 201–213.

Ito, S., Song, Y.H., Josephson-Day, A.R., Miller, R.J., Breton, G., Olmstead, R.G. and Imaizumi, T. (2012) Flowering BHLH transcriptional activators control expression of the photoperiodic flowering regulator CONSTANS in Arabidopsis. *Proc. Natl. Acad. Sci. USA*, 109: 3582–3587.

Jang, S., Marchal, V., Panigrahi, K.C., Wenkel, S., Soppe, W., Den, X.W., Valverde, F. and Coupland, G. (2008). Arabidopsis COP1 shapes the temporal pattern of CO accumulation conferring a photoperiodic flowering response. *EMBO J.*, 27: 1277–1288.

Janzen, D.H. (1976). Why bamboos wait so long to flower. *Annu. Rev. Ecol. Evol. Syst.*, 7: 347–391.

Jiang, C., Gu, J., Chopra, S., Gu, X. and Peterson, T. (2004). Ordered origin of the typical two- and three-repeat Myb genes. *Gene*, 326: 13–22.

Jiao, Y., Hu, Q., Zhu, Y., Zhu, L., Ma, T., Zeng, H., Zang, Q., Li, X. and Lin, X. (2019). Comparative transcriptomic analysis of the fewer induction and development of the Lei bamboo (*Phyllostachys violascens*). *BMC Bioinf.*, 20: 687.

John, C.K. and Nadgauda, R.S. (2002). Bamboo flowering and famine. *Current Science*, 82(3): 261–262.

Kakishima, S., Yoshimura, J., Murata, H. and Murata, J. (2011). Six-year periodicity and variable synchronicity in a mass-flowering plant. *PLoS One*, 6: e28140.

Kendrick, R.E. and Kronenberg, G.H.M. (1994). *Photomorphogenesis in Plants*. Kluwer Academic Publishers, Dordrecht, The Netherlands 2.

Kevei, E., Gyula, P. and Hall, A. (2006). Forward genetic analysis of the circadian clock separates the multiple functions of *ZEITLUPE*. *Plant Physiology*, 140: 933–945.

Khan, S., Rowe, S.C. and Harmon, F.G. (2010). Coordination of the maize transcriptome by a conserved circadian clock. *BMC Plant Biol.*, 10: 126.

Kim, J.Y., Song, H.R., Taylor, B.L. and Carre, I.A. (2003). Light-regulated translation mediates gated induction of the Arabidopsis clock protein LHY. *EMBO Journal*, 22: 935–944.

Kim, W.Y., Fujiwara, S., Suh, S.S., Kim, J., Kim, Y., Han, L., David, K., Putterill, J., Nam, H.G. and Somers, D.E. (2007). *Zeitlupe* is a circadian photo receptor stabilized by *Gigantea* in blue light. *Nature*, 449: 356–360.

Klebs, G. (1913). Über das Verhältnis der Aussenwelt zur Entwicklung der Pflanze. *Sber Akad Wiss Heidelberg*, 5: 1–47.

Kloosterman, B., Abelenda, J.A., del, M.C., Gomez, M., Oortwijn, M., de Boer, J.M., Kowitwanich, K., Horvath, B.M., van Eck, H.J., Smaczniak, C. and Prat, S. (2013). Naturally occurring allele diversity allows potato cultivation in northern latitudes. *Nature*, 495: 246250.

Koornneef, M., Alonso-Blanco, C., Peters, A.J. and Soppe, W. (1998). Genetic control of flowering time in Arabidopsis. *Annual Review in Plant Physiology and Plant Mol. Biol.*, 49: 345–370.

Koshy, K.C. and Pushpangadan, P. (1997). *Bambusa vulgaris* blooms, a leap towards extinction? *Curr. Sci.*, 72: 622–624.

Lariguet, P. and Dunand, C. (2005). Plant photoreceptors: Phylogenetic overview. *J. Mol. Evol.*, 61: 559–569.

Laubinger, S., Marchal, V., Le Gourrierec, J., Wenkel, S., Adrian, J., Jang, S., Kulajta, C., Braun, H., Coupland, G. and Hoecker, U. (2006). Arabidopsis SPA proteins regulate photoperiodic flowering and interact with the floral inducer *CONSTANS* to regulate its stability. *Development*, 133: 3213–3222.

Levy, Y.Y. and Dean, C. (1998). The transition to flowering. *Plant Cell*, 10: 1973–1990.

Li, F., Zhang, X., Hu, R., Wu, F., Ma, J., Meng, Y. and Fu, Y. (2013). Identification and molecular characterization of *FKF1* and *GI* homologous genes in soybean. *PLoS One*, 8: e79036.

Li, Y., Zhang, C., Yang, K., Shi, J., Ding, Y. and Gao, Z. (2019). *De novo* sequencing of the transcriptome reveals regulators of the floral transition in Fargesia macclureana (Poaceae). *BMC Genom.*, 20: 1035.

Lin, C.T. (2000). Photoreceptors and regulation of flowering time. *Plant Physiology*, 123: 39–50.

Linnaeus, C. (1751). Philosophia botanica, in qua explicantur fundamenta botanica cum definitionibus partium, exemplis terminorum, observationibus rariorum, adjectis figuris aeneis. G. Kiesewetter, Stockholm.

Liu, B., Zuo, Z., Liu, H., Liu, X. and Lin, C. (2011). Arabidopsis cryptochrome 1 interacts with SPA1 to suppress COP1 activity in response to blue light. *Genes Dev.*, 25: 1029–1034.

Liu, J., Cheng, Z., Li, X., Xie, L., Bai, Y., Peng, L., Li, J. and Gao, J. (2019). Expression analysis and regulation network identification of the *CONSTANS*-like gene family in Moso bamboo (*Phyllostachys edulis*) under photoperiod treatments. *DNA Cell Biol.*, 38: 607–626.

Liu, J., Yu, J., McIntosh, L., Kende, H. and Zeevaart, J.A. (2001). Isolation of a *CONSTANS* ortholog from *Pharbitis nil* and its role in flowering. *Plant Physiology*, 125: 1821–1830.

Lou, P., Wu, J., Cheng, F., Cressman, L.G., Wang, X. and McClung, C.R. (2012). Preferential retention of circadian clock genes during diploidization following whole genome triplication *in Brassica rapa*. *Plant Cell*, 24: 2415–2426.

Lu, S.X., Knowles, S.M., Andronis, C., Ong, M.S. and Tobin, E.M. (2009). Circadian clock associated1 and late elongated hypocotyl function synergistically in the circadian clock of Arabidopsis. *Plant Physiology*, 150: 834–843.

Marchesini, V.A., Sala, O.E. and Austin, A.T. (2009). Ecological consequences of a massive flowering event of bamboo (*Chusquea culeou*) in a temperate forest of Patagonia, Argentina. *J. Veg. Sci.*, 20(3): 424–432.

Marcolino-Gomes, J., Rodrigues, F.A., Fuganti-Pagliarini, R., Bendix, C., Nakayama, T.J., Celaya, B., Molinari, H.B., de Oliveira, M.C., Harmon, F.G. and Nepomuceno, A. (2014). Diurnal oscillations of soybean circadian clock and drought responsive genes. *PLoS One*, 9: e86402.

McClung, C.R. (2008). Comes a time. *Curr. Opin. Plant Biol.*, 11: 514–520.

McClung, C.R. (2014). Wheels within wheels: New transcriptional feedback loops in the Arabidopsis circadian clock. *F1000 Prime Reports*, 6: 2.

McClung, C.R. and Gutiérrez, R.A. (2010). Network news: Prime time for systems biology of the plant circadian clock. *Curr. Opin. Genet. Dev.*, 20: 588–598.

Michael, T.P., Mockler, T.C., Breton, G., McEntee, C., Byer, A., Trout, J.D., Hazen, S.P., Shen, R., Priest, H.D., Sullivan, C.M., Givan, S.A., Yanovsky, M., Hong, F., Kay, S.A. and Chory, J. (2008). Network discovery pipeline elucidates conserved time-of-day-specific cis-regulatory modules. *PLoS Genet.*, 4: e14.

Miller, T., Muslin, E. and Dorweiler, J. (2008). A maize *CONSTANS*-like gene, *conz1*, exhibits distinct diurnal expression patterns in varied photoperiods. *Planta*, 227: 1377–1388.

Mishra, P. and Panigrahi, K.C. (2015). *GIGANTEA*: An emerging story. *Front. Plant Sci.*, 6: 8.

Mizuno, T. and Nakamichi, N. (2005). Pseudo-response regulators (PRRs) or true oscillator components (TOCs). *Plant Cell Physiology*, 46: 677–685.

Mockler, T.C., Guo, H., Yang, H., Duong, H. and Lin, C. (1999). Antagonistic actions of *Arabidopsis* cryptochromes and phytochrome B in the regulation of floral induction. *Development*, 126: 2073–2082.

Mohan Ram, H.Y. and Harigopal, B. (1981). Some observations on the flowering of bamboos in Mizoram. *Curr. Sci.*, 50: 708–710.

Murakami, M., Ashikari, M., Miura, K., Yamashino, T. and Mizuno, T. (2003). The evolutionarily conserved *OsPRR* quintet: Rice Pseudo-response regulators implicated in circadian rhythm. *Plant Cell Physiology*, 44: 1229–1236.

Murakami, M., Tago, Y., Yamashino, T. and Mizuno, T. (2007). Comparative overviews of clock-associated genes of *Arabidopsis thaliana* and *Oryza sativa*. *Plant Cell Physiology*, 48: 110–121.

Nakamichi, N., Kiba, T., Henriques, R., Mizuno, T., Chua, N.H. and Sakakibara, H. (2010). *Pseudo-response regulators 9, 7*, and *5* are transcriptional repressors in the *Arabidopsis* circadian clock. *Plant Cell*, 22: 594–605.

Nakamichi, N., Kiba, T., Kamioka, M., Suzuki, T., Yamashino, T., Higashiyama, T., Sakakibara, H. and Mizuno, T. (2012). Transcriptional repressor *PRR5* directly regulates clock-output pathways. *Proceedings of the National Academy of Sciences USA*, 109: 17123–17128.

Nakamichi, N., Kita, M., Ito, S., Yamashino, T. and Mizuno, T. (2005). *Pseudo-response regulators, PRR9, PRR7*, and *PRR5*, together play essential roles close to the circadian clock of *Arabidopsis thaliana*. *Plant Cell Physiology*, 46: 686–698.

Nakamichi, N., Kita, M., Niinuma, K., Ito, S., Yamashino, T., Mizoguchi, T. and Mizuno, T. (2007). *Arabidopsis* clock-associated *Pseudo-response regulators PRR9, PRR7*, and *PRR5* coordinately and positively regulate flowering time through the canonical *CONSTANS* dependent photoperiodic pathway. *Plant Cell Physiology*, 48: 822–832.

Navarro, C., Abelenda, J.A., Cruz-Oró, E., Cuéllar, C.A., Tamaki, S., Silva, J., Shimamoto, K. and Prat, S. (2011). Control of flowering and storage organ formation in potato by *Flowering Locus T*. *Nature*, 478: 119–122.

Nemoto, Y., Kisaka, M., Fuse, T., Yano, M. and Ogihara, Y. (2003). Characterization and functional analysis of three wheat genes with homology to the *CONSTANS* flowering time gene in transgenic rice. *Plant J.*, 36: 82–93.

Pao, C.I. and Morgan, P.W. (1986). Genetic regulation of development in *Sorghum bicolor*: I. Role of the maturity genes. *Plant Physiology*, 82: 575–580.

Park, H.J., Kim, W. and Yun, D. (2013). A role for *GIGANTEA*: Keeping the balance between flowering and salinity stress tolerance. *Plant Signal. Behav.*, 8: e24820.

Peng, Z., Lu, Y., Li, L., Zhao, Q., Feng, Q., Gao, Z., Lu, H., Hu, T., Yao, N., Liu, K., Li, Y., Fan, D., Guo, Y., Li, W., Lu, Y., Weng, Q., Zhou, C.C., Zhang, L., Huang, T., Zha, Y., Zhu, C., Liu, X., Yang, X., Wang, T., Miao, K., Zhuang, C., Cao, X., Tang, W., Liu, G., Liu, Y., Chen, J., Liu, Z., Yuan, L., Liu, Z., Huang, X., Lu, T., Fei, B., Ning, Z., Han, B. and Jiang, Z. (2013). The draft genome of the fast-growing non-timber forest species Moso bamboo (*Phyllostachys heterocycla*). *Nat. Genet.*, 45: 456–461.

Pruneda-Paz, J.L. and Kay, S.A. (2010). An expanding universe of circadian networks in higher plants. *Trends Plant Sci.*, 15: 259–265.

Pruneda-Paz, J.L., Breton, G., Para, A. and Kay, S.A. (2009). A functional genomics approach reveals CHE as a novel component of the Arabidopsis circadian clock. *Science*, 323: 1481–1485.

Putterill, J., Robson, F., Lee, K., Simon, R. and Coupland, G. (1995). The *CONSTANS* gene of *Arabidopsis* promotes flowering and encodes a protein showing similarities to zinc finger transcription factors. *Cell*, 80: 847–857.

Qian, H., Han, X., Peng, X., Lu, T., Liu, W. and Fu, Z. (2014). The circadian clock gene regulatory module enantio selectively mediates imazethapyr-induced early flowering in *Arabidopsis thaliana. J. Plant Physiol.*, 171: 92–98.

Quail, P.H. (2002). Phytochrome photosensory signaling networks. *Nat. Rev. Mol. Cell Biol.*, 2: 85–93.

Quail, P.H., Boylan, M.T., Parks, B.M., Short, T.W., Xu, Y. and Wagner, D. (1995). Phytochromes: Photosensory perception and signal transduction. *Science*, 268: 675–680.

Raffaele, E., Kitzberger, T. and Veblen, T.T. (2007). Interactive effects of introduced herbivores and post-flowering die-off of bamboos in Patagonia Nothofagus forests. *J. Veg. Sci.*, 18: 371–378.

Ramanayake, S.M.S.D. (2006). Flowering in bamboo: An enigma! *Ceylon J. Sci.*, 35: 95–105.

Ramanayake, S.M.S.D. and Yakandawala, K. (1998). Incidence of flowering, death, and the phenology of the giant bamboo (*Dendrocalamus giganteus* Wall. Ex Munro). *Ann. Bot.*, 82: 779–785.

Rand, D.A., Shulgin, B.V., Salazar, J.D. and Millar, A.J. (2006). Uncovering the design principles of circadian clocks: Mathematical analysis of flexibility and evolutionary goals. *J. Theor. Biol.*, 238: 616–635.

Redei, G.P. (1962). Super vital mutants of *Arabidopsis. Genetics*, 47: 443–460.

Rhind, D. (1945). *Grasses of Burma*. Baptist Mission Press, Calcutta (bamboo), India, pp. 1–26.

Riboni, M., Galbiati, M., Tonelli, C. and Conti, L. (2013). *GIGANTEA* enables drought escape response via abscisic acid-dependent activation of the florigens and *suppressor of overexpression of CONSTANS. Plant Physiol.*, 162: 1706–1719.

Salome, P.A. and McClung, C.R. (2005). *Pseudo-response regulators 7* and *9* are partially redundant genes essential for the temperature responsiveness of the Arabidopsis circadian clock. *Plant Cell*, 17: 791–803.

Satbhai, S.B., Yamashino, T., Okada, R., Nomoto, Y., Mizuno, T., Tezuka, Y., Itoh, T., Tomita, M., Otsuki, S. and Aoki, S. (2011). Pseudo-response regulator (PRR) homologues of the moss *Physcomitrella patens*: Insights into the evolution of the PRR family in land plants. *DNA Research*, 18: 39–52.

Sawa, M., Nusinow, D.A., Kay, S.A. and Imaizumi, T. (2007). *FKF1* and *Gigantea* complex formation is required for day-length measurement in Arabidopsis. *Science*, 318: 261–265.

Schranz, M.E. and Mitchell-Olds, T. (2006). Independent ancient polyploidy events in the sister families Brassicaceae and Cleomaceae. *Plant Cell Online*, 18: 1152–1165.

Seifriz, W. (1950). Gregarious flowering of *Chusquea. Nature*, 22: 635–636.

Serrano, G., Herrera-Palau, R., Romero, J.M., Serrano, A., Coupland, G. and Valverde, F. (2009). *Chlamydomonas CONSTANS* and the evolution of plant photoperiodic signaling. *Curr. Biol.*, 19: 359–368.

Sertse, D., Disasa, T., Bekele, K., Alebachew, M., Kebede, Y. and Eshete, N. (2011). Mass flowering and death of bamboo: A potential threat to biodiversity and livelihoods in Ethiopia. *J. Biol. Environ.*, 1(5): 16–25.

Sharma, M.L. (1991). The Flowering of Bamboo: Fallacies and Facts. *Proceedings of the 4th International Bamboo Workshop, Chiangmai, Thailand.*

Singh, A.N. and Singh, J.S. (1999). Biomass, net primary production, and impact of bamboo plantation on soil redevelopment in a dry tropical region. *Forest Ecology and Management*, 119: 195–207.

Somers, D.E. (2005). Entrainment of the Circadian Clock. *In: Endogenous Plant Rhythms. Annu. Plant Rev.*, 21: 85–105.

Somers, D.E., Schultz, T.F., Milnamow, M. and Kay, S.A. (2000). *Zeitlupe* encodes a novel clock-associated PAS protein from Arabidopsis. *Cell*, 101: 319–329.

Stearns, S.C. (1980). A new view of life-history evolution. *Oikos*, 35: 266–281.

Suarez-Lopez, P., Robson, W., Onouchi, H., Valverde, F. and Coupland, G. (2001). *CONSTANS* mediates between the circadian clock and the control of flowering in *Arabidopsis. Nature*, 410: 1116–1120.

Sulzman, F.M., Ellman, D., Fuller, C.A., Mooreede, M.C. and Wassmer, G. (1984). *Neurospora* circadian rhythms in space: A reexamination of the endogenous exogenous question. *Science*, 225: 232–234.

Takahashi, M., Furusawa, H., Limtong, P., Sunanthapongsuk, V., Marod, D. and Panuthai, S. (2007). Soil nutrient status after bamboo flowering and death in a seasonal tropical forest in western Thailand. *Ecol. Res.*, 22: 160–164.

Takano, M., Inagaki, N., Xie, X., Yuzurihara, N., Hihara, F., Ishizuka, T., Yano, M., Nishimura, M., Miyao, A., Hirochika, H. and Shinomura, T. (2005). Distinct and cooperative functions of phytochromes A, B, and C in the control of de-etiolation and flowering in rice. *Plant Cell*, 17: 3311–3325.

Takata, N., Saito, S., Saito, C.T., Nanjo, T., Shinohara, K. and Uemura, M. (2008). Molecular phylogeny and expression of poplar circadian clock genes, LHY1 and LHY2. *New Phytol.*, 181: 808–819.

Takata, N., Saito, S., Saito, C.T. and Uemura, M. (2010). Phylogenetic footprint of the plant clock system in angiosperms: Evolutionary processes of Pseudo-response regulators. *BMC Evol. Biol.*, 10: 126.

Tanaka, O. and Takimoto, A. (1972). Effects of the photoperiod on growth of the bamboo grass, *Pleioblastus variegatus. Plant and Cell Physiology*, 13: 187S–189.

Taylor, A.H., Jinyan, H. and Shi-Qiang, Z. (2004). Canopy tree development and undergrowth bamboo dynamics in old-growth Abies-Betula forests in southwestern China: A 12-year study. *For. Ecol. Manag.*, 200: 347–360.

Thapliyal, M., Joshi, G. and Behra, F. (2015). Bamboo: Flowering, seed germination, and storage. pp. 89–108. *In*: Kaushik, S., Singh, Y.P., Kumar, D., Thapliyal, M. and Barthwal, S. (eds.). *Bamboos in India*. ENVIS Centre on Forestry.

Thomas, B. and Vince-Prue, D. (1997). *Photoperiodism in Plants*. Academic Press, New York.

Toledo-Ortiz, G. (2003). The Arabidopsis Basic/Helix-Loop-Helix transcription factor family. *Plant Cell*, 15: 1749–1770.

Tseng, T., Salomé, P.A., McClung, C.R. and Olszewski, N.E. (2004). *SPINDLY* and *GIGANTEA* interact and act in *Arabidopsis thaliana* pathways involved in light responses, flowering, and rhythms in cotyledon movements. *Plant Cell*, 16: 1550–1563.

Valverde, F., Mouradov, A., Soppe, W., Ravenscroft, D., Samach, A. and Coupland, G. (2004). Photoreceptor regulation of *CONSTANS* protein in photoperiodic flowering. *Science*, 303: 1003–1006.

Wang, L., Fujiwara, S. and Somers, D.E. (2010). *PRR5* regulates phosphorylation, nuclear import, and subnuclear localization of *TOC1* in the *Arabidopsis* circadian clock. *EMBO Journal*, 29: 1903–1915.

Wang, W., Franklin, S.B., Lu, Z. and Rude, B.J. (2016). Delayed flowering in bamboo: Evidence from *Fargesia qinlingensis* in the Qinling Mountains of China. *Front. Plant Sci.*, 7: 151.

Wang, X., Wang, Y., Yang, G., Zhao, L., Zhang, X., Li. D. and Guo, Z. (2020). Complementary transcriptome and proteome analyses provide insight into the floral transition in bamboo (*Dendrocalamus latiflorus* Munro). *Int. J. Mol. Sci.*, 21(22): 8430.

Weller, J.L. and Reid, J.B. (1993). Photoperiodism and photocontrol of stem elongation in two photomorphogenic mutants of *Pisum sativum* L. *Planta*, 189: 15–23.

Wilkins, O., Waldron, L., Nahal, H., Provart, N.J. and Campbell, M.M. (2009). Genotype and time of day shape the *Populus* drought response. *Plant J.*, 60: 703–715.

Wu, F., Price, B.W., Haider, W., Seufferheld, G., Nelson, R. and Hanzawa, Y. (2014). Functional and evolutionary characterization of the *CONSTANS* gene family in short-day photoperiodic flowering in soybean. *PloS One*, 9: e85754.

Wysocki, W.P., Ruiz-Sanchez, E., Yin, Y. and Duvall, M.R. (2016). The floral transcriptomes of four bamboo species (Bambusoideae; Poaceae): Support for common ancestry among woody bamboos. *BMC Genom.*, 17: 384.

Xiao, G., Li, B., Chen, H., Chen, W., Wang, Z., Mao, B., Gui, R. and Guo, X. (2018). Overexpression of PvCO1, a bamboo *CONSTANS*- like gene, delays flowering by reducing expression of the FT gene in transgenic Arabidopsis. *BMC Plant Biol.*, 18: 232.

Yanagisawa, S. and Schmidt, R.J. (1999). Diversity and similarity among recognition sequences of DOF transcription factors. *Plant J.*, 17: 209–214.

Yang, S., Weers, B., Morishige, D. and Mullet, J. (2014). *CONSTANS* is a photoperiod regulated activator of flowering in sorghum. *BMC Plant Biol.*, 14: 148.

Yanovsky, M. and Kay, S.A. (2003). Living by the calendar: How plants know when to flower. *Nat. Rev. Mol. Cell Biol.*, 4: 265–276.

Yanovsky, M.J. and Kay, S.A. (2002). Molecular basis of seasonal time measurement in Arabidopsis. *Nature*, 419(6904): 308–12.

Zdepski, A., Wang, W., Priest, H.D., Ali, F., Alam, M., Mockler, T.C. and Michael, T.P. (2008). Conserved daily transcriptional programs in *Carica papaya*. *Trop. Plant Biol.*, 1: 236–245.

Zhang, E.E. and Kay, S.A. (2010). Clocks not winding down: Unravelling circadian networks. *Nat. Rev. Mol. Cell Biol.*, 11: 764–776.

Zhang, X.M., Zhao, L., Larson, R.Z., Li, D.Z. and Guo, Z.H. (2012). *De Novo* sequencing and characterization of the floral transcriptome of Dendrocalamus latiforus (Poaceae: Bambusoideae). *PLoS One*, 7: e42082

Zheng, X., Lin, S., Fu, H., Wan, Y. and Ding, Y. (2020). The bamboo flowering cycle sheds light on flowering diversity. *Front Plant Sci.*, 11: 381.

Zoltowski, B.D. and Imaizumi, T. (2014). Structure and function of the ZTL/FKF1/LKP2 group proteins in Arabidopsis. *Enzymes*, 35: 213–239.

Zuo, Z., Liu, H., Liu, B., Liu, X. and Lin, C. (2011). Blue light dependent interaction of CRY2 with SPA1 regulates COP1 activity and floral initiation in Arabidopsis. *Curr. Biol.*, 21: 841–847.

Chapter 8

Genetics of Abiotic Stress Resistance in Bamboos

Muhamed Adem[1],* and *Oumer Abdie*[2]

8.1 Introduction

Bamboos belong to the family Poaceae and subfamily Bambusoideae hosts approximately 1,670 species within 125 genera which are distributed in Asia, America, and Africa (Basak et al., 2021). Bamboos are one of the plant kingdom's most versatile and renewable resources. Bamboos are known as the "green gold of the forest" because they have benefited rural livelihood while bolstering the urban economy (Ahmad et al., 2020). Bamboos have the potential to be biofuel plants because of their rapid shoot growth (5–20 m within 2–4 months) within a single growing season (Basak et al., 2021). Bamboo is the fastest-growing plant on earth, growing at a rate of 100 cm each day, according to Tao et al. (2018). The most important gigantic bamboo species in Asia are Moso bamboo (*Phyllostachys edulis*) and Ma bamboo (*Dendrocalamus latiflorus* Munro) which may grow up to 20 m tall and around 30 cm in diameter. Bamboo species are mostly evergreen and create abundant forests (Lin et al., 2006). Bamboo forests are undoubtedly one of the most abundant non-timber plants on earth and cover a wide area of tropical and subtropical regions around the world covering over 40 million hectares of land (Lobovikov and Yping, 2012; Qiao et al., 2014).

Bamboos are among the most important non-timber forestry plants. They are playing increasingly important roles in socio-economic development and ecological protection. Around 2.5 billion individuals are believed to directly produce or consume bamboo. The worldwide bamboo market was valued at USD 68.8 billion in 2018,

[1] Department of Forestry, College of Agriculture and Natural Resources, Madda Walabu University, Bale Robe, Ethiopia.
[2] Department of Biotechnology, College of Natural and Computational Science, Wolkite University, Wolkite, Ethiopia..
Email: oumar.abdie@gmail.com
* Corresponding author: muhamed.adem@aau.edu.et/muhamedadem@gmail.com

and the demand for economically important bamboo species is increasing due to their immense benefits to humans (King, 2019). By 2028, the world wide bamboos market size is estimated to be worth USD 82.90 billion. From 2021 to 2028, it is predicted to grow at 5.7%. According to an analysis by Grand View Research, a greater breadth of application of bamboos in home and commercial applications, such as furniture, fabric, fodder, construction, medicine, food, biofuel, pulp and paper, charcoal, handicrafts and textiles is predicted to represent strong development potential for the industry (Emamverdian et al., 2020; Banerjee et al., 2021). China accounts for almost 71% of worldwide bamboo product exports in 2020, Asia Pacific accounted for nearly 80% of global bamboo product commerce. The existence of indigenous bamboo species in Brazil, Mexico, Ecuador, Chile, and the Dominican Republic, North America, Central and South America, and Europe has resulted in their having emerged as key traders of bamboo goods in recent years (https://www.grandviewresearch.com/press-release/global-bamboos-market). Madagascar has most of Africa's bamboo resources (Chaomao et al., 2006; Das et al., 2008). Other African countries with potential bamboo resources include Ethiopia, Tanzania, Malawi, Uganda, and Zambia (Kassahun, 2000; Mekonen et al., 2014; FAO and INBAR, 2018). In general, 80% of bamboo forests are in Asia, 10% in Africa, and 10% in Latin America (Lobovikov et al., 2007). Bamboo attracts global interest due to its unique life form, ecological importance, and vast range of human benefits (Guo et al., 2019).

Despite its outstanding growth performance, bamboo cultivation is hampered by several abiotic and biotic stresses like other plants. Plant development and productivity are threatened by several climate stressors. Environmental pressures are expected to become more severe and span a larger area through time. Plant sciences research has traditionally focused on yield traits, but today's demands necessitate stress resistance enhancement projects (Polle et al., 2019). Because of their sessile nature, plants have had little choice but to evolve diverse mechanisms that allow them to adapt to changing environmental conditions. Drought, high salt, high temperature/heat, waterlogging, cold, nutrient inadequacy, and heavy metals have all forced plants to create a variety of coping mechanisms to deal with the unavoidable problems of environmental stress. Stress-responsive genes control several facets of these adaptation processes, including developmental, physiological, and biochemical alterations (Peng et al., 2013; Huang et al., 2016; Xiang et al., 2021).

Due to its unique species-specific characteristics, such as the fastest growing ability on earth, a long and irregular flowering habit, unique rhizome-dependent systems, and the expansion of gene families because of polyploidization, bamboo could serve as an interesting model for addressing fundamental biological questions (Guo et al., 2019; Basak et al., 2021). However, the current state of passive genomes, transcriptomics, and proteome studies prevents the species from fully maximizing its study potential (Basak et al., 2021). Using in-depth analysis of bamboo genetics, physiology, genomics, and biotechnology to gain a pragmatic perspective on viable solutions may provide possible insights into the genetics of abiotic stress resistance in bamboos (Ahmad et al., 2020).

The release of whole-genome information for many bamboo species leads to genetic manipulations (Zhao et al., 2017; Ma et al., 2021). This shift is expected to harness the research aimed to tackle the natural hindrances of bamboo cultivation. Working on bamboo genetic resilience to environmental stress is one of the feasible options thus far. Stress-resistant bamboo types are likely to emerge because of combined and integrated efforts. Bamboo could be employed in degraded and deserted land rehabilitations, soil and water conservation measures, climate change mitigation, and green development activities if these lofty goals are realized (Ahmad et al., 2021). Now is the moment to create new plants to adapt to the new climate. Putting resources into research to develop abiotic stress tolerant bamboo would be a win-win situation because it would address both economic and environmental concerns (FAO and INBAR, 2018). In general, plant geneticists, physiologists, ecologists, breeders, environmentalists, foresters, economists, and policy makers working together could bring about a large increase in bamboo economic and environmental returns.

8.2 Bamboo and the Environment

Bamboo is a versatile plant that can provide climate-smart solutions to millions of rural communities. Bamboo can offer unparallel economic and environmental benefits if decision makers, planners, national sustainable development policies recognize its benefits (FAO and INBAR, 2018). The immense potential of bamboo in addressing interconnected concerns such as food, energy, degraded land rehabilitation, and climate change mitigation has not been properly realized. Proper investigation of the mechanisms related to the growth, strength, ways to utilize bamboo in industry, employment, climate change mitigation, and soil erosion reduction should be the concern of the time (Emamverdian et al., 2020). Bamboos should be mass-cultivated in marginal, degraded, and adverse areas for proper utilization of their enormous potential (FAO and INBAR, 2018) and to exploit their immune responses against various biotic and abiotic challenges. Bamboo has a large root system and underground biomass, which allows it to survive and regenerate even when the above-ground biomass deteriorates (Yanxia and Frith, 2018).

8.3 Bamboo Genome

The field of genomics has made studying every element of plant biology much easier. Understanding the plant genome at functional, physical, epigenomic and comparative levels has been made possible by the advent of next generation sequencing. Hundreds of model plants have been successfully sequenced, and their genome data are now available to resolve problems of plant biology (Bennetzen et al., 2012; Zhao et al., 2021). So far, scientists have sequenced roughly 328 vascular plants (consisting of 323 angiosperms, five gymnosperms), three non-vascular terrestrial plans, and 60 green algaes. Poaceae has 102 genome assemblies, the highest amongst a total of 104 reported of all angiosperms (Kersey, 2019). The genome data for almost all the important grass lineages has been compiled and made public, but the gigantic grass, bamboo, is one of the neglected species despite having considerable economic and ecological properties. Genome data of only five species are available in public domains out of around 1,670 species reported (Zhang et al., 2012; Li et al., 2020).

New and young genes allow organisms to explore new habitats as drivers of evolutionary progress. Balancing selection as an evolutionary force probably contributes to environmental adaptation (Zhao et al., 2021). According to Jin et al. (2021), 1,622 bamboo-specific orphan genes arose in the last 46 million years, 19 of which evolved from non-coding ancestral sequences with the whole de novo origination process reconstructed (Jin et al., 2021). The adaptive abilities of bamboos are ascribed to their evolutionary process. Bamboos are thought to undergo whole-genome duplication, resulting in the formation of multigene superfamilies, which are groups of genes with substantial similarity and a common ancestor. They are related in decent, but their functions have diverged in terms of position and structure across genomes (Hartl and Clark, 2007; Jin et al., 2021). The modular gene expression of the multigene superfamilies, known as exon duplication and shuffling, is one of their unique characteristics. Exon shuffling supports the switching up of different modules of the functional genes from various regions of the genome in order to create variety in functional proteins that aid in the adaptation process (Kolkman and Stemmer, 2001; Long et al., 2003). Long interval between vegetative and reproduction phases and flowering time among bamboo species has fascinated researchers in quest of understanding the molecular, physiological, and ecological reasons behind this process. Sequenced bamboo genomes enabled the identification of flowering pathway genes linked to photoperiod, vernalization, and hormonal regulation (Biswas et al., 2021). During flowering in *B. Tulda* many of the identified genes were found to be transcriptionally active (Dutta et al., 2018, 2021).

In comparison to other plants, bamboo genome sequencing is technically rather complicated and so is assembling because of its large genome size and high ploidy level (Ahmad et al., 2021). Bioinformatics analysis of sequenced bamboo genomes revealed that many genes that are present in single copies in herbaceous and diploid bamboos are widely multiplied in tetraploid and hexaploid woody bamboos (Guo et al., 2019a; Li et al., 2021). An account of sequenced bamboo genomes is given in Table 8.1.

The unavailability of genome sequences for the 99.67% bamboo species hampered genetic research and development of abiotic stress tolerant bamboos. Moso and Ma bamboos are the most biologically investigated bamboo species to date, owning to their significant socioeconomic and ecological value. *Raddia distichophyll* is another species having draft genome sequences available (Li et al., 2020; Zhao et al., 2021). The expansion of bamboo genome sequencing will aid in the assessment of genetic diversity, phylogenetic analysis, and more crucially, genetic improvement of bamboos for abiotic stress resistances.

Expanding transcriptome sequencing differ to include diverse developmental stages, tissues, and environmental factors would help with expression analysis, which can be used to deduce gene function in other species (Das et al., 2016). Since a promising initiative to sequence the genomes/transcriptomes of 300 species, belonging to 37 different bamboo genera (Genome Atlas of Bamboo and Rattan, GABR, http:// www. gabr- project. com/), the research challenge in bamboo associated with limited

Table 8.1. Summary of fully/partially sequenced bamboo genomes.

Species Scientific Name (Tribe)	Genome Size (Mb)	Ploidy Level	Genome Re/sequencing	No. of Genes	References
Ferrocalamus rimosivaginus (Arundinariinae)	139,467	–	Chloroplast genome	Total gene (TG) - 131. Protein coding gene (PCG) - 84	Zhang et al., 2011
Bambusa albolineata	139,326	–	Chloroplast Genome	TG = 129 PCG = 82	Deng et al., 2021
Chimonobambusa quadrangularis	139,540	–	Chloroplast genome	PCG = 82	Ren et al., 2021
Chimonobambusa sichuanensis	139,594	–	Chloroplast genome	TG = 140 PCG = 93	Zhao et al., 2021
Chimonobambusa hejiangensis	138,911	–	Chloroplast genome	TG = 133 PCG = 86	Liu et al., 2021
Bambusa rigida	139,500	–	Chloroplast genome	TG = 132 PCG = 84	Zheng et al., 2020
P. nigra var. Henonis (Arundinariinae)	139,839	–	Chloroplast genome	TG - 131. PCG - 84	Zhang et al., 2011
P. edulis (Arundinariinae)	139,679	$2n = 4x = 48$	Chloroplast genome	TG - 131. PCG - 84	Zhang et al., 2011
Indocalamus longiauritus (Arundinariinae)	139,668	–	Chloroplast genome	TG - 131. PCG - 84	Zhang et al., 2011
Bambusa emeiensis (Bambuseae)	139,493	–	Chloroplast genome	TG - 131. PCG - 84	Zhang et al. (2011)
Acidosasa purpurea (Arundinariinae)	139,697	–	Chloroplast genome	TG - 131. PCG - 84	Zhang et al. (2011)
Phyllostachys edulis (Arundinariinae)	2.05 Gb	$2n = 4x = 48$	Draft genome	31,987	Peng et al., 2013b
P. edulis (Carr.) Mitford cv. Luteosulcata	–	$2n = 4x = 48$	Chromosome-level reference genome and alternative splicing	–	Zhao et al., 2018a
Bonia amplexicaulis (Bambuseae)	0.849 GB	$2n = 6x = 72$	Draft genome	47,056	Guo et al., 2019a
Guadua angustifolia (Bambuseae)	1.614 GB	$2n = 4x = 46$	Draft genome	38,575	Guo et al., 2019a
Raddia distichophylla	0.608 GB	–	Draft genome	30,763	Li et al., 2021

genomic resources appears to be alleviated. This will considerably reduce the time and effort required for bamboo annotation, DNA barcoding, and conservation (Zhao et al., 2017; Zhao et al., 2021).

8.4 Development of Abiotic Stress Resistance in Bamboo

Bamboo's long and inconsistent flowering patterns, as well as its polyploidy make traditional cross breeding approaches challenging to improve its agronomic properties. Hence, bamboo germplasm enhancements are more likely only possible through molecular breeding approaches. *In situ* conservation, nature reserve, and bamboo gardens may be effective in managing breeding populations to promote accelerations of breeding programs (Ramakrishnan et al., 2020; Zhao et al., 2021). Basic research and genetic breeding of bamboo, on the other hand, have lagged due to a lack of genetic modification tools (Tu et al., 2021). Numerous candidate genes have been identified, but there has been no functional verification of any gene in Moso bamboo due to the lack of a genetic transformation system (Huang et al., 2022). Although effective transformation is routine practice for many plants, it has remained a nightmare for several bamboo species, despite massive and long-term efforts.

The genetic improvement of bamboo is challenging due to the slow pace of breeding associated with the large population size and long flowering intervals. Dozens of agronomic traits have been improved through genetic engineering, but to date it is restricted to only one bamboo species, Ma bamboo (*D. Latiflorus*) (Xiang et al., 2021; Tu et al., 2021). Modern genetic applications may aid in the promotion blooming in controlled environments (*in situ* or *ex situ* gardens), allowing natural genetic recombination to occur, which might be utilized to increase diversity. To achieve this, development of quality germplasm resources is needed (Zhao et al., 2021; Banerjee et al., 2021).

Despite decades of great efforts to improve bamboo agronomic traits, no new bamboo germplasm has been developed using traditional breeding methods (Ramakrishnan et al., 2020; Ahmad et al., 2021). The only feasible option for bamboo improvement appears to be genetic engineering. For most bamboo species however, establishing regeneration platforms, particularly shoot organogenesis, has proven problematic for most bamboo species (Singh et al., 2013). Despite the limitations, a highly efficient Ma bamboo regeneration and transformation procedures ware developed (Ye et al., 2017), allowing researchers to analyze the spatial and temporal gene expression patterns during the regeneration process. Both Agrobacterium tumefaciens and microprojectile bombardment procedures were devised for genetic transformation of *D. hamiltonii* (Sood et al., 2013). Only three reports demonstrated the successful transformation of bamboo utilizing explants from anthers and stem nodesv till April 2022, due to shortage of bamboo genetic modification tools (Qiao et al., 2014; Ye et al., 2017). Only one study found that ectopic expression of *CodA* gene, which produces a choline oxidase enzyme that converts choline to glycine betaine reduced the temperature of Ma bamboo (Qiao et al., 2014). Overexpression of the *leaf color* (*Lc*) gene in Ma bamboo resulted in a significant increase in the anthocyanin accumulation, and improved plant tolerance to cold and drought stressors, more likely due to the increased antioxidant capacity. However, few studies have reported simultaneously improvements of ornamental and multiple agronomic traits in Ma bamboo (Xiang et al., 2021).

Despite the sequencing of the genomes of five bamboo species (Peng et al., 2013; Guo et al., 2019; Zhao et al., 2018; Li et al., 2020; Zhao et al., 2021), none of these species have been successfully regenerated or transformed. Currently, Ma bamboo is the only bamboo species with an efficient gene manipulation strategy (Qiao et al., 2014; Ye et al., 2017, 2020; Xiang et al., 2021; Tu et al., 2021). However, because of its difficult nature (Guo et al., 2019), the genome of Ma bamboo has yet to be sequenced, preventing its molecular research and breeding. To comprehend genomic organization, anticipate protein functions, unravel the evolutionary history of species, and to select appropriate genes to study in comparative research, a phylogenetic analysis was performed (Som, 2014). Accordingly, bamboos have close evolutionary link with the presently sequenced bamboo accessions, as well as a high sequence similarity of orthologs, implying the gene function was retained over time. Further, all the differentially expressed genes (DEGs) found in Ma bamboo shoot organogenesis have orthologs in the other bamboo genomes. These findings suggests that at the genomic level, bamboo species may share a conserved regulatory mechanism that governs bamboo shoot organogenesis (Tu et al., 2021; Jin et al., 2021).

The application of genetic engineering methods to bamboo molecular breeding is particularly crucial for agronomic enhancement of any character of interest in bamboo (Ahmad et al., 2021) including abiotic stress tolerance (Xiang et al., 2021). Xiang et al. (2021) transformed a single transcription factor, bHLH's *Lc* gene, purple bamboo developed with greater ornamental qualities and abiotic stress tolerances (drought and cold stress tolerances) and food nutrient content. Such infrequent results suggest that bamboo genetic improvements are possible in the future with the potentials of abiotic stress tolerance and with improved nutritional qualities. Despite such sporadic but promising results, transformation of bamboo is still in its infant stage. Annotation and functional verification of essential genes linked with bamboo growth, development, and particularly abiotic stress are still needed (Ma et al., 2021). The lack of an efficient and rapid transformation and regeneration process is a major impediment to bamboo functional genomic research, as a result of the species' high polyploidy and genetic redundancy.

8.5 Stress and Bamboo Shoot Organogenesis

Environmental stresses such as drought, salinity, and coldness were previously thought to be hindering factors for plant growth and development, but it has now been discovered that they also play a role in de-differentiation, allowing cells to acquire a new cell fate during plant shoot organogenesis (Ahmad et al., 2021). Bamboo shoot organogenesis is a complicated and poorly understood process. Even though a global perspective of gene expression dynamics during shoot organogenesis in Ma bamboo is offered, much more research is needed to comprehend the operational regulatory networks during bamboo shoot organogenesis. In the future, an unbiased and comprehensive examination of gene expression changes between bamboo species at different stages will provide vital insights into the mechanisms underpinning bamboo organogenesis in the future (Tu et al., 2021). Shoot organogenesis is a valuable biotechnological tool exploited for basic research in plant sciences. The method is commonly used to create transgenic plants, select genetically superior soma clonal variants, and perform enormous plant multiplications (Duclercq et al., 2011). However, exploiting shoot organogenesis in bamboo plants is difficult due to a variety of factors such as the type and age of explant, chemical factors, and environmental situations (Sang et al., 2018; Shin et al., 2020). The genetic control of this extremely dynamic and sophisticated regeneration process, as well as the genetic flow dynamics during bamboo shoot organogenesis, is yet unknown (Tu et al., 2021).

Some study has found a direct link between shoot organogenesis and plant stress response. Bielach et al. (2017) discovered that during bamboo regeneration, the expression levels of the majority of absacic acid (ABA) and abiotic stress-responsive genes were significantly altered. Exogenous treatment of an optimum concentration of ABA or other abiotic stimuli boosted shoot regeneration efficiency considerably, indicating that these genes may play in bamboo organogenesis. Both external stimuli and endogenous plant hormones drive shoot organogenesis during tissue culture. Exogenous variables may act as triggers, affecting endogenous hormone levels and signaling pathways. ABA and various abiotic stresses such as salt, drought, and osmotic stress all have an impact on auxin and cytokinin homeostasis and signaling (Bielach et al., 2017; Sharma et al., 2015).

It is unclear whether exogenous stress factors and endogenous hormones are linked at many levels in bamboos to promote shoot regeneration. Exogenous stress factors and endogenous hormones must be studied further to determine whether they have a direct relationship in the promotion of bamboo shoot regeneration. Stress can cause epigenetic alterations in chromatin structure, which are essential for the activation of genes that drive the cell destiny towards the regenerative state. Stress has been proposed as a signal that induces plant cells to undergo reprogramming for adaptability via stimulus-based cell fate acquisition (Grafi and Barak, 2015). The expression of 34 genes encoding chromatin modification-related proteins, including the chromatin-remodeling complex ATPase chain, chromatin condensation inducer, chromatin structure-remodeling complex protein SYD isoform, and EAF1 isoform, was significantly altered during bamboo regeneration. More research is needed to find out if stress-induced chromatin alteration helps to bamboo shoot organogenesis (Grafi and Barak, 2015; Ahmad et al., 2021).

Many studies have shown that abiotic stressors can cause plant cells to produce reactive oxygen species (ROS) (Huang et al., 2019). ROS are signaling molecules that regulate many plant growth and developmental processes, including plant organogenesis, in addition to causing harm to plant cells (Huang et al., 2019). In de-differentiating cells, ROS can cause up-and-down regulation of regulatory genes (Grafi and Barak, 2015). Oxidative stress-related genes are substantially expressed during bamboo organogenesis, suggesting that ROS may play a role in this process. The specific association between stress and shoot ganogenesis, however, has yet to be determined. Investigating the hypothesis that ROS-mediated signaling promotes genomic reprogramming, leading to stimulus-based acquisition of a novel cell fate in bamboo shoot organogenesis, will be more important in the future. The biological function verification of candidate genes with suspected functions in bamboo *de novo* shoot organogenesis should be the focus of future research (Tu et al., 2021).

8.6 Abiotic Stresses

Based on its environmental and economic characteristics, bamboo is one of the most important non-timer forest plants (Banerjee et al., 2021). It is a widely held truth that if bamboo receives proper attention just from policy to execution, it may play a significant role in combating land degradation and alleviating rural poverty. Despite their importance, bamboos have received little attention in scientific research, notably in the genetics of abiotic stress resistance (IFAD and INBAR, 2019; Hou et al., 2020; Ramakrishnan et al., 2020). Despite the presences of remarkable differences in their adaptation and resilience to environmental stress, all bamboos are affected by environmental stress in their growth and development, which limits their geographic distributions. Understanding the mechanisms that underpin bamboo, water use in response to stress is crucial for forecasting the growth in a changing environment (Wu et al., 2019). From global transcriptome profiling to specific gene expression, profiling has been conducted in bamboos aiming to identify abiotic stress responsive genes (Adem et al., 2019a; Wu et al., 2018; Zhang et al., 2022). Metabolic profiling of Moso bamboo in response to drought stress in a field investigation has been carried out to identify metabolites and their corresponding pathways associated to abiotic stress response and adaptation (Tong et al., 2020). Abiotic stress, which is commonly induced by salt, drought, and high light intensity challenges, can be avoided by introducing abiotic stress-tolerant genes into bamboo transgenic experiments (Sun et al., 2017; Liu et al., 2019; Cheng et al., 2020; Xie et al., 2020; Xiang et al., 2021). As a result, it is critical to investigate all conceivable processes for boosting bamboo's tolerance to environmental stress (Wu et al., 2015; Tu et al., 2021). The field of bamboo basic biological research and the industrial sectors are still grappling with how to improve abiotic stress resistances in bamboo (Xiang et al., 2021). Recent evidence is suggesting that abiotic stress, particularly drought, may act as an important 'cue' to determine flowering time in bamboo species (Biswas et al., 2021). The marker hormone for drought response, ABA, is higher in young leaf from flowering culm than a leaf from non-flowering culm. Fire and pruning have also found to be determinant factors in flowering time in *Bambusa tulda*, *B. balcooa*, *Dendrocalamus hookeri*, and *Melocanna baccifera* (Biswas et al., 2021).

8.6.1 Drought and High Temperature

Drought is a meteorological word that refers to a combination of diminished rainfall, dwindling ground water levels, and limited water supply caused by rising temperatures (Singh et al., 2015). While plants have developed a variety of defense strategies to cope up with high temperatures and other environmental challenges (Rizhsky et al., 2002; Chauhan et al., 2011). Such challenges cause cell injury from ROS, increased cellular temperature, resulting in an increase in the viscosity of cellular contents, changes in the protein-protein interactions, and protein aggregation and denaturation (Farooq et al., 2008). Understanding the fundamental mechanisms of plant water usage in response to drought is crucial for predicting the trend of global ecosystem in a changing climate (Wu et al., 2019). More often high temperatures and dryness have been linked to climate change, lowering the ecological and economic values of bamboos. To improve bamboo stress resistance, it is critical to understand the molecular mechanisms involved in drought and heat stress responses (Liu et al., 2014). Drought occurrences are affecting an increasing number of bamboo habits. The advancement of our knowledge of the process of bamboo drought resistances promotes their genetic improvement and forest sustainability (Tong et al., 2020).

Bamboo's growth is reliant on natural rainfall and is susceptible to high temperatures and drought. High temperatures >40°C and drought for > 10 days during August, according to Liu et al. (2014) caused devastating losses in Moso bamboo forests. Drought in the spring hampered the development, yield, and quality of Moso bamboo. From July to September, high temperatures and drought severely affected the sprouting phase of bamboo. These stresses have an impact on winter shoots yield and quality, as well as new bamboo yield in the following year and wood yield in succeeding years (Zhang et al., 2008). The physiological and biochemical response of four dwarf bamboos (*Pleioblastus kongosanensis, Sasa fortunei, Sasa argenteostriata*, and *Sasa auricoma*) were studied by withholding water. The findings showed that the free proline contents, cell membrane permeability, and the MDA contents of bamboo leaves steadily increased with the prolongation of the stress period. However, the quantity of soluble sugars and the activity of SOD, POD, and CAT increased at the beginning and thereafter decreased (Lan et al., 2010). Heat shock proteins (HSPs) operate as molecular chaperones, inhibiting protein aggregation and protecting the cells from heat and other stresses (Queitsch, 2000; Hart and Hayerhart, 2002). The terminal components of the stress signal transduction chain, and heat shock stress transcription factors (Hsfs) bind to the promoter regions of HSP genes to control transcription in response to stress (Pelham and Bienz, 1982; Döring et al., 2000), particularly high temperature stress (Von et al., 2007).

Metabolic profiles of Moso bamboo in response to drought stress in a field investigation studied by Tong et al. (2020) revealed an important insight. The researchers used liquid chromatography coupled to mass spectrometry (LC-MS) based on untargeted metabolomic profiling to explore metabolic changes in Moso bamboo in the field under drought stress. Results showed that the metabolic profiles induced by drought stress were relatively consistent among the three growth stages (initial vegetative growth stage, vigorous vegetative growth stage, and reproductive growth stage). Specifically, most responsive metabolites exhibited enhanced accumulation under drought stress, including anthocyanins, glycosides, organic acids, amino acids, and sugars and sugar alcohols. The putative metabolism pathways implicated in drought stress response were mostly included in amino acid metabolism and sugar metabolism pathways. Linoleic acid metabolism, ubiquinone and other terpenoid-quinone production, tyrosine metabolism, starch, and sucrose metabolism, and isoquinoline alkaloid biosynthesis were discovered to be common among the three growth stages (Tong et al., 2020).

8.6.2 *Salinity*

Excessive salts in the soil solution cause inhibition of plant growth, which is known as salt stress in plants. No toxic material inhibits plant growth more than salt does on worldwide scale (Zhu, 2007). Many physiological disorders in plants are caused by an increase in Na+ and Cl- ions in the soil (Tavakkoli et al., 2010). Salinity at physiologically dangerous levels causes nutriatentional deficiency and unavailability, altered levels of growth regulators, enzyme inhibition, membrane damage, and metabolic malfunctioning, including photosynthesis, all of which eventually leads to the plant's death (Turkana and Demiral, 2009; Ali et al., 2012). As a result of these issues, it is critical to make every effort feasible to design plants that can thrive in saline environments.

Toxicity caused by excessive salt levels in plants can be divided into two mechanisms: (1) osmotic regulatory disruption and (2) ionic toxicity. Dehydration is produced by osmotic disruption caused by salinity stress, which inhibits water uptake, cell elongation, and leaf development (Gupta and Huang, 2014). In the second phase, ionic stress causes an increase in sodium ions while potassium ions decrease dramatically. As a result of nutrient imbalance, leaves age and die quickly, and enzyme activity, protein synthesis, and photosynthetic ability are all hampered (Munns and Tester, 2008). Signal transduction may occur during the early stages of osmotic and ionic stress, resulting in either osmotic adjustment and ion homeostasis or cell death. The plant's recovery or adaptability is very likely if signal transduction successfully initiates osmotic adjustment and ion homeostasis. However, if cell death occurs, the plant's chance of recovery from salinity-induced toxicity are reduced considerably (Munns and Tester, 2008). Various physiological changes in plants can be observed during the early stages of high salinity stress, including membrane disruption, nutrient imbalance, impaired ability to detoxify ROS, alteration of antioxidant enzymes, reduced and altered photosynthetic activity, and decreased stomatal aperture. The decrease in leaf area, chlorophyll content, and stomatal conductance associated with high salinity stress and reduced photosystem II efficiency caused by high salt levels resulted in impaired photosynthetic performance (Tanou et al., 2009).

The expression of a variety of salt-response genes are critical for adapting hostile strategies to induce biochemical and physiological alterations. Plants' ability to endure the effects of salinity on their growth and development is based on the interaction of salt sensitive genes and proteins. The mitogen-activated protein kinase (MAPK), salt overly sensitive (SOS), and calcium-dependent protein kinase (CDPK) pathways perform adaptive roles to combat stresses (Nakagami et al., 2005; Goyal et al., 2016). Aside from these mechanisms, plant hormones such as salicylic acid, jasmonic acid, and abscisic acid play a key role in stress signaling and adaptation.

Despite the significant progress made in understanding the processes of molecular mechanisms behind salt stress tolerances, the complexity of interactions involved in salt stress tolerance mechanisms in bamboo necessitates further research (Ma et al., 2006). Three bamboo species, *Dendrocalamus strictus*, *Dendrocalamus longispathus*, and *Bambusa bambos* were investigated for salinity tolerance using a hydroponic growth technique. After 14 days of salt treatment, an investigation of vegetative growth metrics revealed no significant effects on the treated species' shoot height, root height, number of leaves, or fresh weight. In salt-tolerant species, proline buildup was more noticeable than glycine betaine (GB). *D. longispathus* salt-sensitive species had 7.57 and 22.85 times more proline than *D. strictus* and *B. bambos* salt-resistant species treated with the same dose. Whereas, with increasing NaCl concentration, GB content showed a completely different trend, its concentration did not fluctuate significantly in *D. longispathus*, but it was observed to decrease in *B. bambos* and increased in *D. strictus* (Pulavarty and Sarangi, 2015). The impact of salt on Moso bamboo has been investigated. Salinity has a significant impact on membrane permeability, proline content, and superoxide dismutase (SOD) activity of seedling leaves, resulting in their death (Huang, 2010; Meng, 2010). Moso bamboo may develop a variety of systems to combat and adapt to the negative impacts of salt stress. For a better understanding of bamboo stress and rising bamboo yields, additional salinity associated genes must be identified and validated, as well as their resistances mechanisms should be investigated (Meng, 2010).

8.6.3 Cold/Chilling

Bamboo distribution is heavily influenced by agroclimatic zones, climatic circumstances, and human intervention (Yeasmin et al., 2015). One of the most important environmental constraints limiting bamboo growth, development, and geographic distribution is the cold. Sudden snow and ice storms are expected to become more common as global change models indicate. Such catastrophes are likely to have a profound influence on all terrestrial ecosystems' inhabitants. In January 2008, for example, the Yangtze River Basin and southern China saw exceptional cold spells caused by persistent rain, snow, or ice storms, resulting in widespread and severe freezing. With its harsh winter weather, this occurrence has impacted many parts of Asia (Ding et al., 2008; Jiang et al., 2008). Among the higher plants, 51 bamboo species were severely damaged among the taller plants, with 14 of them dying. Due to frigid temperatures, the hybrid bamboo (*B. pervariabilis* × *D. grandis*) and *D. latiflorus* could not survive. Surprisingly, both *N. affinis* and *B. rigida* remained relatively unscathed (Zhang et al., 2012a). The following bamboo species are listed from the most susceptible to tolerant bamboos under freezing temperatures: hybrid bamboo > *D. Latiflorus* > *B. Rigida* > *N. affinis* (Jiang et al., 2008; Yang et al., 2008; Zhao et al., 2009).

Cold stress affects signal perception and subsequent transduction pathways, triggering transcriptional control and the activation of many genes encoding for cold-regulated proteins (Zhu, 2016: Guo et al., 2018). Low temperature affects membrane fluidity, initiating the cellular cold response through calcium ($Ca2+$) signaling pathways, as evidenced by an integrated investigation of molecular, physiological, and metabolic profiling in response to cold stress (Sangwan et al., 2001; Zhang et al., 2014). Over the last few decades, significant progress has been made in the study of plants response to freezing (< 0°C) and to chilling (0–15°C) in model and non-model plants. *Arabidopsis thaliana* (Lee et al., 2005), rice (Zhang et al., 2014), cotton (Kargiotidou et al., 2010), soybean (Calzadilla et al., 2016), and tomato are only a few of the useful studies (Weiss and Egea-Cortines, 2009). The underlying processes of bamboo cold stress response, on the other hand, are still poorly understood.

By carefully evaluating transcriptome dynamics under cold stress, Liu et al. (2019) offered an overall summary of the cold-responsive transcriptional patterns in Moso bamboo. Even though molecular changes that occur during bamboo adaptation to cold stress are yet unknown, the data revealed that low temperatures cause significant morphological and biochemical changes in Moso bamboo. The study discovered the expression of a small number of differentially expressed genes (DEGs) at the early stage, but a high number of DEGs at later stage. This demonstrates that the bulk of cold responsive genes in bamboo are late responders. A total of 222 transcription factors were found to be differentially expressed across 24 gene families. The expression of hundreds of well-known C-repeat/dehydration responsive element-binding factors were considerably elevated in response to cold, demonstrating the development of unique cold response networks. Cold stress affected the expression of genes involved in cell wall and fatty acid production, suggesting that they may play a role in the development of bamboo cold resistance. This unusual study discovered that Moso bamboo has both plant kingdom-conserved and species-specific cold response pathways, laying the groundwork for future research into the regulatory mechanisms underlying bamboo cold stress response and providing useful gene resources for the development of cold-tolerant bamboo through genetic engineering (Liu et al., 2019).

8.7 Transgenic Approaches for Enhancing Abiotic Stress Tolerance in Bamboos

Many overexpression experiments of bamboo genes in Arabidopsis and other model plants have proved bamboos ability to withstand stress (Table 8.2). The drought-induced 19 protein, *PeDi19-4*, from Moso bamboo improved plant drought and salt tolerance via the ABA-dependent signaling pathway (Wu et al., 2018). Ribon neuclic acid sequencing confirmed that *PeDi19-4* controlled the expression of a large range of stress-/ABA-responsive differentially expressed genes. *PeDi19-4* directly targets stress-responsive genes (*OsZFP252* and *OsNAC6*) and ABA-responsive genes (*OsBZ8* and *OsbZIP23*) were direct targets. Overexpression of PeDi19-4 in rice and Arabidopsis thaliana boosted drought and salt tolerance via the ABA-dependent signaling, according to phenotypic and stress-related physiological markers. Overexpression of the *leaf color* (*Lc*) gene in Ma bamboo resulted in a significant increase in the anthocyanin accumulation and improved plant tolerance to cold and drought stressors, most likely due to increased antioxidant capacity (Xiang et al., 2021). *PheNAC3* from Moso bamboo overexpression improves abiotic stress tolerance in Arabidopsis demonstrating its importance in the drought stress response (Xie et al., 2020).

PeTIP4;1-1, an aquaporin gene identified from bamboo, has been shown to offer considerable drought and salinity stress tolerances in Arabidopsis, a model plant. Abiotic stresses control *TIP4;1-1*, which plays a key role in bamboo shoot growth. In bamboo shoots and leaves, *PeTIP4;1-1* was dramatically upregulated

Table 8.2. Bamboo genes conferring tolerance to abiotic stresses.

Gene Sources	Gene	Transgenic Species	Phenotype of Stress Tolerance	References
Bamboo	*PeTIP4;1–1*	Arabidopsis	Drought, salinity	Sun et al., 2017
Bamboo	*Pepip2;7*	Arabidopsis	Drought, salinity, high light	Sun et al., 2021
Bamboo	*pheNAC3*	Arabidopsis	Drought	Xie et al., 2020
Maize	*leaf color (Lc) gene*	Ma bamboo	Drought, cold	Xiang et al., 2021
Bamboo	*PeDi19-4*	Rice and Arabidopsis	Drought, salinity	Wu et al., 2018
Bamboo	*PheASR*	Arabidopsis	Drought	Hu et al., 2012
Bamboo	*pheWRKY86*	Arabidopsis and rice	Drought	Wu et al., 2022
Bamboo	*PheDi19-8*	Arabidopsis	Drought	Wu et al., 2020b
Bamboo	PheDof12-1	Arabidopsis	Drought, cold, salinity	Liu et al., 2019
Bacteria	CodA gene	Ma bamboo	Cold	Qiao et al., 2014
Bamboo	*Phehdz1*	Rice	Drought	Gao et al., 2020
Bamboo	*PeVQ28*	Rice	Salinity	Cheng et al., 2020
Bamboo	PheCDPK22	Arabidopsis	Drought	Wu et al., 2020b
Bamboo	PeSNAC1	Rice	Drought, Salinity	Hou et al., 2020
Bamboo	*PheASR2*	Rice	Drought, Salinity	Wu et al., 2020a
Bamboo	*PeWRKY83*	Arabidopsis	Salinity	Wu et al., 2017
Bamboo	*PeC3H74*	Arabidopsis	Drought	Chen et al., 2020
Bamboo	PeTCP10	Arabidopsis and rice	Drought	Liu et al., 2020
Bamboo	*PeZEP*	Arabidopsis	Drought	Lou et al., 2017
Bamboo	*PeUGE*	Arabidopsis	Drought, salinity	Sun et al., 2016
Bamboo	*PheMYB4-1*	Arabidopsis	Drought, salinity and cold	Hou et al., 2018
Bamboo	PeSnRK1a	Arabidopsis	Salinity	Pan, 2020
Bamboo	PheWRKY50-1	Arabidopsis	Drought, salinity	Huang et al., 2021
Bamboo	PheWRKY72-2	Arabidopsis	Drought,	Li et al., 2017
Bamboo	PeLAC10	Arabidopsis	Drought, phenolic acid	Li et al., 2020
Bamboo	PeDi19-4	Arabidopsis and rice	Drought, salinity	Wu et al., 2018

in response to drought and salinity stressors. Transgenic *Arabidopsis* plants overexpressing *PeTIP;1–1* under the control of CaMV *35S* promoter were generated and subjected to morphological and physiological experiments to study the role of investigate the role of *PeTIP4;1–1* in response to drought and salinity conditions. Compared to wild plants, transgenic plants demonstrated improved drought and salinity tolerances and generated longer taproots with greener leaves, higher SOD, POD, and CAT activity, lower MDA levels and higher water content. Three stress-related genes had increased by the aquaporin gene (*AtP5CS*, *AtNHX1*, and *AtLEA*). These data suggest that *PeTIP4;1–1* may play a critical role in drought and salinity stress response. Due to such prominent roles, *PeTIP4;1*—has become a potential gene for stress tolerance in other crops (Sun et al., 2017).

WRKY are among the key transcription factors for stress response in plants, including bamboos. On transgenic Arabidopsis and rice, *PheWRKY86* gene from Moso bamboo (*Phyllostachys edulis*) has provided substantial improvements in drought stress tolerance. Drought and ABA treatments significantly increase *PheWRKY86* expression. The *PheWRKY86* protein is found in the nucleus of the cell and can bind to W-box elements. *PheWRKY86* transgenic plants had improved water retention and lower relative electrolyte leakage (REL) and malondialdehyde (MDA) levels than wild type plants (Wu et al., 2022). Overexpression of *Phehdz1* not only improved drought tolerance of transgenic rice, but also changed its secondary metabolism, according to Gao et al. (2020). *Phehdz1*-overexpressing transgenic rice had a faster germination rate and longer shoots in response to mannitol treatments than the wild-type controls. After a 30% polyethylene glycol 6000 treatment, *Phehdz1*-overexpressing rice plants had a greater survival rate, and higher relative water and proline contents, but a lower malondialdehyde content than the WT plants. This unveils that *Phehdz1* is involved in both drought stress tolerance and the regulation of metabolism-related genes (Gao et al., 2021).

Overexpression of the bamboo *PeVQ28* gene in Arabidopsis influenced the expression of salt/ABA-responsive genes, suggesting that *PeVQ28* regulates salt tolerances and hence may affect ABA production triggered by stress in plants. *PeVQ28*-overexpressing Arabidopsis lines demonstrated greater salt stress resistance and improved ABA sensitivity. *PeVQ28*-transgenic plants displayed lower malondialdehyde and higher proline levels under salt stress than wild-type plants, possibly enhancing stress tolerance. In Arabidopsis, overexpression of *PeVQ28* increased the expression of salt- and ABA-responsive genes. These results suggest that *PeVQ28* is involved in the control of salt tolerance via an ABA-dependent signaling pathway (Cheng et al., 2020).

Hou et al. (2020) discovered that plants with *PeSANC-1* overexpression were more resistant to drought and salt stress. The transgenic rice outperformed its wild counterpart in terms of physiological markers like superoxide dismutase, peroxidase, and catalase activities, as well as malondialdehyde, H_2O_2 and proline. Furthermore, in both yeast and plant cells, protein interaction studies demonstrated that *PeSNAC -1* interacts with stress responsive *PeSNAC-2/4* and *PeNAP-1/4/5*, indicating that those proteins work together to regulat the Moso bamboo stress response. The ability of *PeSNAC-1* to confer salt and drought stress tolerance is linked to its ability to modulate gene regulation in both ABA-dependent and independent signaling pathways. This gene also acts as a positive stress regulator in Moso bamboo, participating in the *PeSNAC-1* and *PeSNAC-2/4* or *PeSNAC-1* and *PeNAP-1/4/5* interaction networks.

Overexpression of *PeWRKY83* caused Arabidopsis to exhibit superior physiological attributes than WT under salt stress conditions, according to Wu et al. (2017). Transgenic plants were less sensitive to ABA and were more endogenous at both the germination and postgermination stages. Under salt stress, overexpression of *PeWRKY83* was shown to affect the expression of ABA biosynthesis genes (*AtAAO3*,

AtNCED2, AtNCED3), signaling genes (*AtABI1, AtPP2CA*), and responsive genes (*AtRD29A, AtRD29B, AtABF1*). The combined findings of investigations show that *PeWRKY83* serves as a novel *WRKY*-associated TF that functions as a positive role in salt tolerance by regulating stress-induced ABA synthesis. Ectopic expression of *PheASR2* has boosted drought stress tolerance in rice, as revealed by physiological assessments of germination rate, plant height, water loss, and survival rate. The *PheASR2* overexpressing transgenic plants showed a significant rise in ROS, electrolyte leakage and malondialdehyde levels, as well as decreased enzyme (CAT and SOD) activities, and greater expression of genes encoding ROS-scavenging enzymes. Transgenic plants were found to be more resistant to oxidative stress than wild-type plants. The expression marker genes such as *OsAREB, OsP5CS1, OsLEA*, and *OsNCED2* were upregulated during drought treatment of the *PheASR2*-overexpressing rice (Wu et al., 2020a).

In a study by Chen et al. (2020), *PeC3H74* (*PeC3H74-OE*) enhanced drought tolerance of transgenic Arabidopsis. Phenotypic and physiological analysis comprising water content, survival rate, electrolyte leakage, and malondialdehyde content confirmed stress tolerances conferring capacity of the *PeC3H74*. Furthermore, transgenic Arabidopsis thaliana seedling roots grew faster than wild-type plants at 10 mm ABA. ABA treatment significantly changed the Arabidopsis stomata over-expressing *PeC3H74*. According to Wu et al. (2020b), under drought stress conditions, *PheDi19-8*-overexpressing lines had a greater survival rate, but smaller stomatal openings compared with the wild-type plants. There was also higher biomass and souble sugar in the *PheDi19-8*-overexpressing lines, but less relative electrolyte leakage and malondialdehyde (MDA) contents. According to the same study, the overexpression of *PheCDPK22* enhanced Arabidopsis sensitivity to drought stress.

The research findings of Liu et al. (2020) revealed that *PeTCP10* plays a specific role in modulating drought-stress responses. *PeTCP10* produced significant drought tolerance in transgenic Arabidopsis and rice, as evidenced by phenotypic and stress-related physiological indicators analysis, including relative water content, root growth, survival rate, germination rate, and MDA content. According to Sun et al. (2016), examination of chlorophyll fluorescence characters, and lateral roots of transgenic *Arabidopsis thaliana* plants indicated PeUGE's important function in enhancing abiotic stress resistances. *PheMYB4*, a hub gene, was shown to be involved in the complicated protein interaction network. Further functional research revealed the ectopic overexpression of its homologous gene, *PheMYB4-1*, conferred cold stress resistance and contributed to transgenic Arabidopsis seedling sensitivity to drought and salt stress. Such discoveries provided a thorough understanding of the MYB family members in Moso bamboo, as well as possible MYB genes for further studies into their involvement in stress tolerance (Hou et al., 2018).

The transgenic plants overexpressing *PeZEP* were created utilizing the *PeZEP* expression vector driven by CaMV *35S*. The transgenic *Arabidopsis thaliana* plants displayed substantial drought stress tolerances as compared with the wild-type plants, according to morphological and physiological indices. It was confirmed that the transgenic plants experienced wilting later, yet had a greater survival percentage (Lou et al., 2017). As the study of Pan (2020) determined, under high-salt stress conditions, transgenic *PeSnRK1a*-overexpressing seeds had a considerably higher germination rate and seedlings with longer roots than wild-type (WT) seeds. The same gene appears to work better in the dark, since transgenic leaves aged more slowly than WT leaves under darkness. According to these findings, *PeSnRK1a* appears to serve as a critical regulatory role in bamboo development and stress tolerance.

The *PheWRKY50-1* gene isolated from bamboo was transformed into Arabidopsis via Agrobacterium infection, and found that the transgenic plants were found to have superior stress tolerances than WT plants (Huang et al., 2021). The overexpression of *PheWRKY72-2* in Arabidopsis showed in a reduced sensitivity to drought stress during the seedling growth. *PheWRKY72-2*'s ability to induce plant stress is linked to its role as a positive regulator of stoma closure (Li et al., 2017). In transgenic Arabidopsis, overexpression of *PeLAC10*, a gene identified and isolated from Moso bamboo, has conferred considerable drought and phenolic acid tolerance in addition to increased lignin content. Plants over-expressing *PeLAC10* in the transgenic Arabidopsis were somewhat smaller and had short petioles than wild plants. They did, however, have a larger lignin content in stems and were more resistant to phenolic acids and dryness. These findings suggested that overexpression of *PeLAC10* improved the tolerances of transgenic Arabidopsis to phenolic acid and drought stress by increasing the quantity of lignin (Li et al., 2020). The prospect of developing abiotic stress tolerant bamboo plants via a transgenic technique appears to be imminent, as potentiality of conferring abiotic stress tolerance has already been in other plant systems mentioned above. A list of the potential abiotic stress tolerances genes is depicted in Table 8.2.

Xie et al. (2020) have demonstrated that under salt stress conditions overexpression of *PheNAC3* induced faster seed germination, better seedling growth, and a greater survival rate than the WT plants. Transcriptome sequencing of Moso bamboo (*Phyllostachys edulis*) demonstrated that cold and dehydration stresses upregulated more than half of *PeTIFY* genes, and certain *PeTIFYs* share a similarity with known TIFY's involved in abiotic stress tolerance (Huang et al., 2016).

Most studies focus on the functional study of genes discovered and isolated from bamboo utilizing Arabidopsis and rice as transformation hosts. In contrast, Qiao et al. (2014) have introduced a foreign bacterial originated *CodA* gene in Ma bamboo. The transgenic Ma bamboo behaved better in cold temperatures than wild plants, demonstrating that *CodA* imparted significant cold stress tolerances. The effect of *CodA* was amplified because the amount of glycine betaine (GB) increased by 140% in *CodA* transgenic Ma bamboo, compared to 83% in wild Ma bamboo. Unlike GB, the activity of superoxide dismutase, peroxidase, and catalase rose in both transgenic and WT plants. But the activity of the aforesaid enzymes were higher in the transgenic lines than in the WT plants under cold stress conditions. MDP buildup and electrolyte leakage (REL) were lower in *CodA* transgenic Ma bamboo plantss than in the control plants.

8.8 Role of Transcription Factors in Abiotic Stress Response

Plant transcription factors (TFs) are regulatory proteins that control gene expression in response to environmental challenges such as drought, salinity, coldness, etc. By regulating defense response and gene regulation networks, TFs play an important role in plant growth and development (Feng et al., 2017). A DNA binding domain of transcription factors binds to cis-acting regions found in the upstream region of all gene promoters (Loredana et al., 2011). Plant genes induced by abiotic stress are separated into classes based on their protein products. The first class of genes are signal transduction network regulators,which comprise transcription factors, molecular chaperones, functional proteins (Song et al., 2013). The second class of genes are those coding for molecules that directly let cells to survive environmental challenges, including MDA, late embryogenesis abundant (LEA) protein, betamine, and other anti-freezing proteins and osmotic regulators (Loredana et al., 2011).

The transcription of a plant gene is controlled directly by a network of plant TFs and their binding sites. Because of their role in inhibiting or activating the activity of RNA polymerase, TFs govern gene regulation. TFs are encoded by almost 10% of genes in the plant genome (Franco et al., 2014). Because abiotic pressures are quantitative traits, they may necessitate the regulation of numerous genes, including TFs and in most situations, single TF may regulate multiple genes involved in abiotic stress responses. Conducting a comprehensive study on all TFs linked to abiotic stress regulation pathways is critical, especially for economic species whose development and output are severely hampered by various abiotic stresses (Kimotho et al., 2019).

Traditional breeding procedures have had limited success, due to the intricacy of stress tolerance features; nevertheless, the transgenic approach is now widely utilized to create stress-tolerant crops. As a result, identifying and characterizing the important genes involved in plant stress responses is a decisive step before developing stress-tolerant plants. Engineering particular regulatory genes has emerged as an effective strategy for influencing the expression of multiple stress-responsive genes, much beyond the manipulation of a single function gene (Xie et al., 2020; Basak et al., 2021). Because of their position as master regulators of many stress-responsive genes, TFs are attractive candidates for genetic engineering to generate stress-tolerant plants (Hou et al., 2020). Many TFs of AP2/ERF, MYB, WRKY, NAC, and bZIP families have been discovered to be involved in various abiotic stresses, and certain TF genes have been engineered to improve stress tolerance in model and crop plants (Li et al., 2016; Wang et al., 2016; Adem et al., 2019a, 2019b; Samo et al., 2019; Gao et al., 2021; Ramakrishnan et al., 2020).

In the abiotic stress signal transduction pathway, TFs play a crucial role. They interact with cis-regulatory regions to alter the expression of a group of related genes, leading to increased stress tolerance. TFs are widely regarded as key targets for botechnological stress tolerance engineering in plants (Milla et al., 2006). Many researchers have used transgenic and bioinformatics analysis methodologies to investigate the role of numerous bamboo TFs in stress tolerances. The biological function of the expansion gene in Moso bamboo, for example, has been confirmed as important in drought stress tolerance (Jin et al., 2020). *R2R3MYBs*, subgroup of MYB TFs, have a significant function in various abiotic stress tolerance and flower development, as demonstrated in co-expression studies (Hou et al., 2018). As listed in Table 8.2, several TFs that are involved in conferring abiotic stress tolerances have been experimentally shown.

Signal transducers (e.g., secondary messengers such as Ca^{2+}, ROS, phytohormones, kinases, and small signaling peptides) trigger the regulatory pathways involving transcriptional, post-transcriptional modifications, translation, and epigenetic regulation. Multiple stress signals activate TFs that govern the stress-inducible gene expression cascade. Some stress-inducible genes encode functional proteins that have a direct impact role in stress tolerance, while others encode regulatory proteins like signal transducers (Lohani, et al., 2022). Figure 8.1 depicts a graphical overview of abiotic stress in plant cells, including sensing, signal transduction, and regulation.

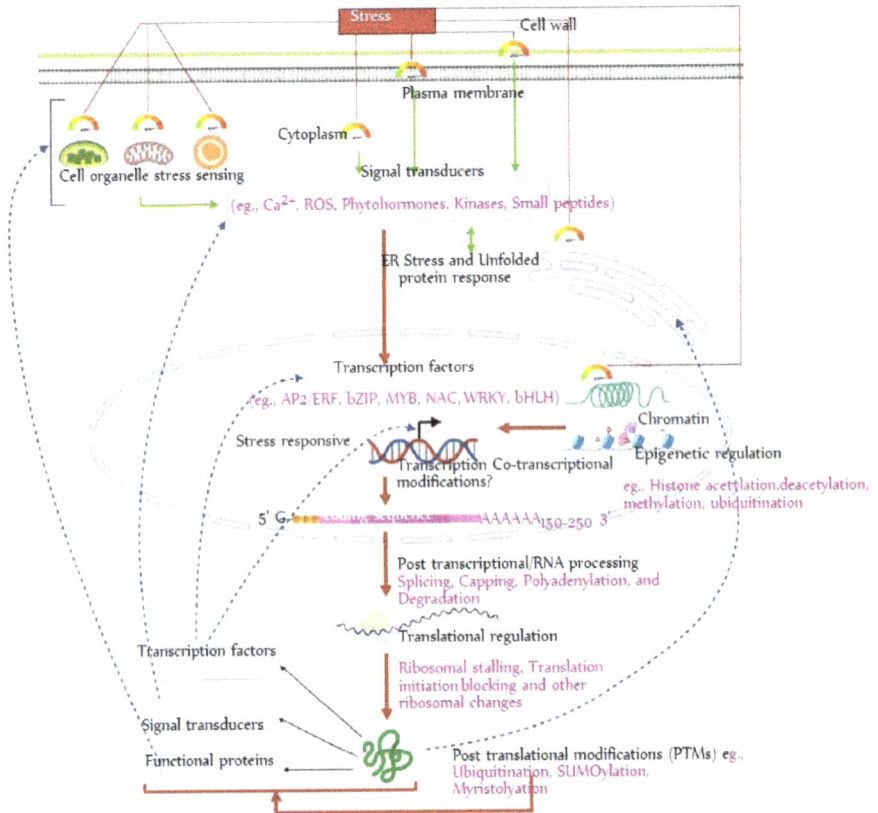

Fig. 8.1. An overview of stress response sensing, signal transduction, and regulation in plant cells.
Source: Redrawn from Lohani et al. (2022). *BioDesign Research.* https://doi.org/10.34133/2022/9819314.

8.9 Effect of Abiotic Stresses on Physiology of Transgenic Lines with Bamboo Originated Genes

Drought, salinity, and cold stress effects on the physiology of the transgenic plants were studied. Accordingly, transgenic plants have shown profound tolerance against abiotic stress. Sugar molecular content was more in the transgenic lines than in wild or mutant plants. The relative electrolyte leakage (REL) content, which is mostly used to study the amount of membrane damage, and MDA, the degree of membrane lipid injury, are the most important measurable parameters used to evaluate the physiological performance of transgenic plants under stress conditions (Smirnoff, 1993). Transgenic Arabidopsis seedlings various bamboo originated genes had lower physiological effect of REL and MDA, which cause membrane damage, than mutants and non-transgenic Arabidopsis seedlings (Wei et al., 2018).

Plants transformed with the *PheDi19-8* and *PheCDPK22* genes had their biochemistry and physiology studied under drought and normal conditions (Wei et al., 2018). To deal with the stress, the transgenic plants accumulated soluble carbohydrates. Under drought stress, transgenic plants overexpressing the *PheDi19-8* gene produced more soluble sugars than non-transgenic plants (Abraham et al., 2003). However, the biochemistry and physiological performance of transgenic plants under stress is not always ideal. Under drought stress, transgenic plants overexpressing *PheCDPK22*, for example, have a very low soluble sugar concentration. Under salt stress, the REL content and MDA levels of transgenic plants also suffered (Hu et al., 2012).

In contrast to Hu's findings, it has been reported that *PheCDPK22*-overexpressing plants have 1.65 and 1.35 times more REL and MDA content than their wild counterparts. The more noteworthy research finding is that transgenic plants carrying the *PheDi19-8* gene increased the expression of stress-responsive genes LEA, RD29A, DREB2A, and RD22, whereas these genes were drastically reduced in *PheCDPK22*-overexpressing Arabidopsis. These abiotic stress sensitive and marker genes LEA, RD29A, RD22, and DREB2A have been extensively researched (Finkelstein and Gampala, 2002). Further research confirmed that *PheDi19-8* and *PheCDPK22* appear to operate as both positive and negative regulators of drought tolerance (Wu et al., 2020b).

8.10 Conclusions

Plant growth and development are threatened by several climate stressors. Environmental pressures are expected to spread their life-threatening influence from location to location and over time. Plant science research has traditionally focused on yield qualities, but today's plant development efforts must pay close attention to stress resistance. Bamboos consistent economic and environmental benefits are jeopardized by ever-changing environmental conditions. Bamboo growth is hampered by abiotic stress, particularly drought, salinity, and cold, despite its vital functions in reversing land degradation, mitigating climate change, and alleviating rural poverty. Due to bamboos possessing a peculiar character of having an unpredictable and extended flowering cycle, as well as a high genome size linked with polyploidy, breeding and creating abiotic stress tolerant bamboos remains a major scientific challenge. Another roadblock to bamboo enhancement is its low genetic transformation efficiency and limited regeneration capabilities. Some bamboo species have undergone successful genetic transformation, but more studies are needed to incorporate various ecological and commercial bamboo variants that are resistant to abiotic stress. Despite the difficulties, current scientific investigations have suggested that producing abiotic stress tolerant bamboos is possible. Drought, salinity, and cold tolerance have been provided by over-expression of various bamboo-derived genes. Despite certain abiotic stress that affects transgenic plants, the majority of cases have shown superior performance in terms of biochemistry and physiological activities, as evidenced by analyses of REL content and malondialdehyde levels (MDA). Despite the difficulties, the prospect for bamboo genetic improvement against abiotic stress is not a nightmare. This is because modern, cost-effective genetic engineering approaches, genome editing, and synthetic biology tools have simplified all elements of complex genetic studies. More importantly, prior and current bamboo research outputs should be effectively gathered in order to convey the major scientific issues surrounding bamboos. This allows for a better understanding of bamboo biology in general and abiotic stress improvement in particular, which leads to conservation and a brighter future for bamboos. The development of new plants for the new climate is critical in this century. As a result, focusing resources on research to develop abiotic stress tolerant bamboo would be a win-win strategy, as it would solve both economic and environmental concerns. Last but not the least, the integrated efforts of several expertise on plant genetics, physiology, ecology, environment, and forestry, and with sound policies may result in a large increase in the economic and environmental returns of bamboos.

References

Abraham, E., Rigo, G., Szekely, G., Nagy, R., Koncz, C. and Szabados, L. (2003). Light-dependent induction of proline biosynthesis by abscise acid and salt stress is inhibited by brassinosteroid in Arabidopsis. *Plant Mol. Biol.*, 51: 363–372.

Adem, M., Zhao, K., Beyene, D., Feyissa, T. and Jiang, T. (2019a). *De novo* assembly and transcriptome profiling of Ethiopian lowland bamboo Oxytenanthera abyssinica (A. Rich) Munro under drought and salt stresses. *Open Biotechnol. J.*, 13: 6–17.

Adem, M., Zhao, K., Beyene, D., Feyissa, T. and Jiang, T. (2019b). Genome-wide analysis and expression profile of bZIPs transcription factor gene family in Oxytenanthera abyssinica (A. Rich) in response to osmotic and salt stress. *Journal of Environmental and Experimental Biology*, 17: 201–208. Doi: 10.22364/eeb.17.21.

Ahmad, Z., Ding, Y. and Shahzad, A. (2020). *Biotechnological Advances in Bamboo the "Green Gold" on the Earth.* Springer, ISBN 978-981-16-1309-8 ISBN 978-981-16-1310-4 (eBook) https://doi.org/10.1007/978-981-16-1310-4.

Ali, Z., Zhang, D.Y., Xu, Z.L., Xu, L. and Yi, J.X. (2012). Uncovering the salt response of soybean by unraveling its wild and cultivated functional genomes using tag sequencing. *PLoS One*, 7: e48819.

Banerjee, S., Basak, M., Dutta, S., Chanda, C., Dey, A. and Das, M. (2021). Ethnobamboology: Traditional uses of bamboos and opportunities to exploit genomic resources for better exploitation. pp. 313–352. *In*: Ahmad, Z., Ding, Y. and Shahzad, A. (eds.). *Biotechnological Advances in Bamboo: The "Green Gold" on the Earth.* Springer, Singapore.

Basak, M., Dutta, S., Biswas, S., Chakraborty, S., Sarka, A., Rahaman, T., Dey, S., Biswas, P. and Das, M. (2021). Genomic insights into growth and development of bamboos: What have we learnt and what more to discover? *Trees*. https://doi.org/10.1007/s00468-021-02197-6.

Bassil, E. and Blumwald, E. (2014). The ins and outs of intracellular ion homeostasis: NHX-type cation/H+ transporters. *Curr. Opin. Plant. Biol.*, 22: 1–6.

Bencina, M., Bagar, T. and Lah, L. (2009). A comparative genomic analysis of calcium and proton signaling/homeostasis in Aspergillus species. *Fungal Genet. Biol.*, 46: S93–S104.

Bennetzen, J.L., Schmutz, J., Wang, H., Percifield, R., Hawkins, J., Pontaroli, A.C. and Jenkins, J. (2012). Reference genome sequence of the model plant Setaria. *Nat. Biotechnol.*, 30: 555–556.

Bielach, A., Hrtyan, M. and Tognetti, V.B. (2017). Plants under stress: Involvement of auxin and cytokinin. *International Journal of Molecular Sciences*, 18: 1427.

Biswas, S., Sarkar, A., Kharlyngdoh, E., Somkuwar, B.G., Biswas, P., Dutta, S., Guha, S. and Das, M. (2021). Evidence of stress induced flowering in bamboo and comments on probable biochemical and molecular factors. *Journal of Plant Biochemistry and Biotechnology*. https://doi.org/10.1007/s13562-021-00719-4(0123).

Calzadilla, P.I., Maiale, S.J., Ruiz, O.A. and Escaray, F.J. (2016). Transcriptome response mediated by cold stress in Lotus japonicus. *Front Plant Sci.*, 7: 374.

Chaomao, H., Weiyi, L., Xiong, Y. and Yuming, Y. (2006). Environmental benefits of bamboo forests and the sustainable development of bamboo industry in Western China. *In: Bamboo for the Environment, Development and Trade.* Proceedings of the Industrial Bamboo Workshop, China, 23 October 2006.

Chauhan, H., Khurana, N., Agarwal, P. and Khurana, P. (2011). Heat shock factors in rice (*Oryza sativa* L.): Genome-wide expression analysis during reproductive development and abiotic stress. *Mol. Genet. Genom.*, 286: 171–187.

Chen, F., Liu, H.L., Wang, K., Gao, Y.M., Wu, M. and Xiang, Y. (2020). Identification of CCCH zinc finger proteins family in Moso bamboo (*Phyllostachys edulis*), and PeC3H74 confers drought tolerance to transgenic plants. *Front. Plant Sci.*, 11: 579255. Doi: 10.3389/fpls.2020.579255.

Cheng, X., Wang, Y. and Xiong, R. (2020). A Moso bamboo gene VQ28 confers salt tolerance to transgenic Arabidopsis plants. *Planta*, 251: 99. https://doi.org/10.1007/s00425-020-03391-5.

Das, M., Bhattacharya, P., Singh, T.S. and Pal, A. (2008). Bamboo taxonomy and diversity in the era of molecular markers. *In: Advances in Botanical Research*, 47: 225–268. Doi: 10.1038/ng.2569. desiccation. *New Phytol.*, 125: 27–58.

Das, M., Haberer, G., Panda, A., Das, M., Laha, S., Ghosh, T.C. and Schaffner, A.R. (2016). Expression pattern similarities support the prediction of orthologs retaining common functions after gene duplication events. *Plant Physiol.*, 171: 2343–2357.

Deng, Z., Chen, Q., Wu, J., Ren, K., Waqqas, M., Tarin, K., He, T., Chen, L., Chen, L., Haq, A.U. and Zheng, Y. (2021). The complete chloroplast genome sequence of Bambusa albolineata (Bambusodae). *Mitochondrial DNA Part B*, 6: 2748–2749. Doi: 10.1080/23802359.2021.1967798.

Ding, Y.H., Wang, Z.Y., Song, Y.F. and Zhang, J. (2008). The unprecedented freezing disaster in January 2008 in southern China and its possible association with the globe warming. *Acta Meteorol Sin.*, 22: 538–558 (in Chinese).

Döring, P., Treuter, E., Kistner, C., Lyck, R., Chen, A. and Nover, L. (2000). The role of AHA motifs in the activator function of tomato heat stress transcription factors HsfA1 and HsfA2. *Plant Cell*, 12: 265–278.

Duclercq, J., Sangwan-Norreel, B., Catterou, M. and Sangwan, R.S. (2011). *De novo* shoot organogenesis: From art to science. *Trends in Plant Science*, 16: 597–606.

Dutta, S., Biswas, P., Chakraborty, S., Mitra, D., Pal, A. and Das, M. (2018). Identification, characterization, and gene expression analyses of important flowering genes related to photoperiodic pathway in bamboo. *BMC Genom*, 19: 190.

Dutta, S., Deb, A., Biswas, P., Chakraborty, S., Guha, S., Mitra, D., Geist, B., Schaffner, A.R. and Das, M. (2021). Identification and functional characterization of two bamboo FD gene homologs having contrasting effects on shoot growth and flowering. *Sci Rep.*, 11: 7849.

Emamverdian, A., Ding, Y., Ranaei, F. and Ahmad, Z. (2020). Application of bamboo plants in nine aspects. *Hindawi Scientific World Journal*. https://doi.org/10.1155/2020/7284203.

FAO and INBAR. (2018). *Bamboo for Land Restoration*. INBAR Policy Synthesis Report 4. INBAR, Beijing, China.

Farooq, M., Basra, S.M.A., Wahid, A., Cheema, Z.A. and Cheema, M.A. (2008). Physiological role of exogenously applied glycinebetaine to improve drought tolerance in fine grain aromatic rice (Oryza sativa L.). *J. Agronomy and Crop Sciences*, 194: 325–333.

Feng, P.F., Wang, Y., Liu, H., Wu, M. and Chu, W. (2017). Genome-wide identification and expression analysis of SBP-like transcription factor genes in Moso Bamboo (Phyllostachys edulis). *BMC Genomics*, 18: 486.

Finkelstein, R.R. and Gampala, S.S. (2002). Abscisic acid signaling in seeds and seedlings. *The Plant Cell*, 15: S45.

Franco, J.M., López, I., Carrasco, J.L., Godoy, M. and Vera, P. (2014). DNA-binding specificities of plant transcription factors and their potential to define target genes. *Proc. Natl. Acad. Sci. U.S.A.*, 111: 2367–2372.

Gao, Y., Liu, H. and Zhang, K. (2021). A Moso bamboo transcription factor, Phehdz1, positively regulates the drought stress response of transgenic rice. *Plant Cell Rep.*, 40: 187–204. https://doi.org/10.1007/s00299-020-02625-w.

Goyal, E., Singh, K., Ravi, S., Singh, and Ajay, K. (2016). Transcriptome profiling of the salt stress response in Triticum aestivum Kharchia local. *Scientific Reports*, 6: 27752.

Grafi, G. and Barak, S. (2015). Stress induces cell dedifferentiation in plants. *Biochimica et Biophysica Acta*, 1849: 378–384.

Guo, X., Liu, D. and Chong, K. (2018). Cold signaling in plants: Insights into mechanisms and regulation. *J. Integr. Plant Biol.*, 60: 745–756.

Guo, Z.H., Ma, P.F., Yang, G.Q., Hu, J.Y., Liu, Y.L., Xia, E.H., Zhong, M.C., Zhao, L., Sun, G.L., Xu, Y.X., Zhao, Y.J., Zhang, Y.C., Zhang, Y.X., Zhang, X.M., Zhou, M.Y., Guo, Y., Guo, C., Liu, J.X., Ye, X.Y., Chen, Y.M., Yang, Y., Han, B., Lin, C.S., Lu, Y. and Li, DZ. (2019a). Genome sequences provide insights into the reticulate origin and unique traits of woody bamboos. *Mol. Plant*, 12: 1353–1365.

Gupta, B. and Huang, B. (2014). Mechanism of salinity tolerance in plants: Physiological, biochemical, and molecular characterization. *International Journal of Genomics*, 1–18. Doi: 10.1155/2014/701596.

Hartl, D. and Clark, A. (2007). *Principles of Population Genetics* (4th Edn). Sinauer Associates Publishing, Sunderland, MA.

Hartl, F.U. and Hayerhartl, M. (2002). Molecular chaperones in the cytosol: From nascent chain to folded protein. *Science*, 295: 1852–1858.

Hashem, A., Abdallah, E.F. and Alqarawi, A.A. (2019). Comparing symbiotic performance and physiological responses of two soybean cultivars to arbuscular mycorrhizal fungi under salt stress. *Saudi Journal of Biological Sciences*, 26(1): 38–48.

Hou, D., Cheng, Z., Xie, L., Li, X., Li, J., Mu, S. and Gao, J. (2018). The R2R3MYB gene family in Phyllostachys edulis: Genome-wide analysis and identification of stress or development-related R2R3MYBs. *Front. Plant Sci.*, 9: 738. Doi: 10.3389/fpls.2018.00738.

Hou, D., Zhongyu, Z., Qiutao, H., Ling, L., Naresh, V., Juan, Z., Wei, Ze., Aimin, W. and Xinchun, L. (2020). PeSNAC-1 a NAC transcription factor from Moso bamboo (Phyllostachys edulis) confers tolerance to salinity and drought stress in transgenic rice. *Tree Physiol.*, 5: 1792–1806. Doi: 10.1093/treephys/tpaa099.

Hu, L., Li, H., Pang, H. and Fu, J. (2012). Responses of antioxidant gene, protein, and enzymes to salinity stress in two genotypes of perennial ryegrass (Lolium perenne) differing in salt tolerance. *J. Plant Physiol.*, 169: 146–156.

Huang, H., Ullah, F., Zhou, D.X., Yi, M. and Zhao, Y. (2019). Mechanisms of ROS regulation of plant development and stress responses. *Frontiers in Plant Science*, 10: 800.

Huang, Y. (2010). *Physiological Responses of Phyllostachys Edulis to Salt stress and Expression Analysis of Salt Tolerance Gene*. Zhejiang Agriculture and Forest University, Hangzhou, China.

Huang, R., Gao, H. and Liu, J. (2022). WRKY transcription factors in Moso bamboo that are responsive to abiotic stresses. *J. Plant Biochem. Biotechnol.*, 31: 107–114. https://doi.org/10.1007/s13562-021-00661-5.

Huang, Z., Jin, S.H., Guo, H.D., Zhong, X.J., He, J., Li, X., Jiang, M.Y., Yu, X.F., Long, H., Ma, M.D. and Chen, Q.B. (2016). Genome-wide identification and characterization of TIFY family genes in Moso bamboo (Phyllostachys edulis) and expression profiling analysis under dehydration and cold stresses. *PeerJ*, 4: e2620. Doi: 10.7717/peerj.2620.

Huang, Z., Zhong, X.J., He, J., Jin, S.H., Guo, H.D. and Yu, X.F. (2016). Genome-wide identification, characterization, and stress-responsive expression profiling of genes encoding LEA (late embryogenesis abundant) proteins in Moso bamboo (*Phyllostachys edulis*). *PLoS One*, 11(11): e0165953. Doi: 10.1371/journal. pone.0165953.

IFAD and INBAR. (2019). South-South knowledge transfer strategies for scaling up pro-poor bamboo livelihoods, income generation and employment creation, and environmental management in Africa. A Brochure.

Jiang, J.M., Li, B.X., Jiangm, N.Q., Zhu, W.S., Yu, Y. and Chen, X.M. (2008). Impact of the snow disaster occurred in 2008 in south China to the clump bamboo in south Sichuan. *Sci. Silvae Sin.*, 44: 141–144 (in Chinese).

Jin, G., Peng-Fei Ma, P.F., Wu, X., Lianfeng Gu, L., Long, M., Zhang, C. and Li, D.Z. (2021). New genes interacted with recent whole-genome duplicates in the fast stem growth of bamboos. *Mol. Biol. Evol.*, 38: 5752–5768. Doi:10.1093/molbev/msab288.

Jin, K.M., Zhuo, R.Y., Xu, D., Wang, Y.J., Fan, H.J., Huang, B. and Qiao, G.R. (2020). Genome-wide identification of the expansion gene family and its potential association with drought stress in Moso bamboo. *International Journal of Molecular Sciences*, 21: 9491. Doi:10.3390/ijms21249491.

Kargiotidou, A., Kappas, I., Tsaftaris, A., Galanopoulou, D. and Farmaki, T. (2010). Cold acclimation and low temperature resistance in cotton: Gossypium hirsutum phospholipase Dα isoforms are differentially regulated bytemperature and light. *J. of Exp. Botany*, 61: 2991–3002.

Kassahun, E. (2000). T indigenous bamboo forests of Ethiopia: An overview. *AMBIO, A Journal of the Human Environment*, 29(8): 518–521.

Kersey, P.J. (2019a). Plant genome sequences: Past, present, future. *Curr. Opin. Plant Biol.*, 48: 1–8.

Kimotho, R.Y., Baillo, E.H. and Zhang, Z. (2019). Transcription factors involved in abiotic stress responses in maize (*Zea mays* L.) and their roles in enhanced productivity in the post genomics era. *PeerJ*, 7: e27549v1. https://doi.org/10.7287/peerj.preprints.27549v1.

King, C. (2019). *Bamboo and Sustainable Development: A Briefing Note*. The International Bamboo and Rattan Organisation, Beijing.

Kolkman, J.A. and Stemmer, W.P.C. (2001). Directed evolution of proteins by exon shuffling. *Nature Biotechnology*, 19: 423–428. https:// doi.org/10.1038/88084.

Kronzucker, H.J. and Britto, D.T. (2011). Sodium transport in plants: A critical review. *New Phytolologist*, 189(1): 54–81.

Lan, Z., Ting, X.X., ZeHui, J., Hua, Y.X., Juan, X.L. and Rong, G.X. (2010). Study on the drought resistance of four dwarf ornamental bamboos. *Forest Research*, 23: 221–226.

Lee, B., Henderson, D.A. and Zhu, J.K. (2005). The Arabidopsis coldresponsive transcriptome and its regulation by ICE1. *Plant Cell*, 17: 3155–3175.

Li, W., Shi, C., Li, K., Zhang, Q.J., Tong, Y. and Zhang, Y. (2020). The draft genome sequence of herbaceous diploid bamboo Raddia distichophylla. *BioRxiv*, 1–29. https://doi.org/10.1101/2020.04.27.064089.

Li, W., Shi, C., Li, K., Zhang, Q.J., Tong, Y., Zhang, Y., Wang, J., Clark, L. and Gao, L.Z. (2021). Draft genome of the herbaceous bamboo Raddia distichophylla. *G3 (Bethesda)*, 11: jkaa049.

Li, L., Mu, S., Cheng, Z., Cheng, Y., Zhang, Y., Miao, Y., Hou, C., Li, X. and Gao, J. (2016). Characterization and expression analysis of the WRKY gene family in Moso bamboo. *Scientific Reports*, 7: 6675. Doi:10.1038/s41598.

Li, L., Yang, K., Wang, S., Lou, Y., Zhu, C. and Gao, Z. (2020). Genome-wide analysis of laccase genes in Moso bamboo highlights PeLAC10 involved in lignin biosynthesis and in response to abiotic stresses. *Plant Cell Reports*, 39: 751–763. https://doi.org/10.1007/s00299-020-02528-w.

Lin, C.S.C., Tseng, M., Hong, I.P. and Chang, W.C. (2006). Albino inflorescence proliferation of Dendrocalamus latiflorus. *Vitro Cell Dev. Biol. Plant*, 42: 331–335.

Lin, Y.T., Tang, S.L., Pai, C.W., Whitman, W.B., Coleman, D.C. and Chiu, C.Y. (2014). Changes in the soil bacterial communities in a cedar plantation invaded by Moso bamboo. *Microb. Ecol.*, 67: 421–429.

Liu, H., Gao, Y., Wu, M., Shi, Y., Wang, H. and Xiang, Y. (2020). TCP10, a TCP transcription factor in Moso bamboo (*Phyllostachys edulis*), confers drought tolerance to transgenic plants. *Environmental and Experimental Botany*, 172: 04002. Doi.org/10.1016/j.envexpbot.2020.104002.

Liu, J., Cheng, Z., Xie, L., Li, X. and Gao, J. (2019). Multifaceted role of PheDof12-1 in the regulation of flowering time and abiotic stress responses in Moso bamboo (*Phyllostachys edulis*). *Int. J. Mol. Sci.*, 20: 424. Doi: 10.3390/ijms20020424.

Liu, L., Dong, D., Li, Y., Li, X. and Bureau, A.M. (2014). Investigation of Moso bamboo forest under high temperature and drought disaster. *World Bamboo Rattan*, 12: 24–27.

Liu, Y., Wu, C., Hu, X., Gao, H., Wang, Y., Luo, H., Cai, S., Li, G., Zheng, Y., Lin, C. and Zhu, Q. (2019). Transcriptome profiling reveals the crucial biological pathways involved in cold response in Moso bamboo (*Phyllostachys edulis*). *Tree Physiology*, 40: 538–556. Doi:10.1093/treephys/tpz133.

Liu, Y.K., Liu, W., Xiao, Q.G., Zeng, Z., Zhuang, L., Zhang, T.H., Jun Zhang, Huang, Q.Y. and Wang, B.X. (2021). The complete chloroplast genome of Chimonobambusa hejiangensis. *Mitochondrial DNA Part B*, 6: 1824–1825. Doi: 10.1080/23802359.2021.1930599.

Lobovikov, M., Guardia, M. and Russo, L. (2007). *World Bamboo Resources: A Thematic Study Prepared in the Framework of the Global Forest Resources Assessment.* Food and Agriculture Organization, Rome, Italy.

Lobovikov, M.Y. and Ping, L. (2012). Bamboo in climate change and rural livelihoods. *Mitigation and Adaptation Strategies for Global Change*, 17: 261–276.

Lohani, N., Singh, M.B. and Bhalla, P.L. (2022). Biological parts for engineering abiotic stress tolerance in plants. *BioDesign Research*. https://doi.org/10.34133/2022/9819314.

Long, M., Betran, E., Thornton, K. and Wang, W. 2003. The origin of new genes: Glimpses from the young and old. *Nature Reviews Genetics*, 4: 865–875. https://doi.org/10.1038/nrg1204.

Loredana, F., Woodrow, C.P., Fuggi, A., Pontecorvo, G. and Carillo, P. (2011). Plant genes for abiotic stress. In: *Abiotic Stress in Plants: Mechanisms and Adaptations*.

Lou, Y., Sun, H., Li, L., Zhao, H. and Gao, Z. (2017). Characterization and primary functional analysis of a bamboo ZEP gene from Phyllostachys edulis. *DNA and Cell Biology*, 747–758. http://doi.org/10.1089/dna.2017.3705.

Ma, S., Gong, Q. and Bohnert, H.J. (2006). Dissecting salt stress pathways. *J. of Exp. Botany*, 57: 1097–1107.

Ma, X., Zhao, H., Yan, H., Sheng, M., Cao, Y., Yang, K., Xu, H., Xu, W., Gao, Z. and Su, Z. (2021). Refinement of bamboo genome annotations through integrative analyses of transcriptomic and epigenomic data. *Computational and Structural Biotechnology Journal*, 19: 2708–2718.

Mekonnen, Z., Worku, A., Yohannes, T., Alebachew, M. and Kassa, H. (2014). Bamboo resources in Ethiopia: Their value chain and contribution to livelihoods. *Ethnobotany Research and Applications*, 12: 511–524.

Meng, Y. (2010). *Effect of NaCl Stress on Electrical Impedance Spectroscopy Parameters and Chlorophyll Fluorescence Characteristics of Moso Bamboo Seedling Leaves.* Hebei Agriculture University, Baoding, China.

Milla, M.A., Townsend, J., Chang, I.F. and Cushman, J.C. (2006). The Arabidopsis AtDi19 gene family encodes a novel type of Cys2/His2 zinc-finger protein implicated in ABA-independent dehydration, high salinity stress and light signaling pathways. *Plant Mol. Biolo.*, 61: 13–30.

Munns, R. and Tester, R. (2008). Mechanisms of salinity tolerance. *The Annual Review of Plant Biology*, 59: 651–681.

Nakagami, H., Pitzschke, A. and Hirt, H. (2005). Emerging MAP kinase pathways in plant stress signaling. *Trends Plant Sciences*, 7: 339–346.

Pan, L. (2020). Role of the Phyllostachys edulis SnRK1a gene in plant growth and stress tolerance. *South African Journal of Botany*, 130: 414–421. https://doi.org/10.1016/j.sajb.2020.01.033.

Pelham, H.R. and Bienz, M.A. (1982). Synthetic heat-shock promoter element confers heat-inducibility on the herpes simplex virus thymidine kinase gene. *EMBO J.*, 1: 1473–1477.

Peng, Z., Lu, Y., Li, L., Zhao, Q., Feng, Q., Gao, Z., Lu, H., Hu, T., Yao, N., Liu, K., Li, Y., Fan, D., Guo, Y., Li, W., Lu, Y., Weng, Q., Zhou, C.C., Zhang, L., Huang, T., Zhao, Y., Zhu, C., Liu, X., Yang, X., Wang, T., Miao, K., Zhuang, C., Cao, X., Tang, W., Liu, G., Liu, Y., Chen, J., Liu, Z., Yuan, L., Liu, Z., Huang, X., Lu, T., Fei, B., Ning, Z., Han, B. and Jiang, Z. (2013a). The draft genome of the fast-growing non-timber forest species Moso bamboo (Phyllostachys heterocycla). *Nature Genetics*, 45: 456–461. Doi: 10.1038/ng.2569.

Peng, Z., Zhang, C. and Zhang, Y. (2013b). Transcriptome sequencing and analysis of the fast-growing shoots of Moso bamboo (Phyllostachys edulis). *PLoS One*, 8: e78944. https:// doi. org/ 10. 1371/ journ al. pone. 00789 44.

Polle, A., Chen, S.L., Eckert, C. and Harfouche, A. (2019). Engineering drought resistance in forest trees. *Frontiers in Plant Sciences*, 9: 1875. Doi: 10.3389/fpls.2018.01875.

Pulavarty, A. and Sarangi, K. (2015). Salt tolerance screening of bamboo genotypes (bamboo sps.) using growth and organic osmolytes accumulation as effective indicators. *Proceedings of the 10th World Bamboo Congress, Korea.*

Qiao, G., Yang, H., Han, X., Liu, M., Jiang, J. and Jiang, Y. (2014). Enhanced cold stress tolerance of transgenic *Dendrocalamus latiflorus Munro* (Ma bamboo) plants expressing a bacterial CodA gene. *In Vitro Cellular and Developmental Biology – Plant*, Doi: 10.1007/s11627-013-9591-z.

Queitsch, C., Hong, S.W., Vierling, E. and Lindquist, S. (2000). Heat shock protein 101 plays a crucial role in thermotolerance in Arabidopsis. *Plant Cell*, 12: 479–492.

Ramakrishnan, M., Yrjala, K., Vinod, K.K., Sharma, A., Cho, J. and Satheesh, V. (2020). Genetics and genomics of Moso bamboo (*Phyllostachys edulis*): Current status, future challenges, and biotechnological opportunities toward a sustainable bamboo industry. *Food and Energy Security*, 9: e229.

Ren, F., Wang, L., Zhuo, W., Zhu, X., Lu, S., Huang, H. and Chen, D. (2021). The first complete plastome of *Chimonobambusa quadrangularis* (Fenzl) Makino: Assembly, annotation, and phylogenetic analysis. *Mitochondrial DNA Part B*, 6: 2762–2763. Doi: 10.1080/23802359.2021.1967808.

Rizhsky, L., Liang, H. and Mittler, R. (2002). The combined effect of drought stress and heat shock on gene expression in tobacco. *Plant Physiol.*, 130: 1143–1151.

Samo, N., Imran, M., Shanglian, H., Xuegang, L., Cao, Y. and Yan, H. (2019). Molecular characterization and expression pattern analysis of a novel stress-responsive gene 'BeSNAC1' in Bambusa emeiensis. *Journal of Genetics*, 98: 52. https://doi.org/10.1007/s12041-019-1098-x.

Sang, Y.L., Cheng, Z.J. and Zhang, X.S. (2018). Plant stem cells and *de novo* organogenesis. *New Phytologist*, 218: 1334–1339.

Sangwan, V., Foulds, I., Singh, J. and Dhindsa, R.S. (2001). Cold-activation of Brassica napus BN115 promoter is mediated by structural changes in membranes and cytoskeleton, and requires Ca2+ influx. *Plant J.*, 27: 1–12. Doi: 10.1046/j.1365-313x.2001.01052.x.

Sharma, E., Sharma, R., Borah, P., Jain, M. and Khurana, J.P. (2015). Emerging roles of auxin in abiotic stress responses. pp. 299–328. *In*: Pandey, G.K. (ed.). Elucidation of Abiotic Stress Signaling in Plants: Functional Genomics Perspectives. Springer, New York, NY.

Shin, J., Bae, S. and Seo, P.J. (2020). *De novo* shoot organogenesis during plant regeneration. *Journal of Experimental Botany*, 71: 63–72.

Singh, B.P., Singh, B., Kumar, V., Singh, P.K. and Jayaswal, P.K. (2015b). Haplotype diversity and association analysis of SNAC1 gene in wild rice germplasm. *Indian J. Genet. Plant Breed*, 75: 157–166.

Singh, S.R., Singh, R., Kalia, S., Dalal, S., Dhawan, A.K. and Kalia, R.K. (2013). Limitations, progress, and prospects of application of biotechnological tools in improvement of bamboo: A plant with extraordinary qualities. *Physiology and Molecular Biology of Plants: An International Journal of Functional Plant Biology*, 19: 21–41.

Smirnoff, N. (1993). The role of active oxygen in the response of plants to water deficit and stress in two genotypes of perennial ryegrass (Lolium perenne) differing in salt tolerance. *Plant Journal.*

Som, A. (2014). Causes, consequences, and solutions of phylogenetic incongruence. *Briefings in Bioinformatics*, 16: 536–548.

Song, X., Li, Y. and Hou, X. (2013). Genome-wide analysis of the AP2/ERF transcription factor super family in Chinese cabbage (Brassicarapa ssp. pekinensis). *BMC Genomics*, 14: 573.

Sun, H., Li, L., Lou, Y., Zhao, H., Yang, Y. and Gao, Z. (2016). Cloning and preliminary functional analysis of PeUGE gene from Moso bamboo (*Phyllostachys edulis*). *DNA and Cell Biology*, 706–714. http://doi.org/10.1089/dna.2016.3389.

Sun, H., Li, L., Lou, Y., Zhao, H., Yang, Y., Wang, S. and Gao, Z. (2017). The bamboo aquaporin gene PeTIP4;1–1 confers drought and salinity tolerance in transgenic Arabidopsis. *Plant Cell Reports*, 36: 597–609. Doi: 10.1007/s00299-017-2106-3.

Sun, H., Wang, S. and Lou, Y. (2021). A bamboo leaf-specific aquaporin gene PePIP2;7 is involved in abiotic stress response. *Plant Cell Reports*, 40: 1101–1114. https://doi.org/10.1007/s00299-021-02673-w.

Tanou, G., Molassiotis, A. and Diamantidis, G. (2009). Induction of reactive oxygen species and necrotic death-like destruction in strawberry leaves by salinity. *Environmental and Experimental Botany*, 65: 270–281.

Tavakkoli, E., Rengasamy, P. and McDonald, G.K. (2010). High concentrations of Na+ and Clions in soil solution have simultaneous detrimental effects on growth of faba bean under salinity stress. *J. of Exp. Botany*, 61: 4449–4459.

Tao, G., Fu, Y. and Zhou, M. (2018). Advances in studies on molecular mechanisms of rapid growth of bamboo species. *Journal of Agricultural Biotechnology*, 26(5): 871–887.

Tong, R., Zhou, B., Cao, Y., Ge, X. and Jiang, L. (2020). Metabolic profiles of Moso bamboo in response to drought stress in a field investigation. *Science of the Total Environment*, 720: 137722.

Tu, M., Wang, W., Yao, N., Cai, C., Liu, Y., Lin, C., Zuo, Z. and Zhu, Q. (2021). The transcriptional dynamics during *de novo* shoot organogenesis of Ma bamboo (Dendrocalamus latiflorus Munro): implication of the contributions of the abiotic stress response in this process. *The Plant Journal*. Doi: 10.1111/tpj.15398.

Turkana, I. and Demiral, T. (2009). Recent developments in understanding salinity tolerance. *Environmental and Experimental Botany*, 67: 2–9.

Von, K.D.P., Scharf, K., and Nover, L. 2007.The diversity of plant heat stress transcription factors. Trends Plant Sci., 12: 452–457.

Wang, H., Wang, H., Shao, H. and Xiaoli, T.X. (2016). Recent advances in utilizing transcription factors to improve plant abiotic stress tolerance by transgenic technology. *Frontiers in Plant Sciences*. https://doi.org/10.3389/fpls.2016.00067.

Wei, W., Cui, M., Hu, Y., Gao, K., Xie, Y., Jiang, Y. and Feng, J. (2018). Ectopic expression of FvWRKY42, a WRKY transcription factor from the diploid woodland strawberry (Fragaria vesca), enhances resistance to powdery mildew, improves osmotic stress resistance, and increases abscisic acid sensitivity in Arabidopsis. *Plant Sci.*, 275: 60–74. Doi: 10.1016/j.plantsci.2018.07.01.

Weiss, J. and Egea-Cortines, M. (2009). Transcriptomic analysis of cold response in tomato fruits identifies dehydrin as a marker of cold stress. *J. of Appl. Genet.*, 50: 311–319.

Wu, H.L., Li, L., Cheng, Z.C., Ge, W., Gao, J. and Li, X.P. (2015). Cloning and stress response analysis of the PeDREB2A and PeDREB1A genes in Moso bamboo (*Phyllostachys edulis*). *Genetics and Molecular Research*, 14(3): 10206–10223.

Wu, M., Liu, H., Gao, Y., Shi, Y., Pan, F. and Xiang, Y. (2020). The Moso bamboo drought-induced 19 protein PheDi19-8 functions oppositely to its interacting partner, PheCDPK22, to modulate drought stress tolerance. *Plant Sciences*. https://doi.org/10.1016/j.plantsci.2020.110605.

Wu, M., Liu, H., Han, G., Cai, R., Pan, F. and Xiang, Y. (2017). A Moso bamboo WRKY gene PeWRKY83 confers salinity tolerance in transgenic Arabidopsis plants. *Scientific Reports*, 7: 11721. Doi: 10.1038/s41598-017-10795-z.

Wu, M., Zhang, K., Xu, Y., Wang, L., Liu, H., Qin, Z. and Xiang, Y. (2022). The Moso bamboo WRKY transcription factor, PheWRKY86, regulates drought tolerance in transgenic plants. *Plant Physiolo. Biochem.* Doi: 10.1016/j.plaphy.2021.10.024.

Wu, X.P., Liu, S., Luan, J., Wang, Y. and Cai, C. (2019). Responses of water use in Moso bamboo (*Phyllostachys heterocycla*) culms of different developmental stages to manipulative drought. *Forest Ecosystems*, 6: 31. https://doi.org/10.1186/s40663-019-0189-8.

Xiang, M., Ding, W.S., Wu, C., Wang, W., Ye, S., Cai, C., Hu, X., Wang, N., Bai, W., Tang, X., Zhu, C., Yu, X., Xu, Q., Zheng, Y., Ding, Z., Lin, C. and Zhu, Q. (2021). Production of purple Ma bamboo (*Dendrocalamus latiflorus Munro*) with enhanced drought and cold stress tolerance by engineering anthocyanin biosynthesis. *Planta*, 254: 50. https://doi.org/10.1007/s00425-021-03696-z.

Xie, L., Cai, M., Li, X., Zheng, H., Xie, Y., Cheng, Z., Bai, Y., Li, J., Mu, S. and Gao, J. (2020). Overexpression of PheNAC3 from Moso bamboo promotes leaf senescence and enhances abiotic stress tolerance in Arabidopsis. *PeerJ*, 8: e8716. Doi: 10.7717/peerj.8716.

Yamaguchi, T. and Blumwald, E. (2005). Developing salt-tolerant crop plants: Challenges and opportunities. *Trends Plant Sci.*, 10(12): 615–620.

Yang, G.Y., Xu, L., Yang, L.S. and He, X.B. (2008). Bamboo forest damage caused by snow storm in Sichuan province in 2008 and silvicultural reestablishment measures. *Sci. Silvae Sin.*, 44: 96–100 (in Chinese).

Yanxia, L. and Frith, O. (eds.). (2018). Chishui. *In*: INBAR. *Bamboo for Land Restoration. INBAR Policy Synthesis Report 5*. INBAR, Beijing, China.

Ye, S., Cai, C., Ren, H., Wang, W., Xiang, M., Tang, X., Zhu, C., Yin, T., Zhang, L. and Zhu, Q. (2017). An efficient plant regeneration and transformation system of Ma bamboo (*Dendrocalamus latiflorus Munro*) started from young shoot as explant. *Frontiers in Plant Sciences*, 8: 1298.

Yeasmin, L., Ali, M.N., Gantait, S. and Chakraborty, S. (2015). Bamboo: An overview on its genetic diversity and characterization. *3 Biotech*, 5: 1–11.

Zhang, F., Wan, X.Q., Zhang, H.Q., Liu, G.L., Jiang, M.Y., Pan, Y.Z. and Chen, Q.B. (2012a). The effect of cold stress on endogenous hormones and CBF1 homolog in four contrasting bamboo species. *Journal for Research*, 17: 72–78.

Zhang, Q., Chen, Q., Wang, S., Hong, Y. and Wang, Z. (2014). Rice and cold stress: Methods for its evaluation and summary of cold tolerance related quantitative trait loci. *Rice*, 7: 24.

Zhang, X.M., Zhao, L., Larson-Rabin, Z., Li, D.Z. and Guo, Z.H. (2012). *De novo* sequencing and characterization of the floral transcriptome of *Dendrocalamus latiflorus* (Poaceae: Bambusoideae). PloS One, 7: e42082.

Zhang, Y.J., Ma, P.F. and Li, D.Z. (2011). High-throughput sequencing of six bamboo chloroplast genomes: Phylogenetic implications for temperate woody bamboos (Poaceae: Bambusoideae). *PloS One*, 6: e20596.

Zhang, P., Wang, J. and Zhang, H. (2008). Measures of water management and increasing drought resistance of moso forests in Anji County, Zhejiang Province. *World Bamboo Rattan*, 6: 23–24.

Zhang, Z., Huang, B., Chen, J., Jiao, Y., Guo, H., Liu, S., Ramakrishnan, M. and Qi, G. (2022). Genome-wide identification of JRL genes in moso bamboo and their expression profiles in response to multiple hormones and abiotic stresses. *Front. Plant Sci.*, 12: 809666. Doi: 10.3389/fpls.2021.809666.

Zhao, H., Gao, Z., Wang, L., Wang, J., Wang, S. and Fei, B. (2018). Chromosome-level reference genome and alternative splicing atlas of Moso bamboo (*Phyllostachys edulis*). *Gigascience*, 7: giy115.

Zhao, H., Gao, Z., Wang, L., Wang, J., Wang, S., Fei, B., Chen, C., Shi, C., Liu, X., Zhang, H., Lou, Y., Chen, L., Sun, H., Zhou, X., Wang, S., Zhang, C., Xu, H., Li, L., Yang, Y., Wei, Y., Gao, Q., Yang, H., Zhao, S. and Jiang, Z. (2018a). Chromosome-level reference genome and alternative splicing atlas of Moso bamboo (*Phyllostachys edulis*). *GigaScience*, 7: giy115.

Zhao, H., Zhao, S., Fei, B., Liu, H., Yang, H., Dai, H., Wang, D., Jin, W., Tang, F., Gao, Q., Xun, H., Wang, Y., Qi, L., Yue, X., Lin, S., Gu, L., Li, L., Zhu, T., Wei, Q., Su, Z., Wan, T.A., Ofori, D.A., Muthike, G.M., Mengesha, Y.M., deCastro, E., Silva, R.M., Beraldo, A.L., Gao, Z., Liu, X. and Jiang, Z. (2017). Announcing the genome atlas of bamboo and rattan (GABR) project: Promoting research in evolution and in economically and ecologically beneficial plants. *Gigascience*, 6: 1–7.

Zhao, J.F., Dong, W.Y., Mao, W.J. and Wang, L. (2009). Investigation on snow damage on four kinds of bamboo forests in Daguan County. *Journal of West China for Science*, 38: 96–100 (in Chinese).

Zhao, W., Cao, B., Yang, G., Wengen Zhang, W. and Yu, F. (2021). Complete chloroplast genome sequence and phylogenetic analysis of Chimonobambusa sichuanensis (Bambusoideae). *Mitochondrial DNA Part B*, 6: 824–825. Doi: 10.1080/23802359.2021.1884017.

Zheng, Y., Hou, D., Zhuo, J., Zheng, R., Wang, Y., Li, B., Yu, X. and Lin, X. (2020). Complete chloroplast genome sequence of *Bambusa rigida* (Bambuseae). *Mitochondrial DNA Part B*, 5: 2972–2973. Doi: 10.1080/23802359.2020.1793699.

Zhu, J.K. (2007). Plant salt stress. *Encyclopedia of Life Sciences*.

Zhu, J.K. (2016). Abiotic stress signaling and responses in plants. *Cell*, 167: 313–324.

Chapter 9

Current Understanding on Major Bamboo Diseases, Pathogenicity, and Resistance Genes

Sonali Dey,[1] *Subhadeep Biswas,*[1] *Anirban Kundu,*[2]
Amita Pal[3] *and Malay Das*[1,*]

9.1 Introduction

Bamboo is the one of the fastest developing grasses mainly growing in the tropical and sub-tropical regions of Asia, America, and Africa (Das et al., 2008; Shu and Wang, 2015; Basak et al., 2021; Biswas et al., 2022). It is an economically important plant group having a wide range of utilities such as food, fodder, medicine, construction materials, fishing materials, handicrafts, clothes, paper along with its use in landscaping and soil conservation (Banerjee et al., 2021). Such multipurpose usage of bamboo rapidly increases its demand in major Asian economies such as China, Japan, and India (Akinlabi et al., 2017). Therefore, it is imminent to gather knowledge on major pathogens and pests that affect bamboo growth and development to design appropriate mitigation strategies.

A range of pathogens and pests such as bacteria, fungi, virus, and insects can inflict diseases to bamboo.

It has been reported that more than 195 species of bamboos belonging to 35 genera are affected by various diseases. A total of 1,200 insect species, 580 fungi, five bacteria, three viruses, one phytoplasma (mycoplasma-like organism), and one bacterium-like organism have been identified that can be pathogenic to bamboos (Mohanan, 1997; Shu and Wang, 2015; Xu and Wang, 2004; Xu et al., 2006). Despite that, detailed information on pathogenicity of these pathogens and the extent of

---

[1] Department of Life Sciences, Presidency University, Kolkata 700073.
[2] Department of Botany, Ramakrishna Mission Vivekananda Centenary College, 700118.
[3] Division of Plant Biology, Bose Institute, Kolkata 700054.
* Corresponding author: malay.dbs@presiuniv.ac.in

economic loss inflicted to different species are very scanty, although they are very important for proper management of bamboo bio-resources (Xu and Wang, 2004). It is always challenging for any forest-grown plants to gather such information, which will be dependent on extensive survey works. Therefore, the main objective of this chapter is to collate all available information and to highlight future research directions.

9.2 Distribution of Major Bamboo Diseases and Pests in Asia

Since culm is the major economically important part of the vegetative plant body, substantial work has been done on diseases observed on young and matured culms. For instance, rot of emerging culms is one of the most common bamboo diseases and had been reported from India, Bangladesh, China, Pakistan, Philippines, and Thailand (Mohanan, 1997). It is caused by fungi such as *Fusarium moniliforme* and *Fusarium flocciferum*. In most instances, maximum economic losses have been reported in unmanaged natural stands than in plantations. Emerging culms, which are 15–30 cm in height, are the most vulnerable one. The average mortality rate of emerging bamboo culms ranges from 5.5 to 25.5%. However, a few species are more susceptible to the disease. For instance, mortality rate of *Bambusa bambos* is approximately 34% (Mohanan, 1997). Rot of emerging culms caused by *Pterulicium xylogenum*, has been recorded in different bamboo species that grow in the northern and northeastern states of India. It particularly affects shoot growth in the edible bamboo *Melocanna baccifera*. Young, 2–4 years old clumps of *B. bambos*, *Dendrocalamus longispathus*, and *D. strictus* were worst affected by this disease in China and India. Bamboo blight had been reported from India and Bangladesh, affecting *B. Bambos*, *B. balcooa*, *B. tulda*, *B. vulgaris*, and *B. nutans*. Culm brown rot had been reported from *Phyllostachys viridis* and *P. glauca*. The disease was first recorded in 1974 in Nanjing, China, and later spread to stands located in Jangsu and Zhejiang provinces. *Ceratosphaeria phyllostachydis* Zhang and *Coccostroma arundinariae* fungi were frequently observed in *P. edulis* forests of China during 2005–2006 (Xu et al., 2020). Occurrence of bamboo top blight of *P. edulis* (Moso bamboo) and *P. pubescens* was reported from Jiangsu, Zhejiang, Anhui, Jiangxi, Fujian, and Shanghai Provinces of China. Witches-broom disease has been reported from India, Indonesia, Japan, China, and Vietnam. It affects many bamboo species such as *Phyllostachys*, *Ochlandra*, *Bambusa*, *Gigantochloa*, and *Sasa*. Witches-broom caused by *Aciculosporium take* has been extensively observed in Japan. Branch die-back disease and little leaf disease has been reported from India only. Bamboo wilt, caused by *Fusarium oxysporum*, was reported from a hybrid bamboo, *B. Pervariadilix Grandis Nin* (*B. pervariabilis* x *D. daii*), growing in Guangxi and Sichuan provinces in China. This hybrid bamboo is used for paper production and hence the disease poses a challenge to the paper industry. Culm rust caused by *Stereostratum corticioides* affect *P. glauca* and *P. meyeri* stands in Jiangsu, Hunan, Zhejiang, and Anhui provinces in China. Bamboo mosaic disease, caused by the BaMV has been reported in two major cultivated species of bamboos, which are *D. latiflorus* and *B. oldhamii*. The disease had been reported from 13 species of bamboos having pachymorph rhizomes and they primarily grow in China and India.

Shoots of the diseased plant appear hard in texture and have compromised edibility and canning quality (Mohanan, 1997). In China, more than 800 insect species including locusts, leaf rollers, puss moths, tussock moths, and sawflies cause huge economic loss (Shu and Wang, 2015). Pests like bamboo shoot wireworms caused severe losses in bamboo production in Zhejiang. Primarily, two insects, *Ceracris* spp. and *Rivula biatomea* (Moore) caused enormous damage to *P. edulis* population spreading over 135 ha area in Anji Zhejiang, China. These insects were also detected in the Anhui, Fujian, and Sichuan provinces (Xu et al., 2020).

9.3 Bamboo Diseases Caused by Fungi

Several pathogenic fungi have been identified that can cause disease to the bamboo culm, although the disease severity may vary. A few among them can cause significant damage such as culm rot caused by *Pterulicium xylogenum*, culm wilt caused by *Fusarium incarnatum*, rot of emerging (Figs. 9.1A, B) and growing culm (Fig. 9.1C) generated by *Fusarium* spp. and *Pterulicium* sp., culm blight engendered by *Sarocladium oryzae*, culm rust affected by *Steroestratum corticioides*, top blight of *Phyllostachys* spp., produced by *Ceratosphaeria phyllostachydis*, witches-broom (Fig. 9.1D), brought about by *Balansia* spp., *Aciculosporium* sp. (Table 9.1). Culm purple blotch (Fig. 9.1E) caused by *F. stilboides* is another major disease of bamboos. Besides this, necrosis of culm internode affected by *Curvularia lunata*, sooty stripe

Fig. 9.1. Symptoms of a few selected diseases that commonly occur in bamboos.
Note: (A, B) Emerging culm rot of *B. tulda*, (C) Rot of Growing Culms of *B. vulgaris*, (D) Witch's Broom of *B. tulda*, (E) Culm Purple Blotch of *B. balcooa*, (F) Puccinia Leaf Rust of *B. balcooa*, (G) Phakopsora Leaf Rust of *D. hookeri*, (H) Phoma Leaf Spot of *Arundinaria mailing*.

Table 9.1. Summary of important pathogenic fungi, their symptoms, and host bamboo species.

Disease	Causal Organisms	Symptoms	Bamboo Host	Transmission	Control Measure	References
Rot of emerging culms	*Fusarium moniliforme* var. *intermedium*	Dark brown spots on outermost sheath of emerging culm, gradually cover the entire sheath, emerging shoot becomes discoloured, succulent, strong smell of molasses.	*B. balcooa, B. bambos, B. polymorpha, B. vulgaris, D. longispathu, D. strictus, Thyrsostachys oliveri*	Insects, wild animals like monkeys, porcupines, and squirrels	Biocontrol using *Trichoderma,* leaf and bark extracts of *Cleistanthus collinus* and *Prosopis juliflora*	Mohanan, 1997, 2017
Rot of growing culm	*F. moniliforme, F. equiseti, F. fujikuroi*	Greyish brown, spindle-shaped lesions at the base of culm sheaths.	*B. balcooa, B. bambos, B. polymorpha, D. strictus*	Sap sucking insect like *Purohitha cervina* Distant	Carbendazim, mancozeb, and Monocrotophos	Mohanan, 2017
Culm rot	*Pterulicium xylogenum*	Profuse, white mycelial growth at clump base, mycelial patches on culm sheath, necrotic lesions on culm.	*B. vulgaris* var. *waminii, D. giganteus, Gigantochloa sp. M. baccifera, B. pallida, D. longispicula, D. asper, B. tulda*	Air current	Copper oxychloride and carbendazim	Harsh et al., 2005; Mohanan, 1997, 2010, 2017
Culm brown rot	*F. solani, F. equiseti*	Pale yellow spots on the lower culm, spread vertically, upward direction, later form violet-brown to black-brown streaks	*P. viridis, P. viridis, P. aureosulcata, P. glauca*	=	No control measure suggested	Mohanan, 2017; Zhou et al., 2011
Culm base rot	*Arthrinium phaeospermum, A. alternata, Arthrinium sp., F. oxysporum, F. fujikuroi*	Browning and necrosis occur at culm base, culm withering and dying in severe infection.	*P. pubescens*	Etiology of the disease has not been well studied	Bayleton	Mohanan, 2017
Culm purple blotch	*F. stilboides*	Pale yellow mottled spots and stripes on culm, yellow leaves	*P. viridis, P. pubescens, P. viridis, P. dulcis, P. praecox, B. balcoa*	–	Limestone powder on the floor around the clumps	Mohanan, 2017

Table 9.1 contd. ...

...Table 9.1 contd.

Disease	Causal Organisms	Symptoms	Bamboo Host	Transmission	Control Measure	References
Bamboo blight	*Sarocladium oryzae, Paraconiothyrium fuckelii, Fusarium* spp., *S. strictum, Pteroconium* sp., *Arthrinium* sp.	Premature death of culm sheaths, partial collapse of the fragile apical regions, wet rotten patches developed on the internodes.	*B. bambos, B. balcooa, B. tulda, B. vulgaris, B. nutans*	—	Carbendazim combined with mancozeb or fytolan	Mohanan, 2017
Bamboo top blight	*Ceratosphaeria phyllostachydis*	Browning and necrosis of the culm and subsequent withering of branches.	*P. pubescens, P. heterocycla, P. edulis*	Wind or rain splashes	Carbendazim or bordeaux mixture	Lin, 2001; Mohanan, 2017
Branch die-back	*F. pallidoroseum*	Small, grayish magenta, linear lesions occur on top three to five internodes of young culm, pale yellowish, linear lesions occur on foliage.	*B. bambos, B. vulgaris, D. strictus, B. bambos, Phyllostachys* sp.	Air current	Mancozeb	Mohanan, 2017
Witches-broom	*Aciculosporium take, Heteroepichloe bambusae, Heteroepichloe sasae, Epichloe bambusae, Balansia take, Linearistroma lineare, Loculistroma bambusae, Aciculosporum sasicola, Heteroepichloe bambusae, Heteroepichloae sasae, Phaeosphaeria bambusae, Aciculosporum sasicola*	Numerous highly shortened abnormal shoots at the nodes of mature culms further develop into highly reduced shoots, shortening of culm sheaths, boat shaped shoots, with a prominent ligule.	*P. viridis, P. glauca, P. praecox, P. heteroclada, P. aurosulcata, P. incarnate, P. nuda, Semiarundinarina fastuosa, B. multiplex, G. apus kurz, G. robusta, P. bambusoides, P. nigra, Sasa berealisvar, Sasa* spp., *Ochlandra travancorica, O. scriptoria, O. ebracteata*	Rain drop transmission	Tetracycline, oxytetracycline, carbendazim and triazolone	Mohanan, 2017

Disease	Pathogen	Symptoms	Host	Transmission	Control	Reference
Thread blight	*Erythricium salmonicolor*, *Venkatanarayanan*, *Corticium koleroga*	Large, water-soaked, irregular lesions with greyish green center and grayish white margins on leaves, appear on leaf base grow towards the tip.	*B. balcooa, B. bambos, B. multiplex, B. polymorpha, B. tulda, B. tuldoides, B. vulgaris, D. brandistii, D. longispathus, D. strictus, Thyrsostachys siamensis, O. scriptoria, O. ebracteata*	—	—	Mohanan, 2017
Bamboo wilt	*F. oxysporum*, *F. incarnatum*, *Gliocladium* sp., *Erwinia sinocalami*	Yellowing of the foliage, premature defoliation, discoloration and shrivelling of culm basal, browning and necrosis of root and rhizome.	*D. latiflorus, B. pervariadilisx Grandis (Bambusa pervariabilis* X *Dendrocalamopsis daii), P. edulis*	Wind and rain splashes	—	Ma et al., 2008; Mohanan, 2017
Bamboo culm rust	*Stereostratum corticioides*	Rusty patches on the lower parts of young culms, emerging shoots.	*Bambusa* sp., *Chimonobambusa* sp., *Pleioblastus* sp., *Pseudosasa* sp., *Sasa* sp., *Arundinaria* sp., *Semiarundinaria* sp., *P. glauca, P. dulcis, P. viridis, P. praecox, P. heteroclada, P. incarnate, P. bambusoides, P. nidularia, P. aureosulcata, P. vivax, P. glabrato, P. propinqua, P. congesta, P. meyeri, Pleioblastus vaginatus, P. higoensis*	Through wind	Coal tar and diesel oil mixture and cresoli saponatus solution	Mohanan, 2017

Table 9.1 contd....

...Table 9.1 contd.

Disease	Causal Organisms	Symptoms	Bamboo Host	Transmission	Control Measure	References
Bamboo culm smut	*Bambusiomyces shiraianus*	Black patches on culms and shoot tips.	*Phyllostachys sulphurea, P. glauca, P. pubescens, P. nigra, P. flexuosa, P. bambusoides, P. incarnata, P. congesta, P. aurea, Pleioblastus amarus, P. makinoi, Fargesia sp., Arundinaria spp., Sasa ramose, S. nana*	–	–	Mohanan, 2017
Culm staining and die-back	*Apiospora* sp.	Pale purple to dark brown, linear lesions around the bore hole wounds of growing culm, later spread to the entire culm, internodes become necrotic, black.	*B. vulgaris, D. longispathus*	–	Endosulphan	Mohanan, 1997, 2017
Foliage blight	*Bipolaris maydis* anamorph of *Cochliobolus heterostrophus,* *B. bambusae*	Small, water-soaked, grayish brown, spindle-shaped lesions found both young and mature leaves, yellowish orange haloes develop around this lesion.	*B. bambos, D. brandisii, D. longispathus, D. strictus, Pseudoxytenanthera ritcheyi*	Wind or rain splashes	Difolatan or fytolan	Mohanan, 2017

| Kweilingia leaf rust | *Kweilingia divina, Puccinia inflexa* | Water-soaked pin-head flecks found at lower surface of the foliage, along with yellowish orange to rust brown linear urediniosor. | *B. balcooa, B. bambos, B. multiplex, B. polymorpha, B. tulda, B. tuldoides, B. vulgaris, D. brandisii, D. hamiltonii, D. lojngispathus, D. strictus, Pseudoxytenanthera ritcheyi, Thyrsostachys oliveri, T. siamensis, Ochlandra travancorica, O. scriptoria, B. oldhamii, D. latiflorus, B. multiplex, B. oldhamii, B. shimadai, D. latiflorus, Bambusa sp., Sasa sp., D. membranaceus* | — | — | Mohanan, 2017 |
| Puccinia leaf rust | 25 species of *Puccinia* like *P. aduncta, P. bambusicola, P. cymbiformis, P. flammuliformis, P. gracilenta, P. hikawaensis, P. kusanoi, P. longicornis, P. melanocephala, P. mitiioformis, P. nigroconoidea, P. melanocephala, P. phyllostachydis, P. sasicola, P. sasae, P. scabrida, P. sinarundinaria, P. tenella, P. xanthosperma* | Yellowish brown to dark-brown, linear urediniosor, scattered on the infected leaves. | *Phyllostachys sp., Bambusa spp., B. tessellata, B. oldhami, B. vulgaris, Ischurchola spinosa, Sasa sp., Arundinaria sp., A. atropurpurea, A. okadana, Disporopsis arisanensis, D. latiflorus, D. oldhami, P. nigra, P. pubescens, P. aurea, Sinarundinaria nitida, Schizostachyum sp.* | Etiology and disease cycle are not available | Triadimefon and kejunning | Ma et al., 2005; Mohanan, 2017 |

Table 9.1 contd. ...

...*Table 9.1 contd.*

Disease	Causal Organisms	Symptoms	Bamboo Host	Transmission	Control Measure	References
Uredo leaf rust	Seven species of *Uredo* like *U. arundinariae, U. sasae, U. arundinis-donacis, U. bambusae-nanae, U. dendrocala, U. inflexa, U. ochlandrae*	Orange to brown urediniosori arranged on lower leaf surface, leaf blotch and defoliation in severe infection.	*Arundinaria* sp., *Sasa* sp., *Bambusa* spp., *Phyllostachys* sp., *Sasa septentrionalis, Sinarundinaria* spp., *Sasamorpha amabilis, B. nana, D. strictus, D. latiflorus, Physopella inflexa, O. stridula*	–	–	Mohanan, 2017
Exserohilum leaf spot	*Exserohilum rostratum* anamorph of *Setosphaeria rostrata* Leonard, *Exserohilum holmii* (anamorph of *Setosphaeria holmii* (Luttr.)	Small, water-soaked, grayish black, linear to irregular lesions on mature leaves, spread to the entire leaf lamina.	*D. strictus, D. longispathus, B. polymorpha*	–	Difolatan or fytolan	Mohanan, 2017
Colletotrichum leaf spot	*Colletotrichum gloeosporioides* anamorph of *Glomerella cingulata, Colletotrichum septorioides, Colletotrichum* sp.	Water-soaked, small grayish brown spots developed on juvenile and mature leaves.	*B. vulgaris, B. bambos, D. strictus, D. asper, D. giganteus, D. pendulus, Gigantochloa levis, Gigantochloa ligulata, G. latifolia, G. rostrata, G. Scortechinii, P. pubescens, P. ebracteata, Phyllostachys* spp. *Arundinaria* sp., *O. travancorica, O. travancorica, O. Scriptoria*	Wind or rain splash	Thiram, benlate or captan	Mohanan, 1997, 2017

Smut (Inflorescence)	*Claviceps* sp., *Bambusiomyces shiraianus*, *Tilletia bambusae*	Developing spikelets affected during infection.	*B. bambos, Bambusa* sp., *P. heterocycla, S. nana, S. ramosa, P. sulphurea, P. glauca, P. pubescens, P. nigra, P. flexuosa, P. bambusoides, P. incarnata, P. congesta, P. aurea, Pleioblastus amarus, P.makinoi, Fargesia* sp., *Arundinaria* spp.	–	Carboxin, thiabendazole, and etaconazole	Mohanan, 2017
Ergot (Inflorescence)	*Claviceps purpurea, Claviceps* sp., *Hypocrella semiamplexa*	Developing spikelets are affected during infection and seeds are also replaced with fungal fructifications.	*Phyllostachys* sp., *D. hamiltonii*	‖	–	Mohanan, 2017
Rhizome dud rot	*Globisporangium proliferatum*	Yellowing of the entire foliage, complete defoliation within 15–20 days, browning and rot of the rhizome buds and tender buds.	*B. bambos*	‖	–	Mohanan, 2017
Rhizome and root rot	*Amylosporus campbellii; Serpula similis, S. eurocephala, Sphaerostilbe bambusae, Polyporus* sp., *Poria* sp.	Yellowing of the leaves followed by die-back of culms, white fibrous or white spongy rot of root and rhizome.	*D. strictus, Thyrsostachys oliveri, D. longispathus, B. Bambos, B. balcooa, O. travancorica*	–	Copper oxychloride	Mohanan, 2017
Decay of rhizome, root, and basal culm	*Ganoderma lucidum, Amylosporus campbellii, Coltricia bambusicola, Antrodia rhizomorpha, Rigidoporus microcarpus, Rosellinia emergens, Rosellinia* sp.	White spongy-fibrous or brown cubical rot in root, rhizome, and culm.	*B. bambos, M. baccifera, D. giganteus*	‖	–	Mohanan, 2017

disease produced by *Arthrinium arundinis*, and *A. phaeospermum*, culm sheath spot generated by *Shiraia bambusicola*, *Myriangium haraeanum*, *Pestalozziella bambusae*, and *Sarocladium* sp., sooty mould disease, a result of *Myriangium* sp., *Capnodium* spp., *Spiropes scopiformis* are considered as a regular pathogen of bamboo genera like *Ochlandra*, *Dendrocalamus*, *Phyllostachys*, *Bambusa*, *Pleioblastus*, and *Fargesai* are other minor diseases that impact commercial exploitation of bamboos to some extent (Mohanan, 2010, 2017).

Not only culm, but also foliage is affected by fungal infection. A total of 245 species of fungi are known to be associated with foliage infections. Among them, 113 are associated with major and 132 species are associated with minor infections (Mohanan, 2010, 2017). Leaf rust disease caused by *Kweilingia divina* (Gautam and Avasthi, 2018), *Puccinia inflexa*, *Uredo* spp., *Puccinia* spp. (Ma et al., 2005) (Fig. 9.1F) *Tunicopsora bagchii*, leaf spot caused by *Exserohilum holmii*, *Dactylaria bambusina*, *Ascochyta arundinariae*, and leaf blight caused by *Bipolaris maydis* are the most commonly occurring diseases of leaves. Another frequently occurring, minor disease of bamboos is Phakopsora leaf rust (Fig. 9.1G) caused by *Phakopsora loudetiae*. Various leaf spot diseases (Fig. 9.1H) observed in bamboo are also considered as minor pathogens because they are not associated with severe damage and economic losses (Mohanan, 2017; Table 9.1).

Apart from culm and leaf, there are several fungal species that can infect root and rhizome. A few prominent candidates among them include *Globisporangium proliferatum*, *Amylosporus campbellii*, *Serpula similis*, *S. eurocephala*, *Sphaerostilbe bambusae*, *Polyporus* sp., *Poria* sp. (Mohanan, 2017; Table 9.1). Fungal pathogens primarily disperse through wind or rain splashes. However, a few are also transmitted via sap-sucking insects, wild animals like monkeys, porcupines, and squirrels (Mohanan, 2010, 2017).

9.4 Bamboo Diseases Caused by Bacteria

Compared to fungi, much smaller number of pathogenic bacteria has been identified in bamboo. The wilt disease of bamboo, which is caused by the bacteria *Erwinia sinoaalami* can naturally infect *D. latiflorus*. Common symptoms of this disease include yellowing of foliage, premature defoliation, discolouration of culm base, and browning of rhizomes (Mohanan, 1997, 2002, 2017; Qiao et al., 2011). It has been observed that bamboos are also susceptible to Phytoplasma, a mycoplasma-like organism, responsible for little leaf disease. They are mainly transmitted by insect vectors like aphids, and leafhoppers. This disease, also known as 'witches-broom disease', is caused by *Candidatus Phytoplasma asteris*. It has been identified in *P. nigra* var. *henonis* and its occurrence has been reported from Korea (Jung et al., 2006). Later studies found that the disease can also occur on other bamboo hosts such as *D. strictus*. The main symptoms of this disease include formation of numerous, highly reduced abnormal bushy shoots emerging from the nodes of newly emerged culms, reduced leaves with needle-like appearance, and stunted culm growth (Mohanan, 2017). It would be interesting to investigate in future whether lesser incidents of disease caused by bacteria is due to lesser number of studies or bamboo's enhanced resistance to biotrophic organisms (e.g., bacteria) than that of necrotrophic (e.g., fungi) ones.

9.5 Bamboo Diseases Caused by Virus

BaMV belongs to the genus *Potexvirus* of the family Alphaflexiviridae (Lin et al., 1994) and was first isolated from *Bambusa multiplex* (Lour) Raeusch., and *B. vulgaris* Schrad. The disease has already been reported from multiple countries like Brazil, China, USA, Australia, and Hawaii (Hull, 2013; Table 9.2). The BaMV is mainly associated with mosaic symptoms on bamboo leaves. BaMV virus particles are flexuous, filamentous rods, 480–500 nm in length and 15 nm in diameter. It can be thermally inactivated in the temperature range between 75°–80°C. Electron microscopy observations of the infected tissues revealed that virus appears in the form of electron-dense crystalline bodies (EDCBs) within infected cells. The fully sequenced RNA genome of BaMV is approximately 6.4 kb in length, possesses a total of 6,366 nucleotides, along with a 5'-cap and a 3' poly (A) tail and six conserved open reading frames (ORFs) (Chen et al., 2007).

In contrast to BaMV, satellite bamboo mosaic virus (satBaMVs) possesses small, single-stranded positive-sense 836-nt long, linear RNA molecules that encode a 20-kDa non-structural protein called P20, which is flanked by a 159-nt 5'-UTR and 125-nt 3'-UTR (Lin and Hsu, 1994). A study on BaMV co-infected *Nicotiana benthamiana* revealed that P20 is necessary for long-distance transport of satBaMV (Chang et al., 2016). Three clades of satBaMV have been identified revealing wide sequence diversity (Wang et al., 2014). Clade I consists of satBaMV isolates collected from most of the bamboos species; clade II hosts isolates obtained from Ma bamboo, whereas clade III harbours isolates obtained from *B. vulgaris*. Assessment of nucleotide diversity conferred that isolates placed in clades II and III possess greater sequence diversity compared to clade I (Wang et al., 2014). Sequence analysis of satBaMV isolates showed a hyper-variable region with high sequence variation in the 5'-UTR of satBaMV but possesses a conserved secondary RNA structure (Yeh et al., 2004). SatBaMV is a sub-viral agent and hence totally dependent on BaMV for replication and encapsidation (Lin and Hsu, 1994). It has been reported that BaMV can infect 13 species of bamboo and is found to be a major deterrent of *B. edulis* production in many bamboo growing regions of the world (Singh et al., 2013).

Cherry necrotic rusty mottle virus (CNRMV) is one of the most important viruses that infect sweet cherries. Interestingly, it has been found that 21 different species of bamboos like *Arundinaria falconerii*, *B. balcooa*, *B. bambos*, *B. multiplex*, *B. nutans*, *B. pallida*, *B. tulda*, *B. entricose*, *D. asper*, *D. asper* (Chinese), *D. bambusoides*, *D. bannaenensis*, *D. barbatus*, *D. dianxiensis*, *D. giganteus*, *D. hamiltonii* (local maggar), *D. hamiltonii* (north east variety), *D. sinicus*, *D. yunnanensis*, *Froesiochloa somnigensis*, and *M. baccifera* can be infected by CNRMV (Awasthi et al., 2014). It is an unassigned member of the family Betaflexiviridae, flexuous filamentous plant viruses (Adams et al., 2012). The genome of CNRMV consists of a positive stranded ssRNA, which is 8,432 nucleotides long (excluding the 3' poly (A) tails) sequence that encodes for seven significant ORFs. The genome of CNRMV consists of a single-stranded positive-sense RNA, which is approximately 8.4 kb in size. The genome possesses seven ORFs (Awasthi et al., 2014). Five of these ORFs are found conserved among all foveaviruses, allexiviruses, potexviruses, and carlaviruses. The ORF1 encodes for the replicase. ORFs 2a and ORFs 5a, are nested completely within

Table 9.2. Summary of important pathogenic viruses, their symptoms, and available information on genome structures and sequences.

Pathogen	Family	Symptoms	Bamboo Host	Transmission Vector	Genome Organisation	Genome information	Control Measure	References
Bamboo Mosaic virus (BaMV)	Genus Potexvirus, Family Alpha flexivirideae	Foliar mosaic pattern, stripes, brown internal streaking of both shoot and culm, aborted culm, poorly developed culms, short internodes, and new emerging shoots are hard in texture.	*B. mutabilis, B. beecheyana, B. dolichoclada, B. edulis, B. multiplex, B. oldhamii, B. pachinensis, B. utilis, B. ventricosa, B. vulgaris, D. giganteous, D. latiflorus, M. baccifera*	Dipterans species (*A. orientalis, G. fasciventris*)	Single stranded, (+) RNA	6366 nucleotides long, six conserved ORFs with 5' cap structure and 3' poly (A) tail	–	Bhardwaj et al., 2017; Mohanan, 2017
Cherry Necrotic Rusty Mottle Virus (CNRM)	Family Betaflexiviridae	Mosaic pattern on leaves, chlorosis, yellow streaks, necrotic lesions, leaf curling.	*A. falconerii, B. balcooa, B. bambos, B. multiplex, B. nutans, B. pallida, B. tulda, B. ventricosa, D. asper, D. asper,* (Chinese) *D. bambusoides, D. bannaenensi, D. barbatus, D. dianxiensis, D. giganteous, D. hamiltonii* (local maggar) *D. hamiltonii* (northeast verity) *D. sinicus, D. yunmanensis, Fargesia somnigensis, M. baccifera*	-	Single stranded, (+) RNA	Approximately, 8.4 kb long genome, seven conserved ORFs without 5' cap structure and 3' poly (A) tail.	–	Bhardwaj et al., 2017; Mohanan, 2017

| Apple stem grooving virus (ASGV) | Genus Capillovirus Family Beta flexivirideae | Mosaic pattern, chlorosis, yellow streaks on leaves, necrotic lesions, and curling of leaves. | *D. bannaenensi, Gigantochloa takserah, D. strictus, D. barbatus, D. latiflorus, D. membranaceus, D. bambusoide, D. dianxiensis, B. ventricosa, D. sinicus, D. yunnanensis, B. bambos, B. distegia, B. multiplex, B. nutans, B. pallid, B. nana, B. balcooa, A. falconerii, B. tulda, P. nigra, Guadua angustifoli, Gigantochloa apus, Sasa auricoma, F. somnigensis* | – | Single stranded, (+) RNA | Possess near about 6.5 kb genome two ORFs without 3' poly (A) tail, and a single coat protein (CP) of 27 kDa. | – | Bhardwaj et al., 2017 |

ORFs 2 and 5, respectively and the function of other two ORFs are still unknown (Rott and Jelkmann, 2001).

Apple stem grooving virus (ASGV) is another important, globally distributed, latent virus causing infection to apples. Later studies found that it can also infect 27 bamboo species such as *D. bannaenensis, D. hamiltonii (Local maggar), Gigantochloa takserah, D. strictus, D. barbatus, D. latiflorus, D. membranaceus, D. bambusoides, D. dianxiensis, B. ventricosa, D. sinicus, D. yunnanensis, B. bambos, B. distegia, B. multiplex, B. nutans, B. pallid, B. nana, B. balcooa, A. falconerii, B. tulda, P. nigra, Guadua angustifoli, Gigantochloa apus, Sasa auricoma, F. somnigensis* (Bhardwaj et al., 2017). It is an unassigned member of the genus *Capillovirus*, family Betaflexiviridae. The virus particle is flexuous, filamentous, 600–700 nm in length and 12 nm wide (Lister, 1970; Table 9.2). The ASGV genome comprises of a positive stranded ssRNA. Its ORF1 encodes a multifunctional polyprotein having replication associated protein plus the coat protein (CP) at its C-terminal end, which is necessary for the viral infection. The ORF2 codes for a putative movement protein, nested within ORF1 (Komatsu et al., 2012; Table 9.2).

9.6 Bamboo Diseases Caused by Insects

Multiple herbivorous insects also cause huge damage to bamboos. Insects mainly feed on foliage, bore holes on shoots, culm and finished products, or suck sap from bamboo. There are some insects, which attack live bamboos (attack seeds, foliage, and culms), while the other group attack post-harvested bamboos (Verma and Sajeev, 2012). Insects can damage bamboos in diverse ways such as by sucking plant sap, inflicting mechanical injury, introducing toxic compounds into the plant, and transmitting diseases as vector (Verma and Sajeev, 2012). It has been reported that in China more than 400 insects' species feed bamboo leaves, and near about 39 different species like locusts, leaf rollers, puss moths, tussock moths, and sawflies are cause a huge economic loss (Xu and Wang, 2004). It has been also found that in India defoliators, sapsuckers, and culm and shoot borers are the potential threats for bamboos. The seed pest, *Udonga montana* found cause huge damage in seed production in bamboos and borers *Dinoderus* spp. main threat for stored bamboos. In the case of Asian countries, 'ghoon' borer (*Dinoderus* spp.) causes maximum damage to bamboos strands.

Insects that attack live bamboos mainly belong to the orders Orthoptera, Hemiptera, Lepidoptera, Hymenoptera, and Coleoptera. The nature of damage caused by them includes foliage feeding, sucking the sap, and making bore holes on culms and shoots (Koshy et al., 2001). Among all the insects, shoot and culm borers cause maximum damage to bamboo. Although lesser in frequency and population size, infestation by defoliating insects may lead to epidemics and total defoliation of bamboo stands (Verma and Sajeev, 2012). It has been reported that leaf feeders reduce the surface area available for photosynthesis and thereby affect the vigour, growth, and survival of plants. The finished products made of bamboo are mainly prone to attack by insect borers (Mathew and Nair, 1988). The post-harvest pests are from the order Coleoptera and are members of the families Bostrychidae,

Lyctidae, and Anobidae. Among them Bostrychidae causes maximum damage to bamboo. It has been reported that approximately 16 species of Bostrychids attack post-harvest and finished bamboo products (Verma and Sajeev, 2012). In India, termites like *Odontotermes feae* (Wasmann), *O. horni* (Wasmann), and *O. obesus* (Rambur) may attack rhizomes and culms of bamboo species like *D. stocksii, D. strictus, D. asper, Guadua angustifolia, B. bambos*, and *B. balcooa* (Shanbhag et al., 2010). In another study, it has been found that *Bambusa hirose, B. oldhamii, Dendrocalamus brandisii, D. latiflorus, Gigantocholoa pseudoarundinacea, Guadua angustifolia* growing in Hawaii were predominantly invaded by termites such as *C. gestroi* (Wasmann), *C. formosanus* (Shiraki) (Hapukotuwa and Grace, 2011).

9.7 Physical and Chemical Methods of Bamboo Preservation

Bamboo plants contain a large amount of starch and sugar and hence are highly susceptible to various pathogens, which may cause extensive damage to culm biomass during storage (Mohanan, 2017). Untreated bamboo culms if stored, demonstrate approximately five years of life span. Therefore, it is very important to pretreat bamboo culms before storage. Both traditional/physical and non-traditional/chemical methods are used for the treatment of bamboo to prolong their durability (Omyama et al., 1995). Traditional methods include torching, smoking, curing, whitewashing, and soaking. In contrast to traditional methods, non-traditional methods rely on chemical treatments, confer long-term protection to bamboos (Kaur et al., 2016). Chemical preservatives that are used for bamboo preservation can be classified into four groups (Kumar et al., 1994), which are natural toxicants, waterborne preservatives (fixing and non-fixing types), oil-based preservatives (Creosote), and organic solvent- based preservatives like pentachlorophenol (PCP) (trichlorophenol (TCP), and copper/zinc soaps). Fungal diseases can be controlled by applying various fungicides like carbendazim, mancozeb, copper oxychloride, carbendazim, thiophanate methyl, fytolan, copper oxychloride, mancozeb, thiophanate methyl carboxin, thiabendazole, etaconazole, while insects can be controlled by insecticides like endosulphan and dimethoate.

Since, chemicals are toxic and expensive, biocontrol agents are being used for bamboo disease management. For instance, emerging culms can be controlled by bio-control agent *Trichoderma*, and leaf and bark extracts of *Cleistanthus collinus* and *Prosopis juliflora* (Sharada et al., 2013). In China, bamboo leaf rust disease caused by *Puccinia* sp. could be controlled up to 64.8% with the help of hyper-parasite *Acremonium salmoneum* (Ye et al., 2011, 2012). Similarly, the bacterial wilt of bamboo could be successfully controlled by *Pseudomonas aeruginosa* ZB27 (Qiao et al., 2011). Besides this, *Metarhizium anisopliae* has been successfully used to control bamboo shoot wireworms in China (Wang et al., 2010). Similarly, *Beauveria bassiana* and *Bacillus thuringiensis* were used to manage bamboo leaf defoliators such as tussock moths, puss moths, and leaf rollers (Li, 2006; Sun et al., 2007; Wu, 2009).

Due to environmental and human health hazards of conventional chemical preservatives (Xu et al., 2013), few nontoxic chemicals have been developed and successfully applied in the field. For instance, bio-oil, hydrogels, organic acids, and boron complexes are also very effective ingredients for bamboo culm storage. It has been found that the application of hot neem seed oil on *B. vulgaris* could improve water resistance and resistance against *Pycnoporus sanguineus* (Erakhumen, 2012a, b). Another report suggested that bamboo samples soaked with resin mixed with camphor leaf extract could provide reasonable level of protection against *Phanerochaete chrysosporium* and *Gloeophyllum trabeum* (Xu et al., 2013). Many such treatments such as mesoporous aluminosilicate (Wu et al., 2019) and electrochemical silver treatment (Ju et al., 2021) were also found effective against fungal diseases.

9.8 Biochemical and Genetic Defence in Bamboo

Our understanding on defence response of bamboo plants at the biochemical and genetic levels are currently limiting due to shortage of sequenced bamboo genomes, mutant lines, and effective transformation procedures. Only few available studies indicate involvement of known stress-induced hormones in this process. For instance, salicylic acid (SA) confers plant resistance to virus infection via R gene resistance and regulates a few genes in the RNA-silencing pathway (Alazem et al., 2017). Recent findings suggest that Abscisic Acid (ABA) treatment decreases titres of BaMV in inoculated leaves of Arabidopsis. It also found that ABA induces resistance against BaMV, which is mediated by argonaute (AGO) gene family, specifically the AGO 2 and 3 (Alazem et al., 2017). In a recent study conducted on Moso bamboo, Jacalin-related lectins (JRLs) have been identified. JRLs are responsive to a broad group of biotic and abiotic stress hormones such as SA, abscisic acid (ABA), and methyl jasmonate (MeJA) treatments (Zhang et al., 2022). Similarly, lateral organ boundaries domain (LBD) gene family is another candidate, which may play an important role in molecular defence responses of bamboo. Promoters of *P. edulis* LBD genes contain numerous motifs responsive to ABA, MeJA, and IAA (Huang et al., 2021). Studies demonstrate that transcription factors also contribute to bamboo disease resistance. For instance, overexpression of the Moso bamboo transcription factor PheWRKY1 enhances disease resistance against *Pseudomonas syringae* pv. tomato DC3000 (Pst DC3000) in transgenic *A. thaliana* (Cui et al., 2013). A few important studies have been done to characterize viral genome, which are pathogenic to bamboos. For instance, BaMV, which was first isolated from *B. multiplex*, poses a major threat to many bamboo species and thus seriously affects their commercial exploitation (Elliott and Zettler, 1996; Hsu et al., 2000). Complete genome sequences of the Indonesian isolates of BaMV infecting *B. oldhamii* revealed occurrence of genomic recombination events (Abe et al., 2018). BaMV infection had been induced on *Nicotiana benthamiana* to identify the responsive defence genes. Total eight out of the nine plants demonstrated increased BaMV coat protein. Transcripts associated with cell rescue, defence, death, aging, signal transduction, and energy production were found elevated in all these plants (Cheng et al., 2010).

To combat BAMV mediated loss of bamboo production, plant genetic engineering had been introduced to acquire virus resistance and is mostly based on pathogen derived resistance. For instance, when interfering satBaMV was incorporated into

N. benthamiana and *A. thaliana*, all transgenic plants showed resistance to BaMV (Lin et al., 2013). Another study found that expression of genes such as GDSL-like lipase (Gds 1-ll), Myb4-like (Myb4l), pectinesterase 53-like (Pec53), cinnamyl alcohol dehydrogenase 5 (Cad5), and peroxidase 16 (Pod16) were higher after being infected with *Arthrinium phaeospermum* and these genes may be involved in the resistance response of *B. pervariabilis* × *D. grandis* (Luo et al., 2021).

9.9 The Bamboo Resistance (R)-gene Family and their Potential in Marker Assisted Selection

Plants are often exposed to pathogenic threats and, in response, have evolved an advanced innate immune system that enables them to resist pathogenic invasion. This immune reaction is a function of gene interaction between the host resistance (R) genes and cognate pathogenic avirulence (AVR) genes. Successful recognition of the pathogenic AVR results in induction of a plethora of downstream immune reactions triggering effector-triggered immunity (ETI) (Jones and Dangl, 2006). The nucleotide binding site-leucine rich repeat (NBS-LRR) proteins are the largest family of plant R-proteins, with the LRR domain engaged in pathogen recognition and the NBS domain initiate downstream signaling cascades (DeYoung and Innes, 2006). Activation of the R-genes eventually culminates in defence responses through induction of hypersensitive reactions and programmed cell death. Therefore, characterization of the R-genes could provide valuable insights into the complex molecular events underlying host pathogen interaction. Although characterization of the monocot and eudicot R-genes revealed a great diversity in the NBS-LRR gene family, however their characterization in the Bambusoideae subfamily is still warranted. Among the woody bamboo species, the Moso bamboo genome revealed a total of 344 NBS-LRR genes that comprise 343 CNL (CC-NBS-LRR) and a sole RNL (RPW8-NBS-LRR) member (Shao et al., 2016). Interestingly, TNL (TIR-NBS-LRR), the third group of NBS-LRR genes, is completely absent in the Moso bamboo genome. This is consistent with other monocot species such as rice and maize, suggesting an ancient gene-loss event following the divergence of monocot from the dicots (Shao et al., 2016). However, the recently sequenced genomes of four other bamboo species, *Bonia amplexicaulis*, *Guadua angustifolia*, *Olyra latifolia*, and *Raddia guianensis* may reveal the interspecific variability of the R-gene repertoire. Moreover, the polyploid genomes of *Bonia amplexicaulis* (hexaploid) and *Guadua angustifolia* (tetraploid) may unravel the complex pattern of R-gene expansion and reticulate evolution within the Bambusoideae subfamily.

In a separate study, analyses of 427 Moso bamboo genomes revealed high-frequency heterozygous SNPs to be the greatest contributor underpinning the variation of the R-gene member RPM1 (Zhao et al., 2021). As evidenced in other plants, a unique balancing selection of the R-gene members was observed that explains their diversity within a bamboo population and may serve as an evolutionary basis for the parallel hosts-pathogen coevolution.

Gao et al. (2011) characterized EST sequences of Ma bamboo (*Dendrocalamus latiflorus*) through construction of a cDNA library. Functional annotation of the 891 unigenes through GO and KEGG analyses revealed a significant share

(67 unigenes, 7.5%) to be involved in the host defence system. Members of the RPP (recognition of *Phytophthora parasitica*) group, RGA (resistance gene analogues), and RDL/RF (resistance to dieldrin/resistant factor) are among the most prominent R-gene members. Genotype-specific pathogenic recognition is a critical feature of the RPP group, which is represented by 24 members having a putative RPP like protein features. Additionally, members of the RGAs and RDLs groups were also enriched in the unigene collection that could serve as a valuable resource for marker-assisted resistance breeding. Considering the susceptibility of bamboo species to various pathogenic viruses including bamboo mosaic virus (BaMV), identification of putative R-genes might lead to a permanent solution and stabilize its commercial exploitation (McDowell and Woffenden, 2003).

Li-Xia et al. (2018), in their attempt to clone and characterize the RGA members, reported 33 RGA members in the black bamboo, *Phyllostachys nigra*. As per maximum likelihood phylogeny, the members were categorized into 10 distinct clades revealing the variation within the R-gene family. Analysis of selection pressure upon the 33 identified members demonstrated that most of them underwent positive selection. Interestingly, the positively selected sites appeared to be clustered more in the LRR-domain compared to the NBS-domain which can be functionally correlated with the LRR domain indicating specificity of pathogenic recognition. Conversely, amino acid substitutions within the NBS-domain, might lead to a loss of disease resistance resulting in gene elimination.

Whole genome sequencing provides a wealth of genomic data that not only allows improvement of novel features, but also aids in breeding operations through genomic selection (GS) or marker-assisted selection (MAS). Molecular markers based on genetic composition are distinguishing features of an organism that can be employed to isolate them from the rest of the population for a target trait and can be used to manipulate plant species for higher yields, better quality, and tolerance to biotic and abiotic stresses. With the availability of bamboo genome sequences, appropriate analysis of this massive resource may provide new insights into the development of improved cultivars. Despite prominence of the Bambusoideae subfamily as an important forest resource from econometric point of view, improvement of varieties through GS or MAS has largely been ignored. Considering the huge repertoire of R-genes harbored by the bamboo species, large scale selection programs can now be implemented for their resistance against biotic threats.

9.10 Conclusion and Future Perspective

Bamboos are one of the important lignocellulosic-biomass resources, known for their rapid vegetative growth. They are important grass family members which popularly cultivated for their extensive values in different sectors like food, fodder, pulp, bioenergy production, etc. Due to the susceptibility of various major fungal, bacterial, and pest pathogens, different parts of the bamboo-like root, rhizome, shoot, leaf, etc., are severely damaged, affecting their growth rate and development, which ultimately leads to huge economic growth losses throughout the world. Chemical control is mainly effective against major fungal and bacterial diseases, whereas bio-control measures have been found effective against pests. Beside this, many minor fungal and viral diseases can also be controlled by adopting good cultivation practices as well as better silviculture practices. Phytohormones like SA, MeJA, and ABA were found active against bamboo mosaic virus and thus constitute an important part in the biochemical defense of bamboo (Huang et al., 2021). In addition, gene candidates such as LBD (Huang et al., 2021), RGA (Li-Xia et al., 2018), Gdsl-ll, Myb4l, Pec53, Cad5, and Pod16 (Luo et al., 2021) have been found involved in the molecular defense responses of bamboo. Despite the enormous importance of R-genes in plant disease resistance, their characterization in the Bambusoideae subfamily remains broadly neglected. Considering the enormous collection of R-genes present in polyploid bamboo genomes, multi-institutional research programmes need to be implemented to harvest the full potential of this enormous genetic resource.

Acknowledgments

The authors of this book chapter acknowledge research fundings funded by the Council of Scientific and Industrial Research (CSIR), India [38(1386)/14/EMR-II], [38(1493)/19/EMR-II], Department of Biotechnology, India (BT/PR10778/PBD/16/1070/2014, BT/PR28859/FCB/125/3/2018, BT/INF/22/SP45088/2022), and FRPDF grant of Presidency University and Alexander von Humboldt Foundation, Germany.

References

Abe, S., Neriya, Y., Noguchi, K., Hartono, S., Sulandari, S., Somowiyarjo, S., Ali, A., Nishigawa, H., and Natsuaki, T. (2018). First report of the complete genomic sequences from Indonesian isolates of bamboo mosaic virus and detection of genomic recombination events. *J. Gen. Plant Pathol.*, 85: 158–161.

Adams, M.J., Candresse, T., Hammond, J., Kreuze, J.F., Martelli, G.P., Namba, S., Pearson, M.N., Ryu, K.H., Saldarelli, P. and Yoshikawa, N. (2012). Family betaflexiviridae. pp. 920–941. *In*: King, A.M.Q., Adams, M.J., Carstens, E.B. and Lefkowitz, E.J. (eds.). *Virus Taxonomy: Ninth Report of the International Committee on Taxonomy of Viruses.* Elsevier Academic Press, London.

Akinlabi, E.T., Anane-Fenin, K. and Akwada, D.R. (2017). *Bamboo: The Multipurpose Plant.* Springer.

Alazem, M., He, M.H., Moffett, P. and Lin, N.S. (2017). Abscisic acid induces resistance against bamboo mosaic virus through Argonaute 2 and 3. *Plant Physiol.*, 174(1): 339–355.

Awasthi, P., Ram, R., Zaidi, A.A., Prakash, O., Sood, A., Hallan, V. and Hantula, J. (2014). Molecular evidence for bamboo as a new natural host of cherry necrotic rusty mottle virus. *Pathol.*, 45(1): 42–50.

Banerjee, S., Basak, M., Dutta, S., Chanda, C., Dey, A. and Das, M. (2021). Ethnobamboology: Traditional uses of bamboos and opportunities to exploit genomic resources for better exploitation. pp. 313–352. *In*: Ahmad, Z., Ding, Y., Shahzad, A. (eds.). *Biotechnological Advances in Bamboo: The "Green Gold" on the Earth.* Springer, Singapore.

Basak, M., Dutta, S., Biswas, S., Chakraborty, S., Rahaman, T., Sarkar, A., Dey, S., Biswas, P. and Das, M. (2021). Genomic insights into growth and development of bamboos: What have we learnt and what more to discover? *Trees*, 35: 1771–1791.

Bhardwaj, P., Awasthi, P., Prakash, O., Sood, A., Zaidi, A.A. and Hallan, V. (2017). Molecular evidence of natural occurrence of apple stem grooving virus on bamboos. *Trees*, 31(1): 367–375.

Biswas, S., Rahaman, T., Gupta, P., Mitra, R., Dutta, S., Kharlyngdoh, E., Guha, S., Ganguly, J., Pal, A. and Das, M. (2022). Cellulose and lignin profiling in seven, economically important bamboo species of India by anatomical, biochemical, FTIR spectroscopy and thermogravimetric analysis. *Biomass Bioenergy*, 158: 106362.

Chang, C.H., Hsu, F.C., Lee, S.C., Lo, Y., Wang, J.D., Shaw, J. et al. (2016). The nucleolar fibrillarin protein is required for helper virus-independent long-distance trafficking of a subviral satellite RNA in plants. *Plant Cell*, 28: 2586–2602.

Chen, H.C., Hsu, Y.H. and Lin, N.S. (2007). Downregulation of *Bamboo mosaic virus* replication requires the 5' apical hairpin stem loop structure and sequence of satellite RNA. *Virology*, 365: 271–284.

Cheng, S.F., Huang, Y.P., Wu, Z.R., Hu, C.C., Hsu, Y.H. and Tsai, C.H. (2010). Identification of differentially expressed genes induced by Bamboo mosaic virus infection in *Nicotiana benthamiana* by cDNA-amplified fragment length polymorphism. *BMC Plant Biol.*, 10(1): 1–12.

Cui, X.W., Zhang, Y., Qi, F.Y., Gao, J., Chen, Y.W. and Zhang, C.L. (2013). Overexpression of a Moso bamboo (*Phyllostachys edulis*) transcription factor gene PheWRKY1 enhances disease resistance in transgenic *Arabidopsis thaliana*. *Botany*, 91(7): 486–494.

Das, M., Bhattacharya, S., Singh, P., Filgueiras, T.S. and Pal, A. (2008). Bamboo taxonomy and diversity in the era of molecular markers. *Adv. Bot. Res.*, 47: 225–268.

Deng, S., Shu, J.P. and Wang, H.J. (2010). Investigation of host range of wireworms (*Melanotus cribricollis*) and their spatial distribution in soil. *Chin. Bull. Entomol.*, 47(5): 983–987 (in Chinese with English abstracts).

DeYoung, B.J. and Innes, R.W. (2006). Plant NBS-LRR proteins in pathogen sensing and host defence. *Nat. Immunol.*, 7: 1243–1249.

Elliott, M.S. and Zettler, F.W. (1996). Bamboo mosaic virus detected in ornamental bamboo species in Florida. *Proc. Fla. State. Hort. Soc.*, 109: 24–25.

Erakhrumen, A.A. (2012a). Evaluating the efficacy of neem (Azadirachta indica A. Juss) seed oil treatment for Bambusa vulgaris Schrad. against *Pycnoporus sanguineus* (L. ex Fr.) Murr. using static bending strength properties. *Forest Pathology*, 42: 191–198.

Erakhrumen, A.A. (2012b). Absorption of neem (*Azadirachta indica*) seed oil by split-bamboo (Bambusa vulgaris) at different temperature regimes and treatment durations. *Floresta*, 42(2): 231–241.

Gao, Z.M., Li, C.L. and Peng, Z.H. (2011). Generation and analysis of expressed sequence tags from a normalized cDNA library of young leaf from Ma bamboo (*Dendrocalamus latiflorus* Munro). *Plant. Cell. Rep.*, 30: 2045.

Gautam, A.K. and Avasthi, S. (2018). A new record to rust fungi of Northwestern Himalayas (Himachal Pradesh), India. *Studies in Fungi*, 3(1): 234–240.

Hapukotuwa, N.K. and Grace, J.K. (2011). Comparative study of the resistance of six Hawaii-grown bamboo species to attack by the subterranean termites *Coptotermes formosanus* Shiraki and *Coptotermes gestroi* (Wasmann) (Blattodea: Rhinotermitidae). *Insects*, 2(4): 475–485.

Harsh, N.S.K., Singh, Y.P., Gupta, H.K., Mushra, B.M., McLaughlin, D.J. and Dentinger, B. (2005). A new culm rot disease of bamboo in India and its management. *Journal of Bamboo and Rattan*, 4(4): 387–398.

Hsu, Y.H., Annamalai, A.P., Lin, C.S., Chen, Y.Y., Chang, W.C. and Lin, N.S. (2000). A sensitive method for detecting bamboo mosaic virus (BaMV) and establishment of BaMV-free meristem tip cultures. *Plant. Pathol.*, 49: 101–107.

Huang, B., Huang, Z., Ma, R., Ramakrishnan, M., Chen, J., Zhang, Z. and Yrjälä, K. (2021). Genome-wide identification and expression analysis of LBD transcription factor genes in Moso bamboo (*Phyllostachys edulis*). *BMC Plant Biol.*, 21(10): 1–22.

Hull, R. (2013). *Plant Virology*. (5th edn.). Academic Press, New York, 1118 pp.

Jones, J.D. and Dangl, J.L. (2006). The plant immune system. *Nature*, 444: 323–329.

Ju, Z., Zhan, T., Cui, J., Brosse, N., Zhang, H., Hong, L. and Lu, X. (2021). Eco-friendly method to improve the durability of different bamboo (*Phyllostachys pubescens*, Moso) sections by silver electrochemical treatment. *Industrial Crops and Products*, 172: 113994.

Jung, H.Y., Chang, M.U., Lee, J.T. and Namba, S. (2006). Detection of 'Candidatus Phytoplasma asteris' associated with henon bamboo witches-broom in Korea. *J. Gen. Plant Pathol.*, 72(94): 261–263.

Kaur, P.J., Satya, S., Pant, K.K. and Naik, S.N. (2016). Eco-friendly preservation of bamboo species: Traditional to modern techniques. *BioRes.*, 11(4): 10604–10624.

Komatsu, K., Hirata, H., Fukagawa, T., Yamaji, Y., Okano, Y. et al. (2012). Infection of capilloviruses requires subgenomic RNAs whose transcription is controlled by promoter-like sequences conserved among flexiviruses. *Virus Res.*, 167: 8–15.

Koshy, K.C., Harikumar, D. and Narendran, T.C. (2001). Insect visits to some bamboos of the Western Ghats, India. *Curr. Sci.*, 81(7): 833–838.

Kumar, S., Shukla, K.S., Dev, I. and Dobriyal, P.B. (1994). *Bamboo Preservation Techniques: A Review. INBAR Technical Report No. 3*. INBAR, New Delhi and ICFRE, Dehradun, 59 pp.

Li, K.S. (2006). Effects of Sendebao and *Bacillus thuringiensis* on controlling *Algedonia coclesalis*. *J. Zhejiang For Coll.*, 23(4): 445–448 (in Chinese with English abstracts).

Lin, K.Y., Hsu, Y.H., Chen, H.C. and Lin, N.S. (2013). Transgenic resistance to bamboo mosaic virus by expression of interfering satellite RNA. *Mol. Plant Pathol.*, 14(7): 693–707.

Lin, N.S. and Hsu, Y.H. (1994). A satellite RNA associated with bamboo mosaic potexvirus. *Virology*, 202(2): 707–714.

Lin, N.S., Lin, B.Y., Lo, N.W., Hu, C.C., Chow, T.Y. and Hsu, Y.H. (1994). Nucleotide sequence of the genomic RNA of bamboo mosaic potexvirus. *J. Gen. Virol.*, 75: 2513–2518.

Lin, Q.Y. (2001). Technology of integrated measures to control die-back of *Phyllostachys edulis*. *J. Nanjing Univ. Nat.*, 25(1): 39–43.

Lister, R.M. (1970). Apple stem grooving virus. *In: CMI/AAB Descriptions of Plant Viruses No. 31*. ATCC, Virginia, U.S.A.

Li-Xia, Y., Jun-Jun, X., Qiong, B., Ya-ping, Z., Bin, L. and Bo, Y. (2018). Cloning and molecular evolution analysis of NBS class resistance gene analogs in black bamboo (*Phyllostachys nigra*). *Silvae Genet.*, 67(1): 117–123.

Luo, F., Fang, X., Liu, H., Zhu, T., Han, S., Peng, Q. and Li, S. (2021). Differential transcriptome analysis and identification of genes related to resistance to blight in three varieties *of Bambusa pervariabilis × Dendrocalamopsis grandis*. *PeerJ.*, 9: e12301.

Ma, L.J., Chi, W.Y., Huang, D.R., Zheng, W.J. and Xu, Y.B. (2005). Pathogenesis of leaf rust of *Dendrocalamopsis oldhami [Bambusa oldhamii]* and selection of its control chemicals. *Journal of Zhejiang Forestry College*, 22(1): 66–69.

Ma, Y.L., Wei, C.Y. and Li, J.Z. (2008). Study on infection of wilt to hybrid bamboo of *Bambusa pervariabilis × Dendrocalamopsis daii*. *Journal of Zhejiang Forestry Science and Technology*, 28(5): 29–32.

Mathew, G. and Nair, K.S.S. (1988). Storage pests of bamboos in Kerala. pp. 212–214. *In*: Rao, I.V.R., Gnanaharan, R. and Sastry, C.B. (eds.). *Bamboos: Current Research*. International Bamboo Workshop, Cochin 14–18 November 1988.: Proceedings edited by. Peechi, KFRI.

McDowell, J.M. and Woffenden, B.J. (2003). Plant disease resistance genes: Recent insights and potential applications. *Trends Biotechnol.*, 21(4): 178–83.

Mohanan, C. (1997). *Diseases of Bamboos in Asia: An Illustrated Manual*. INBAR, New Delhi. 228 pp.

Mohanan, C. (2002). *Diseases of Bamboos in Asia: An Illustrated Manual. INBAR Technical Report; 10*. BRILL, California, 228 pp.

Mohanan, C. (2010). *Rust Fungi of Kerala. KFRI Handbook No. 26*. Peechi, KFRI, 148 pp.

Mohanan, C. (2017). *Diseases of Bamboos in Asia: An illustrated Manual*. International Network for Bamboo and Rattan (INBAR).

Omyama, A.-R., Mohamd, A.-L., Walter, L. and Norini, H. (1995). *Planting and Utilization of Bamboo in Peninsular Malaysia, Research Pamphlet No. 118*. Forest Research Institute of Malaysia (FRIM), 11 pp.

Qiao, T.M., Zhu, T.H. and Li, S.J. (2011). Colonization of pseudomonas aeruginosa ZB27 and its control effect on hybrid bamboo blight. *Acta. Phytophylacica. Sin.*, 38(2): 133–138.

Rott, M.E. and Jelkmann, W. (2001). Complete nucleotide sequence of cherry necrotic rusty mottle virus. *Arch. Virol.*, 146: 395–401.

Shao, Z.Q., Xue, J.Y., Wu, P., Zhang, Y.M., Wu, Y., Hang, Y.Y., Wang, B. and Chen, J.Q. (2016). Large-scale analyses of angiosperm nucleotide-binding site-leucine-rich repeat genes reveal three anciently diverged classes with distinct evolutionary patterns. *Plant Physiol.*, 170(4): 2095–2109.

Shanbhag, R.R., Jagadish, M.R., Dhanya, B. and Viswanath, S. (2010). Assessment of termite infestation in six industrially important bamboo species in semiarid tracts of Karnataka, India. *Journal of Bamboo and Rattan*, 9(3-4): 109–113.

Sharada, P., Nagaveni, H.C., Remadevi, O.K. and Jain, S.H. (2013). Toximetric studies on some major bamboo pathogens. *Indian For.*, 139(9): 814–820.

Shu, J. and Wang, H. (2015). Pests and diseases of bamboos. *In*: Liese, W. and Köhl, M. (eds.). *Bamboo: Tropical Forestry*, Vol. 10. Springer, Cham. https://doi.org/10.1007/978-3-319-14133-6_6.

Singh, S.R., Singh, R., Kalia, S., Dalal, S., Dhawan, A.K. and Kalia, R.K. (2013). Limitations, progress, and prospects of application of biotechnological tools in improvement of bamboo: A plant with extraordinary qualities. *Physiol. Mol. Biol. Plants*, 19: 21–41.

Sun, P.L., Lu, G., Chen, W.M., Huang, Z.G. and Xue, W.Y. (2007). Field trials of emulsifiable suspensions of Beauveria bassinaconidia for control aphid in shoot bamboo. *J. Zhejiang Univ. (Agric. Life Sci.)*, 33(2): 197–201 (in Chinese with English abstracts).

Verma, R.V. and Sajeev, T.V. (2012). Insect pest of bamboo in India. *INVIS Bulletin*, 227–246.

Wang, I.N., Hu, C.C., Lee, C.W., Yen, S.M., Yeh, W.B. and Hsu, Y.H. et al. (2014). Genetic diversity and evolution of satellite RNAs associated with the bamboo mosaic virus. *PLoS One*, 9: e108015.

Wang, P., Zhang, Y.B., Shu, J.P., Deng, S. and Wang, H.J. (2010). Virulence of Metarhizium anisopliae var. anisopliae to the larvae of *Melanotus cribricollis* (Coleoptera: Elateridae). *Chin. J. Biol. Control*, 26(3): 274–279 (in Chinese with English abstracts).

Wu, J.Q. (2009). Experimentation of various application methods of *Beauveria bassiana* on controlling *Pantana phyllostachysae*. *Sci. China. Tech.*, 23(4): 101–105.

Wu, Z., Huang, D., Wei, W., Wang, W., Wang, (Alice) X., Wei, Q., Niu, M., Lin, M., Rao, J. and Xie, Y. (2019). *Mesoporous aluminosilicate* improves mildew resistance of bamboo scrimber with Cu B P anti-mildew agents. *Journal of Cleaner Production*, 209: 273–282.

Xu, G.Q., Wang, L.H., Liu, J.L. and Hu, S.H. (2013). Decay resistance and thermal stability of bamboo preservatives prepared using camphor leaf extract. *Int. Biodeterior. Biodegrad.*, 78: 103e107.

Xu, M., Fan, S., Jin, L., Lu, Q., Tian, G. and Wang, L. (2006). Records of bamboo diseases and the taxonomy of their pathogens in China (I). *Forest Research, Beijing*, 19(6): 692–699.

Xu, Q.F., Liang, C.F., Chen, J.H., Li, Y.C., Qin, H. and Fuhrmann, J.J. (2020). Rapid bamboo invasion (expansion) and its effects on biodiversity and soil processes. *Global Ecology and Conservation*, 21: p.e00787.

Xu, T.S. and Wang, H.J. (2004). *Main Pests of Bamboo in China*. Chinese Forest Publish House, Beijing, pp. 45–46.

Yeh, W.B., Hsu, Y.H., Chen, H.C. and Lin, N.S. (2004). A conserved secondary structure in the hypervariable region at the 5' end of *Bamboo mosaic virus* satellite RNA is functionally interchangeable. *Virology*, 330: 105–115.

Ye, L.Q., Wu, X.Q. and Wang, L. (2012). Study *Acremonium salmoneum* of bamboo leaf rust. *Journal of Nanjing Forestry University (Natural Sciences Edition)*, 36(2): 64–68.

Ye, L.Q., Wu, X.Q. and Ye, J.R. (2011). Mycoparasitic activities of *Acremonium salmoneum* against bamboo leaf rust. *Chinese Journal of Biological Control*, 27(4): 528–534.

Zhang, Y.M., Chen, M., Sun, L., Wang, Y., Yin, J., Liu, J., Sun, X.Q. and Hang, Y.Y. (2020). Genome-wide identification and evolutionary analysis of NBS-LRR genes from *Dioscorea rotundata*. *Front. Genet.*, 11: 484.

Zhang, Z., Huang, B., Chen, J., Jiao, Y., Guo, H., Liu, S., Ramakrishnan, M. and Qi, G. (2022). Genome-wide identification of JRL genes in Moso bamboo and their expression profiles in response to multiple hormones and abiotic stresses. *Front. Plant Sci.*, 12.

Zhao, H.S., Sun, S., Ding, Y.L., Wang, Y., Yue, X.H., Du, X., Wei, Q., Fan, G.Y., Sun, H.Y., Lou, Y.F. et al. (2021). Analysis of 427 genomes reveals Moso bamboo population structure and genetic basis of property traits. *Nat. Commun.*, 12: 5466.

Zhou, C.L., Wu, X.Q., Ji, J. and Ye, J.R. (2011). Occurrence situation and control countermeasures of bamboo diseases in Nanjing. *J. Nanjing Forestry Univ. Nat. Sci.*, 35(1): 127–131.

Chapter 10

Genome Annotation, *In silico* Tools and Databases with Special Reference to Moso Bamboo

Hansheng Zhao,[1], Lianfu Chen,[1] Yu Wang,[2] Yinguang Hou,[2]
Lei Sun,[2] Junwei Gan,[2] Zeyu Fan[2] and Shanying Li[2]*

10.1 Introduction

The first complete plant genome sequence, *Arabidopsis thaliana*, was reported and published in 2000, marking the beginning of the plant genome era. Over the past 20 years, draft or reference genomes of approximately 800 plant species have been sequenced and the number continues to grow exponentially. The release of these plant genomes has dramatically advanced studies in various disciplines of plant biology. With tremendous advances in sequencing technology and the initiation of many genome sequencing and updating projects, improving the accuracy of the genome annotation stage remains challenging. This is because only a few genomes are available as annotation standards achieved through enormous investments in human curation efforts. Therefore, high-quality genome annotation, as one way of summarizing the existing knowledge of the genomic characteristics of an organism, will be one of the principal research focuses for the plant genome. In addition, genome annotation will play an essential role in some challenging fields in the future, i.e., dissecting information from plant pan-genomes. Here, we summarize the progress of plant genome annotation and the challenges of annotating more complex plant genomes and present state-of-the-art tools and databases for this process.

[1] Professor, International Centre for Bamboo and Rattan, Beijing, China.
[2] International Centre for Bamboo and Rattan, Beijing, China.
* Corresponding author: zhaohansheng@icbr.ac.cn

10.2 Genome Annotation

Genome annotation refers to characterizing the genome, which includes annotations of gene structure and functions at the genome-wide scale. Structural annotation mainly involves identifying repeating sequences on a genome and identifying coding genes (exons, introns, non-coding regions, etc.). Based on the structural annotation, functional annotation involves predicting the functions of genes based on the sequences of the coding genes. Therefore, structural annotation is a prerequisite for functional annotation, and structural annotation serves the purpose. Functional annotation is the basis and reference for functional validation and knowledge discovery. As a result of genome annotation, a connection is established between the sequence of a genome and its biological characteristics. Genome annotation is the first step in analyzing a sequenced species' genome. Data and methods for genome annotation have been improving in recent years, and researchers are increasingly relying on genome annotation information in all aspects of biological research. Therefore, any subsequent biological studies can become inaccurate if the genome annotation results are incomplete. Thus, genomic annotation information for some species (humans, mice, rice, bamboo, etc.) is regularly updated. Accurate and reliable genome annotation information is crucial to a variety of genome-based research studies.

10.2.1 Evaluation of Genome Assembly Quality

A high-quality genome is a prerequisite for genome annotation. Therefore, after the genome assembly has been completed, the first step is to assess the quality of the assembly. We anticipate that each contig will have enough length to contain more complete gene structures. Therefore, this principle may be summarized as the '3Cs':

- Continuity: the obtained contigs should be sufficiently long.
- Correctness: the error rate for assembled contigs is relatively low.
- Completeness: containing as much of the complete gene as possible.

At present, multiple evaluative indicators can be used to describe the integrity and continuity of genome assembly, among which the best known is the contig N50 value. The N50 is like the mean or median but with greater weight than the longer contigs. It is generally accepted that the longer the N50 is, the better the assembly result. Particularly, when the genome assembly is based on short reads, and the N50 length of the assembly is greater or equal to the average gene length of the species, this indicates good assembly quality, and further annotation work can be undertaken. If genomes are assembled based on long enough reads (e.g., PacBio HiFi), chromosome-level N50 data can usually be obtained.

Additionally, some metrics (e.g., BUSCO (Seppey et al., 2019), LAI (Ou et al., 2018)) can be utilized as complementary methods to the N50 to evaluate the quality of genome assembly. For example, BUSCO compares the assembled genome sequence with a set of lineage-specific single-copy genes to determine the percentage of single-copy or duplicated genes that comprise the sequence, thereby assessing the quality of the genome assembly and continuity. As a result, the genome assembly may not be complete, or the N50 may be too low, which means more reads must be added to improve genome assembly results to facilitate follow-up work, such as genome annotation.

10.2.2 Genome Structure Annotation

Genome-wide structural genome annotation refers to identifying all the gene's positions and structures within the genome using bioinformatics methods. We obtain final annotation results by integrating multiple methods (*ab initio* prediction, homology annotation, transcriptome annotation, etc.).

10.2.2.1 Repetition Sequence Identification

After confirming that the genome quality meets the annotation requirements, the first step is to identify repetitive sequences. Repeat sequences can be classified into two categories: tandem repeat and interspersed repeat. Tandem repeats include microsatellite sequences, satellite sequences, etc., and are often used to create genetic biomarkers. Interspersed repeats, also known as transposons, are divided into DNA transposons and RNA retrotransposons and play an essential role in regulating gene expression. Repeated sequence identification can be divided into sequence alignment and *ab initio* prediction. The sequence alignment method identifies homologous repeats within the genome based on their degree of similarity. The results predicted by this method are frequently accurate but not comprehensive. Among the most widely used alignment prediction software is Repeatmasker (Tempel, 2012). In addition, the *ab initio* method utilizes structural characteristics to predict repeats within the genome. A high level of performance is achieved when the *ab initio* method is applied to predict sequences with prominent structural elements, such as MITEs (miniature inverted-repeat transposable elements) and LTRs (long terminal repeats). The software Repeatmodeler (Flynn et al., 2020) is often used for *ab initio* methods-based prediction. RepeatModeler2 integrates the LTR prediction module to facilitate the enhancement of predictive accuracy. Repeats are identified by combining homology alignment and *ab initio* prediction.

Critical technical challenges in repeat sequence annotation: (1) Due to the short reads of next-generation sequencing and the k-mer algorithm employed in genome assembly, highly similar repeat sequences may be compressed together and interfere with the identification of subsequent repeat sequences. (2) Some highly repeat sequences are difficult to assemble using existing assembly methods and result in unassembled reads, making it more challenging to identify their repetitive regions.

10.2.2.2 Annotation of the Coding Gene

A gene prediction identifies the positions of genes in batches throughout the genome, i.e., identifying the DNA sequences containing information about genes, such as promoters, start/stop codons, transcription splice sites, etc. Currently, there are two primary methods for identifying the genetic structure information within the genome. As the first type of method, homologous or transcriptome sequences from the same species, or closely related species, are aligned to the newly assembled genome sequence to perform gene identification. The second method uses an advanced mathematical model to *ab initio* predict gene structure. It is based on the built-in parameters of the software (e.g., codon usage frequency, exon-intron length distribution, and other features) to distinguish genes from intergenic regions and determine the exon-intron structure within genes. Using *ab initio* prediction methods, genome annotation can still be performed even if a newly sequenced genome does not contain sufficient evidence.

Nevertheless, the parameters that come with the software are species-specific, and most of them come from the genomes of classic model species. Therefore, gene prediction based on mathematical models will be less accurate if the genome annotated species are distantly related to these model species. As a result, the state-of-the-art approach combines the first and second methods for gene annotation separately and then selects the 'best' annotation after integrating both ways.

10.2.2.3 Release of Data

When genome annotation is complete, the first step is to provide the most comprehensive annotation datasets (including gene exon-intron structure, gene start codon, gene stop codon, gene alternative splicing, etc.) in an appropriate file format (e.g., GFF3/GTF format). Once the genome annotation information has been gathered, submit it to large bioinformatics public databases (such as GenBank and Ensembl), or set up a small database (such as bambooGDB.org (Zhao et al., 2014)) where the results can be shared. As a result, genome annotation can be made available to a broader range of researchers to facilitate research in related areas.

10.2.3 Annotation of Gene Functions

Consequent to obtaining gene structure information, we hope to be able to obtain gene function information. The genome-wide functional annotation is achieved by inferring the function of all genes in a batch. Currently, we mainly use the homologous protein alignment method, in which an unknown protein is compared to proteins retrieved from the database to find homologous proteins. These databases, which provide annotation of protein functions, classifications, and biological pathways, are the most widely used. They include NCBI Nr (Pruitt et al., 2005), UniProtKB/Swiss-Prot (Poux et al., 2017), COG (Tatusov et al., 2003), eggNOG (Huerta-Cepas et al., 2019), InterPro (Finn et al., 2017), Pfam (Mistry et al., 2021), KEGG (Kanehisa et al., 2021), and Gene Ontology (The Gene Ontology resource: enriching a GOld mine 2021). Currently, we make use of four commonly used databases for functional annotation.

These databases include UniProtKB/Swiss-Prot protein sequence databases, KEGG biological pathway databases, Interpro protein family databases, and Gene Ontology gene function annotation databases.

(1) Align the sequences with the UniProtKB/Swiss-Prot databases to obtain preliminary sequence information.
(2) Compare with the KEGG database to predict the biological pathways that the protein might represent.
(3) Aligning the protein sequence with the Interpro database enables the identification of conserved sequences, motifs, and domains.

(4) Predict the function of the protein. Interpro has established an interaction system with Gene Ontology, which records the correspondence between each protein family and the function nodes in Gene Ontology. Through this system, we can predict the biological function of the protein.

Nevertheless, the functional annotation relies on alignment, which will result in two significant challenges. Firstly, this approach relies on external data and may be too restrictive for some species with less-studied evolution. Secondly, sequence similarity does not necessarily translate into a similar biological function. Consider introducing methods other than sequence alignment to improve functional annotation further.

10.3 Moso Bamboo Genome Annotation

In 2013, our team reported that the genome of Moso bamboo, the first species of the Bambusoideae subfamily, has been completed and published (Peng et al., 2013). In 2018, we released the second version of the Moso bamboo genome (Zhao et al., 2018). Currently, our team works on the assembly and annotation of Moso bamboo using PacBio HiFi reads. We also made an annotation update to the second version of the genome. The results are summarized in Fig. 10.1.

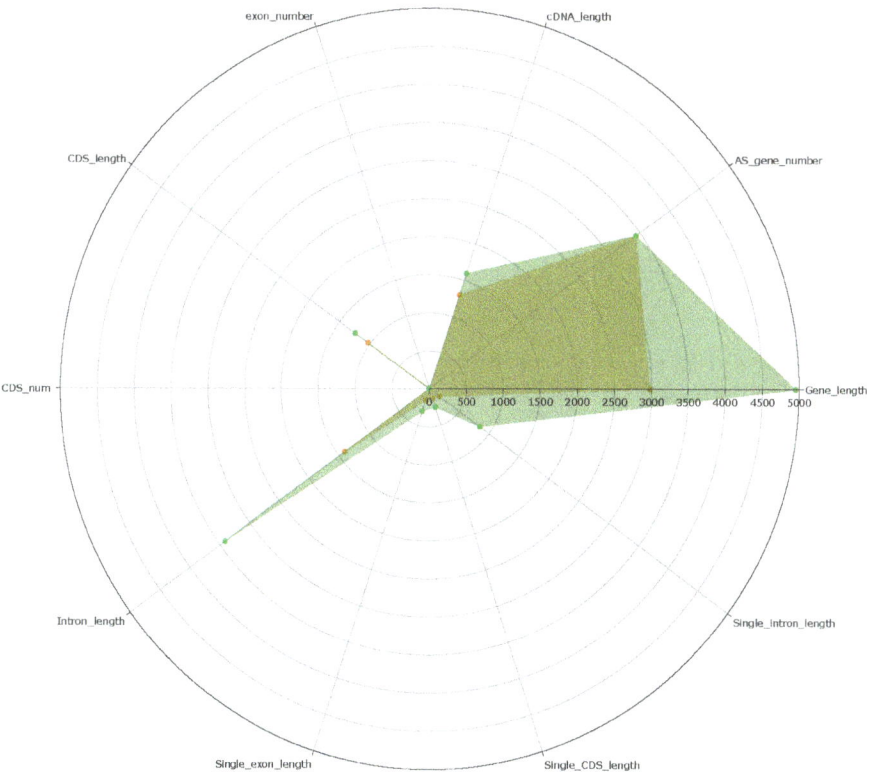

Fig. 10.1. Main feature statistics of genome annotation.

According to the latest Moso bamboo annotation results (which will be released as soon as possible), we have identified 10 aspects, including gene length, single intron length, single CDS length, single exon length, intron length, CDS numbers, CDS length, exon number, cDNA length, AS gene number.

In addition, we have achieved some results in gene structure annotation. Our team has developed a pipeline for genome annotation. In Fig. 10.2, we found that our pipeline performed better than mainstream annotation software (BRAKER2 (Brůna et al., 2021), MAKER2 (Holt and Yandell, 2011), AUGUSTUS (Stanke et al., 2006), etc.). We will continue improving and updating our pipeline to meet the release requirements.

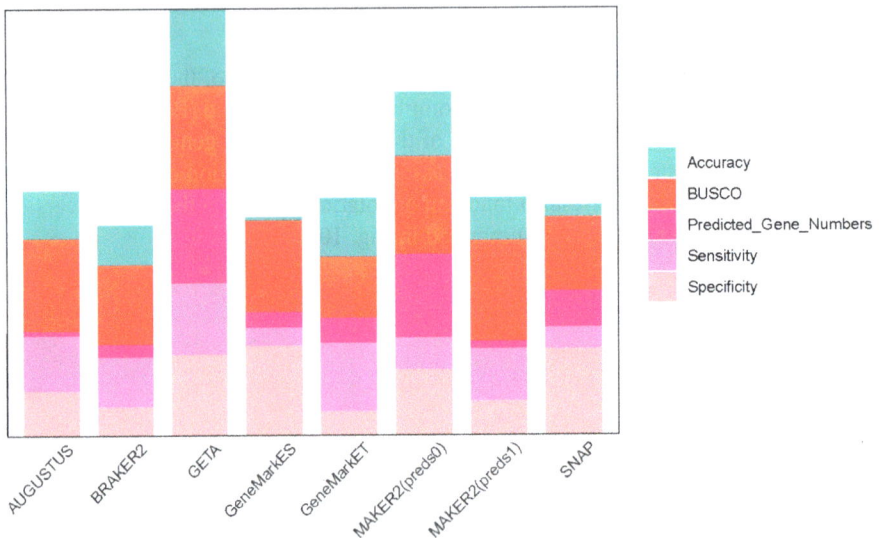

Fig. 10.2. An evaluation of the accuracy of the in-house pipeline and other gene prediction software for rice genome annotation.

We compared seven major software or pipelines for plant genome assembly. These are GETA (software developed by us), BRAKER2, MAKER2 (two different parameters), AUGUSTUS, GeneMarkET, GeneMarkES, SNAP. We used default parameters to assemble the rice genome and evaluated the assembly quality using Sensitivity, Specificity, and BUSCO. Based on the results of the above three levels, GETA shows the best results. For example, according to Sensitivity, the first GETA (46.40%) is 17.18% higher than the second AUGUSTUS (29.52%). According to Specificity, the first GETA (57.43%) is 18.11% higher than the second MAKER2 (preds0) (39.32%). Regarding BUSCO, the first GETA achieved a near-full score of 98.7%, which was 11.1% better than the second MAKER2 (preds1) (87.6%).

Finally, high-quality genome annotation is a crucial piece of information for discovering critical functional genes, identifying essential genes, and breeding different varieties. However, eukaryotic genome annotation cannot be completed once and for all since, with the ongoing development of annotating tools and sequencing technologies, it is necessary to update the existing genome annotations periodically. Consequently, there is a great deal of work to be done regarding the annotation of eukaryotes such as bamboo.

References

Brůna, Tomáš, Katharina J. Hoff, Alexandre Lomsadze, Mario Stanke and Mark Borodovsky. (2021). BRAKER2: Automatic eukaryotic genome annotation with GeneMark-EP+ and AUGUSTUS supported by a protein database. *NAR Genomics and Bioinformatics*, 3(1): lqaa108.

Finn, Robert D., Teresa K. Attwood, Patricia C. Babbitt et al. (2017). InterPro in 2017: Beyond protein family and domain annotations. *Nucleic Acids Research*, 45(D1): D190–D199.

Flynn, Jullien M., Robert Hubley, Clément Goubert et al. (2020). RepeatModeler2 for automated genomic discovery of transposable element families. *Proceedings of the National Academy of Sciences*, 117(17): 9451–9457.

Holt, Carson and Mark Yandell. (2011). MAKER2: An annotation pipeline and genome-database management tool for second-generation genome projects. *BMC Bioinformatics*, 12(1): 1–14.

Huerta-Cepas, Jaime, Damian Szklarczyk, Davide Heller et al. (2019). eggNOG 5.0: A hierarchical, functionally and phylogenetically annotated orthology resource based on 5090 organisms and 2502 viruses. *Nucleic Acids Research*, 47(D1): D309–D314.

Kanehisa, Minoru, Miho Furumichi, Yoko Sato, Mari Ishiguro-Watanabe and Mao Tanabe. (2021). KEGG: Integrating viruses and cellular organisms. *Nucleic Acids Research*, 49(D1): D545–D551.

Mistry, Jaina, Sara Chuguransky, Lowri Williams et al. (2021). Pfam: The protein families database in 2021. *Nucleic Acids Research*, 49(D1): D412–D419.

Ou, Shujun, Jinfeng Chen and Ning Jiang. (2018). Assessing genome assembly quality using the LTR Assembly Index (LAI). *Nucleic Acids Research*, 46(21): e126–e126.

Peng, Zhenhua, Ying Lu, Lubin Li et al. (2013). The draft genome of the fast-growing non-timber forest species Moso bamboo (Phyllostachys heterocycla). *Nature Genetics*, 45(4): 456–461.

Poux, Sylvain, Cecilia N. Arighi, Michele Magrane et al. (2017). On expert curation and scalability: UniProtKB/Swiss-Prot as a case study. *Bioinformatics*, 33(21): 3454–3460.

Pruitt, Kim D., Tatiana Tatusova and Donna R. Maglott. (2005). NCBI Reference Sequence (RefSeq): A curated non-redundant sequence database of genomes, transcripts, and proteins. *Nucleic Acids Research*, 33(suppl_1): D501–D504.

Seppey, Mathieu, Mosè Manni and Evgeny M. Zdobnov. (2019). BUSCO: Assessing genome assembly and annotation completeness. *In: Gene Prediction*. Springer.

Stanke, Mario, Oliver Keller, Irfan Gunduz, Alec Hayes, Stephan Waack and Burkhard Morgenstern. (2006). AUGUSTUS: *Ab initio* prediction of alternative transcripts. *Nucleic Acids Research*, 34(suppl_2): W435–W439.

Tatusov, Roman L., Natalie D. Fedorova, John D. Jackson et al. (2003). The COG database: An updated version includes eukaryotes. *BMC Bioinformatics*, 4(1): 1–14.

Tempel, Sébastien. (2012). Using and understanding RepeatMasker. *In: Mobile Genetic Elements*. Springer.

The Gene Ontology resource: Enriching a GOld mine. (2021). *Nucleic Acids Research*, 49(D1): D325–D334.

Zhao, Hansheng, Zhimin Gao, Le Wang et al. (2018). Chromosome-level reference genome and alternative splicing atlas of Moso bamboo (Phyllostachys edulis). *Gigascience*, 7(10): giy115.

Zhao, Hansheng, Zhenhua Peng, Benhua Fei et al. (2014). BambooGDB: A bamboo genome database with functional annotation and an analysis platform. *Database*, 2014.

Chapter 11

Functional Genomics Study of Bamboo Shoot and Rhizome Development
An Overview

Jiaxiang Zhang,[1,2] *Han Li,*[1,2] *Huiming Xu,*[1] *Xingyan Fang,*[1,2] *Gangjian Cao*[1] *and Liuyin Ma*[1,*]

11.1 Introduction

Bamboo is an important non-timber forest plant, accounting for about 1% of the world forest area, and is known as the "second largest forest" in the world (Mao et al., 2017). Bamboo provides important resources for food, construction, paper and fiber, and thus has important ecological, economic, cultural, and industrial value (Wang et al. 2019). Bamboo can reach a height of 20 m within 45–60 days (Wang et al., 2019). Therefore, deciphering the regulation mechanism of rapid growth of bamboo is one of the most important scientific questions in the field of bamboo research.

Recently, the area of bamboo forests has been increasing at a rate of 3% per year despite the sharp decline in global forest area (Wang et al., 2017). This is partly due to the widespread rhizome development of monopodial bamboo such as Moso bamboo (Wang et al., 2017). The rhizome provides a fast and efficient response system for nutrient transport among different bamboo individuals (Wang et al., 2017; Song et al., 2016). Therefore, widespread rhizome development is another interesting feature of bamboo development. This chapter will systematically summarize the current research progress on gene expression regulation of rapid growth and rhizome development in bamboos.

[1] College of Forestry, School of Future Technology, Fujian Agriculture and Forestry University, Fuzhou, China.
[2] College of Life Sciences, Fujian Agriculture and Forestry University, Fuzhou, China.
* Corresponding author: lma223@fafu.edu.cn

11.2 The Rapid Growth of Bamboo Culms

Since the sequencing of several bamboo genomes completed (Peng et al., 2013; Guo et al., 2019; Zhao et al., 2018; Zheng et al., 2022), dozens of literatures report the rapid growth of bamboo culms (Wang et al., 2019; Wei et al., 2019; Chen et al., 2021). In terms of methodology, these studies mainly focus on introducing multi-omics methods such as transcriptomics and proteomics to decipher the regulatory network of rapid growth (Cui et al., 2012; Li et al., 2018; Wang et al., 2019; Wang et al., 2021). Few literatures have identified and validated the function of bamboo's rapid growth regulatory genes by reverse genetics or dwarf mutants (Wang et al., 2019; Wang et al., 2020; Wei et al., 2017). Therefore, the first section of this chapter will focus on summarizing the transcriptional regulation (hormone responses, transcription factors) and post-transcriptional regulatory networks (kinases, miRNAs, alternative splicing, and *cis*-nature antisense transcripts (*cis*-NATs)) for rapid bamboo growth.

11.3 The Characteristics of Rapid Growth in Bamboos

11.3.1 Rapid-growth Stages

Bamboos take about two months to complete from unearthed to the height of mature bamboo culms (Wang et al., 2019). The growth rate of bamboos is unevenly distributed (Cui et al., 2012). Understanding the definition of bamboo's rapid-growth stages is of great significance to the study of its rapid-growth. To better understand the rapid-growth stages of bamboos, we first summarize the stage classification of rapid-growth. According to the existing research literature on bamboo rapid-growth studies (Table 11.1), the rapid growth of bamboo culms can be divided into three stages: initial, intermediate, and late stages (Cui et al., 2012; Wang et al., 2019). The first stage is the initial stage of rapid growth of bamboos, with a height of bamboo culm less than 1.00 m. The intermediate stage is when the bamboo culms are between

Table 11.1. Sampling height distribution for bamboo rapid-growth studies.

Sampling Height for Rapid Growth	References
0.05, 0.20, 0.50, 1.00, 2.00, 3.00, 6.00, 8.50 and 12.00 m	(Cui et al., 2012; He et al., 2013)
winter bamboo shoot, 0.10, 0.50, 1.00, 3.00, 6.00, 9.00, 12.00 m, and culm after leaf expansion	(Li et al., 2018; Peng et al., 2013)
0.15, 0.50, 1.60, 4.20, 9.00 and 12.00 m	(Wang et al., 2019; Wang et al., 2020; Lin et al., 2021)
0.05, 0.25, 0.35, 0.50 and 0.9 m	(Gamuyao et al., 2017)
0.20, 1.00, 3.00 and 6.70 m	(Xu et al., 2019)
0.35 m	(Guo et al., 2020)
0.20, 0.50, 1.00, 2.00, 3.00, 5.00, 6.00 and 7.00 m	(Wang et al., 2019)
2.30 m	(Chen et al., 2021)
1.60–1.70 m	(Wang et al., 2021)
0.03, 0.06, 0.15, 0.25 m	(Wei et al., 2019)

1.00–6.00 m in height. Bamboo culms higher than 6.00 m are considered to be the later stage of rapid growth (Cui et al., 2012).

11.3.2 Cell Division or Cell Elongation

Plant growth begins from cell division to cell elongation, and finally forms mature cells (Wang et al., 2021). Therefore, cell division and cell elongation are two crucial processes of cell growth. How cell division and cell elongation contribute to the explosive rapid growth of bamboos is an interesting scientific question to be answered? In the initial stage of rapid-growth, the dwarf variant-*Pseudosasa japonica var. tsutsumiana* significantly decreased cell number and reduced the cell elongation in internodes compared to that from the wild-type-*Pseudosasa japonica* based on anatomical analyses (Wei et al., 2017). This result indicates that the rapid growth of bamboo culm requires cell division and cell elongation.

11.3.3 Dynamics of Cell Division and Cell Elongation

The next question is whether the abundance of cell division and cell elongation change dynamically during the rapid growth of bamboos? According to the histological observation of Moso bamboo, a large number of nuclei were present in the parenchyma and fiber cells in the initial stage, and the number of nuclei gradually decreased until they almost disappeared in the late stage (Cui et al., 2012). In contrast, the length of parenchyma and fiber cells gradually increased from the initial to the late stages (Cui et al., 2012). Since the number of nuclei is related to cell division, and cell length is a direct indicator of cell elongation, cell division is the major contributor to the initial stage of bamboo culm development, while cell elongation is dominant in the intermediate and late stages of bamboo culm development. Furthermore, cells are lignified at late stage (Cui et al., 2012; Li et al., 2018). Therefore, the rapid growth of bamboo culm is the result of dynamic changes among cell division, cell elongation, and cell wall modification. In summary, cell division plays a major role in the early stage, cell elongation mainly occurs in the intermediate and late stages, and cell wall modification function on cell maturation in the late stage. Therefore, cell division, cell elongation, and cell wall modification all play important roles in the rapid growth of bamboo.

11.4 Transcriptional Regulation of Rapid Growth in Bamboos

11.4.1 Auxin and BR Responses

Auxin influences stem elongation and regulates the formation, activity, and fate of meristems, and has therefore been recognized as a major hormone shaping plant architecture (Gallavotti, 2013). The expression of auxin signaling genes was upregulated during the rapid growth of bamboos (Li et al., 2018). *AUX/LAX* (Auxin transporter-like protein) family genes encode carrier proteins involved in proton-driven auxin influx and are key regulatory genes in auxin pathway (Péret et al., 2012). Five *AUX/LAX* genes were up-regulated in the rapid growth of bamboo culms, and increased *AUX/LAX* activity enhances auxin-dependent cell separation

(Li et al., 2018). The small auxin upregulated RNA (*SAUR*) family genes has been proposed as a bridge linking auxin-mediated acid growth and cell elongation (Stortenbeker and Bemer, 2019). Therefore, *SAUR* expression has long been used as an indicator of auxin-dependent cell elongation (Stortenbeker and Bemer, 2019). In Moso bamboo, the expression of *SAUR* was lower in the initial stage, but highly expressed in the intermediate and late stages of rapid growth of bamboo culms (Li et al., 2018). Therefore, similar to Arabidopsis, the expression pattern of bamboo *SAUR* is positively correlated with its function on enhancing cell elongation (Li et al., 2018). These results suggest that auxin signaling regulates rapid growth by controlling the expression of *AUX/LAX* and *SAUR* genes. Brassinosteroids (BRs) are essential steroid hormones that control cell division, elongation, and differentiation, and are therefore required for plant growth and development (Planas-Riverola et al., 2019). The poaceae-specific Brassinosteroids response gene 1 (*PSBR1*) is identified in Moso bamboo, the expression of *PSBR1* is negatively regulated by BR, and over-expression of *PSBR1* in Arabidopsis significantly inhibits the height of transgenic Arabidopsis (Guo et al., 2020). Therefore, bamboo BRs may be involved in the regulation of plant height.

11.4.2 Hormones in SAM Region

Hormone-responsive gene expression was assessed in the shoot apical meristem (SAM) region during the initial stages of rapid-growth bamboos (Gamuyao et al., 2017). The SAM region has higher concentrations of total gibberellins (GA) and cytokinins (especially trans-zeatins, iPRPs) (Gamuyao et al., 2017). GA promotes the shoot elongation and mutant of GA biosynthesis genes such as *CPS* (ent-copalyl diphosphate synthase, also named *GA1*), *KS-A* (ent-kaurene synthase, also named *GA2*), *KO2* (ent-kaurene oxidase, also named *GA3*), and *KAO-B* (ent-kaurenoic acid oxidase) leads to dwarf or semidwarf phenotype of Arabidopsis (Talon et al., 1990; Yamaguchi et al., 1998; Regnault et al., 2014). Interestingly, these gibberellin biosynthesis genes were highly expressed in the SAM of Moso bamboo (Gamuyao et al., 2017). Cytokinins are involved in cell division by controlling the size of shoot apical meristem and thus affecting shoot morphology (Kuroha et al., 2009; Miyawaki et al. 2006). Mutation of cytokinin biosynthesis gene *IPT3* (isopentenyltransferase 3 for production of *iPRPs*) reduces the shoot apical meristem size and reduces the inflorescence stems in Arabidopsis (Miyawaki et al., 2006). Mutation of trans-zeatins biosynthesis genes-*LOGs* delays inflorescence growth and thus decreases plant height of Arabidopsis (Kuroha et al., 2009). In Moso bamboo, cell division biosynthetic genes such as *IPT3*, LOG-like genes (*LOGL1, LOG2A, LOGL7A*, and *LOGL10B* corresponding to the production of trans-zeatins) were also highly expressed in the SAM (Gamuyao et al., 2017). Auxin plays positive roles in regulating plant height. Mutation of auxin biosynthesis genes-*TAA1* and *YUCCA1* inhibit the growth of plant height in Arabidopsis (Stepanova et al., 2008; Zhao et al., 2001). In Moso bamboo, the rate-limiting transaminase and flavin monooxygenase encoding genes *TAA1;4* and *YUCCA7* for tryptophan-derived auxin are also dominantly expressed in the SAM (Gamuyao et al., 2017). These results proved that auxin, gibberellin, and cytokinin are involved in the rapid growth of bamboo culm in the initial stage.

11.4.3 Crosstalk Between Hormones

Aux/IAA proteins are short-lived transcription factors to repress expression of primary auxin response genes (Tiwari et al., 2004). During the rapid growth of Moso bamboo, co-expression analysis revealed that the auxin signaling gene *PeIAA1* is a hub gene in the crosstalk of multiple hormones signaling (Li et al., 2018). *PeIAA1* interacts with BR, cytokinin, and JA (jasmonic acid) signaling genes at 13 edges of the co-expressed regulatory network (Li et al., 2018). This phenomenon was also observed in *PeIAA2* and other *AUX/IAA*-mediated co-expression networks (Li et al., 2018). Four *AUX/IAA* genes link BR signaling, and this interaction has been demonstrated in Arabidopsis (Li et al., 2018). Therefore, auxin plays a central role in mediating the hormonal crosstalk during the rapid growth of bamboo.

11.4.4 Transcription Factors

Gene expression analysis of transcription factor families has been widely used to identify transcription factors that regulate the rapid growth of bamboo culms. Fifteen *NAC*s showed higher expression in the intermediate and late stages, suggesting that NAC transcription factors may also be involved in cell elongation and cell wall modification of bamboo culms (Shan et al., 2019). The expression of *PeMYB26* and *PeMYB33* is significantly higher in 6.7 m tall bamboo culms than that in 0.2 m shoots (Yang et al., 2019), suggesting that MYB transcription factors may be involved in cell wall modification. Nine *MYBs* genes (*PeMYB 3, –10, –14, –22, –29, –37, –50, –64*, and *–74*) may be involved in the cell elongation of bamboo, and their relative expression is highest at 3 m height in bamboo culms compared to that at 0.2 m and 6.7 m (Yang et al., 2019). The expression of six *LBD*s (*PeLBD 06, –10, –20, –29*, and *–46*) were consistently upregulated during the rapid growth of Moso bamboo (Huang et al., 2021). *PeMYB35* regulates rapid cell elongation and cell wall accumulation of Moso bamboo by binding to the promoter of *PeGT43-5* (glycosyltransferases 43-5) (Li et al., 2021). The expression of *PeBBXs* (*PeBBX 17, –20*, and *–25*) were positively correlated with the height of bamboo shoots (Ma et al., 2021). Similar observations have been reported in *PheE2F/DP* gene family (Li et al., 2021). These results indicate that transcription factors play important roles in the rapid growth of bamboo culms.

11.4.5 Crosstalk Between TF and Hormones

Transcription factors (TFs) and hormones are two well-known regulators of gene expression. How do TFs and hormones interact during the rapid growth of bamboo? GATA transcription factors regulate plant growth through hormone signal transduction. It was reported that about 61.29% of Moso bamboo *GATA* genes (19/31) showed significant differential expression under exogenous GA treatment (Wang et al., 2020). Among them, *PeGATA26* showed a 56% reduction in expression under GA treatment (Wang et al., 2020). In addition, overexpression of *PeGATA26* in Arabidopsis significantly suppressed the plant height of transgenic Arabidopsis (Wang et al., 2020). *PeGATA26* reduces GA levels by activating the expression of the GA turnover genes (Wang et al., 2020). Since the expression of *PeGATA26* also gradually decreased during the rapid growth of bamboo, the GATA transcription factor regulates cell growth by downregulating *PeGATA26*, resulting in increased GA levels.

11.5 Post-transcriptional Regulation of Rapid Growth in Bamboos

11.5.1 Kinase

The BR signaling is negatively regulated by the kinase family-GLYCOGEN SYNTHASE KINASE 3. GSK can interact with BZR1 to phosphorylate BZR1 and inactivate BR-BZR1-dependent cell elongation (Wang et al., 2019). Bamboo GSK3/ shaggy-like kinase 1 (PeGSK1) was shown to be an important post-transcriptional regulator of BR signaling. PeGSK1 rescues *atgsk1* growth-arrested phenotypes including dwarfism (Wang et al., 2019). Overexpression of *PeGSK1* in Arabidopsis significantly suppressed Arabidopsis height. *PeGSK1* interacts with BZR1 to downregulate the expression of BR-dependent cell growth genes, including *EXPs* (*EXP1, EXPA1, –4, –8*), *PREs* (*PRE1, 5*), *IBH1, BEE1, THE1*, and *HERK1* (Wang et al., 2019). The expression pattern of *PeGSK1* was negatively correlated with the expression of BR-dependent cell elongation genes during the rapid growth of bamboo culms (Wang et al., 2019). Therefore, *PeGSK1* is an important post-transcriptional regulator in rapid-growth control of bamboos.

Sucrose non-fermenting kinase 1 (SNF1) is a key energy sensor in Arabidopsis cell growth regulation. *SNF1* inhibits plant growth and development under low-energy conditions (Wei et al., 2019). In bamboo, *SNF1* was significantly downregulated in both cell division and cell elongation regions of fast-growing internodes, ultimately promoting rapid-growth by increasing soluble sugar content (Wei et al., 2019). Therefore, sucrose is also involved in the rapid growth regulation of bamboo culms.

11.5.2 Small RNAs

miRNAs from 12 families are involved in the internode growth of Moso bamboo, and they act as negative regulators of their target genes. Six miRNAs, including miR_N47, miR_N64, miR167, miR528, miR164, and miR396, are involved in cell division of basal internodes (Wang et al., 2021). Eight miRNAs, including miR164a, miR396a, miR_N21, miR529, miR2257, miRN39, miR397, and miR444, act on the upper internodes and play roles in cell elongation and cell wall modification (Wang et al., 2021). In particular, miRNA396 targets *GRF1* and miR164 targets *GA2ox1*, both of which are involved in the GA signaling pathway (Wang et al., 2021).

The expression of *PeNAC36, –42* and *45* in the intermediate and late stages was higher than that in the initial stage. More importantly, the expression patterns of these three *PeNACs* were negatively correlated with those from miRNA164 (Shan et al., 2019). Therefore, the miRNA164-*PeNAC* module may promote the rapid growth of Moso bamboo by regulating cell wall modifications at the post-transcriptional level. Thus, these results suggest that the expression of fast-growing hormone signaling genes is highly regulated at the post-transcriptional level.

11.5.3 Alternative Splicing, cis-NATs, and Circular RNAs

In addition to miRNAs, *GRF1* has also been shown to be regulated by alternative splicing and contains intron retention (Zhang et al., 2018). Transcriptome analysis of exogenous GA-treated Moso bamboo results report that genes associated with internode elongation (*ABAR*) and cell wall modification (*CESA6, CESA7, CSLF8, CSLG1, CSLE1, CSLA1, CSLF2, CCR*, and *LAC4*) are post-transcriptionally regulated by alternative splicing and *cis*-natural antisense transcripts (*cis*-NATs) (Zhang et al., 2018). Furthermore, increased *CESA5* intron retention was positively correlated with circular RNA production (*PH01000040:442256-443044*) (Wang et al., 2019). Two cell wall modification related genes: *PH01001529G0090* (hemicellulose metabolism) and *PH01003669G0180* (lignin metabolism), which produce circular RNAs, are also targeted by miRNAs (Wang et al., 2019). Therefore, alternative splicing, miRNAs, circRNAs, and *cis*-NATs are all involved in the rapid growth of bamboo by regulating cell wall modification.

11.6 Rhizome Development of Bamboos

11.6.1 Characteristics of Bamboo Rhizome

The rapid growth of bamboo culms requires enormous energy. How does the bamboo supply energy continuously and quickly is crucial to understand for the growth and development of bamboo culms. During evolution, monopodial bamboos developed a complex rhizome system that can spread and communicate horizontally between young and mature bamboos, providing energy storage and transport (Wang et al., 2017; Song et al., 2016). Moso bamboo rhizomes can develop new shoot tips, lateral bud, rhizome tips and roots (Wang et al., 2017). New shoot tip could develop to Moso bamboo culms, while the lateral bud is dormant. Rhizome tip spread the rhizome horizontally, and the root absorbs nutrients from the soil (Wang et al., 2017). Therefore, understanding how to control rhizome development is another important scientific question in bamboo development. Here, we systematically analyze the current progress of rhizome and root development at the transcriptional and post-transcriptional levels.

11.6.2 Transcriptional and Post-transcriptional Regulation of Rhizome Development

11.6.2.1 Transcriptional Regulation

Phosphatidylethanolamine-binding protein (PEBP) is known for regulating plant flowering, and FT (flowering locus T) is one of subfamilies of PEBP. In bamboo rhizome development, *PeFT 6, –9*, and *–10* play key roles in the transition between dormant buds to new shoot tips (Zhao et al., 2019). The expression of *PeFT6* and *PeFT10* was highly expressed in lateral buds, but significantly suppressed in new shoot tips (Zhao et al., 2019). In contrast, *PeFT9* was expressed at a lower level in lateral buds, while its expression was increased 25.4-fold in new shoot tips (Zhao et al., 2019). Therefore, *PeFT6, PeFT10*, and *PeFT9* may play important roles in activation of dormant buds to new shoot tips.

11.6.2.2 *Post-transcriptional Regulation*

At the post-transcriptional level, 820, 503, and 729 differentially alternative splicing events (DEAS) were identified in pairwise comparisons of rhizome tips, new shoot tips, and lateral buds, respectively (Wang et al., 2017). These DEAS are rich in 3'-end processing, which is determined by another RNA processing pathway: alternative polyadenylation (Wang et al., 2017). Polyadenylation is a step that cleaves pre-mRNA and adds a poly(A) tail to the 3'-end of mRNA. However, more than 50% of plant genes contain multiple poly(A) sites, thus, a gene can generate different RNA isoforms by alternatively using poly(A) sites. This event is called alternative polyadenylation (APA) (Wang et al., 2017). A total of 1,224 APA genes were identified in different rhizome tissues, including rhizome tips, lateral buds, and new shoot tips (Wang et al., 2017). More importantly, compared with rhizome tips, new shoot tips tend to selectively use proximal poly(A) sites, while lateral buds tend to select distal poly(A) sites (Wang et al., 2017). Therefore, new shoot tips and lateral buds have different poly(A) site selection, and it is speculated that APA is involved in controlling the development of new shoot tips and lateral buds.

11.7 Nitrate Signaling in Bamboo Development

The rapid growth of bamboo culms and the development of rhizomes require huge amount of nutrients, and nitrate is one of the two major nitrogen sources for plant development. Unveiling the relationship between bamboo development and nitrate is important for understanding the nutrients supply and regulatory network in bamboos. NIN-like proteins (NLPs) are central transcription factors which involves in regulating nitrate signaling, assimilation, and utilization. A total of eight PeNLPs were identified in Moso bamboo (Lin et al., 2021). By introducing the protoplast system and reverse genetics approach, for group I or II PeNLPs (PeNLP1, –2, –5, and –8) can orchestrate nitrate signaling, and they have different regulatory patterns from NLPs in Arabidopsis. In Arabidopsis, the group III NLP-AtNLP7 is a central regulator of nitrate signaling, whereas in bamboo group I and II, NLPs act as central regulators of nitrate signaling instead of group III NLPs (Lin et al., 2021). Nitrate induces the cytoplasmic-to-nuclear shuttling of PeNLP8, suggesting that nitrate signaling is highly regulated in Moso bamboo (Lin et al., 2021). Over-expression of *PeNLP8* in Arabidopsis significantly improves plant vigor, dry mass and accumulates total amino acids in transgenic Arabidopsis (Lin et al., 2021). Overall, this study suggested that nitrate acts as one layer of regulator to the development of bamboos.

Overall, the rapid growth of bamboo culm and the development of rhizomes are tightly regulated at the transcriptional or post-transcriptional level. Uncovering the complexity of these regulatory networks and identifying the pivotal regulatory genes is of great significance for comprehensively unraveling the mystery of bamboo development. However, most bamboo functional genetics work relies heavily on transcriptome analysis. Since gene expression at the RNA level does not always correlate with its expression at the protein level, it is expected that in the future an increasing number of regulated genes can be functionally verified by reverse or forward genetics approaches.

References

Chen, M., Ju, Y., Ahmad, Z. et al. (2021). Multi-analysis of sheath senescence provides new insights into bamboo shoot development at the fast growth stage. *Tree Physiology*, 41(3): 491–507.

Cui, K., He, C.Y., Zhang, J.G., Duan, A.G. and Zeng, Y.F. (2012). Temporal and spatial profiling of internode elongation-associated protein expression in rapidly growing culms of bamboo. *Journal of Proteome Research*, 11: 2492–507.

Gallavotti, A. (2013). The role of auxin in shaping shoot architecture. *Journal of Experimental Botany*, 64(9): 2593–2608.

Gamuyao, R., Nagai, K., Ayano, M. et al. (2017). Hormone distribution and transcriptome profiles in bamboo shoots provide insights on bamboo stem emergence and growth. *Plant and Cell Physiology*, 58(4): 702–716.

Guo, Z.H., Ma, P.F., Yang, G.Q. et al. (2019). Genome sequences provide insights into the reticulate origin and unique traits of woody bamboos. *Molecular Plant*, 12(10): 1353–1365.

Guo, Z., Zhang, Z., Yang, X. et al. (2020). PSBR1, encoding a mitochondrial protein, is regulated by brassinosteroid in Moso bamboo (*Phyllostachys edulis*). *Plant Molecular Biology*, 103: 63–74.

He, C.Y., Cui, K., Zhang, J.G., Duan, A.G. and Zeng, Y.F. (2013). Next-generation sequencing-based mRNA and microRNA expression profiling analysis revealed pathways involved in the rapid growth of developing culms in Moso bamboo. *BMC Plant Biology*, 13(1): 119.

Huang, B., Huang, Z., Ma, R. et al. (2021). Genome-wide identification and expression analysis of LBD transcription factor genes in Moso bamboo (Phyllostachys edulis). *BMC Plant Biology*, 21(1): 296.

Kuroha, T., Tokunaga, H., Kojima, M. et al. (2009). Functional analyses of LONELY GUY cytokinin-activating enzymes reveal the importance of the direct activation pathway in Arabidopsis. *The Plant Cell*, 21(10): 3152–3169.

Li, L., Cheng, Z., Ma, Y. et al. (2018). The association of hormone signalling genes, transcription and changes in shoot anatomy during moso bamboo growth. *Plant Biotechnolgy Journal*, 16: 72–85.

Li, L., Shi, Q., Li, Z. and Gao, J. (2021). Genome-wide identification and functional characterization of the PheE2F/DP gene family in Moso bamboo. *BMC Plant Biology*, 21(1): 158.

Li, Z., Wang, X., Yang, K. et al. (2021). Identification and expression analysis of the glycosyltransferase GT43 family members in bamboo reveal their potential function in xylan biosynthesis during rapid growth. *BMC Genomics*, 22(1): 867.

Lin, Z., Guo, C., Lou, S. et al. (2021). Functional analyses unveil the involvement of Moso bamboo (Phyllostachys edulis) group I and II NIN-like proteins in nitrate signaling regulation. *Plant Science*, 306: 110862.

Ma, R., Chen, J., Huang, B., Huang, Z. and Zhang, Z. (2021). The BBX gene family in Moso bamboo (*Phyllostachys edulis*): Identification, characterization and expression profiles. *BMC Genomics*, 22(1): 533.

Mao, F., Li, X., Du, H. et al. (2017). Comparison of two data assimilation methods for improving MODIS LAI time series for bamboo forests. *Remote Sensing*, 9(5): 401.

Miyawaki, K., Tarkowski, P., Matsumoto-Kitano, M. et al. (2006). Roles of Arabidopsis ATP/ADP isopentenyltransferases and tRNA isopentenyltransferases in cytokinin biosynthesis. *Proceedings of the National Academy of Sciences*, 103(44): 16598–16603.

Peng, Z., Lu, Y., Li, L. et al. (2013). The draft genome of the fast-growing non-timber forest species Moso bamboo (Phyllostachys heterocycla). *Nature Genetics*, 45(4): 456–61, 461e1–2.

Peng, Z., Zhang, C., Zhang, Y. et al. (2013). Transcriptome sequencing and analysis of the fast growing shoots of Moso bamboo (*Phyllostachys edulis*). *PloS One*, 8: e78944.

Péret, B., Swarup, K., Ferguson, A. et al. (2012). AUX/LAX Genes encode a family of auxin influx transporters that perform distinct functions during Arabidopsis development *The Plant* Cell, 24(7): 2874–2885.

Planas-Riverola, A., Gupta, A., Betegón-Putze, I., Bosch, N., Ibañes, M. and Caño-Delgado, A. (2019). Brassinosteroid signaling in plant development and adaptation to stress. *Development*, 146(5): dev151894.

Regnault, T., Davière, J.M., Heintz, D., Lange, T. and Achard, P. (2014). The gibberellin biosynthetic genes AtKAO1 and AtKAO2 have overlapping roles throughout Arabidopsis development. *The Plant Journal : For cell and molecular biology*, 80(3): 462–474.

Shan, X., Yang, K., Xu, X., Zhu, C. and Gao, Z. (2019). Genome-wide investigation of the NAC gene family and its potential association with the secondary cell wall in Moso bamboo. *Biomolecules*, 9(10): 609.

Song, X., Peng, C., Zhou, G., Gu, H., Li, Q. and Zhang, C. (2016). Dynamic allocation and transfer of non-structural carbohydrates, a possible mechanism for the explosive growth of Moso bamboo (Phyllostachys heterocycla). *Scientific Reports*, 6(1): 25908.

Stepanova, A.N., Robertson-Hoyt, J., Yun, J. et al. (2008). TAA1-mediatedaAuxin biosynthesis is essential forhormone crosstalk and plant development. *Cell*, 133(1): 177–191.

Stortenbeker, N. and Bemer, M. (2019). The SAUR gene family: The plant's toolbox for adaptation of growth and development. *Journal of Experimental Botany*, 70(1): 17–27.

Talon, M., Koornneef, M. and Zeevaart, J.A. (1990). Endogenous gibberellins in Arabidopsis thaliana and possible steps blocked in the biosynthetic pathways of the semidwarf ga4 and ga5 mutants. *Proceedings of the National Academy of Sciences*, 87(20): 7983–7987.

Tiwari, S., Hagen, G. and Guilfoyle, T. (2004). Aux/IAA proteins contain a potent transcriptional repression domain. *The Plant Cell*, 16: 533–43.

Wang, K.L., Zhang, Y., Zhang, H.M. et al. (2021). MicroRNAs play important roles in regulating rapid growth of the *Phyllostachys edulis* culm internode. *New Phytologist*, 231: 2215–2230.

Wang, T., Li, Q., Lou, S. et al. (2019). GSK3/shaggy-like kinase 1 ubiquitously regulates cell growth from Arabidopsis to Moso bamboo (*Phyllostachys edulis*). *Plant Science*, 283: 290–300.

Wang, T., Wang, H., Cai, D. et al. (2017). Comprehensive profiling of rhizome-associated alternative splicing and alternative polyadenylation in Moso bamboo (*Phyllostachys edulis*). *The Plant Journal*, 91(4): 684–699.

Wang, T., Yang, Y., Lou, S. et al. (2020). Genome-wide characterization and gene expression analyses of GATA transcription factors in Moso Bamboo (*Phyllostachys edulis*). *International Journal of Molecular Sciences*, 21: 14.

Wang, Y., Gao, Y., Zhang, H. et al. (2019). Genome-wide profiling of circular RNAs in the rapidly growing shoots of Moso bamboo (*Phyllostachys edulis*). *Plant and Cell Physiology*, 60: 1354–1373.

Wei, Q., Guo, L., Jiao, C. et al. (2019). Characterization of the developmental dynamics of the elongation of a bamboo internode during the fast growth stage. *Tree Physiology*, 39: 1201–1214.

Wei, Q., Jiao, C., Ding, Y. et al. (2017). Cellular and molecular characterizations of a slow-growth variant provide insights into the fast growth of bamboo. *Tree Physiology*, 38: 1–14.

Xu, X., Lou, Y., Yang, K., Shan, X., Zhu, C. and Gao, Z. (2019). Identification of Homeobox genes associated with lignification and their expression patterns in bamboo shoots. *Biomolecules*, 9(12): 862.

Yamaguchi, S., Sun, T.P., Kawaide, H. and Kamiya, Y. (1998). The GA2 locus of Arabidopsis thaliana encodes ent-Kaurene synthase of gGibberellin biosynthesis. *Plant Physiology*, 116(4): 1271–1278.

Yang, K., Li, Y., Wang, S. et al. (2019). Genome-wide identification and expression analysis of the MYB transcription factor in Moso bamboo (*Phyllostachys edulis*). *PeerJ*, 6: e6242.

Zhang, H., Wang, H., Zhu, Q. et al. (2018). Transcriptome characterization of Moso bamboo (*Phyllostachys edulis*) seedlings in response to exogenous gibberellin applications. *BMC Plant Biology*, 18(1): 125.

Zhao, H., Gao, Z., Wang, L. et al. (2018). Chromosome-level reference genome and alternative splicing atlas of Moso bamboo (*Phyllostachys edulis*). *GigaScience*, 7: giy115.

Zhao, J., Gao, P., Li, C., Lin, X., Guo, X. and Liu, S. (2019). PhePEBP family genes regulated by plant hormones and drought are associated with the activation of lateral buds and seedling growth in Phyllostachys edulis. *Tree Physiology*, 39(8): 1387–1404.

Zhao, Y., Christensen, S.K., Fankhauser, C. et al. (2001). A role for flavin monooxygenase-like enzymes in auxin biosynthesis. *Science*, 291(5502): 306–309.

Zheng, Y., Yang, D., Rong, J. et al. (2022). Allele-aware chromosome-scale assembly of the allopolyploid genome of hexaploid Ma Bamboo (Dendrocalamus latiflorus Munro). *Journal of Integrative Plant Biology*, 64: 649–670.

Chapter 12

In vitro Propagation and Genetic Transformation in Bamboo
Present Status and Future Prospective

Adla Wasi,[1] Zishan Ahmad,[2,3] Anwar Shahzad,[1,]* Sabaha Tahseen,[1]
Yulong Ding,[2,3] Abolghassem Emamverdian[1,2] and
Muthusamy Ramakrishnan[1,2]

12.1 Introduction

Bamboos belong to the Poaceae family and are one of the fastest-growing plants on the earth. Bamboo comprises 120 genera and 1,641 species (Polesi et al., 2019; Soreng et al., 2015). These are early maturing plants that cover most of tropical part of the world. Bamboos provide environmental, economic, and social benefits to the millions of people living in tropical areas. The stem of bamboos is very useful, and it provides raw material for housing, handicraft, agriculture applications, paper, and pulp industry, etc. (Singh, 2008). Thus, commonly, bamboo is referred as the "friend of people", "poor man's timber", "green gasoline", and "the cradle to coffin timber" (Singh, 2008; Lobovikov et al., 2012; Singh and Reeta, 2020). Bamboos contain low fat and high fibres potassium vitamins and several other mineral nutrients, thus, has been in people's diet in various forms. It also contains antioxidants, glycosides, flavones and can be used for medicinal purposes. The sap obtained from bamboo

[1] Plant Biotechnology Section, Department of Botany, Aligarh Muslim University, Aligarh 202002, U.P., India.
[2] Bamboo Research Institute, Nanjing Forestry University, Nanjing, Jiangsu, 210037, People's Republic of China.
[3] Co-Innovation Centre for Sustainable Forestry in Southern China, Nanjing Forestry University, Nanjing, 210037, China.
* Corresponding author: ashahzad.bt@amu.ac.in

shoots has the potential to cure jaundice. Similarly, *Bambusa arundinacea* consists of bamboo manna which is a siliceous concretion of shoot of bamboo effective against respiratory problems (Soni et al., 2013). Leaf juice of bamboo provides strength to the cartilage of those suffering from osteoporosis (Vanithakumari et al., 1989). The rapid growth of bamboos is exploited for various applications (Nayak and Mishra, 2016; Khalil et al., 2012; Sen and Reddy, 2011). Therefore, the natural bamboo resources are getting depleted due to overexploitation. In addition, the long monocarpic reproductive cycle of 120 years and recalcitrant seed production make it difficult to replenish natural bamboo resources. Furthermore, vegetative propagation of bamboos through cutting is a difficult procedure (Sood et al., 2013). The limited availability of propagule, difficulty in transportation, inadequate roots of propagules, and low survival rate make vegetative propagation of bamboos much more cumbersome (Mudoi et al., 2013). There are many pests which destroy bamboo nursery and adult bamboo plants (Singh et al., 2013). The growing demand of bamboo can be accomplished by applying tissue culture technique. Tissue culture is able to produce 0.5 million plants per year starting from a single explant (Gielis, 1995). Bamboo tissue culture is useful in understanding the reproductive organ and its phenomenon through *in vitro* flowering because in natural conditions after flowering, the bamboo's clump dies. Tissue culture is also used to preserve hybrid seeds because obtaining superior traits through conventional breeding is very difficult due to peculiar flowering in bamboos (Singh et al., 2013). Similarly, adding biotechnological advancement to this technique is helpful in obtaining better varieties within a short duration. Biotechnological advancements include somatic hybrid formation through protoplast fusion, use of DNA marker to choose gene of interest, and formation of the genetically engineered organism. Transformation in bamboo is a challenging task. However, efforts have been made to obtain genetically transformed bamboo species, namely, *Dendrocalamus hamiltonii*, *D. farinosus*, and *D. Latiflorus* (Jiang and Zhou, 2014; Qiao et al., 2014; Sood et al., 2014). In *D. Latiflorus* genetic transformation has been done through anther culture. Poor *in vitro* regeneration efficiency and slow growth of bamboo regenerants makes genetic transformation much more challenging. However, genetic transformation is the need of the time to obtain stress tolerant bamboo species. Hence, it is necessary to invest several resources and extensive research to obtain genetically transformed bamboo. Sustainable utilization and conservation of bamboo resources can bring more profits to humankind all over the world. There are limited reviews available on *in vitro* propagation, genetic transformation, and transgenic approaches of bamboo (Singh et al., 2013; Mudoi et al., 2013; Qiao et al., 2014; Shahzad et al., 2021; Ganie et al., 2021; Chen et al., 2021). However, there is no comprehensive report nor is there a ready reference available on *in vitro* propagation, genetic transformation, and application of biotechnological tools for the bamboo advancement. Therefore, the present chapter focuses mainly on the available information on micropropagation protocols for direct or indirect organogenesis, disinfection of explants, and genetic transformation of bamboo species.

12.2 Micropropagation

12.2.1 Explant Selection and Disinfection

Selection of the explant and procedure involved in disinfection are the keys to attain success during tissue culture (Shahzad et al., 2017). In 1968, Alexander and Rao did the *in vitro* propagation of the bamboo for the first time. They were able to germinate mature embryos of *Bambusa arundinacea* on a nutrient medium successfully. Systemic and surface contaminations are the major reason for the failure of *in vitro* propagation of bamboos (Oprins et al., 2004). Explants, such as nodes, consist of intercellular space and vessels which provide a cavity for contaminating agents to settle in and escape the sterilization process (Thakur and Sood, 2006). Although *in vitro* propagation of different bamboo species has been done through seeds, node, internode, axillary bud, etc. (Chowdhury et al., 2004; Das and Pal, 2005a, b; Maiya et al., 2021; Kaladhar et al., 2017; Saini et al., 2016; Jat et al., 2016). It is very difficult to establish a regeneration protocol by using vegetative tissue as explant in bamboos, but it is necessary to produce true-to-type plants (Ye et al., 2017). Table 12.1 is dealing with the available information on the use of explants and sterilization procedures applied during bamboo *in vitro* propagation.

12.2.2 Nutrient Medium

Different *in vitro* growing tissues have different nutritional requirements. Hence, specific culture mediums should be investigated for tissues, organs, and different developmental stages of various species of plant. The chemical composition of each culture medium has its own benefits or disadvantages on bamboo shoot induction and multiplication. In comparison to solid media, liquid media were found more beneficial in the growth and multiplication of bamboo as suggested in several reports (Saxena, 1990; Sood et al., 2002; Das and Pal, 2005a, b; Arya et al, 2006; Ogita et al., 2008; Kabade, 2009). Table 12.2 represents the developed protocols for *in vitro* propagation of different bamboo species. In *B. balcooa*, stunting in shoots was observed when the solidifying agent was added to nutrient medium (Negi and Saxena, 2011a). Similarly, in *B. nutans*, the gelling agent reduced the multiplication rate and shoot length (Negi and Saxena, 2011a). In another study, Gantait et al. (2016) found semi-solid media are best for the establishment of culture, while for proliferation and rooting the liquid medium gives more satisfactory results. It is also attributed that the gelling agent present in solid medium binds the water, absorbs PGRs and nutrients and makes it less available to cultured tissue hence it reduces growth and shoot multiplication in some bamboos (Singh et al., 2013). MS (Murashige and Skoog, 1962) medium are well known for initiation of the shoot in bamboos. Arya et al. (2008a) reported in *Bambusa balcooa* and *Drepanostachyum falcatum*, wherein high-frequency bud breaks have occurred in MS medium when compared to B5 medium (Gamborg et al., 1968) or Woody Plant Medium (WPM) (McCown and Lloyd, 1981). Similarly, Kabade (2009) reported in *B. bambos* that MS medium is best suited for shoot initiation from nodal segment when compared to

Table 12.1. Source of explant and sterilization procedures in *Bamboo species.*

Bamboo Species	Nature of Explant	Procedure of Sterilization	References
Bambusa balcooa	Nodal segment	Nodal segment → 70% Ethanol → 0.1% HgCl$_2$ (5-min) → Tween 20 sol (1–2 drop) → DW	Sharma and Sarma (2011)
Phyllostachys heterocycla	Seeds	Seeds → 75% ethanol (1 min) 4–5 ×DW → 2% NaOCl → 0.1% Tween-80 (15–20 min) → 6–8 × DW	Yuan et al. (2013)
Bambusa bambos	Nodal segment	Node → 0.1% Bavistin and teepol → 0.1% HgCl$_2$	Anand et al. (2013)
Bambusa tulda	Nodal segment	Node → 70% ethanol → 0.1 HgCl2 sol (5 min) → 5% tween 20 sol (2 h) → RTW (30 min) → DW	Pratibha and Sarma (2013)
Bambusa arundinacea	Seeds	Seeds → Soaked overnight in Bavistin → 10% Tween 20 → DW → 0.1% HgCl2 (15 min) → 5 × DW	Kalaiarasi (2014)
Bambusa nutan	Nodes with axillary bud	Nodal segment → 70% Ethanol → 0.1% HgCl$_2$ (5-min) → 3 × DW	Pratibha and Sarma (2014)
Bambusa ventricosa	Axillary buds	Axillary buds → 70% Ethanol (30s) → 2 × DW → 0.1% HgCl2 → 2% NaOCl (15, 20 and 25 min) 10% H2O2 → 5 × DW	Wei et al. (2015)
Dendrocalamus strictus	Nodes	Nodes → Dip in 1% extran (10 min) → Wash DDW 1% extran → 0.1% HgCl2 (5 min) → Rinsed DDW → Dip 70% Ethanol (1 min) → Wash DDW	Goyal et al. (2015)
Drepanostachyum falcatum	Axillary buds	Axillary buds → 5% Cetrimide solution (5min) → Clean with 90% Ethanol → 0.1% HgCl2 (10–12 min) → Wash 4 × DW	Saini et al. (2016)
Bambusa bamboos	Nodal segment	Nodes → wash in RTW → Liquid detergent (15 min) 4 × DW → Bavistin (5–7 min) → 3 × DW 0.05%, 0.1% and 0.2% HgCl2 (3–10 min) → Wash in DW	Raju and Roy (2016)
Dendrocalamus strictus	Seeds	Seed → washed in 2% Teepol sol (5 minute) → Wash with RTW (15–20 min) → Rinse in DW → 79% alcohol → 3 × DW → 0.1% HgCl2 sol 5 × DW	Jat et al. (2016)
Dendrocalamus latiflorus	Newly emerged shoots	Shoots → 70% ethanol (1min) → 2 × SW → 0.1% HgCl$_2$ (8 min) → 5–7 × SW	Ye et al. (2017)

Table 12.1 contd. ...

...Table 12.1 contd.

Bamboo Species	Nature of Explant	Procedure of Sterilization	References
Bambusa vulgaris	Intermodal segment	Internodes → RTW (10 min) → Tween 20 (5 min) → 2–3 × DW → Soaked in 1% Bavistin (10 min) → DW → 70% Ethanol (1 min) → 2–3 DW → 0.1% HgCl2 (5 min) 4–5 × DW	Kaladhar et al. (2017)
Dendrocalamus strictus	Nodal segment	Nodes → washed in fungicide (50% Copper Chloride) + Bactericide (Streptocycline) + Tween-20 (15 min) → 3 × DW → 0.1% HgCl2 (8 min) → Dip in 70% Ethanol (60 sec) → 5 × DW	Rajput et al. (2019)
Thyrsostachys siamensis	Nodal segment	Nodes → 1% Teepol → 70% ethanol (1 min) → surface-sterilized (10% Haiter 6% commercial NaOCl solution) 0.02% Tween 20 (10 min) → 3 × DW	Obsuwan et al. (2019)
Bambusa balcooa	Nodes	Nodal cuttings → Teepol (5–10 min) → 3 × DDW Bavistin (5 min) → 2 × DDW → 0.1% HgCl2 (5 min) → 5 × SDW	Rajput et al. (2020)
Bambusa balcooa	Nodal segment	Nodes → DW → Bavistin (10 min) → tween-20 (5 min) → 3 × DW → 70% Alcohal → (30 sec) → 3 × DW	Chavan et al. (2021)
Bambusa nutans	Nodal segment	Nodes → 5% Tween 20 + 1% Gentamicin + 5% Bavistin → 70% Ethanol → 0.1% HgCl2	Maiya et al. (2021)

Abbrreviations: DDW, double distilled water; DW, distilled water; HgCl2, mercuric chloride; NaOCl, sodium hypochlorite; RTW, running tap water; EtOH, ethyl alcohol (ethanol); s. second(s); minute (min); SDW, sterilized distilled water; SW, sterilized water; H2O2, hydrogen peroxide.

Table 12.2. Developed protocols for *in vitro* propagation in *Bamboo* species.

Species	Culture Medium (Shoot and Rooting Medium, pH, Sucrose, and Agar Concentration)	Culture Conditions	Experimental Output	References
Bambusa nutan	MS + BA 1 mg/L (SIM). MS + BA 1 mg/L (SMM). MS + IAA 5.0 mg/L + IBA 8.0 mg/L + NAA 5.0 mg/L (RIM). Sucrose 3% agar 2.5% pH 5.7.	Temp. $25 \pm 2°C$. PP. 16 h of 45 μmol m^{-2} s^{-1} CFWT.	In SMM, 11.06 shoot produced per culture and mean shoot length were obtained 4.0 cm. In RIM, 90% of the culture produced root and mean length and no of root per culture were 6.08 and 8, respectively.	Pratibha and Sarma (2014)
Dendrocalamus strictus	Ms + 4 mg/L (SIM). MS + 4 mg /l (SMM). Ms + 3 mg/l NAA (RIM). Sucrose 3% agar 0.8% pH 5.7.	Temp. $25°C \pm 2°C$ PP. 16 h of 2,000–3,000 lux CFWT.	In 49% of the culture shoot bud formation occur and average no of shoots produced 3.40. In SMM, the highest rate of shoot multiplication was 3.68 per explant. The maximum root regenerated were 1.36 per explant in 87% of culture. Only 70% of the plant survives during acclimatization and this plant is transferred to the field. Genetic stability was confirmed by RAPD and ISSR marker.	Goyal et al. (2015)
Drepanostachyum falcatum	MS + 4.5 BAP mg/L (SIM). MS + 3.5 mg/L BAP (SMM). MS + 6.5 mg/L IBA (RIM). Subculture every 4 weeks. Sucrose 3% agar 0.8% pH 5.6.	Temp. $25°C \pm 2°C$ PP. 16 h of 2500 lux CFWT. RH. NS	In SIM, 68.75% of culture produces shoot with average no of shoot 4.15 and mean shoot length 1.98 ± 0.12. In SMM, the average no of shoot per culture were 41.49 with average length 2.15. Rooting was observed in 99% of the culture with average root length of 2.18 cm. After hardening and acclimatization 90%–95% plantlets survive in field condition.	Saini et al. (2016)
Bambusa bamboos	MS + 2.0 mg/L BAP + 1.0 mg/L TDZ (SIM). MS + 2.0 mg/L BAP + 1.0 mg/l TDZ (SMM). ½ MS + 2.5 mg/l IBA and 2.5 mg/l NAA (RIM). Sucrose 3% agar 0.8% pH 5.8.	Temp. $24 \pm 2°C$ PP. 16 h CFWT. RH NS	In SIM, maximum 3.14 no of shoot produces per explant. In SMM, 16.58 shoots obtained per culture and average shoot length reported 9.21 cm. Rooting were observed in 86.7% of the culture with average 8.72 root per shoot. Acclimatization had been done for 1 month and after field transfer 100% survivability was reported.	Raju and Roy (2016)

Table 12.2 contd. ...

...Table 12.2 contd.

Species	Culture Medium (Shoot and Rooting Medium, pH, Sucrose, and Agar Concentration)	Culture Conditions	Experimental Output	References
Thyrsostachys siamensis	MS + 11.3 µM 2,4-D + 4.65 µM kn and 1.96 µM + IBA (CIM). MS + 11.1 µM BA + 3.43 µM IBA (SIM). MS + 26.85 µM NAA (RIM). Sucrose 3% agar 0.5% pH 5.7.	Temp. 24 ± 1°C. PP16 h of 35–40 µmol $m^{-2} s^{-1}$ provided by CWFT. RH (NS).	In CIM, an average diameter 0.76 cm of callus were produced. In SIM, medium average no. of shoots produced were 13.5 and average length of shoot was 2.53. In SMM medium maximum 27.0 shoot were reported. Roots were produced efficiently in 80% of culture. Maximum 2.80 no. of roots produced per shoot.	Obsuwan et al. (2019)
Bambusa balcooa	MS + 4.0 mg/L BAP + 50 mg/L ascorbic acid + 25 mg/L arginine + 25 mg/L citric acid + 25 mg/L adenine sulphate SIM. MS + 4.0 mg/L BAP + 1.0 mg/L NAA (SMM). ½ MS + 6.0 mg/L NAA + 100 mg/L activated charcoal (RIM). Sucrose 3% agar 0.8% pH 5.8.	Temp. 25 ± 2°C. PP 14 h of 50 µmol $m^{-2} s^{-1}$ provided by CWFT. RH (NS).	In SIM, 96% shoot sprouting were observed. In SMM, 62.0 shoots per node per culture were produced. In RIM 8.3 roots were produced per shoot within four weeks of incubation in 100% of culture. The plants were hardened in soilrite® for 4 weeks and display 100% survival when transfer to field condition. Genetic stability was confirmed by ISSR and SCoT.	Rajput et al. (2020)
Bambusa mutans	MS +1 mg/L BAP + 0.5 mg/L KN in (SIM). MS + 1 mg/L + BAP and 0.5 mg/L KN in (SMM). Subculture each 3 weeks. MS + 3 mg/L NAA + 3 mg/L IBA (RIM). 3% sucrose, 0.8% agar pH 5.8.	Temp 25 ± 2°C. PP 16 h, of 3000 Lux provided by CWFT. RH 75–85%.	In Liq. SIM, average length of shoot produced are 2.5 ± 0.19 cm in 100% of cultures. In SMM, shoot multiply rapidly. In rooting media, 84.67% roots were achieved. 100% survival rate was obtained after field transfer.	Maiya et al. (2021)
Bambusa balcooa	MS + 4 mg/L (SIM). 3% sucrose, 0.8% agar pH 5.8.	Temp 25 ± 2°C. PP 16 h provided by CWFT RH 50–60%.	In SIM average no of shoots produced are 7.67 and average length of shoot is 7.67 cm.	Chavan et al. (2021)

| *Bambusa nutans* | MS + 8 mg/L 2,4-D + 0.5 mg/L IBA (CIM). MS + BAP 2 mg/L + NAA 0.5 mg/L (SIM). 1/2 MS + 1 mg/L IAA (RIM). Sucrose 3% agar 0.42% pH 5.8. | Temp. 25 ± 2°C. PP 16 h of 60–70 mmol/m²/s CWFT. RH. 60–65%. | In CIM, 53% of Culture produce calluses after 2.5 month of induction. Shoot induction reported in 50% of induction. The frequency of root regeneration was reported 72.8% within one month of culture. | Shawan et al. (2021) |

Abbreviations: 2,4-D, 2,4-dichlorophenoxyacetic acid; BAP, 6-Benzylaminopurine; Ba, benzyl adenine; TDZ, thidiazuron; kn, kinetin; IBA, indole-3-yl-butyric acid IBA; NAA, α-Naphthalene acetic acid; cm, centimetre; no, number; h, hour; wk; weeks ; mg, milligram; L, liter; mmol, milimolar; μM, micromolar CIM, callus induction medium; SIM, shoot induction medium; SMM, shoot multiplication medium; RIM, root induction medium; RAPD, Random amplified polymorphic DNA; ISSR, inter simple sequence repeats; SCoT, start codon targeted ; NS, not stated; SPFD, Spectral Photon Flux Density ; CWFT, cool white fluorescent tube; PP, photoperiod; RH, relative humidity; Temp., temperature; AGR, absence of growth regulator; SGM, Seed germination medium; Liq, liquid.

WPM, B5, and SH (Schenk and Hildebrandt, 1972) and WPM was recognized least effective. Carbohydrate is one of the crucial factors throughout tissue culture as it is a source of energy and maintains the osmotic balance. In culture vessels, explant phase photo mixotrophic condition, i.e., low CO_2 lower gaseous exchange and light intensity hence depend upon medium carbohydrate. Sucrose is commonly used as a carbohydrate source because of its easy availability low cost, easy translocation and free from enzymatic degradation due to its non-reducing property (Jiménez et al., 2021). In some bamboo species other carbohydrates were tested but sucrose gave a more satisfactory result, e.g., in *B. pallida* sucrose-containing medium caused the highest bud break (Beena et al., 2012). Similarly, in *B. balcooa* and *B. arundinacea* 1% and 4% sucrose were found most effective (Brar et al., 2014; Venkatachalam et al., 2015). When sucrose was replaced with any other sugar, as in the case of *D. asper* culture, it did not give a satisfactory result and multiplication rate of shoots declined (Singh et al., 2011).

12.2.3 Regeneration Process

Alexander and Rao (1968) initiated the bamboo micropropagation for the first time by using zygotic embryo as explant in *D. strictus* species of bamboo. Several techniques have been put forward for the regeneration of bamboo species such as adventitious shoot formation, somatic embryogenesis, and organogenesis are being utilized successfully. Regeneration through organogenesis involves two types of procedures referred to as direct and indirect organogenesis. Plants produced during directed organogenesis through shoot tips or nodal segment are found more genetically stable and devoid of soma-clonal variations (Das and Pal, 2005b; Singh et al., 2011). Hence, many studies have been reported where axillary meristems were used for *in vitro* propagation of bamboo species using juvenile and mature tissues and only limited studies are available on indirect organogenesis of bamboos (Agnihotri et al., 2009; Cheah and Chaille, 2011; Hu et al., 2011).

12.2.3.1 Direct Organogenesis

Axillary buds, and zygotic embryos, are mainly used as explants to induce *in vitro* propagation of bamboos. However, in most of the *in vitro* propagation studies of bamboo, nodal segment was reported to give best result (Chowdhury et al., 2004; Das and Pal, 2005a, b; Raju and Roy, 2016). In many species of bamboo lower concentrations of 6-Benzylaminopurine (BAP) was used for the induction of bud break opening of axillary buds and shoot multiplication enhancement while its higher concentration reduces the shoot multiplication bud break and causes restricted shoot growth (Chowdhury et al., 2004; Das and Pal, 2005a, b; Lee et al., 2019). For the rooting in bamboos, mainly auxin (Saxena and Bhojwani, 1993; Chowdhury et al., 2004; Das and Pal, 2005a, b) or a combination of auxin with cytokinin was reported in most of the cases (Mudoi et al., 2013). In *Bambusa nutans*, when nodal segments were inoculated in the combination of BAP and Kinetin (Kn) it causes sprouting in 100% of the culture (Maiya et al., 2021). Similarly, the nodal segment was found to produce multiple shoots in *Thyrsostachys siamensis* when cultured in a combination of Benzyl adenine (BA) and Indole butyric acid (IBA) (Obsuwan et al., 2019). In a different study, 90% of the nodal explants produced shoots directly when treated with BAP along with Thidiazuron (TDZ) (Raju and Roy, 2016). In *Dendrocalamus strictus*, 4–5 shoots produced per culture on the MS medium containing BAP 4.0 mg/L and nearly 26 shoots produced in the shoot multiplication medium (Rajput et al., 2019). BA, BAP, and KN are found to be the most common cytokinins used for shoot induction and shoot multiplication (Chowdhury et al., 2004; Das and Pal, 2005a, b). Seeds of *Dendrocalamus giganteus* germinated in combination of BAP, KN, and Gibberellic acid (GA3) (Waikhom et al., 2012). Besides plant growth hormone, various other adjuvant or additives were used to improve the development and growth of the culture. In *D. hamiltonii*, nodal explant was reported to produce flowers and multiple shoots in BAP containing MS medium. The best result of shoots regeneration per explant was obtained on 4.4 µM BAP after 12 weeks of culture. *In vitro* flowering was first observed after 13 weeks with the highest flowering of 47% in nodal explant after seedlings were transferred to hormone-free basal medium. It was a successful attempt to overcome unpredictable flowering and post-flowering consequences in bamboos (Chambers et al., 1991). It has been reported that spring and summer seasons are best suited for shoot initiation and enhanced frequency of bud break because in these seasons contamination rate is very low and auxin production is very high in young buds of bamboos (Funada et al., 2001; Singh et al., 2012b). On the contrary, *in vitro* bud break was maximum in October and September in *B. balcooa* and *B. tulda* (Das and Pal, 2005b) when phenolic contents were lowest (Das and Pal, 2005a). There are several phenolic compounds which release during culture and cause lethal browning and blackening of cultures. Hence, antioxidant or polyphenol adsorbents or adjuvant such as activated charcoal, adenine sulphate, and amino acids are used along with PGRs. In *D. hamiltonii*, culture ascorbic acid gives a good result compared to PVP and AC, while in *D. strictus*, PVP was found to be more effective for shoot multiplication (Saxena and Dhawan, 1999). Similarly, glutamine is also a source of reduced nitrogen which promotes growth of shoot and reduces the problem of leaching in *D. giganteus* culture (Sanjaya et al., 2005). In *B. arundinacea*, silver nitrate (AgNO3) enhanced the multiplication of shoots (Venkatachalam et al., 2015). In *B. arundinacea*, *B. bambos*, and *B. balcooa*, coconut water was used for the improvement of shoot multiplication (Das and Pal, 2005a, b; Venkatachalam et al., 2015; Negi and Saxena, 2011b). Citric acid and cysteine were found to be useful in improving shoot multiplication frequency in *B. balcooa* (Choudhary et al., 2017). Similarly, in *B. balcooa*, adenine sulphate, L-arginine, citric acid, and ascorbic acid combinations enhance healthy shoot formation while in the absence of the additive, vitrified shoots were formed (Sandhu et al., 2018). Hence, it has been observed that in bamboos, mostly additives were reported to be useful for the enhancement of shoot multiplication.

12.2.3.2 Indirect Organogenesis

12.2.3.2.1 Induction of Callus and Embryo

Calli were produced from pericycle of explant adjacent to xylem under the effect of exogenously provide auxin and cytokinin (Atta et al., 2009; Ikeuchi et al., 2013). The calli were then differentiated into shoot and root on shoot and root induction medium, respectively (Obsuwan et al., 2019). There are several reports available on micropropagation of bamboo via indirect organogenesis including *Otatea acuminate* subsp. *aztecorum* (Woods et al., 1992), *B. ventricosa*, *D. farinosus* (Cheah and Chaille, 2011; Hu et al., 2011), *D. giganteus* (Devi et al., 2012), *D. hamiltonii* (Agnihotri et al., 2009), and *P. nigra* (Ogita, 2005). For callus induction in bamboos usually a higher concentration of 2, 4-Dicholorophenoxyacetic acid (2, 4-D) has been used while embryogenic calli were induced at the lower concentration of the hormones and simultaneous germination of embryos was reported (Yuan et al., 2013). Obsuwan et al. (2019) obtained undifferentiated callus from the node and internode of *Thyrsostachys siamensis* Gamble on 2, 4-D 11.30 μM, 4.65 μM kinetin, and 1.96 μM IBA. Similarly, Ogita (2005) found that half-strength MS medium along with 3 μM 2, 4-D was effective to produce whitish yellow calli in *P. nigra*. Wei et al. (2015) suggested that TDZ could also play a crucial role in inducing callus in bamboos. In *B. ventricosa*, 27 μM 2, 4-D, 2.7 μM NAA, and 0.0045 μM TDZ effectively produced callus. These calli can amplify on the medium containing 22.6 μM 2, 4-D, 2.2 μM 6-BA, and 5.4 μM NAA. Plantlets were able to regenerate on 13.3 μM 6-BA and 2.7 μM Napthalene acetic acid (NAA) and finally get acclimatized and transfered to experimentation pots (Wei et al., 2015). In *D. asper*, 77.7% of the calli were produced from the base of leaf sheath on 30 μM 2, 4-D (Ojha et al., 2009). Similarly, in *Phyllostachys bambusoides*, yellow globular and friable calli were produced from the nodal segments on MS medium containing 8.0 mg L^{-1} picloram. However, only friable calli were able to show morphogenetic activity and could produce shoot and root (Komatsu et al., 2011).

12.2.3.2.2 Root Induction

Root induction is a crucial step both for direct and indirect *in vitro* regeneration of bamboos (Anand et al., 2013). Root number, length of root, and frequency of rooting are affected by the type and concentration of nutrient media used. Usually, high frequency of rooting was obtained on any medium containing lower concentration of hormones (Sandhu et al., 2018). Similarly, in a bamboo cluster of at least three, shoots are more responsive for induction of roots than a single isolated shoot (Table 12.2). In *B. bambos*, a cluster of three to four shoots induce root formation in 60% of the cultures (Anand et al., 2013). It was reviewed by Scott (1972) and Torrey (1976) that auxins play an important part in the development of root. Depending upon mobilization capability, auxin differ in their physiological role. Generally, cytokinin is not required for induction of root because it is present in shoots in sufficient quantity (Singh et al., 2013). In most of the previous studies, IBA alone or a combination of IBA with NAA was the preferred rooting hormone in bamboos (Singh et al., 2013). The effect of IBA and NAA on induction of root had been stated earlier in other bamboos like *B. vulgaris*, *B. oldhamii*, and *D. asper* (Rout and Das, 1997; Lin et al., 2005; Rathore, et al., 2009). Better rooting was observed on half-strength MS medium in *D. asper* and *D. hamiltonii* (Singh et al., 2011, 2012b). Similarly, in *B. balcooa* and *B. bambos*, half-strength MS medium was used for rooting (Rajput et al., 2020; Raju and Roy, 2016). This response occurs due to a reduction in nitrogen requirement during root induction (Ajithkumar and Seeni, 1998). In *B. nutans* subsp.

Cupulata, 84.67% rooting was achieved in rooting media containing full-strength MS medium along with NAA and IBA (Maiya et al., 2021). Rooting was observed in 86.7% of the culture of *Bambusa bambos* with an average 8.72 root per shoot in MS medium containing IBA and NAA (Raju and Roy, 2016). Adding choline chloride in rooting medium along with IBA improved rooting up to 89% in *D, hamiltonii* (Singh et al., 2012b) which is higher than the report of Sood et al. (2002) where they found only 25–30% shoots were able to develop root. Similarly, adding activated charcoal in MS medium containing NAA produced 8.3 roots per shoot within four weeks of incubation in 100% of culture (Rajput et al., 2020) (Table 12.2).

12.2.4 Somatic Embryogenesis

Somatic embryos are like zygotic ones and have a bipolar axis which is not directly attached to the source tissue. Somatic embryogenesis (SE) has been less emphasized in the past due to the difficulty associated with the process and its modulation. In bamboo, most of the SE reported so far, are in genus *Dendrocalamus* and *Bambusa* (Polesi et al., 2021). The first report of SE in bamboo was published by Mehta et al. (1982). SE was observed in *B. arundinacea* induced from the culture of the zygotic embryo. Most of the SE was reported in *D. strictus* and *B. bamboos* because of the availability of seeds at regular intervals. But the main disadvantage associated with SE differentiated from seeds was because of the high heterogeneity that occurs in the seed population of bamboos. Moreover, it took at least 10 years to identify elite plants. Gillis et al. (2007) reported the SE in *B. balcooa* by using pseudo spikelets which come out to be a cost-effective and efficient method for the multiplication of mature bamboo.

To obtain somatic embryos selection of specific explant is very important. Embryogenic calli have been produced by using seeds, zygotic embryos, root, shoot, and leaf explants. Similarly, exposure to auxin plays an important role in the determination of the embryogenic fate of the cells. Nowadays for large-scale propagation, SE is one of the preferred methods because root and shoot primordia formation is a single-step procedure. Also, it is a reliable technique for mass multiplication of bamboo as it is economically feasible and involves minimum labor and consumes less time. Similarly, the formation of the synthetic seeds via somatic embryo formation offers great opportunity in micropropagation of bamboos (Singh et al., 2013). SE is useful for genetic transformation and producing stress resistance bamboos through transgenic technology. It is also helpful to induce *in vitro* flowering and to understand the mechanism of this complex process in bamboos. The dense organized embryogenic units can be clearly screened, multiplied, and easily germinated (Sood et al., 2013). A detail description of somatic embryogenesis is given in Table 12.3.

12.2.5 Acclimatization and Field Transfer

Transfer of *in vitro* raised plants to field condition is one of those major challenges faced during micropropagation of plants. This is due to the shock faced by the tissue culture-raised plants when exposed to a natural environment of low humidity and

Table 12.3. Somatic embryogenesis in *Bamboo* species.

Species	Nature of Explant	Culture Medium and PGRs, Subcultures	Result of Experiment	References
Dendrocalamus strictus	Seed embryo for callus induction	MS + 2, 4-D; 3 × 10⁻⁵ M (ECIM). MS + 1 × 10⁻⁵ M 2, 4-D + 5 × 10⁻⁶ M Kn + 2 × 10⁻⁶ M IBA (SEMM). MS + 5 × 10⁻⁶ M NAA + 5 × 10⁻⁶ M Kn (SEGM). ½ MS+ 3 × 10⁻⁶ M NAA + 2.5 × 10⁻⁶ M IBA + sucrose 3% + agar 0.2% (RIM).	In ECIM, 40% of the callus were embryogenic in nature. On average 5–15 somatic embryos were formed per culture. When the initial embryo transferred to SEMM the multiplication has been taken place in 70% of the culture. Frequency of germination were 90% and healthy plant were obtained in nearly 85% of the culture. 25–40 shoots of length of 7–12 cm were produced per culture. 92% of the plant survived in glasshouse.	Saxena and Dhawan (1999)
Bambusa edulis	Nodes and internode	MS + 0.046 μM TDZ + 13.6 μM 2,4-D (CIM). Ms + 0.455 μM TDZ (SEGM).	At lower conc. of TDZ sufficient of embryogenic callus were formed and browning of callus was reduced. In SEGM, 81% 84% of somatic embryos were germinated.	Lin et al. (2004)
Bambusa balcooa	Pseudo spikelets	MS + 4.5 μM 2,4-D (ECIM). MS + 22.2 μM BAP (SEMM).	In ECIM, 98% of the callus were formed and 90% of the embryogenic calli have the capacity to regenerate into somatic embryo. Actual regeneration occurred in 46% of the SE, rooting and hardening in 100% of the SE and overall conversion of SE in to established plant were obtained in 40% of the culture.	Gillis et al. (2007)
Dendrocalamus hamiltonii	Zygotic embryos	MS + 3.0 mg/L 2,4-D (ECIM). MS + 2 mg/L BA + 1 mg/L. Kn + and 1 mg/L NAA (SIM). MS + 5 mg/L IBA (RIM).	At optimum concentration of 2,4-D compact embryogenic calli were formed. In 89.5% of the culture shoot differentiation and elongation were observed. At higher concentration of IBA optimum root formation were observed and regenerated plantlet shows 100% survival in field condition.	Zhang et al. (2010)
Bambusa nutans	Sprouted buds	MS + 5 mg/L 2,4-D (ECIM). MS + 1 mg/L BAP + 1 mg/L 2,4-D (SEMM). MS + 1 mg/L BAP + 1 mg/L 2,4-D + 20 mg/L ascorbic acid. (SEGM).	In 40% of the culture embryogenic calli were formed. These embryogenic calli were turned in to mature embryo after 2 to 3 subcultures. After germination, at least 50 SE derived plantlet transfer to field and they showed 90% survival rate.	Mehta et al. (2011)

| *Phyllostachys heterocycle* | Seed embryo for callus induction. | MS + 4.0 mg/L 2, 4-D and 0.1 mg/L ZT (ECIM). MS + 5.0 mg/L ZT (SIM). MS + 2.0 mg/L NAA (RIM). | Callus induction occur in 50% of the culture and 15% of the calli were embryogenic in nature. Somatic embryoid containing 5% embryogenic calli were regenerated into plantlets directly. Successful rooting was observed. Transfer of plantlet to greenhouse had been done after 30 d of subculture. | Yuan et al. (2013) |

Abbreviations: ZT, zeatin; d, days; ECIM, embryogenic callus induction medium; SEMM, Somatic embryo multiplication medium; SEGM, somatic embryo germination medium; CM, coconut milk; M, molar; SIM, shoot induction medium; RIM, shoot induction medium.

high irradiance from a high humid and low irradiance *in vitro* condition. Plants grown *in vitro* conditions are devoid of cuticular wax on leaves, absence of a well-developed stomatal mechanism and vascular system, and presence of a lower number of proteins, carbohydrates, phenols, and inefficient photosynthetic activity. Hence, effective hardening and acclimatization procedures are required before transferring the culture regenerants to natural conditions. During acclimatization it is necessary to decrease the nutrient supply and humidity gradually to make the regenerants capable of developing strong photosynthetic and defence mechanisms for the survival in the field conditions (Singh et al., 2011). Plantlets with a healthy root system are usually transferred to disposable cups or polybags containing soil rite, vermiculite, compost, or manure singly or in the mixture and kept under high humid conditions and lower amount of MS basal medium (Mishra et al., 2008; Singh et al., 2012b). Mixing of sand with vermicompost enhanced sand porosity and root aeration which in turn increased the survivability of bamboos (Singh et al., 2012b). After acclimatization, when different bamboo species are transferred to the field conditions, their rate of survival is also different. Survivability was recorded in *G. angustifolia* (85%), and 80–85% in *B. bambos* var. *gigantea* (Jiménez et al., 2006; Kapoor and Rao, 2006). Similarly, in *D. hamiltonii*, 85% survival was reported in field conditions (Agnihotri et al., 2009) and 80–85% and 78% field survival was reported by Sood et al. (2002), Godbole et al. (2002), respectively. Sixty thousand culture regenerants of *D. asper* were transferred successfully in the field condition (Arya et al., 1999). Negi and Saxena (2011b) achieved 95.83% acclimatization in *B. nutans*, and they were able to transfer 12 plants in the field with 100% of survivability and 91% of the *B. tulda* survival was reported in greenhouse condition (Mishra et al., 2011). Similarly, 90–100% field survival was reported in *D. strictus* (Rajput et al., 2019), in *B. tulda* (Das and Pal, 2005b), and in *B. bambos* (Raju and Roy, 2016). Recently, 100% survival rate in *B. nutans* subsp. *Cupulata* was recorded after field transfer (Maiya et al., 2021). Monopodial type of microrhizome formation of *D. strictus* (Chowdhury et al., 2004) and sympodial type of microrhizome development in *B. tulda* (Das and Pal, 2005) under *in vitro* condition further helped establishing culture regenerants in the field condition.

12.3 Genetic Transformation and Editing of the Genome in Bamboos

Work on practicing genetic transformation of various crop plants is going on since the last three decades for their qualitative and quantitative improvements. There are numerous approaches to transfer foreign genes such as particle bombardment, polyethylene glycol (PEG), electroporation, and use of *Agrobacterium*. Establishment of efficient tissue regeneration system is a prerequisite for the successful genetic transformation in plants. Also, genetic fidelity test is an important step to confirm regeneration efficiency and expression of foreign genes (Saeed and Shahzad, 2016). Genetic transformation efforts have been made in various bamboo species such as *D. hamiltonii*, *D. farinosus*, and *D. Latiflorus*, but with limited success (Jiang and Zhou, 2014; Qiao et al., 2014; Sood et al., 2014). A positive transformation was accomplished in *D. latiflorus* through anther culture. Attempt was made to transfer bacterial gene CodA which code for choline oxidase, hence helpful in achieving cold tolerant bamboo (Qiao et al., 2014). Agrobacterium-mediated transformation was done in *D. latiflorus*. However, it took an eight-month period to get sufficient callus and a further eight months for positive transformation results. Genetic transformation is a difficult procedure but in bamboo it is much more difficult because of low efficiency of regeneration and slow development of plantlets. Hence, transient transformation can be a suitable approach in bamboos. It refers to the expression of a target gene prominently into host cells within a limited time (Shen et al., 2014). Transient expression technology is advantageous than the stable genetic transformation in many aspects since it is less costly, simple, rapid, safe, and can give a high expression level (Li et al., 2017). Various gene models have been recognized through the sequencing of the whole genome of Moso bamboo. Hence, efficient transient transformation technology is required to investigate the function of a particular gene. While doing protoplast transformation, explant tissue selection, growth of the protoplasts, and culture environment affects the efficiency of transformation (Zhang et al., 2011; Zhao et al., 2019). Chen et al. (2021) reported the transformation ability of the protoplast of Moso bamboo and Ma bamboo through PEG-mediated transformation. They found etiolated seedlings were the best explants generating 44.7 and 35.2% transformation in Moso bamboo and Ma bamboo, respectively. Similarly transient transformation through particle bombardment was achieved in suspension cells of *Phyllostachys* bamboo using a construct with hygromycin phosphotransferase gene and a fluorescent protein (FP) gene, namely *AcGFP1* and *mCherry* 9 (Ogita et al., 2011). The transformed cells showed presence of the FP gene under fluorescent stereo microscope and it showed hygromycin resistance under selection medium (Ogita et al., 2011).

Gene editing technology using clustered regularly interspaced short palindromic repeat (CRISPR) and CRISPR associated endonuclease (Cas9) are the most important breakthroughs in the recent past. This CRISPR editing was applied to the hexaploid *D. Latiflorus* to alter the plant height, targeting either a single allele or total homo-alleles of a gene (Ye et al., 2020). It is expected that the recently released bamboo genome further boosts the gene transformation and gene editing in Moso bamboo. However, intensive work is important to delineate the molecular mechanism of somatic embryogenesis and regeneration for attaining an effective transformation (Ramakrishnan et al., 2020) (Table 12.4).

12.4 Conclusion and Future Prospective

The human population depends on bamboo for many aspects of their life. The future of bamboo seems very promising due to an increasing concern about the environment and forest protection. The bamboo market is growing enormously. Micropropagation is the most appropriate and suitable technique to fulfil the increasing demand of the bamboos as it can produce millions of bamboos starting from a single explant within a short period of time. Although a tissue culture-raised plant is better in vigor and quality as compared to plants produced through conventional methods

Table 12.4. Genetic transformation in *Bamboo* species.

Bamboo Species	Method of Transformation	Target Gene	Aim of the Transformation	Result/Remark	References
Dendrocalamus giganteus	*Agrobacterium* mediated transformation	*Luc*	Cloning of luciferase gene of *Photinus pyralis* which is a reporter gene	Gene successfully transferred to *D. giganteus* and yield a fluorescent product that can be quantified by measuring released light.	Wiersma 2008
Phyllostachys nigra	Particle bombardment	*AcGFP1*, *mCherry* gene	To enhance expression of fluorescent protein	Achieved stable transformation of bamboo cells. These transformed cells were able to express *hpt* gene along with enhanced fluorescent protein genes *AcGFP1 and mCherry*.	Ogita et al. (2011)
Dendrocalamus latiflorus	*Agrobacterium* mediated transformation	*CodA*	To enhance production of cold tolerant enzyme cholin oxidase	Cold tolerance improved.	Qiao et al. (2014)
Dendrocalamus latiflorus	*Agrobacterium* mediated transformation	*Lc* gene of maize	Overexpression of *Lc* gene of maize	Achieved over-accumulation of anthocyanin.	Ye et al. (2017).
Phyllostachys nigra	Particle bombardment	*HvACT1* gene of barley	Switching of pathways of metabolite production	Pathways of agmatine production were become more active in transformed cells than previously active pathway of putrescine production.	Nomura et al. (2018)
Dendrocalamus latiflorus	*Agrobacterium* mediated transformation	*Lc (leaf color)* gene of maize	Overexpression of *Lc* gene of maize	Expression of *Lc* gene improved accumulation of anthocyanin conferring purple color also enhanced resistant in bamboo against heat and drought. This resistance increases due to production of antioxidant comparably higher quantity then the wild bamboo.	Xiang et al. (2021)
Phyllostachys edulis, *Dendrocalamus latiflorus*	*Agrobacterium* mediated transformation	*RUBY*	To achieve overexpression of external protein and to develop *RUBY* reporter system	After 3 days of transformation red color of root was observed due to expression of RUBY reporter gene.	Chen et al. (2021)

Abbreviations: RUBY, *ruby particles in mucilage*; LC, leaf color; HvACT1, hordeum vulgare agmatine coumaroyl transferase; CodA, cholin oxidase; *AcGFP1*, aequorea coerulescens green fluorescent protein; *mCherry*, monomeric red fluorescent protein; *LUC*, luciferase gene; *HPT*, hygromycin phosphotransferse gene.

but tissue culture applications are limited or restricted to some species. To improve the quality and genetic trait of any plant or for producing plant resistance against various abiotic and biotic stresses, genetic transformation is the best approach these days. In bamboos, genetic transformation is reported but confined to a few species; hence further emphasis and refinement is required to obtain genetically transformed bamboo on a larger scale (Shahzad et al., 2021). Presence of inadequate embryogenic tissue, low regeneration efficiency, and time-taking procedures limit the genetic transformation procedure in bamboos. Soon, if a researcher can overcome all these problems mentioned, it would be a major success in the field of genetic transformation of bamboo. Similarly, CRISPR Cas9 technology is a breakthrough discovery and has been applied in plants recently (Ye et al., 2017) but in bamboos, its application is limited and is only used by a few researchers. In future its application would widen along with genetic transformation to improve genetic modification and to obtain improved bamboo variety within a limited period (Ye et al., 2020). Also, recently the release of few bamboo genomes would help the researcher to perform gene transformation more efficiently. Understanding the molecular mechanism behind somatic embryogenesis and regeneration is also necessary to exploit gene editing technology at its best. Monocarpic or the long flowering cycle of bamboo is also a hindrance in the breeding of bamboos. This can be overcome by the induction of flowering in a controlled *in situ* or *ex situ* condition by various biotechnological interventions and this opportunistic flowering could be exploited for obtaining further variability. Further, an efficient anther culture system needs to be developed for several bamboo species to obtain haploid and double haploid. Bamboo cultivation is threatened by numerous adverse factors such as drought, salinity, nutrient deficiency, and land degradation. These problems of bamboo cultivation can be further overcome in the future by combining the classical approach of breeding along with biotechnological intervention such as CRISPER and genetic transformation. In this sense, future advances in the knowledge of bamboo micropropagation and genetic transformation can boost both fundamental studies and large-scale commercial production of the bamboos.

Author Contribution

Conceptualization: A.W., Z.A., A.S.; Writing – original draft and revise preparation: A.S., Z.A., A.S., S. T., A.E., M.R.; Supervision: A.S.

Conflict of Interest

Authors declare no conflict of interest.

Funding

The study is supported by the Council of Scientific and Industrial research (CSIR), India [Award No. 09/112(0677)/2020-EMR-I] and Department of Science and Technology (DST), India (Award No. IF; 190296) for providing financial assistance in the form of junior research fellowship (JRF) to the author Adla Wasi and Sabaha Tahseen, respectively.

References

Agnihotri, R.K. and Nandi, S.K. (2009). *In vitro* shoot cut: A high frequency multiplication and rooting method in the bamboo *Dendrocalamus hamiltonii*. *Biotech.*, 8: 259–263.

Ajithkumar, D. and Seeni, S. (1998). Rapid clonal multiplication through *in vitro* axillary shoots proliferation of *Aegle marmelos* (L.) Corr., a medicinal tree. *Plant Cell. Rep.*, 17: 422–426.

Alexander, M.P. and Rao, T.C. (1968). *In vitro* culture of bamboo embryo. *Curr. Sci.*, 415: 37.

Anand, M., Brar, J. and Sood, A. (2013). *In vitro* propagation of an edible bamboo Bambusa Bambos and assessment of clonal fidelity through molecular markers. *J. Med. Bioen.*, 2: 4.

Arya, S.A. Kant, Sharma, D. and Arya, I.D. (2008a). Micropropagation of two economically important bamboos: *Drepanostachyum falcatum* (Nees) Keng and *Bambusa balcooa* Roxb. *Indian Forest*, 134: 1211–1221.

Arya, S., Rana, P.K., Sharma, R. and Arya, I.D. (2006). Tissue culture technology for rapid multiplication of *Dendrocalamus giganteus* Munro. *Indian Forest*, 132: 345–357.

Arya, S., Sharma, S., Kaur, R. and Arya, I.D. (1999). Micropropagation of *Dendrocalamus asper* by shoot proliferation using seeds. *Plant Cell Rep.*, 18: 879–882.

Atta, R., Laurens, L., Boucheron-Dubuisson, E., Guivarc'h, A., Carnero, E., Giraudat-Pautot, V., Rech, P. and Chriqui, D. (2009). Pluripotency of Arabidopsis xylem pericycle underlies shoot regeneration from root and hypocotyl explants grown *in vitro*. *Plant J.*, 57: 626–644.

Beena, D.B., Rathore, T.S. and Rao, P.S. (2012). Effects of carbohydrates on *in vitro* axillary shoot initiation and multiplication of *Bambusa pallida* Munro. *J. Phytol.*, 4: 55–58.

Brar, J., Shafi, A., Sood, P., Anand, M. and Sood, A. (2014). *In-vitro* propagation, biochemical studies, and assessment of clonal fidelity through molecular markers in *Bambusa balcooa*. *J. Trop. For. Sci.*, 26: 115–124.

Chambers, S.M., Heuch, J.H.R. and Pirrle, A. (1991). Micropropagation and *in vitro* flowering of the bamboo *Dendrocalamus hamiltonii* Munro. *Plant Cell Tissue Organ Cult.*, 27: 45–48.

Chavan, N.S., Kale, S.S. and Deshmukh, V.S. (2021). Effect of different concentrations of BAP on *in vitro* shoot multiplication of bamboo. *J. Pharm. Innov.*, 10: 161–166.

Cheah, K.T. and Chaille, L.C. (2011). Somatic embryogenesis from mature *Bambusa ventricosa*. *Biotechnology*, 11: 1–5.

Chen, K., Hu, K., Xi, F., Wang, H., Kohnen, M.V., Gao, P. and Gu, L. (2021). High-efficient and transient transformation of Moso bamboo (*Phyllostachys edulis*) and Ma bamboo (*Dendrocalamus latiflorus* Munro). *J. Plant Biol.*, https://doi.org/10.1007/s12374-020-09294-y.

Chowdhury, P., Das, M., Sikdar, S.R. and Pal, A. (2004). Influence of the physiological age and position of the nodal explants on micropropagation of field-grown *Dendrocalamus strictus* Nees. *Plant Cell Biotechnol. Mol. Biol.*, 5: 45–50.

Choudhary, A.K., Priyanka, K. and Ashish, R. (2017). Refinement of protocol for rapid clonal regeneration of economical bamboo, *Bambusa balcooa* in the agroclimatic conditions of Bihar, India. *African J. Biotechnol.*, 16: 450–462.

Das, M. and Pal, A. (2005a). *In vitro* regeneration of *Bambusa balcooa* Roxb.: Factors affecting changes of morphogenetic competence in the axillary buds. *Plant Cell Tissue Organ Cult.*, 81: 109–112.

Das, M. and Pal, A. (2005b). Clonal propagation and production of genetically uniform regenerants from axillary meristems of adult bamboo. *J. Plant Biochem. Biotech.*, 13: 185–188.

Devi, W.S., Bengyella, L. and Sharma, G.J. (2012). *In vitro* seed germination and micropropagation of edible bamboo *Dendrocalamus giganteus* Munro using seeds. *Biotechnol. J.*, 11: 74–80.

Funada, R.T., KuboTabuchi, M., Sugiyama, T. and Fushitani, M. (2001). Seasonal variations in endogenous indole-3-acetic acid and abscisic acid in the cambial region of *Pinus densiflora* Sieb. Et Zucc. Stems in relation to earlywood/latewood transition and cessation of tracheid production. *Holzforschung*, 55: 128–134.

Gamborg, O.L., Miller, R.A. and Ojima, K. (1968). Nutrient requirements of suspension cultures of soybean root cells. *Exp. Cell Res.*, 50: 151–158.

Ganie, I.B., Shahzad, A., Ahmad, Z., Bukhari, N.A. and Parveen, K. (2021). Transgenic approaches in bamboo. pp. 251–273. *In*: Ahmad, Z., Ding, Y. and Shahzad, A. (eds.). *Biotechnological Advances in Bamboo*. Springer, Singapore.

Gantait, S.B., Pramanik, R. and Banerjee, M. (2016). Optimization of planting materials for large scale plantation of *Bambusa balcooa* Roxb.: Influence of propagation methods. *J. Saudi Soc. Agric. Sci.*, 17: 79–87.

Gielis, J. (1995). Bamboo and biotechnology. *European Bamb. Soc. J. Ser.*, 6: 27–39.

Gillis, K., Gielis, J., Peeters, H., Dhooghe, E. and Oprins, J. (2007). Somatic embryogenesis from mature *Bambusa balcooa* Roxburgh as basis for mass production of elite forestry bamboos. *Plant Cell, Tissue, and Organ Cult.*, 91: 115–123.

Godbole, S., Sood, A., Thakur, R., Sharma, M. and Ahuja, P.S. (2002). Somatic embryogenesis and its conservation into plantlets in a multipurpose bamboo, *Dendrocalamus hamiltonii* Nees et Arm Ex. Munro. *Curr. Sci.*, 83: 885–889.

Goyal, A.K., Pradhan, S., Basistha, B.C. and Sen, A. (2015). Micropropagation and assessment of genetic fidelity of *Dendrocalamus strictus* (Roxb.) Nees using RAPD and ISSR markers. *3 Biotech.*, 5: 473–482.

Hu, S.L., Zhou, J.Y., Cao, Y., Lu, X.Q., Duan, N., Ren, P. and Chen, K. (2011). *In vitro* callus induction and plant regeneration from mature seed embryo and young shoots in a giant sympodial bamboo, *Dendrocalamus farinosus* (Keng et Keng f.) Chia et H.L. Fung. *Afr. J. Biotechnol.*, 10: 3210–3215.

Ikeuchi, M., Sugimoto, K. and Iwase, A. (2013). Plant callus: Mechanisms of induction and repression. *Plant Cell*, 25: 3159–3173.

Jat, B.L., Panwar, R., Gena, D., Mir, M.A. and Rawat, R.S. (2016). *In vitro* micropropagation of *Dendrocalamus strictus* (solid bamboo). *World J. Pharm. Res.*, 5: 838–864.

Jiang, K. and Zhou, M. (2014). Recent advances in bamboo molecular biology. *J. Trop. Subtrop. Bot.*, 22: 632–642.

Jiménez, V.M., Castillo, J., Tavares, E., Guevara, E. and Montiel, M. (2006). *In vitro* propagation of the neotropical giant bamboo, *Guadua angustifolia* Kunth, through axillary shoot proliferation. *Plant Cell Tissue Organ Cult.*, 86: 389–395.

Jiménez, V.M., Holst, A., Carvajal-Campos, P. and Guevara, E. (2021). Standard protocols for *in vitro* propagation of bamboo with emphasis on axillary shoot proliferation. *In*: *Biotechnological Advances in Bamboo*. Springer, Singapore, pp. 63–84.

Kabade, A.U. (2009). *Studies on Refinement of Protocols for Rapid and Mass in Vitro Clonal Propagation, Evaluation of Genetic Fidelity and Growth Performance of Bamboo Species—Bambusa bambos (L.) Voss and Dendrocalamus strictus (Roxb.) Nees.* Forest Research Institute, Dehradun.

Kaladhar, D.S.V.G.K., Tiwari, P. and Duppala, S.K. (2017). A rapid *in vitro* micropropagation of *Bambusa vulgaris* using inter-node explant. *Int. J. Life. Sci. Scienti. Res.*, 1052–1054.

Kalaiarasi, K., Sangeetha, P., Subramaniam, S. and Venkatachalam, P. (2014). Development of an efficient protocol for plant regeneration from nodal explants of recalcitrant bamboo (Bambusa arundinacea Retz. wild) and assessment of genetic fidelity by DNA markers. *Agrofor. Syst.*, 88: 527–537.

Kapoor, P. and Rao, I.U. (2006). *In vitro* rhizome induction and plantlet formation from multiple shoots in *Bambusa bambos* var. *Gigantea* Bennet and Gaur by using growth regulators and sucrose. *Plant Cell Tissue Organ Cult.*, 85: 211–217.

Khalil, H., Bhat, A., Jawaid, I.U.H., Zaidon, M., Hermawan, A., Hadi, D. and Hadi, Y.S. (2012). Bamboo fibre reinforced biocomposites: A review. *Materials & Design*, 42: 353–368.

Komatsu, Y.H., Batagin-Piotto, K.D., Brondani, G.E., Gonçalves, A.N. and de Almeida, M. (2011). *In vitro* morphogenic response of leaf sheath of *Phyllostachys bambusoides*. *J. For. Res.*, 22: 209–215.

Lee, P.C., Muniandi, S.K. and Ab Shukor, N.A. (2019). *In vitro* regeneration of bamboo species. *Pertanika Journal of Scholarly Research Reviews*, 4: 3.

Li, S., Cong, Y., Liu, Y., Wang, T., Shuai, Q., Chen, N., Gai, J. and Li, Y. (2017). Optimization of *Agrobacterium*-mediated transformation in soybean. *Front Plant. Sci.*, 8: 246.

Lin, C.S., Lin, C.C. and Chang, W.C. (2004). Effect of thidiazuron on vegetative tissue-derived somatic embryogenesis and flowering of bamboo *Bambusa edulis*. *Plant Cell Tissue Organ Cult.*, 76: 75–82.

Lin, C.S., Lin, C.C. and Chang, W.C. (2005). Shoot regeneration, re-flowering, and post flowering survival in bamboo inflorescence culture. *Plant Cell Tissue Organ Cult.*, 243–249.

Lobovikov, M.D. Schoene and Yping, L. (2012). Bamboo in climate change and rural livelihoods. *Mitigation and Adaptation Strategies for Global Change*, 17: 261–276.

Maiya, S.M., Janardan, L. and Prasad, G.D. (2021). The impact of various factors of *in vitro* culture on shoot multiplication and plant production of the *Bambusa nutans* subsp. cupulata in *in vitro* propagation through nodal segments. *Int. J. Res. Anal. Rev.*, 8: 766–776.

McCown, B.H. and Lloyd, G. (1981). Woody plant medium (WPM)—a mineral nutrient formulation for microculture of woody plant-species. *Hort. Science*, 16: 453–453.

Mehta, U., Rao, I.V.R. and Mohan Ram, H.Y. (1982). Somatic embryogenesis in bamboo. pp. 109–110. *In*: Fujiwara, A. (ed.). *Plant Tissue Culture. Jpn. Assoc. Proc. 5th Intl Cong Plant Tissue Cell Cult.*

Mehta, R., Sharma, V., Sood, A., Sharma, M. and Sharma, R.K. (2011). Induction of somatic embryogenesis and analysis of genetic fidelity of *in vitro*-derived plantlets of *Bambusa nutans* Wall., using AFLP markers. *Eur. J. For. Res.*, 130: 729–736.

Mishra, Y., Patel, P. and Ansari, S.A. (2011). Acclimatization and macroproliferation of micropropagated plants of *Bambusa tulda* Roxb. *Asian J. Exp. Biol. Sci.*, 2: 498–550.

Mishra, Y., Patel, P.K., Yadav, S., Shirin, F. and Ansari, S.A. (2008). A micropropagation system for cloning of *Bambusa tulda* Roxb. *Sci. Hortic.*, 115: 315–318.

Mudoi, K.D., Saikia, S.P., Goswami, A., Gogoi, D., Bora, D. and Borthakur, M. (2013). Micropropagation of important bamboos: A review. *Afr. J. Biotechnol.*, 12: 20.

Murashige, T.F. Skoog. (1962). A revised medium for rapid growth and bioassays with tobacco tissue cultures. *Physiol. Plant.*, 15: 473–497.

Nayak, L. and Mishra, S.P. (2016). Prospect of bamboo as a renewable textile fiber; historical overview, labeling, controversies, and regulation. *Fashion and Textiles*, 3: 1–23.

Negi, D. and Saxena, S. (2011a). Micropropagation of *Bambusa balcooa* Roxb. through axillary shoot proliferation. *In vitro Cell. Dev. Biol. Plant.*, 47: 604–610.

Negi, D. and Saxena, S. (2011b). *In vitro* propagation of *Bambusa nutans* Wall. ex Munro through axillary shoot proliferation. *Plant Biotechnol. Rep.*, 5: 35–43.

Nomura, T., Ogita, S. and Kato, Y. (2018). Rational metabolic flow switching for the production of exogenous secondary metabolites in bamboo suspension cells. *Sci. Rep.*, 8: 1–11.

Obsuwan, K., Duangmanee, A. and Thepsithar, C. (2019). *In vitro* propagation of a useful tropical bamboo, *Thyrsostachys siamensis* Gamble, through shoot-derived callus. *Hortic. Environ. Biotechnol.*, 60: 261–267.

Ogita, S. (2005). Callus and cell suspension culture of bamboo plant, *Phyllostachys nigra*. *Plant Biotechnol.*, 22: 119–125.

Ogita, S., Kashiwagi, H. and Kato, Y. (2008). *In vitro* node culture of seedlings in bamboo plant, *Phyllostachys meyeri* McClure. *Plant Biotechnol.*, 25: 381–385.

Ogita, S.N., Kikuchi, T. Nomura and Kato, Y. (2011). A practical protocol for particle bombardment-mediated transformation of *Phyllostachys* bamboo suspension cells. *Plant Biotechnol.*, 28: 43–50.

Ojha, A., Verma, N. and Kumar, A. (2009). *In vitro* micropropagation of economically important edible bamboo (*Dendrocalamus asper*) through somatic embryos from root, leaves, and nodal segments explants. *Res. Crops*, 10: 430–436.

Oprins, J., Grunewald, W., Gillis, K., Delaere, P., Peeters, H. and Gielis, J. (2004). Micropropagation: A general method for commercial bamboo production. *In*: *7th World. Bamboo Congress*, 7: 1–11.

Polesi, L.G.L., Vieira, D.N., Guerra, M.P. and Fraga, H.P.D.F. (2021). Somatic embryogenesis in bamboos. pp. 85–105. *In*: Ahmad, Z., Ding, Y. and Shahzad, A. (eds.). *Biotechnological Advances in Bamboo*. Springer, Singapore.

Polesi, L.G., Fraga, H.P.D.F., Vieira, L.D.N., Heringer, A.S., Ornellas, T.S., dos Santos, H.P. and Pescador, R. (2019). Chloroplast ultrastructure and hormone endogenous levels are differently affected under light and dark conditions during *in vitro* culture of Guadua chacoensis (Rojas) Londoño & PM Peterson. *Acta. Physiologiae. Plantarum*, 41: 1–12.

Pratibha, S. and Sarma, K.P. (2013). *In vitro* propagation of *Bambusa tulda*: An important plant for better environment. *J. Environ. Res. and Develop.*, 7: 1216–1223.

Pratibha, S. and Sarma, K.P. (2014). *In vitro* propagation of *Bambusa pallida* on commercial scale in Assam, India. *J. Environ. Res. and Develop.*, 8: 4.

Qiao, G., Yang, H., Zhang, L., Han, X., Liu, M., Jiang, J. and Zhuo, R. (2014). Enhanced cold stress tolerance of transgenic *Dendrocalamus latiflorus* Munro (Ma bamboo) plants expressing a bacterial CodA gene. *In Vitro Cell Dev. Biol. Plant*, 50: 385–391.

Rajput, B.S., Jani, M.D., Gujjar, M.R. and Shekhawat, M.S. (2019). Effective and large-scale *in vitro* propagation of *Dendrocalamus strictus* (Roxb.) Nees using nodal segments as explants. *World Sci. News*, 130: 238–249.

Rajput, B.S., Jani, M., Ramesh, K., Manokari, M., Jogam, P., Allini, V.R. and Shekhawat, M.S. (2020). Large-scale clonal propagation of *Bambusa balcooa* Roxb.: An industrially important bamboo species. *Ind. Crops Prod.*, 157: 112–905.

Raju, R.I. and Roy, S.K. (2016). Mass propagation of *Bambusa bambos* (L.) Voss through *in vitro* culture. *Jahangirnagar University Journal of Biological Sciences*, 5: 15–26.

Ramakrishnan, M., Yrjälä, K., Vinod, K., Sharma, A., Cho, J., Satheesh, V. and Zhou, M. (2020). Genetics and genomics of Moso bamboo (*Phyllostachys edulis*): Current status, future challenges, and biotechnological opportunities toward a sustainable bamboo industry. *Food and Energy Secur.*, 9: 229.

Rathore, T.S., Kabade, U., Jagadish, M.R., Somashekar, P.V. and Viswanath, S. (2009). Micropropagation and evaluation of growth performance of the selected industrially important bamboo species in southern India, in *Proc. 8th World Bamboo Cong.*, 6: 41–55.

Rout, G.R. and Das, P. (1997). *In vitro* plant regeneration via callogene-sis and organogenesis in *bambusa vulgaris*. *Biol. Plant.*, 515–522.

Saeed, T. and Shahzad, A. (2016). Advances in molecular approaches for the integrative genetic transformation of highly important climbers. pp. 367–385. *In*: Shahzad, A., Sharma, S. and Siddiqui, S. (eds.). *Biotechnological Strategies for the Conservation of Medicinal and Ornamental Cimbers*. Springer, Netherlands.

Saini, H., Arya, I.D., Arya, S. and Sharma, R. (2016). *In vitro* micropropagation of Himalayan weeping bamboo, *Drepanostachyum falcatum*. *Am. J. Plant Sci.*, 7–1317.

Sandhu, M., Wani, S.H. and Jiménez, V.M. (2018). *In vitro* propagation of bamboo species through axillary shoot proliferation: A review. *Plant Cell. Tissue Organ Cult.*, 132: 27–53.

Sanjaya, T., Rathore, S. and Rai, V.R. (2005). Micropropagation of *Pseudoxytenanthera stocksii* Munro. *In Vitro Cell. Dev. Biol. Plant.*, 41: 333–337.

Saxena, S. (1990). *In vitro* propagation of the bamboo (*Bambusa tulda* Roxb.) through shoot proliferation. *Plant Cell Rep.*, 9: 431–434.

Saxena, S. and Bhojwani, S.S. (1993). *In vitro* clonal multiplication of 4-year-old plants of the bamboo, *Dendrocalamus longispathus* Kurz. *In Vitro Cellular Developmental Biology-Plant.*, 29: 135–142.

Saxena, S. and Dhawan, V. (1999). Regeneration and large-scale propagation of bamboo *(Dendrocalamus strictus* Nees) through somatic embryogenesis. *Plant Cell Rep.*, 18: 438–443.

Schenk, R.U. and Hildebrandt, A.C. (1972). Medium and techniques for induction and growth of monocotyledonous and dicotyledonous plant cell cultures. *Can. J. Bot. Ser.*, 50: 199–204.

Scott, T.K. (1972). Auxins and roots. *Annu. Rev. Plant Physiol.*, 23: 235–258.

Shahzad, A., Sharma, S., Parveen, S., Saeed, T., Shaheen, A., Akhtar, R., Yadav, V., Upadhyay, A. and Ahmad, Z. (2017). Historical perspective and basic principles of plant tissue culture. pp. 1–36. *In*: Abdin, M.Z., Kiran, U. and Ali, A. (eds.). *Plant Biotechnology: Principles and Applications*. Springer, Singapore.

Shahzad, A., Tahseen, S., Wasi, A., Ahmad, Z. and Khan, A.A. (2021). Application of biotechnological tool in bamboo improvement. pp. 291–312. *In*: Ahmad, Z., Ding, Y. and Shahzad, A. (eds.). *Biotechnological Advances in Bamboo*. Springer, Singapore.

Sharma, P. and Sarma, K.P. (2011). *In vitro* propagation of *Bambusa balcooa* for a better environment. *In*: *International Conferences on Advances in Biotechnology and Pharmaceutical Sciences (ICABPS'11)*. Bangkok, pp. 248–252.

Sen, T. and Reddy, H.J. (2011). Application of sisal, bamboo, coir, and jute natural composites in structural upgradation. *Int. J. Innov. Technol. Manag.*, 2: 186.

Shawan, S.M., Janardan, L. and Prasad, G.D. (2021). The impact of various factors of *in vitro* culture on shoot multiplication and plant production of the *Bambusa nutans* subsp. cupulata in *in vitro* propagation through nodal segments. *Int. J. Res. Anal. Rev.*, 8: 766–777.

Shen, J., Fu, J., Ma, J., Wang, X., Gao, C., Zhuang, C., Wan, J. and Jiang, L. (2014). Isolation, culture, and transient transformation of plant protoplasts. *Curr. Protoc. Cell Biol.*, 63: 1–7.

Singh, G. and Reeta, V. (2020). "Green gold" as superpower potential in for green India and mystery behind bamboo blossom. *J. Med. Plants*, 8: 112–117.

Singh, O. (2008). Bamboo for sustainable livelihood in India. *Indian Forester*, 134: 1193–1198.

Singh, S.R., Dalal, S., Singh, R., Dhawan, A.K. and Kalia, R.K. (2011). Micropropagation of *Dendrocalamus asper* (Schult. & Schult. F. Backer ex K Heyne): An exotic edible bamboo. *J. Plant Biochem. Biotechnol.*, 21: 220–228.

Singh, S.R., Dalal, S.U.N.-I.T.A., Singh, R.O.H.T.A.S., Dhawan, A.K. and Kalia, R.K. (2012b). Seasonal influences on *in vitro* bud break in *Dendrocalamus hamiltonii* Arn. Ex Munro nodal explants and effect of culture microenvironment on large-scale shoot multiplication and plantlet regeneration. *Indian J. Plant Physiol.*, 17: 13.

Singh, S.R., Singh, R., Kalia, S., Dalal, S., Dhawan, A.K. and Kalia, R.K. (2013). Limitations, progress, and prospects of application of biotechnological tools in improvement of bamboo—a plant with extraordinary qualities. *Physiol. Mol. Biol. Plants*, 19: 21–41.

Sood, A., Ahuja, P.S., Sharma, M., Sharma, O.P. and Godbole, S. (2002). *In vitro* protocols and field performance of elites of an important bamboo *Dendrocalamus hamiltonii* Nees et Arn. Ex Munro. *Plant Cell Tissue Organ Cult.*, 71: 55–63.

Sood, P., Bhattacharya, A., Joshi, R., Gulati, A., Chanda, S. and Sood, A. (2014). A method to overcome the waxy surface, cell wall thickening, and polyphenol induced necrosis at wound sites—the major deterrents to *Agrobacterium*-mediated transformation of bamboo, a woody monocot. *J. Plant Biochem. Biotechnol.*, 23: 69–80.

Sood, A., Bhattacharya, A., Sharma, M., Sharma, R.K., Nadha, H.K., Sood, P. and Ahuja, P.S. (2013). Somatic embryogenesis and *Agrobacterium* mediated genetic transformation in bamboos. *Somatic Embryogenesis and Genetic Transformation in Plants*, 166–178.

Soni, V., Jha, A.K., Dwivedi, J. and Soni, P. (2013). Traditional uses, phytochemistry and pharmacological profile of *Bambusa arudinacea* Retz. *CELLMED.*, 3: 20–21.

Soreng, R.J., Peterson, P.M., Romaschenko, K., Davidse, G., Zuloaga, F.O., Judziewicz, E.J., Filgueiras, T.S., Davis, J.I. and Morrone, O. (2015). A worldwide phylogenetic classification of the Poaceae (Gramineae). *J. Syst. Evol.*, 53: 117–137.

Thakur, R. and Sood, A. (2006). An efficient method for explant sterilization for reduced contamination. *Plant Cell Tissue Organ Cult.*, 84: 369–371.

Torrey, J.G. (1976). Root hormones and plant growth. *Annu. Rev. Plant Physiol.*, 27: 435–459.

Vanithakumari, G., Manonayagi, S., Padma, S. and Malini, T. (1989). Antifertility effect of *Bambusa aruninacea* shoot extract in male rat. *J. Ethanopharmocol.*, 25: 173–180.

Venkatachalam, P., Kalaiarasi, K. and Sreeramanan, S. (2015). Influence of plant growth regulators (PGRs) and various additives on *in vitro* plant propagation of *Bambusa arundinacea* (Retz.) Wild: A recalcitrant bamboo species. *J. Genet. Eng. Biotechnol.*, 13: 193–200.

Waikhom, S.D., Bengyella, L. and Sharma, G.J. (2012). *In vitro* seed germination and micropropagation of edible bamboo *Dendrocalamus giganteus* Munro using seeds. *Biotechnol.*, 11: 74–80.

Wei, Q., Cao, J., Qian, W., Xu, M., Li, Z. and Ding, Y. (2015). Establishment of an efficient micropropagation and callus regeneration system from the axillary buds of *Bambusa ventricosa*. *Plant Cell Tissue Organ Cult.*, 122: 1–8.

Wiersma, R. (2008). Bioluminescent Bamboo. *Newsl. South Calif. Chap. Am. Bamboo Soc.* 18: 2–5.

Woods, S.H., Philips, G.C., Woods, J.E. and Collins, G.B. (1992) Somatic embryogenesis and plant regeneration from zygotic embryo explants in Mexican weeping bamboo, Otatea acuminata Aztecorum. *Plant Cell Rep.*, 11: 257–261.

Xiang, M., Ding, W., Wu, C., Wang, W., Ye, S., Cai, C. and Zhu, Q. (2021). Production of purple Ma bamboo (*Dendrocalamus latiflorus* Munro) with enhanced drought and cold stress tolerance by engineering anthocyanin biosynthesis. *Planta.*, 254: 1–17.

Ye, S., Cai, C., Ren, H., Wang, W., Xiang, M., Tang, X. and Zhu, Q. (2017). An efficient plant regeneration and transformation system of ma bamboo (*Dendrocalamus latiflorus* Munro) started from young shoot as explant. *Front. Plant Sci.*, 8: 1298.

Ye, S., Chen, G., Kohnen, M.V., Wang, W., Cai, C., Ding, W. and Lin, C. (2020). Robust CRISPR/Cas9 mediated genome editing and its application in manipulating plant height in the first generation of hexaploid Ma bamboo (*Dendrocalamus latiflorus* Munro). *Plant Biotechnol. J.*, 18: 1501–1503.

Yuan, J.L., Yue, J.J., Wu, X.L. and Gu, X.P. (2013). Protocol for callus induction and somatic embryogenesis in Moso bamboo. *PloS One*, 8–81954.

Zhang, Y., Su, J., Duan, S., Ao, Y., Dai, J., Liu, J., Wang, P., Li, Y., Liu, B., Feng, D., Wang, J. and Wang, H. (2011). A highly efficient rice green tissue protoplast system for transient gene expression and studying light chloroplast-related processes. *Plant Methods*, 7: 30.

Zhang, N., Fang, W., Shi, Y., Liu, Q., Yang, H., Gui, R. and Lin, X. (2010). Somatic embryogenesis and organogenesis in *Dendrocalamus hamiltonii. Plant Cell Tissue Organ Cult.*, 103: 325–332.

Zhao, J.P., Gao, C., Li, X., Lin, X., Guo, X. and Liu, S. (2019). *PhePEBP* family genes regulated by plant hormones and drought are associated with the activation of lateral buds and seedling growth in *Phyllostachys edulis. Tree Physiol.*, 39: 1378–1404.

Chapter 13

Current Status of Bamboo Tissue Culture and Genetic Transformation Technology

Vidya R Sankar[1] and *Muralidharan Enarth Maviton*[2,*]

13.1 Introduction

With about 1,400 species worldwide, bamboos are a large and diverse group within the grass family. This wonderful grass, most of the species of which are woody, can adapt to almost any climate and grow in most of the continents.

Bamboos, which played a very important role in the economy of many rural communities for centuries are now finding applications in diverse areas such as housing, furniture, handicrafts, pulp, paper, charcoal, vinegar, vegetable (the bamboo shoot), and in recent times at an industrial scale for manufacture of engineered bamboo products (panels, boards, veneer, flooring, roofing, lumber, etc). The main advantage of bamboo over other woody plants lies in its fast growth since bamboo plant establishes and attains a harvestable age as early as in four to six years. The nature of growth in bamboo where new culms develop every year and mature in about two years is another unique advantage of bamboo. As a biofuel crop, because of its low ash content and alkali index (Sharma et al., 2018) it has been considered as a good resource for applications in renewable energy. For climate change mitigation and for restoration of degraded landscapes, this plant has been recognized as the best solution due to its excellent carbon sequestration potential and ability to grow in poor soils and help conserve soil and water due to its extensive underground rhizome and root system. Bamboo can also help ease the pressure off natural forests by providing for fuel and timber alternatives. Therefore, a rapid increase in demand

[1] Former Research Associate, Bamboo Technical Support Group, Kerala Forest Research Institute, Peechi, Thrissur, Kerala, India; Expert Member, Task Force on Sustainable Bamboo Management (TFSBM), International Bamboo and Rattan Organization (INBAR), Beijing.

[2] Chief Scientist (Retired), Kerala Forest Research Institute, Peechi, Kerala 680653 India;Consultant, INBAR, Beijing; Chair, Task Force on Sustainable Bamboo Management, INBAR.

Email: vidyarsankar@gmail.com

* Corresponding author: emmurali@gmail.com

for fast grown bamboo biomass in the form of poles for construction, furniture and timber alternatives, and lignocellulosic feedstock for bioenergy and other industrial uses such as paper and viscose, is anticipated in the coming years. Only through a rapid expansion in cultivation of bamboo can the demand be expected to be met since exploitation of natural bamboo forests is neither sustainable nor environmentally sound.

Not surprisingly, there has been a great deal of interest in establishment of bamboo plantations around the world not only in traditional bamboo growing areas but also in other regions where bamboo can adapt well. This generates a large-scale requirement of good quality planting material and necessitates a re-evaluation of the different propagation methods available for bamboo. Despite all the advantages, the wider cultivation of bamboo is currently restricted to only a few dozen species based on their adaptability to climatic factors and the physico-chemical and mechanical properties that are suitable for various applications.

Bamboos have a relatively short history of domestication except for a few species that have been cultivated in agricultural land and fringes of forests. Conventional breeding and tree improvement in bamboo is severely limited by several factors. Most bamboos are semelparous mast flowering species in which the next flowering year is unpredictable. The absence of simultaneous flowering in potential parent clumps is a hurdle to undertake conventional breeding programmes. Identification of genotypes with superior characteristics early in their flowering cycles and clonal propagation of these selections to generate sufficient planting material remains the most promising way ahead for tree improvement in bamboo. Genetic transformation which seeks to overcome such barriers and directly manipulate the plants at the DNA level also holds much promise.

13.2 Propagation

For long, all the bamboo required for traditional uses were collected from the forests and from the clumps growing in homesteads and the demand could be met in a sustainable manner. Seeds are produced in copious quantities after each gregarious flowering event, but seed viability in bamboo is unfortunately very short. Even with the best of storage conditions, seeds lose viability in about a year. The other great disadvantage of using seeds for propagation is that the progeny is genetically heterogeneous being predominantly outcrossing group. While this is an advantage for maintaining long-term genetic diversity in the wild and enables selection of desirable characteristics, in plantations it is desirable to maintain uniformity of form and properties for various applications.

13.2.1 *Vegetative Propagation*

Conventionally, propagation of bamboo has been carried out by a variety of vegetative propagation methods since seed availability is not assured when the need arises. Vegetative parts of the bamboo clump such as rhizomes offsets, culms, and branch cuttings are widely used for this purpose (Banik, 1995; Akinlabi et al., 2017). Species vary widely in their ability to be clonally propagated with the thick-walled species being most amenable. One to three-node culm sections or part of the branches with viable buds sprout shoots and root when provided with a suitable medium and moisture. Success is improved significantly with the aid of rooting hormones and use of mist chamber. Rooted culm and branch cuttings are ready for planting out after a hardening phase in the nursery during which rhizomes are formed.

An alternative method is layering such as ground or simple layering, stump layering, air layering, or marcotting and seedling layering (Banik, 2016) in which the nodal region is brough in contact with the soil or a suitable medium to encourage rooting while still attached to the clump. While success rates are relatively good, the method is not suitable for practice on a large scale due to limitations of space, time, and labour. The technique of macroproliferation has been developed as a means of continuing the multiplication process in the nursery from plants that has been produced through germination of seeds or any of the vegetative propagation methods (Kumar, 1991). The method takes advantage of the formation of new rhizomes every growing season at the base of each shoot (tiller). Separation of the tillers with at least a rhizome attached to each, allows the establishment of a propagule and the process can be repeated every six months and continued over three years (Banik, 2016).

The main shortcomings of the vegetative propagation methods for a large-scale planting programme are that they are time consuming, with low multiplication rates and are limited by the quantity of suitable vegetative material that can be collected from a clump. Extraction of rhizome offsets of medium and large sized species from mature clumps is laborious and prone to damage. Due to the large size of the propagule, handling in the nursery and transportation is cumbersome. The method has a high rate of success and is practical for small farms with small requirements for propagules every year or for botanical collections and germplasm accessions.

13.2.2 *Micropropagation*

Micropropagation has now been well accepted as a worldwide practice for the large-scale propagation of bamboo. The advantages of the method over conventional methods are high multiplication rates, disease-free nature, small size and uniformity of propagules and these increase the acceptance of the technology for large-scale multiplication. There are three pathways to *in vitro* regeneration of whole plants from cultured tissues. The most common method is to induce enhanced axillary bud proliferation in the presence of plant growth regulators in explants that contain a meristem, like the shoot tips and nodal explants. Multiple shoot formation results and the microshoots are in turn rooted *in vitro* or *ex vitro* to obtain whole plantlets. In the next alternative, *de novo* organogenesis is induced in explants without a meristem such as leaves, internodes, etc., and the shoot meristems proliferate into multiple shoots followed by root induction, as in the first method. In the third method, somatic embryogenesis, which is akin to organogenesis except that instead of shoot meristems being induced, bipolar embryos of somatic origin are generated. Somatic embryos follow a path of development which are like zygotic embryos with which they are morphologically similar and are capable of germination (conversion) to give plantlets without the need for a separate rooting step. While the first method of enhanced axillary bud proliferation is the most accepted method for commercial micropropagation for many ornamental species and forest tree species including bamboo, somatic embryogenesis is of relevance due to its potential for scaling-up, and for automation. In bamboo, as in many grasses, *de novo* organogenesis is not a common pathway for plant regeneration.

The first report on regeneration of bamboo plantlets in tissue culture was as early as in 1960s by Alexander and Rao (1968) in *Dendrocalamus strictus*, where embryos were used as the explant. This procedure laid a keystone for bamboo tissue culture and were followed by reports in *Bambusa arundinacea* (synonym: *B. bambos*) by several researchers such as Mehta et al. (1982), Nadgauda et al. (1990), Joshi and Nadgauda (1997). Subsequently seed and seedlings were used as explants for many species such as *D. hamiltonii* (Sood et al., 2002a, 2002b; Arya et al., 2012), *D. strictus* (Reddy, 2006), *D. giganteus* (Devi et al., 2012), *B. oldhamii* (Thiruvengadam et al., 2011), and in *B. bambos* (Vamil et al., 2010). The potential of zygotic embryo has been largely exploited in the species *D. strictus* (Nadgir et al., 1984; Mascarenhas et al., 1988; Zamora et al., 1988; Rout and Das, 1994; Maity and Ghosh, 1997; Ravikumar et al., 1998; Saxena and Dhawan, 1999) for clonal propagation. The uncertain genetic background of seeds, their limited and unpredictable availability, and short period of viability were the reasons that restricted the widespread application of tissue culture using juvenile material in bamboo.

It was only in the early 1990s that tissue culture of bamboo from material taken from adult clumps became successful on a consistent basis. The application of axillary buds as explants became a breakthrough in the field of clonal propagation and commercial production of bamboo. Nadgir et al. (1984) had reported success in *D. strictus* and *B. vulgaris*, and Vongvijitra (1988) in *Bambusa nana*, *D. asper*, and *Thyrsostachys oliveri* by using small branch cuttings. The other early reports are by Chaturvedi et al. (1993), Saxena and Bhojwani (1993). This system has been well established with species like *B. balcooa*, *D. giganteus*, *D. asper*, *D. hamiltonii*, *Guadua angustifolia*, etc. (as reviewed by Mudoi et al., 2013) and huge number of plantlets were produced at the commercial level too. The advantage of this technique is that the generation of true-to-type population eliminates the variability inherent in the sexually produced progeny. It is ideally suited for large-scale generation of quality planting material from selections (Plus Clumps) of bamboo made at the adult stage.

The alternate pathway of somatic embryogenesis permits a diversity in explants such as seeds, shoot tips, inflorescence, young leaf, anther, etc., that can be used to initiate cultures. Conversion of somatic embryos, in which the root is an integral component, to plantlets, overcomes the common constraint of poor rooting in shoot cultures. Somatic embryogenesis is greatly influenced by various factors like species, age of the explant, type of basal media, and plant growth regulators (Godbole et al., 2002) and carries an increased risk of genetic variability especially when a callus stage is involved. The main advantage of the somatic embryogenesis pathway lies in its application for genetic transformation. Somatic embryos are typically of single cell origin and a good example of totipotency of plant cells. This aspect is of importance since genetic engineering of plants involves the introduction of a foreign DNA first into cells and it is desirable to have plant regeneration from the transformed cells alone, so that chimeras are avoided in the regenerant plants.

Regeneration through somatic embryogenesis has been reported in species such as *D. strictus* (Rout and Das, 1994; Saxena and Dhawan, 1999), *D. giganteus* (Rout and Das, 1994), and *D. hamiltonii* (Godbole et al., 2002) and in many other species. Embryogenesis from vegetative tissue such as nodes, internodes, roots, leaf sheath, and floral parts has been found to be successful in many bamboo species. *D. hamiltonii* (Godbole et al., 2002), *D. longispathus* (Saxena and Bhojwani, 1993), *G. angustifolia* (Jiménez et al., 2006), and *B. edulis* (Lin et al., 2004) were successfully propagated with somatic embryos induced on the nodal region. Chang and Lan (1995) developed plantlets from roots in *B. beecheyana*. Floral tissues have been successfully used for embryo induction in *B. balcooa* (Gillis et al., 2007), *B. beecheyana* (Yeh and Chang, 1986b), *B. oldhamii* (Yeh and Chang, 1986a; Supaibulwattana, 1991), *D. asper* and *D. latiflorus* (Supaibulwattana, 1991), and *S. latiflora* (Tsay et al., 1990), etc. *Phyllostachys* species such as *P. nigra* (Ogita, 2005) and *P. viridis* (Hassan and Debergh, 1987) have been propagated from young and tender shoots. Sood et al. (2002) produced plantlets of *D. hamiltonii* on a large scale from seedling explants through somatic embryogenesis.

13.3 Limitations

The well accepted protocol, i.e., axillary proliferation in bamboo led to the emergence of many commercial tissue culture units throughout the world. Even though it is a well-established method, the wider application of this technique has been limited in many species when initiated from field-grown adult bamboo species. Success with adult tissues is limited by several factors such as the prevalence of superficial and systemic contamination, hyperhydricity, browning, *in vitro* flowering, variations in multiplication rates, and low *in vitro* rooting response, coupled with reduced survival during hardening and acclimatization (Gielis et al., 2001; Sandhu et al., 2017).

13.3.1 Tissue Browning

Among the various constraints in bamboo *in vitro* culture, tissue browning was found to be the most important one during the shoot multiplication and rooting stages. According to Compton and Preece (1986) and Oprins et al. (2004), different factors influence this phenomenon such as species, age and position of tissue, age of mother plant and season of explant excision, nutrient medium, sterilizing agent used, etc., and all these are accompanied with enhanced production of polyphenol oxidases (PPO) instigated by wounding either during culture initiation or subculturing (Huang et al., 2002). Presence of certain cytokinins like benzyl adenine (BA) can intensify the tissue browning depending on the pH of the media (Huang et al., 1989). Nutrient media with standard pH of 5.7 (acidic) gave a relatively low browning rate, whereas browning was higher in media having pH values of 7 and 8. Sankar (2019) found that an acidic pH (4.5) of tissue culture media was optimum for the shoot culture of *Pseudoxytenanthera ritcheyi*. Frequent subculturing and addition of polyvinylpyrrolidone (PVP), citric acid or ascorbic acid, and activated charcoal to the nutrient medium have been recommended to prevent browning in *D. giganteus* (Mudoi et al., 2013; Ramanayake and Yakandawala, 1997) and *D. strictus* (Saxena and Dhawan, 1999), respectively. The effectiveness of these treatments was also species specific and PVP has been reported as ineffective in *D. hamiltonii* (Singh et al., 2012) and *Phyllostachys nigra* (Ogita, 2005). These results were supported by findings of Huang et al. (2002) who have also shown PVP, activated charcoal or PPO inhibitors, such as ascorbic acid, cysteine, ferulic acid, kojic acid, and thiourea, to be ineffective in *B. oldhamii*, *D. latiflorus*, and *P. nigra*. Regular transferring to the fresh media in short duration was found to be most effective against tissue browning in most of the species along with the use of antioxidants in some cases for control of browning.

13.3.2 Microbial Contamination

The other major issue in micropropagation of bamboo when using explants from adult mother clumps is that of systemic microbial contamination (Oprins et al., 2004). The initiation and maintenance of sterile cultures from tissues collected from adult plants growing in the field is often one of the most challenging tasks in plant tissue culture of woody perennials in general. An extensive array of microbial communities resides on the surfaces of the plants and consequently only a thorough disinfection procedure using antimicrobial agents will ensure a contamination-free culture. Regular surface sterilization procedures reduced the exogenous contamination of bamboo explants to a range of 8.64%–9.76% depending upon factors such as season, area of collection, and the mother plant (Sankar et al., 2017).

13.4 Conventional Strategies for Control of Contamination

Common surface sterilization procedures such as with mercuric chloride ($HgCl_2$) and sodium hypochlorite (NaOCl) can easily control the phyllosphere flora in bamboo. A sequence of treatments with NaOCl for specific durations was reported as successful in *G. angustifolia* (Borges-García et al., 2004). Pretreatment with antimicrobial agents such as agrimycin and benomyl along with detergents following the use of NaOCl and PPM was also found to be very effective in this species (Jimenez et al., 2006).

A combination of different antimicrobials such as 0.01% Antiseptal (benzethonium chloride 10% and alkyl-arylpolyether alcohol 10%) (Yeh and Chang, 1986a, b), Bavistin with different antibiotics like streptomycin, kanamycin, rifampicin, ciprofloxacin, tetracycline, gentamycin, bacteriomycin, etc., followed by $HgCl_2$ has been reported as successful in many species (Sandhu et al., 2017). Dip sterilization with ethanol after surface sterilization is another strategy for reducing the contamination in bamboo shoot culture (Jha et al., 2013). Prophylactic treatment of the mother plants with different antifungal agents is another method to improve the efficiency of the surface sterilization.

Amendment of tissue culture media with antibiotics is the most common procedure for suppressing endophytic bacteria and Nadha et al. (2012) eliminated the bacterial contaminants *Pantoea agglomerans* and *P. ananatis* from shoot cultures of *G. angustifolia* by adding kanamycin (10 µg/ml) in culture media for 10 days. Axenic cultures of *D. longispathus* harbouring the endophytic Gram-positive bacteria, *Sporosarcina pasteruii* has been obtained by treating with gentamicin (250 µg/ml) (Sankar et al., 2017).

A strategy for control of bacteria and fungi in plant tissue culture that is occasionally in use involves isothiazolones. Plant Preservative Mixture™ (PPM), a technology patented by Plant Cell Technology, Washington, is a broad-spectrum biocide that consists of 5-chloro-2-methyl-3(2H)-isothiazolone and 2-methyl-3(2H)-isothiazolone. The advantage of PPM is that it is a microbicide against both bacteria and fungi, is heat stable and autoclavable with media but as a routine measure to control contamination is not likely to be cost effective and is known to reduce the *in vitro* response as reported by Rihan et al. (2012) and George and Tripepi (2001).

13.5 Latent Contamination

Conventional surface sterilization procedures do not give an assurance of establishment and maintenance of aseptic *in vitro* cultures of bamboo. Leifert et al. (1991) reported the presence of endophytic bacteria in *in vitro* cultures in commercial laboratories. Presence of the endophytic flora which are localized in the cell junctions and intercellular spaces of cortical parenchyma (Thomas, 2004) cannot be removed through this procedure and Thakur and Sood (2006) reported that the anatomical features of bamboo shoots provide space for several of the microbes. This endophytic population which consists of bacteria and fungi, remain dormant for a period of culture due to the conditions such as sucrose concentration, salt concentration, pH, and temperature (Cooke et al., 1992; Danby et al., 1994; Leifert et al., 1994). Since these microorganisms may survive in hidden form, culture indexing using bacteriological media for their detection is warranted to detect them in cultures (Leifert and Woodward, 1998; Thomas, 2004). Thomas and Sekhar (2014) reported the intercellular association of non-cultivable endophytic bacteria from the cytoplasmic and periplasmic spaces of *in vitro* culture of banana. The minor changes in media, its pH or culture conditions triggers the endophytic population into active multiplication which reduces growth rate, retards rooting or result in death of cultures (Leifert and Waites, 1992; Ewald et al., 1997; Leifert and Cassells, 2001). Incidence of latent contamination in bamboo was always observed at the range of 50%–53% and above in *B. balcooa*, *P. stocksii*, and *P. ritcheyi* shoot cultures (Sankar, 2019).

Endophytes which are useful to the host plant in their natural conditions can break the mutualistic association under stressed conditions (Mwamba, 1995). Holland and Polacco (1994) have shown their impact in *in vitro* culture through the hormone-mediated response which influences the reproducibility of the protocol. Their role in different physiological pathways such as biosynthesis of growth-promoting phytohormones, intensification of nutrient availability, and enhanced resistance to pathogens, etc. (Vendan et al., 2010; Goh and Vallejos, 2013; Hassan, 2017) has been reported. The altered morphology of the *in vitro* raised endophyte-freed *Pinus sylvestris* had been reported by Pirttilä et al. (2004) which could be restored by the introduction of the endophytic product to the medium. Dias et al. (2009) related the positive impact of the endophytic bacteria during acclimatization of *in vitro* developed strawberry plantlets. Significant loss in shoot multiplication and growth rates has been reported in shoot cultures of *D. longispathus*, which was treated with antibiotics to remove the endophytic bacterial contamination (Sankar et al., 2017).

13.6 Systemic Acquired Resistance to Control Latent Contamination

Activation of plant defence system was found to be an excellent method to control the exogenous as well as endophytic contamination. Defensive chemicals and non-protein compounds produced in plants during the various stresses such as insect attack can act against a wide range of pathogens. Plant defence activation was found to be a successful strategy for the control of latent contamination in shoot cultures of *B. balcooa*, *P. stocksii*, and *P. ritcheyi* (Sankar, 2019). Both prophylactic treatment of explants with chitosan and the amendment of media with chitosan have been found to be very effective in controlling the latent contamination in bamboo shoots cultures. Application of exo-polysaccharides isolated from endophytic bacteria on mother plants 24 hrs before explant collection reduced the latent contamination in all three bamboo species studied. Treatment of secondary branches with axillary buds with 3% hydrogen peroxide for 5 hrs reduced the initial contamination to between 0.98%–1.11% and latent contamination between 0.57%–1.88% in *B. balcooa*, *P. stocksii*, and *P. ritcheyi*, respectively. Application of jasmonic acid (JA) (100 µM) and beta-amino benzoic acid (BABA) (30 µg/mL) on the leaves of the mother plant one day prior to the collection of nodes was effective against all the exogenous and endogenous contaminants in all the three bamboo species. Both these chemicals were found to be more effective against fungal contaminants. An advantage of JA and BABA pretreatment was that it facilitated the use of sprouted buds for culture initiation especially in the case of *P. ritcheyi*. Conventionally, it is always explants with dormant buds that are used since the chemicals used for surface sterilization damage the softer tissues of sprouted buds. The treatments reduced the contamination to < 3% in all the species during all phases of culture (Sankar, 2019).

13.7 Rooting

A major constraint in micropropagation of bamboo is the difficulty in induction of rooting in the microshoots obtained in the multiplication phase and thereby reducing the efficiency of the overall process. Root induction frequency varied with the age of the mother plant, and it is easier in shoot cultures of seedling origin (Ramanayake et al., 2006). Addition of glucose along with IBA during the root induction phase gave 85% rooting success in *B. nutans* (Yasodha et al., 2008).

Ex vitro rooting, which involves treating the microshoots with rooting hormones to induce roots in the greenhouse as is done in the case of rooting stem or branch cuttings in vegetative propagation, is widely adopted in micropropagation. Not only is an expensive *in vitro* step avoided, but the shift to *ex vitro* rooting has been shown to be beneficial in terms of functional root development and consequently improved hardening of plantlets. Success with *ex vitro* rooting of the bamboo tissue culture plants is rare. Ravikumar et al. (2008) reported the rooting in 85–90% of *D. strictus* shoots originating from seedlings within 20–25 days of incubation at 85–90% RH after pulse treatment with IBA.Similarly, 10 min. treatment in 1000 mg/l NAA induced 99% *ex vitro* rooting in shoots of *P. stocksii* (Somashekar et al., 2008). NAA at 2000 mg/l induced rooting in *B. bambos* shoot clusters (Kabade, 2009). Yan et al. (2010) reported that the plantlets rooted *ex vitro* were more similar in morphology as well as with a similar root system with lateral roots, to their mother plants whereas *in vitro* rooted plants were significantly different during acclimatization.

13.8 *In vitro* Rhizome Induction

The underground rhizome plays an important role in the development of the bamboo plant and the shoots that emerge in the growing season every year grow out from the rhizome that develop during the previous year. Kapoor and Rao (2006) had success in inducing *in vitro* rhizomes in *B. bambos var. gigantea* shoot cultures in media containing 5% sucrose. Earlier Shirgurkar et al. (1996) had demonstrated the phenomenon in *D. strictus*. In the culture regenerants of *D. strictus*, a monopodial type microrhizome developed in the presence of Indole-3-butyric acid (IBA), which helped to transplant the regenerants in the soil (Chowdhury et al., 2004) and develop into adult plants In the experimental garden. It is also expected that plantlets with a perennating organ will help in easy transportation of the micropropagated plants since they can tolerate desiccation to quite an extent.

13.9 *In vitro* Flowering

The phenomenon of *in vitro* flowering in bamboo cultures has been reported in more than a dozen species (Yuan et al., 2017). Like the flowering behavior in bamboo clumps, the phenomenon is not easily amenable to elucidation. When different factors were evaluated in shoot cultures at the multiplication stage of five different bamboo species, *B. balcooa*, *B. tulda*, *B. nutans*, *D. longispathus*, and *P. stocksii*, it was found that photoperiod played an important role (Sankar, 2019). Among the photoperiods of 8/16 to 16/8 tested, the highest incidence in *in vitro* flowering occurred under the 16/8 light period between 40 d–52 d. The next best photoperiod for flowering was 14/10 for all these species, but only after 70 days of incubation. The shoot cultures maintained in photoperiod lower than 12 hrs were found to be optimum for vegetative growth.

Sankar (2019) also evaluated the effect of nutrient stress in shoot cultures on flowering. Nutrient stress was induced either by increasing the length of passage of the shoot culture or by sudden shifting of the shoots into low salt media. Prolonged incubation of shoot cultures without subculturing resulted in the depletion of all nutrients in the media which resulted in flowering in eight weeks under the 16/8 photoperiod and in 10 weeks under the 8/16 photoperiod. Shorter subculture periods induced active shoot growth in all the species tested and therefore maintaining this condition was suggested as the strategy for overcoming the incidence of flowering *in vitro*. When shoot cultures were grown in full strength MS medium for several subculture periods and then transferred to a diluted medium, the time taken for flowering to appear was significantly reduced which was also influenced by photoperiod and the age of the mother plant from which the explants were derived. Flower bud formation was found early in cultures established from axillary buds of adult field grown bamboo which were incubated under shorter photoperiod than in seedling cultures.

The type of cytokinins or their levels, particularly when high, at the shorter photoperiods also influenced flowering. Longer photoperiod (16/8) induced flowering in all the five species tested even at lower concentration of BAP (15 μM) with a delay of up to 70 days, whereas at 90 μM flowering occurred within 30 days of inoculation. Similarly, TDZ also induced flowering under 16/8 h at 1 μM and 2.5 μM within 45 and 30 days, respectively. Application of individual auxins at different concentrations had no influence on *in vitro* flowering in bamboo whereas, a combination of NAA and IBA (10 μM and 13 μM, respectively) with the photoperiod of 10/14 induced flowering in all shoot cultures within 45 days.

13.10 Application of Modern Technology

Micropropagation is today an industry that is expanding globally, and the number of species being commercially propagated through tissue culture is growing (Patil et al., 2021). A perusal of literature will show that the list of species for which laboratory protocols are available is much higher than those that eventually are successful at the commercial level. The case of bamboo is not very different. Propagation of almost all the economically important species of the world have been reported, but the list of species for which commercial micropropagation is successful is far too few. Clearly, scale up of lab procedures requires the focus on some of the steps that effect the overall efficiency of the process. Low multiplication rates and poor rooting are typically the constraints faced in bamboo micropropagation.

13.11 Potential for Photoautotrophic Micropropagation (PAM)

Low survival of plantlets during rooting and hardening stages is linked to the physiology of the shoots that undergoes a transition from heterotrophic to autotrophic conditions (Pérez et al., 2015; Ha et al., 2017). The heterotrophic conditions provided by sugar-containing media and low light conditions typical of plant tissue culture growth rooms are conducive to rapid multiplication but result in impaired anatomy and physiology of the shoots. To overcome many of the problems associated with heterotrophic tissue culture and to improve the efficiency of the process, photoautotrophic micropropagation (PAM) has been advocated (Kozai, 1991).

PAM is carried out under conditions that allow growth, multiplication, and rooting of chlorophyllous explants with photosynthetic ability on sugar-free media and an environment characterized by low relative humidity in the culture vessel headspace, high levels of light and carbon dioxide (Nguyen et al., 2016). Tissues under PAM have reduced incidence of morphological as well as physiological disorders along with much reduced microbial contamination. Plants developed through PAM exhibit higher survival rate during hardening and acclimatization.

In bamboo, Nguyen and Kozai (2005) have reported photoautotrophic micropropagation. Increased fresh weight and number of new, unfolded leaves were seen in treatment under photoautotrophic conditions, at 45 d. Root formation also increased under this condition. After being transferred to the *ex vitro* stage for 21 days, bamboo shoots produced under photoautotrophic conditions had a 20% higher survival rate.

The other advantages of PAM are the enhanced year-round production of plantlets with a simplified micropropagation system. With well-designed culture vessels, this will provide conditions to develop an automated system which results in the reduction of the labour cost (Nguyen et al., 2016).

13.12 Liquid Cultures/Bioreactors

Use of bioreactors and liquid media for micropropagation has been shown in several species to enhance the multiplication rates both through axillary bud proliferation as well as somatic embryogenesis. Muralidharan (2009) found the potential for use of bioreactors for multiplication and development of propagule delivery systems particularly using somatic embryogenesis is high since bamboo is amenable to culture in liquid media.

Sood et al. (2002) compared agar-solidified medium and liquid medium with respect to shoot multiplication and root formation of *D. hamiltonii* and found that liquid culture is more suitable. Ara (2020) has shown that it is possible to control shoot and root development in 11 species of bamboo in liquid culture. Gutiérrez et al. (2016) were successful in growing *G. angustifolia* shoots and rhizomes in a temporary immersion bioreactor (TIM) in which cultures were intermittently exposed to liquid media and results were better than that obtained on semisolid media.

13.13 Genetic Transformation for Bamboo Improvement

The limitations of conventional breeding are especially severe in bamboo that makes any planned crossing between selected plants almost impossible due to the unpredictability of flowering. Selection of superior genotypes capture the unique combinations of traits that may come up in nature, but the breeder's ability to bring together useful traits in planting material is defeated. Biotechnological tools that enable genetic engineering and introduction of desirable traits into bamboo and resulting in rapid and efficient genetic improvement (Limera et al., 2017) therefore attain importance.

Genetic transformation and breeding studies in bamboo are still limited (Ramakrishnan et al., 2020). *Agrobacterium* mediated transformation are found to be ineffective in bamboo because they are not a natural host for this bacterium (De Cleene and De Ley, 1976). Ogita et al. (2011) developed a particle bombardment protocol for suspension cell cultures in *Phyllostachys nigra* Munro var. *henonis* with constructs expressing hygromycin phosphotransferase gene and enhanced fluorescent protein genes namely, *AcGFP1* and *mCherry*.

Successful genetic transformation by *Agrobacterium* has been restricted in bamboo due to the rapid differentiation of the monocot cells/tissues (Graves et al., 1988), lignification/sclerification during the *vir* induction process, lack of/or reduced cell division (Kahl, 1982) and limited transgene integration rates (Frame et al., 2002), etc. Infection evokes different plant responses, therefore, success rates of producing transgenic monocots is highly variable (Sood and Sood, 2013). Sood et al. (2013) working with the somatic embryogenesis system of *D. hamiltonii* found that necrosis due to polyphenol oxidation, lack of differentiation due to cell wall thickening at wound sites, and the waxy surfaces of the somatic embryos were the main reasons that prevent *Agrobacterium* attachment and infection. They developed an effective method for transforming somatic embryos with the use of 0.01% Tween-20 as surfactant during infection. Lack of an efficient regeneration procedure, explant materials that are used for the embryogenesis, ploidy level of the callus derived from the explants, etc., are the reasons for the failure of these techniques in bamboo. Besides, the factors such as tissue type, strains of *Agrobacterium* used for infection, concentration of inoculum, tissue culture media used for culture development, selection agents and markers, as well as the type of vectors play a role in genetic transformation (Sood et al., 2011; Huang et al., 2022). Protocols were developed for genetic transformation of *D. hamiltonii* both by *Agrobacterium tumefaciens* as well as micro projectile bombardment mediated approach (Sood et al., 2013). Ye et al. (2017) reported that the calli derived from young vegetative tissues are ideal for genetic transformation even though it is found to be very difficult. They reported a plantlet regeneration and *Agrobacterium*-mediated transformation protocol for *D. latiflorus* with young shoots as explants. Cold stress tolerance was enhanced *D. latiflorus* by the accumulation of glycine betaine in through the introduction of a bacterial CodA gene encoding choline oxidase by *Agrobacterium*-mediated transformation (Qiao et al., 2014). Genomic studies and transcriptome analysis of bamboo are providing interesting insights that could be of potential application for genetic transformation. Several of these are related to environmental stresses. Sun et al. (2016) analysed the PeUGE gene from Moso Bamboo (*Phyllostachys edulis*) that was induced by abiotic stresses such as drought, salinity, and water stress, involved in biosynthesis of cell wall polysaccharides and conferred tolerance to drought. Sun et al. (2017) demonstrated that the bamboo aquaporin gene PeTIP4;1–1 confers drought and salinity tolerance in transgenic *Arabidopsis*. Gao et al. (2021) reported that a Moso bamboo transcription factor (Phehdz1) positively regulates the drought stress response in transgenic rice. Among other traits, the possibilities of downregulating lignin pathway genes, modulating the abiotic stress responses, or control over flowering would be of interest from the industrial point of view.

Biotechnological tools like RNA interference (RNAi), trans-grafting, cis genesis/intragenesis and genome editing tools, like zinc-finger and CRISPR/Cas9, have been developed recently that permits fast and precise genetic modifications and overcomes the limitations of *Agrobacterium*-mediated transformation. Ye et al. (2020) induced homozygote mutations in protoplasts of *D. latiflorus* through CRISPR/Cas9 technology and they obtained the transgenic lines with altered plant height. Huang et al. (2022) established the first reported immature embryo plant regeneration system along with genome editing in *P. edulis* using CRISPR/Cas9. The advent of CRISPR/Cas9 technologies that allow genome editing of specific sites in the genome and controlling the expression of target genes, is promising for bamboo.

Bamboos are grown as a long-term forestry crop. As with any tree crop, the implications of genetic transformation especially for traits that impart resistance or tolerance against pest and diseases and other abiotic and biotic stress factors, would give the GMO a selective advantage and increase the risk of weediness. Many of the monopodial bamboos with their running rhizomes already are treated with caution when planted in a new locality. The risk of genetically modified bamboo with improved vigor is to be critically examined to avoid invasiveness.

The threat to biodiversity in bamboo planted in forests is also to be viewed with caution. The ease with which bamboo establishes itself in most situations is the reason it is considered ideal for forest landscape restoration. The bamboo root system is profuse in nature and a thick mat of fibrous roots around the clumps discourages the growth of other plants as does the leaf litter that is relatively slow to decompose due to the higher levels of silica. Hence, introduction of GM bamboo to forest areas is to be regulated with discretion.

13.14 Conclusions

The advances made in tissue culture and genetic engineering in bamboo have been encouraging since several important species are now being commercially micropropagated around the world on a large scale and several examples of bamboo species showing the proof of concept of genetic transformation is now available. However, for a group of plants that is poised to become a major source of fast renewable and sustainably produced woody biomass for the emerging industries, there are challenges ahead to improve the efficiency of the techniques to a level comparable to that of the agricultural and forestry species. Bamboo, with its unique biology and relatively short domestication history that imposes a handicap on genetic improvement and production of quality planting material, requires intensive research. Several leads have been obtained in recent years that promise to overcome many of the hurdles and spearhead the genetic improvement of bamboo and help make available planting material of the highest quality for ensuring high productivity and quality of produce from plantations of the future.

References

Akinlabi, E.T., Anane-Fenin, K., Akwada, D.R. and Richard, D. (2017). Regeneration cultivation and sustenance of bamboo. pp. 39–86. *In*: Akinlabi, E.T., Anaen-Fenin, K. and Akwada, D.R. (eds.). *Bamboo: The Multipurpose Plant.* Springer.

Alexander, M.P. and Rao, T.C. (1968). *In vitro* culture of bamboo embryo. *Current Science*, 37: 415.

Ara, M.T., Nomura, T., Kato, Y. and Ogita, S. (2020). A versatile liquid culture method to control the *in vitro* development of shoot and root apical meristems of bamboo plants. *American Journal of Botany*, 11: 262–275. https://doiorg/104236/ajps2020112020.

Arya, I.D., Kaur, B. and Arya, S. (2012). Rapid and mass propagation of economically important bamboo *Dendrocalamus hamiltonii*. *Indian Journal of Energy*, 1(1): 11–16.

Banik, R.L. (1995). A manual of vegetative propagation of bamboos. *INBAR Technical Report* No. 6. INBAR FORTIP and Bangladesh Forest Research Institute, 66 pp.

Banik, R.L. (2016). *Silviculture of South Asian Priority Bamboos.* Springer. Doi: 101007/978-981-10-0569-5.

Borges-García, M., Ros-Araluce, C., Castellanos-Rubio, Y., Milanes-Rodríguez, S. and Velásquez-Feria, R. (2004). Efecto de diferentes métodos de desinfección en el establecimiento *in vitro* de *Guadua angustifolia* Kunth. *Biotecnologia Vegetal*, 4: 237–242.

Chang, W.C. and Lan, T.H. (1995). Somatic embryogenesis and plant regeneration from roots of bamboo (*Bambusa beechayana* Munro Var beechayana). *Journal of Plant Physiology*, 145: 535–538.

Chaturvedi, H.C., Sharma, M. and Sharma, A.K. (1993). *In vitro* regeneration of *Dendrocalamus strictus* Nees through nodal segment taken from field grown culm. *Plant Science*, 91: 97–101.

Chowdhury, P., Das, M., Sikdar, S.R. and Pal, A. (2004). Influence of the physiological age and position of the nodal explants on micropropagation of field-grown *Dendrocalamus strictus* Nees. *Plant Cell Biotechnology and Molecular Biology*, 5: 45–50.

Compton, M.E. and Preece, J.E. (1986). Exudation and explants establishment. *Newsletter of International Association of Plant Tissue Culture*, 50: 9–18.

Cooke, D.L., Waites, W.M. and Leifert, C. (1992). Effect of *Agrobacterium tumefaciens Erwinia carotovora Pseudomonas syringae* and *Xanthomonas campestris* on plant tissue cultures of *Aster Cheiranthus Delphinium Iris* and *Rosa* disease development *in vivo* as a result of latent infection *in vitro*. *Journal of Plant Disease and Protection*, 99: 469–481.

Danby, S., Berger, F., Howitt, D.J., Wilson, A.R., Dawson, S. and Leifert, C. (1994). Fungal contaminants of *Primula Coffea Musa* and *Iris* tissue cultures. pp. 397–403. *In*: Lumsden, P.J., Nicholas, J.R. and Davies, W.J. (eds.). *Physiology Growth and Development of Plants in Culture*. Kluwer Academic Publishers, Dordrecht.

De Cleene, M. and De Ley, J. (1976). The host range of crown gall. *The Botanical Review*, 42(4): 389–466. http://wwwjstororg/stable/4353907.

Devi, W.S., Bengyella, L. and Sharma, G.J. (2012). *In vitro* seed germination and micropropagation of edible bamboo *Dendrocalamus giganteus* Munro using seeds. *Biotechnology*, 11(2): 74–80.

Dias, A.C.F., Costa, F.E.C., Andreote, F.D., Lacava, P.T., Teixeira, M.A., Assumpção, L.C., Araújo, W.L. and Azevedo, J.L. (2009). Isolation of micropropagated strawberry endophytic bacteria and assessment of their potential for plant growth promotion. *World Journal of Microbiology and Biotechnology*, 25(2): 189–195.

Ewald, D., Naujoks, G., Zaspel, I. and Szczygiel, K. (1997). Occurrence and influence of endogenous bacteria in embryogenic cultures of Norway spruce. pp. 149–154. *In*: Cassells, A.C. (ed.). *Pathogen and Microbial Contamination Management in Micropropagation*. Kluwer Academic Publishers, Dordrecht.

Frame, B.R., Shou, H., Chikwamba, R.K., Zhang, Z., Xiang, C., Fonger, T.M. and Wang, K. (2002). *Agrobacterium tumefaciens*-mediated transformation of maize embryos using a standard binary vector system. *Plant Physiology*, 129(1): 13–22.

Gao, Y., Liu, H., Zhang, K., Li, F., Wu, M. and Xiang, Y.A. (2021). Moso bamboo transcription factor *Phehdz*1 positively regulates the drought stress response of transgenic rice. *Plant Cell Reports*, 40(1): 187–204. Doi: 101007/s00299-020-02625-w Epub 2020 Oct 24 PMID: 33098450.

George, M. and Tripepi, R. (2001). Plant Preservative Mixture™ can affect shoot regeneration from leaf explants of chrysanthemum, European birch, and rhododendron. *HortScience*, 36: 768–769.

Gielis, J., Peeters, H., Gillis, J. and Debergh, P.C. (2001). Tissue culture strategies for genetic improvement of bamboo. *Acta Horticulturae*, 552: 195–204. https://doi.org/10.17660/ActaHortic.2001.552.22.

Gillis, K., Gielis, J., Peeters, H., Dhooghe, E. and Oprins, J. (2007). Somatic embryogenesis from mature *Bambusa balcooa* Roxb as basis for mass production of elite forestry bamboos. *Plant Cell Tissue and Organ Culture*, 91: 115–123.

Godbole, S., Sood, A., Thakur, R., Sharma, M. and Ahuja, P.S. (2002). Somatic embryogenesis and its conversion into plantlets in a multipurposs bamboo *Dendrocalamus hamiltonii* Nees et Arn Ex Munro. *Current Science*, 83(7): 885–889.

Goh, C.H., Veliz Vallejos, D.F., Nicotra, A.B. and Mathesius, U. (2013). The impact of beneficial plant-associated microbes on plant phenotypic plasticity. *Journal of Chemical Ecology*, 39(7): 826–839.

Graves, A.E., Goldman, S.L., Banks, S.W. and Graves, A.C.F. (1988). Scanning electron microscope studies of *Agrobacterium tumefaciens* attachment to *Zea mays*, Gladiolus sp. and *Triticum aestivum*. *Journal of Bacteriology*, 170(5): 2395–2400.

Gutiérrez, L.G., López-Franco, R. and Morales-Pinzón, T. (2016). Micropropagation of *Guadua angustifolia* Kunth (Poaceae) using a temporary immersion system RITA®. *African Journal of Biotechnology*, 15(28): 1503–1510. Doi: 105897/AJB201615390.

Ha, J.H., Han, S.H., Lee, H.J. and Park, C.M. (2017). Environmental adaptation of the heterotrophic to autotrophic transition: The developmental plasticity of seedling establishment. *Critical Reviews in Plant Science*, 36(2): 128–137. Doi: 101080/0735268920171355661.

Hassan, A.E. and Debergh, P. (1987). Embryogenesis and plantlet development in bamboo *Phyllostachys viridis*. *Plant Cell Tissue and Organ Culture*, 10(1): 73–77.

Hassan, S.E.D. (2017). Plant growth-promoting activities for bacterial and fungal endophytes isolated from medicinal plant of *Teucrium polium* L. *Journal of Advanced Research*, 8(6): 687–695.

Holland, M.A. and Polacco, J.C. (1994). PPFMs and other covert contaminants: is there more to plant physiology than just plant? *Annual Review of Plant Biology* 45(1): 197–209. http://wwwjstororg/stable/24099837.

Huang, L.C., Huang, B.L. and Chen, W.L. (1989). Tissue culture investigations of bamboo IV: Organogenesis leading to adventitious shoots and plants in excised shoot apices. *Environmental and Experimental Botany*, 29: 307–315.

Huang, L.C., Lee, Y.L., Huang, B.L., Kuo, C.I. and Shaw, J.F. (2002). High polyphenol oxidase activity and low titratable acidity in browning bamboo tissue culture. *In Vitro Cellular and Developmental Biology – Plant*, 38: 358–365.

Huang, B., Zhuo, R., Fan, H., Wang, Y., Xu, J., Jin, K. and Qiao, G. (2022). An efficient genetic transformation and CRISPR/Cas9-based genome editing system for Moso Bamboo (*Phyllostachys edulis*). *Frontiers in Plant Science*, 13: 822022. Doi: 103389/fpls2022822022.

Jha, A., Das, S. and Kumar, B. (2013). Micropropagation of *Dendrocalamus hamiltonii* through nodal explants. *Global Journal of Bioscience and Biotechnology*, 2: 580–582.

Jimenez, V.M., Jhamna, C., Elena, T., Eric, G. and Mayra, M. (2006). *In vitro* propagation of the neotropical giant bamboo *Guadua angustifolia* Kunth through axillary shoot proliferation. *Plant Cell Tissue and Organ Culture*, 86: 389–395.

Joshi, M. and Nadgauda, R.S. (1997). Cytokinins and *in vitro* induction of flowering in bamboo: *Bambusa arundinacea* (Retz) Willd. *Current Science*, 523–526.

Kabade, U.A. (2009). *Studies on Refinement of Protocols for Rapid and Mass in vitro Clonal Propagation Evaluation of Genetic Fidelity and Growth pperformance of Bamboo Species – Bambusa bambos* (L) Voss and *Dendrocalamus strictus* (Roxb) Nees. PhD Thesis. Forest Research Institute University, Dehra Dun, India.

Kahl, G. (1982). Molecular biology of wound healing: The conditioning phenomenon. pp. 211–267. *In*: Kahl, G. and Schell, J.S. (eds.). *Molecular Biology of Plant Tumors*. Academic Press.

Kapoor, P. and Rao, I. (2006). *In vitro* rhizome induction and plantlet formation from multiple shoots in *Bambusa bambos var gigantea* Bennet and Gaur by using growth regulators and sucrose. *Plant Cell Tissue Organ Culture*, 85: 211–217.

Kozai, T. (1991). Photoautotrophic micropropagation. *In Vitro Cellular & Developmental Biology –Plant*, 27(2): 47–51. https://doiorg/101007/BF02632127.

Kumar, A. (1991). Mass production of field planting stock of *Dendrocalamus strictus* through macroproliferation: A technology. *Indian Forester*, 117(12): 146–152.

Leifert, C. and Cassells, A.C. (2001). Microbial hazards in plant tissue and cell cultures. *In Vitro Cell Development Biology - Plant*, 37: 133–138.

Leifert, C. and Waites, W.M. (1992). Bacterial growth in plant tissue culture media. *Journal of Applied Bacteriology*, 72(6): 460–466. https://doi.org/10.1111/j.1365-2672.1992.tb01859.x.

Leifert, C. and Woodward, S. (1998). Laboratory contamination management: The requirement for microbiological quality assurance. *Plant Cell Tissue and Organ Culture*, 52: 83–88.

Leifert, C., Morris, C. and Waites, W.M. (1994). Ecology of microbial saprophytes and pathogens in tissue-cultured and field-grown plants. *Critical Reviews in Plant Sciences*, 13(2): 139–183. Doi: 10.1080/07352689409701912.

Leifert, C., Ritchie, Y.J. and Waites, W. (1991). Contaminants of plant-tissue and cell cultures. *World Journal of Microbiology and Biotechnology*, 7: 452–469. https://doi.org/10.1007/BF00303371.

Limera, C., Sabbadini, S., Sweet, J.B. and Mezzetti, B. (2017). New biotechnological tools for the genetic improvement of major woody fruit species. *Frontiers in Plant Science*, 8: 1418. Doi: 103389/fpls201701418.

Lin, C.S., Lin, C.C. and Chang, W.C. (2004). Effect of thidiazuron on vegetative tissue-derived somatic embryogenesis and flowering of Bamboo—*Bambusa edulis*. *Plant Cell Tissue and Organ Culture*, 76: 75–82.

Maity, S. and Ghosh, A. (1997). Efficient plant regeneration from seeds and nodal segments of *Dendrocalamus strictus* using *in vitro* technique. *Indian Forester*, 123(4): 313–318.

Mascarenhas, A.F., Nadgir, A.L., Thengane, S.R., Phadke, C.H., Khuspe, S.S., Sdhirgurkar, M.V., Parasharami, V.A. and Nadgauda, R.S. (1988). Potential application of tissue culture for propagation of *Dendrocalamus strictus*. pp. 159–166. *In*: Ramanuja Rao, I.V.R., Gnanaharan, R. and Sastry, C.B. (eds.). *Bamboos Current Research*. Proceedings of the International Bamboo Workshop, 14–18 November, Cochin India.

Mehta, U., Rao, I.V.R. and Mohanram, H.Y. (1982). Somatic embryogenesis in bamboo. pp. 109–110. *In*: Fujiwara, A. (ed.). *Proceedings of 5th International Congress on Plant Tissue and Cell Culture*. Tokyo.

Mudoi, K.D., Saikia, S.P., Goswami, A., Gogoi, A., Bora, D. and Borthakur, M. (2013). Micropropagation of important bamboos: A review. *African Journal of Biotechnology*, 12(20): 2770–2785.

Muralidharan, E.M. (2009). Achievements and challenges in micropropagation of bamboo. *Proceedings of National Workshop on Global Warming and Its Implications for Kerala*. Thiruvananthapuram, 19–21 January 2009.

Mwamba, C.K. (1995). Variations in fruit of *Uapaca kirkiana* and effects of *in situ* silvicultural treatments on fruit parameters. pp. 27–38. *In*: Maghembe, J.A., Ntupanyama, Y. and Chirwa, P.W. (eds.). *Improvement of Indigenous Fruit Trees of the Miombo Woodlands of Southern Africa*. Primex Printers, ICRAF Nairobi.

Nadgauda, R.S., Parasharami, V.A. and Mascarenhas, A.F. (1990). Precocious flowering and seedling behaviour in tissue-cultured bamboos. *Nature*, 344: 335–336. https://doi.org/10.1038/344335a0.

Nadgir, A.L., Phadke, C.H., Gupta, P.K. and Parasharami, V.A. (1984). Rapid multiplication of bamboo by tissue culture. *Silva Genetica*, 33: 219–223.

Nadha, H., Salwan, R., Kasana, R., Anand, M. and Sood, A. (2012). Identification and elimination of bacterial contamination during *in vitro* propagation of *Guadua angustifolia* Kunth. *Pharmacognzy Magazine*, 8(30): 93–97. Doi: 10.4103/0973-1296.96547.

Nguyen, Q. and Kozai, T. (2005). Photoautotrophic micropropagation of woody species. pp. 123–146. *In*: Kozai, T. et al. (eds.). *Photoautotrophic (Sugar-free medium) Micropropagation as a New Propagation and Transplant Production System*. Doi:101007/1-4020-3126-2_8.

Nguyen, Q.T., Xiao, Y. and Kozai, T. (2016). Photoautotrophic micropropagation. *Plant Fact*. Elsevier, Netherlands, pp. 271–283.

Ogita, S. (2005). Callus and cell suspension culture of bamboo plant *Phyllostachys nigra*. *Plant Biotechnology*, 22(2): 119–125. Doi:10.5511/plantbiotechnology.22.119.

Ogita, S., Kikuchi, N., Nomura, T. and Kato, Y. (2011). A practical protocol for particle bombardment-mediated transformation of *Phyllostachys* bamboo suspension cells. *Plant Biotechnology*, 28(1): 43–50. https://doi.org/10.5511/plantbiotechnology.10.1101a.

Oprins, J., Grunewald, W., Gillis, K., Delaere, P., Peeters, H. and Gielis, J. (2004). Micropropagation: A general method for commercial bamboo production. *In*: 7th World Bamboo Congress, New Delhi. http://hdl.handle.net/1854/LU-675271.

Patil, A.M., Gunjal, P.P. and Das, S. (2021). *In vitro* micropropagation of *Lilium candidum* bulb by application of multiple hormone concentrations using plant tissue culture technique. *International Journal for Research in Applied Sciences and Biotechnology*, 8(2): 244–253. https://doi.org/10.31033/ijrasb.8.2.32.

Pérez, L.P., Montesinos, Y.P., Olmedo, J.G., Sánchez, R.R., Montenegro, O.N., Rodriguez. R.B., Ribalta, O.H., Escriba, R.C.R., Daniels, D. and Gómez-Kosky, R. (2015). Effects of different culture conditions (photoautotrophic photomixotrophic) and the auxin indole-butyric acid on the *in vitro* acclimatization of papaya (*Carica papaya* L. var Red Maradol) plants using zeolite as support. *African Journal of Biotechnology*, 14(35): 2622–2635. DOI: 105897/AJB201514814.

Pirttila, A.M., Joensuu, P., Pospiech, H., Jalonen, J. and Hohtola, A. (2004). Bud endophytes of Scots pine produce adenine derivatives and other compounds that affect morphology and mitigate browning of callus cultures. *Physiologia Plantarum*, 121(2): 305–312. Doi: 10.1111/j.0031-9317.2004.00330.x.

Qiao, G., Yang, H., Zhang, L., Han, X., Liu, M., Jiang, J., Jiang, Y. and Zhuo, R. (2014). Enhanced cold stress tolerance of transgenic *Dendrocalamus latiflorus* Munro (Ma bamboo) plants expressing a bacterial *CodA* gene. *In Vitro Cell Developmental Biology—Plant*, 50(4): 385–391. Doi:101007/s11627-013-9591-z.

Ramakrishnan, M., Yrjälä, K., Vinod, K.K., Sharma, A., Cho, J., Satheesh, V. and Zhou, M. (2020). Genetics and genomics of Moso bamboo (*Phyllostachys edulis*): Current status, future challenges, and biotechnological opportunities toward a sustainable bamboo industry. *Food Energy Security*, 1–36. https://doi.org/10.1002/fes3.229.

Ramanayake, S.M.S.D. and Yakandawala, K. (1997). Micropropagation of the giant bamboo (*Dendrocalamus giganteus* Munro) from nodal explants of field grown culms. *Plant Science*, 129: 213–223.

Ramanayake, S.M.S.D. (2006). Flowering in bamboo: An enigma! *Ceylon Journal of Science (Biological Sciences)*, 35: 95–105.

Ravikumar, R., Ananthakrishnan, G., Kathiravan, K. and Ganapathi, A. (1998). *In vitro* propagation of *Dendrocalamus strictus* Nees. *Plant Cell Tissue and Organ Culture*, 52: 189–192.

Reddy, G.H. (2006). Clonal propagation of bamboo (*Dendrocalamus strictus*). *Current Science*, 11: 14642–1464.

Rihan, H., Al-issawi, M., Al-Swedi, F. and Fuller, M. (2012). The effect of using PPM (plant preservative mixture) on the development of cauliflower microshoots and the quality of artificial seed produced. *Scientia Horticulturae*, 141: 47–52. Doi: 101016/jscienta201203018.

Rout, G.R. and Das, P. (1994). Somatic embryogenesis and *in vitro* flowering of three species of bamboo. *Plant Cell Reports*, 13: 683–686.

Sandhu, M., Wani, S.H. and Jiménez, V.M. (2017). *In vitro* propagation of bamboo species through axillary shoot proliferation: A review. *Plant Cell Tissue and Organ Culture*, 132(1): 27–53.

Sankar, V.R. (2019). *Studies on the Constraints in Efficient Micropropagation of Bamboo*. Ph.D. Thesis. Cochin University of Science and Technology, 287 pp.

Sankar, V.R., Thomas, G. and Muralidharan, E.M. (2017). Host-specific endophytic bacteria *Sporosarcina pasteruii* enhances growth in *in vitro* shoot cultures of the bamboo *Dendrocalamus longispathus* an economically important bamboo. *Journal of Bamboo and Rattan*, 16(2): 47–64.

Saxena, S. and Bhojwani, S.S. (1993). Clonal multiplication of 4-year-old plants of bamboo *Dendrocalamus longispathus* Kurz. *In Vitro Cell Developmental Biology—Plant*, 29: 135–142.

Saxena, S. and Dhawan, V. (1999). Regeneration of large-scale propagation of bamboo (*Dendrocalamus strictus* Nees) through somatic embryogenesis. *Plant Cell Reports*, 18: 438–444.

Sharma, R., Wahono, J. and Baral, H. (2018). Bamboo as an alternative bioenergy crop and powerful ally for land restoration in Indonesia. *Sustainability*, 10: 4367. Doi:103390/su10124367.

Shirgurkar, M.V., Thengane, S.R., Poonawala, I.S., Jana, M.M., Nadgauda, R.S. and Mascarenhas, A.F. (1996). A simple *in vitro* method of propagation and rhizome formation in *Dendrocalamus strictus* Nees. *Current Science*, 70(10): 940–943.

Singh, S.R., Dalal, S., Singh, R., Dhawan, A.K. and Kalia, R.K. (2012). Seasonal influences on *in vitro* bud break in *Dendrocalamus hamiltonii* Arn ex Munro nodal explants and effect of culture microenvironment on large-scale shoot multiplication and plantlet regeneration. *Indian Journal of Plant Physiology*, 17: 9–21.

Somashekar, P.V., Rathore, T.S. and Shashidhar, K.S. (2008). Rapid and simplified method of micropropagation of *Pseudoxytenanthera stocksii*. pp. 165–182. *In*: Ansari, S.A., Narayanan, C. and Mandal, A.K. (eds.). *Forest Biotechnology in India*. Satishi Serial Publishing House, Delhi.

Sood, A., Ahuja, P.S., Sharma, M., Sharma, O.P. and Godbole, S. (2002). *In vitro* protocols and field performance of elites of an important bamboo *Dendrocalamus hamiltonii* Nees et Arn Ex Munro. *Plant Cell Tissue and Organ Culture*, 71: 55–63.

Sood, P. and Sood, A. (2013). *Development of genetic transformation system for Dendrocalamus hamiltonii Nees et Arn ex Munro*. PhD Thesis. Guru Nanak Dev University Amritsar.

Sood, P., Bhattacharya, A. and Sood, A. (2011). Problems and possibilities of monocot transformation. *Biologia Plantarum*, 55: 1–15. Doi: 10.1007/s10535-011-0001-2.

Sood, P., Bhattacharya, A., Joshi, R., Gulati, A., Chanda, S. and Sood, A. (2013). A method to overcome the waxy surface cell wall thickening and polyphenol induced necrosis at wound sites—The major deterrents to *Agrobacterium* mediated transformation of bamboo a woody monocot. *Journal of Plant Biochemistry and Biotechnology*, 23: 69–80. Doi: 101007/s13562-013-0189-7.

Sun, H., Li, L., Lou, Y., Zhao, H., Yang, Y. and Gao, Z. (2016). Cloning and preliminary functional analysis of PeUGE gene from Moso Bamboo (*Phyllostachys edulis*). *DNA and Cell Biology*, 35(11): 706–714. Doi: 101089/dna20163389 Epub 2016 Aug 15 PMID: 27525704.

Sun, H., Li, L., Lou, Y., Zhao, H., Yang, Y., Wang, S. and Gao, Z. (2017). The bamboo aquaporin gene PeTIP4; 1–1 confers drought and salinity tolerance in transgenic Arabidopsis. *Plant Cell Reports*, 36: 1–13. Doi: 101007/s00299-017-2106-3.

Supaibulwattana, K. (1991). *In vitro* culture of some economic bamboos. *agris.fao. org.*

Thakur, R. and Sood, A. (2006). An efficient method for explant sterilization for reduced contamination. *Plant Cell Tissue and Organ Culture*, 84: 369–371.

Thiruvengadam, M., Rekha, K.T. and Chung, I.M. (2011). Rapid *in vitro* micropropagation of *Bambusa oldhamii* Munro. *Philippine Agricultural Scientist*, 94(1): 7–13.

Thomas, P. (2004). A three-step screening procedure for detection of covert and endophytic bacteria in plant tissue cultures. *Current Science*, 87: 67–72.

Thomas, P. and Sekhar, A.C. (2014). Live cell imaging reveals extensive intracellular cytoplasmic colonization of banana by normally non-cultivable endophytic bacteria. *AoB Plants*, 6: plu002. Doi: 101093/aobpla/plu002.

Tsay, H.S., Yeh, C.C. and Hsu, J.Y. (1990). Embryogenesis and plant regeneration from anther culture of bamboo *Sinocalamus latiflora* (Munro) McClure. *Plant Cell Reports*, 9: 349–351.

Vamil, A., Aniat-ul-Haq and Agnihotri, R.K. (2010). Plant growth regulators as effective tool for germination and seedling growth for *Bambusa arundinaceae*. *Research Journal of Agricultural Sciences*, 1: 233–236.

Vendan, R.T., Yu, Y.J., Lee, S.H. and Rhee, Y.H. (2010). Diversity of endophytic bacteria in ginseng and their potential for plant growth promotion. *The Journal of Microbiology*, 48(5): 559–565. https://doi.org/10.1007/s12275-010-0082-1.

Vongvijitra, R. (1988). Traditional vegetative propagation and tissue culture of some Thai bamboos. pp. 148–150 *In*: Rao, I.V.R., Gnanaharan, R. and Sastry, C.B. (eds.). *Bamboos: Current Research*. Proceedings of International Bamboo Workshop Cochin, 14–18 Nov. 1988. Peechi KFRI.

Yan, H., Liang, C., Yang, L. and Li, Y. (2010). *In vitro* and *ex vitro* rooting of *Siratia grosvenorii*, a traditional medicinal plant. *Acta Physiologiae Plantarum*, 32(1): 115–120. Doi: 10.1007/s11738-009-0386-0.

Yasodha, R., Kamala, S., Kumar, S.P., Kumar, P. and Kalamegam, K. (2008). Effect of glucose on *in vitro* rooting of mature plants of *Bambusa nutans*. *Scientia Horticulturae*, 116: 113–116. 101016/jscienta200710025.

Ye, S., Cai, C., Ren, H., Wang, W., Xiang, M., Tang, X., Zhu, C., Yin, T., Zhang, L. and Zhu, Q. (2017). An efficient plant regeneration and transformation system of Ma bamboo (*Dendrocalamus latiflorus* Munro) started from young shoot as explant. *Frontiers in Plant Science*, 8: 1298. Doi: 103389/fpls201701298.

Ye, S., Chen, G., Kohnen, M.V., Wang, W., Cai, C., Ding, W., Wu, C., Gu, L., Zheng, Y., Ma, X., Lin, C. and Zhu, Q. (2020). Robust CRISPR/Cas9 mediated genome editing and its application in manipulating plant height in the first generation of hexaploid Ma bamboo (*Dendrocalamus latiflorus* Munro). *Plant Biotechnology Journal*, 18(7): 1501–1503. Doi:101111/pbi13320.

Yeh, M. and Chang, W.C. (1986a). Plant regeneration through somatic embryogenesis in callus culture of green bamboo (*Bambusa oldhamii* Munro). *Theoretical and Applied Genetics*, 73(2): 161–163.

Yeh, M. and Chang, W.C. (1986b). Somatic embryogenesis and subsequent plant regeneration from inflorescence callus of *Bambusa beecheyana* Munro var *beecheyana*. *Plant Cell Reports*, 5: 409–411.

Yuan, J.L., Yue, J.J., Gu, X.P. and Lin, C.S. (2017). Flowering of woody bamboo in tissue culture systems. *Frontiers in Plant Science*, 8: 1589. https://doi.org/10.3389/fpls.2017.01589.

Zamora, A.B., Gruezo, S.S. and Damasco, O.P. (1988). Tissue culture of *Dendrocalamus, Bambusa, Gigantochloa*, and *Schizostachyum* species of bamboo. *Philippine Forest Research Journal (Philippines)*, 13: 55–60.

Chapter 14

CRISPR/Cas Based Genome Editing and its Possible Implication in Bamboo Research

Tsheten Sherpa,[1,2] *Khushbu Kumari,*[1,2] *Deepak Kumar Jha,*[1,2]
Manas Kumar Tripathy[1] and *Nrisingha Dey*[1,*]

14.1 Introduction

Bamboo is an essential economic commodity and an important alternative to conventional timber. Bamboo also holds a significant cultural and culinary value in many parts of the world (Singhal et al., 2013). Moso bamboo (*Phyllostachys edulis*) is widely used in constructions (scaffolding, flooring, and roofing) because of its thickness and hardiness. Worldwide, around 1225–1500 species of bamboo have been found, among which China is one of the largest growers of bamboo, having around 500 species. India is the second-largest producer, with about 9.57 million hectares of forest covered with bamboo plantations (Yeasmin et al., 2015). Approximately 80% of all the bamboos grown are present in Asia, and the rest, 20%, are found in Latin America and Africa. With around 2.5 billion US dollars in trade related to the bamboo industry, it provides massive employment for the masses. But the improvement of the bamboo industry lags far behind in comparison to other valuable plants mainly because of the slow rate of breeding (Ramakrishnan et al., 2020). Therefore, for improving bamboo with better traits, applying modern techniques for precise genetic modification is an imminent requirement.

[1] Division of Plant and Microbial Biotechnology, Institute of Life Sciences, NALCO Square, Chandrasekharpur, Bhubaneswar, Odisha 751023.
[2] Regional Centre for Biotechnology, National Capital Region Biotech Science Cluster, Faridabad, Haryana (NCR Delhi) 121001.
Emails: Tshetensherpa70@gmail.com; khusikumari357@gmail.com; deepakjha1515@gmail.com; mktripathy@gmail.com
* Corresponding author: ndey@ils.res.in, nrisinghad@gmail.com

With a remarkable increase in genome sequencing technologies in the last decade, many genetics and genomics data have been gathered from hundreds of plants (Sun et al., 2022). This enormous data should be utilized appropriately for better resource sustainability and security. For this reason, targeted gene manipulating techniques have been gaining increased attention. With the discovery of different genome modifying nucleases such as transcription activator-like effector nucleases (TALENs), zinc finger nuclease (ZFNs), and clustered regularly interspaced short palindromic repeats (CRISPR)-associated protein (CRISPR/Cas), targeted gene manipulation has become more straightforward (Nasti and Voytas, 2021). Among all these techniques, the recently developed CRISPR/Cas gene editing (GE) technique has gained significant popularity primarily because of its ease of target design, increased specificity, multiplexing, undemanding delivery methods, and the availability of many *in silico* and *in vitro* tools for generating and testing its components (Zhu et al., 2020). These genetic editing enzymes can scan, bind and change the DNA sequence in a genome. TALEN and ZFN use a FokI nuclease to induce a double-strand break (DSB) into the target site, whereas CRISPR/Cas has innate nuclease activities that cause DSBs. The cell's endogenous repair machinery repairs these DSBs induced by the nucleases through non-homologous end joining (NHEJ) or homologous-directed repair (HDR) (Gaj et al., 2013). NHEJ is prone to error and usually leads to random insertion or deletion of nucleotide sequences, usually knocking out or changing a gene function. In the case of HDR, homologous sequences are inserted into the cleavage site through homologous recombination. The HDR pathway can be utilized for precise DNA insertion and gene adjustment experiments (Nambiar et al., 2019). Genome editing techniques provide a platform for precisely manipulating a plant's genotype and changing the phenotype as per our desire.

Bamboo belongs to the grass family *Poaceaea* and comes under the subfamily of Bambusoideae. It is mainly divided into two types; woody and herbaceous. Most woody bamboos have n = 12 chromosomes, whereas most herbaceous bamboos have n = 11 (Yeasmin et al., 2015). Many tropical woody bamboos are hexaploid (6n = 6*12 = 72 chromosomes), while temperate woody bamboo are tetraploid (4n = 4*12 = 48 chromosomes), and herbaceous bamboos are diploid (2n = 2*11 = 22) (Ramakrishnan et al., 2020). The release of draft genome sequences from the Moso bamboo by Peng et al., 2013, marked the first whole-genome sequencing from the Bambusoideae subfamily (Peng et al., 2013). Recently, five more draft genome sequences of bamboo, namely, *Raddia guianensis*, *Olyra latifolia*, *R. distichophylla*, *Guadua angustifolia*, and *Bonia amplexicaulis* have been published (Guo et al., 2019; Li et al., 2020). This vast sequence of information has tremendous potential to improve bamboo cultivation through genetic manipulation using gene-editing techniques. Therefore, this chapter briefly mentions different gene-editing techniques successfully used in plant improvement programs with special emphasis on the potential of CRISPR/Cas as the next-generation gene-editing technique.

14.2 Different Genome Editing Techniques and Advantages of CRISPR/Cas9

In plants like bamboos, where breeding for superior traits is challenging because of the long juvenile phase, developing micro propagation-based genetic manipulation is highly desirable. The micropropagation-based gene editing provides a better alternative to traditional breeding because it is less time-consuming, less laborious, and specific (Chen et al., 2019). Some essential genome editing techniques that have been successfully applied in plants are discussed below. These techniques have been thoroughly standardized and can improve bamboo cultivations.

14.2.1 Zinc Finger Nuclease (ZFN)

Zinc finger nucleases are artificially designed restriction enzymes consisting of DNA binding and DNA cleaving domain (Fig. 14.1a). The DNA binding domain contains around 4–6 zinc finger protein, where each protein recognize about three base pairs (bps) of DNA (Petolino, 2015). These DNA binding domains are arranged in a two-finger module that recognizes six bps of DNA sequence; therefore, the domain junction between each module can be optimized for maximum binding and specificity (Mohanta et al., 2017). Presently recognition specificity of each codon by zinc finger protein has been appropriately standardized, and all the 64 possible tri-

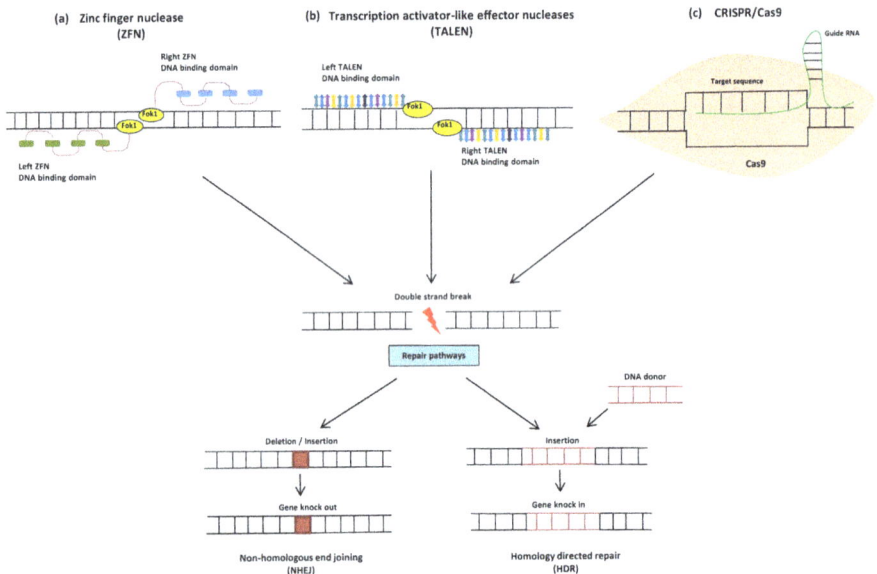

Fig. 14.1. Schematic representation of (a) ZFN, (b) TALEN, and (c) CRISPR/Cas9 mediated double-stranded break (DSBs) on the target site. ZFN and TALEN use FokI nuclease to create DSBs, and, in the case of CRISPR/Cas9, Cas protein induces DSB. These breaks are repaired by non-homology end joining (NHEJ), leading to deletion/insertion of DNA, or homology-directed repair (HDR), and leading to insertion from homologous donor DNA.

nucleotide codon sequences can be targeted by zinc finger moiety. For DNA cleavage, a nuclease enzyme called FokI is used. These two domains, DNA binding zinc finger domain and DNA cleavage nuclease, form ZFN. One of the critical properties of FokI-mediated DNA cleavage is that the catalytic domain must form a dimer for DNA cleavage. So for DNA cleavage, two ZFNs should be designed flanking the target site and must be present adjacent to one another (Fig. 14.1a). Also, correct spacing between these two ZFNs is required to form the FokI enzyme dimer. This property of ZFN makes the editing highly specific. Some variants of FokI have also been developed where dimerization is not required, further simplifying this method (Gaj et al., 2013). These DSBs caused by the FokI will then be repaired by NHEJ and HDR, leading to gene mutation or DNA integration into the target cleavage site (Fig. 14.1). The biggest drawback of ZFN-mediated gene editing is that it requires an expert to design and assemble a particular zinc finger domain, which is strenuous work (Gupta and Musunuru, 2014).

14.2.2 *Transcription Activator-like Effector Nucleases (TALEN)*

Transcription activator-like effector nuclease, or TALEN, is a gene-editing technique similar to ZFN. Both rely on the FokI nuclease to cleave the target DNA sequence. But unlike ZFN, TALEN comprises TALE molecules as DNA binding and recognition domain (Fig. 14.1b) (Joung and Sander, 2013). TALE proteins are found in *Xanthomonas* bacteria and function as a DNA binding effector molecule to regulate host genes (Boch and Bonas, 2010). The DNA binding domain of TALE molecules consists of around 30 copies of 33–35 amino acid sequences. These sequences are mostly conserved, except the 12th and 13th positions. These 12th and 13th variable regions between different TALE proteins are called repeat-variable diresidue (RVD) and are the main component behind different nucleotide recognition. Each RVD can recognize a single nucleotide, and therefore it can be arranged in a specific way to target DNA sequences. The DNA cleavage region is similar to ZFN, where the FokI needs to form a dimer for DSBs to occur. Therefore for DNA cleavage, two different TALEN present in the opposite strand and separated by specific base pairs are required (Gaj et al., 2013). A comparative study between TALEN and ZFN has found that TALEN showed a higher DSB capability. Designing and producing TALE targeting DNA sequence was also comparatively easier than ZFN as TALE required recognizing single bases instead of triple bases in ZF, hence providing better flexibility (Joung and Sander, 2013). But TALEN technique does come with some disadvantages. Because of many amino acids in TALE protein, the size of the DNA sequence to encode these TALE proteins becomes very large. These provide a technical challenge in cloning and transforming. Also, many repeats present in TALE arrays provide another technical challenge in cloning as these repeats may deter some viral vectors' efficiency (Gupta and Musunuru, 2014). Overall, TALEN is a better gene-editing technique than ZFN. But as both techniques involve protein engineering, it is a tedious job and usually requires an expert.

14.2.3 Clustered Regularly Interspaced Short Palindromic Repeats (CRISPR)-associated Protein (CRISPR/Cas)

The Japanese research group in 1987 first discovered CRISPR (Clustered Regular Interspaced Palindromic Repeats) inside the *E. coli* genome when they were studying the genes involved in phosphate metabolism (Ishino et al., 1987). Similar repeats were found in *Haloferax mediterranei* by Francisco Mojica in 1989 (Mojica et al., 1993). These sequences were described as unusual short palindromic repeats of 20–40 bp separated by spacers of unique 20–60 bases. In 2005, Mojica and Pourcel discovered that CRISPR sequences were present in bacteria and archaea along with bacteriophage, prophage, and plasmids. They also realized that if the bacterium and its phage have similar CRISPR sequences, the bacterium was resistant to that phage. These led to the realization that CRISPR sequences may function as a defense mechanism against foreign genetic materials (Mojica et al., 2005; Pourcel et al., 2005).

Siksnys in 2011 successfully transferred the whole CRISPR system from *S. thermophiles* to *E. coli*. They performed CRISPR-Cas9 cleavage *in vitro*, using a reprogrammable CRISPR array, which could cleave the sequence of choice (Gasiunas et al., 2012). Shortly, Charpentier and her colleague were successful in developing a custom-made sgRNA (single guide RNA) by combining two RNA, CRISPR RNA (crRNA) and trans-activating CRISPR RNA (tracrRNA) (Jinek et al., 2012). Finally, Zhang and the group were able to edit multiple genes using three-component systems consisting of Cas9 from *S. pyrogenes* or *S. thermophiles*, tracrRNA, and CRISPR array in both human and mouse genomes with high efficiency and specificity (Cong et al., 2013). Soon afterward, CRISPR/Cas9 was established in plants too, where genes in *Arabidopsis thaliana*, *Nicotiana benthamiana*, and rice were edited successfully (Nekrasov et al., 2013; Shan et al., 2013; Xie and Yang 2013).

Various CRISPR/Cas systems have been found in archaeal and bacterial genomes. These have been designed into two classes, Class I and II, where Class I is further divided into type I, II, and IV, and Class II into type II, V, and VI. This classification is based primarily on the function of Cas protein, the presence of "signature genes" of CRISPR-Cas types, and the organization, sequence similarity, and phylogenetic analysis of CRISPR-Cas genes (Makarova et al., 2020). Mainly, Class I CRISPR-Cas systems have Cas protein which constitutes multiple subunits. In contrast, Class II Cas proteins are more straightforward, with only one subunit (Fig. 14.1c). Type II CRISPR-Cas system, which consists of Cas9 endonuclease, has become the most commonly used in genome editing technology (Montecillo et al., 2020).

In a type II system, the host integrates a short fragment of the invading pathogen's DNA sequence into its CRISPR array site in the form of spacers (Barrangou et al., 2007). When the same pathogen infects again, this CRISPR array is transcribed and forms a pre-crRNA which binds with tracrRNA, forming an RNA duplex. This duplex between pre-crRNA and tracrRNA is catalyzed by RNase III, leading to the formation of stem-loop structured RNA called mature crRNA. These mature crRNA (crRNA/tracrRNA hybrid) then forms a complex with Cas9, leading to the activation of Cas protein for targeted DNA cleavage (Wright et al., 2016). Cas9 has a double-lobed structure. One lobe constitutes an alpha-helical lobe and another nuclease. The nuclease lobe harbors two nuclease domains, HNH and RuvC, responsible for the double-stranded cleavage of target DNA (Nishimasu et al., 2014). For the DSBs to occur, a short DNA sequence in the host's genome called protospacer adjacent motif (PAM) is an essential component. PAM is a 3–6 bp of DNA sequence present after the crRNA target. This site acts as a cleavage site for the Cas protein and also to differentiates between foreign genetic material and the innate CRISPR loci. For Cas9, 5'-NGG-3' PAM sequence occurs, where "N" stands for any nucleobase and "G" for guanine nucleobase (Anders et al., 2014).

Other CRISPR/Cas proteins such as Cas12a are also frequently used for gene editing. Cas12a belongs to the same class as Cas9, i.e., Class II, but is of a different type, type V. Unlike Cas9, which makes a blunt-ended cut on the cleavage site, Cas12a can make sticky end cuts. These sticky end cuts can be advantageous when doing knock-in or gene insertion studies. Also, Cas12a requires a different PAM site, increasing the targeting flexibility of CRISPR/Cas mediated gene editing (Bandyopadhyay et al., 2020).

CRISPR genome editing provides a few significant advantages over TALEN and ZFN. Effector design for each target site should be meticulously designed from scratch for TALEN and ZFN as it involves re-engineering each site's new protein. But in the case of CRISPR, target effector design is straightforward, as it only involves designing a 20 nucleotide sequence. Therefore developing plasmid constructs and expressing these sgRNAs are undemanding (Nemudryi et al., 2014). This short and direct target design of CRISPR sgRNA also gives it a considerable advantage in multiplexing experiments over other techniques (Endo et al., 2015). But in the case of TALEN and ZFN, different modules have to be inserted for different target sites, vastly increasing the experiments' complexity and cost (Armario Najera et al., 2019).

14.3 Different Methods in CRISPR-mediated Gene Editing

14.3.1 Base Editing through CRISPR

Base editors provide highly efficient targeted substitutions of different nucleotides without relying on HDR or DSB formation. The cytosine base-editor (CBE) system and the adenine base-editor (ABE) system are two main classes of base editors. These base editors can impart all the four possible substitution mutations, i.e., from A→G, G→A, C→T, and T→C. It consists of an impaired Cas nuclease known as nCas9 (nickase Cas9) fused with deaminase enzyme. The nCas9 is a product of mutation in one of the nuclease domains of Cas9, which causes the Cas9 to make just a single-strand break (Gasiunas et al., 2012). Fused with nCas9 are different deaminase enzymes which change the nucleotide sequences as per our needs.

The cytosine base-editor (CBE) is a type of a base editor that changes cytosine to thymine (Fig. 14.2a). This system consists of a cytidine deaminase fused with nCas9 and uracil glycosylase inhibitor (Komor et al., 2016). When CBE binds to the target DNA sequence, the nCas9 nicks a single DNA strand, and the cytidine

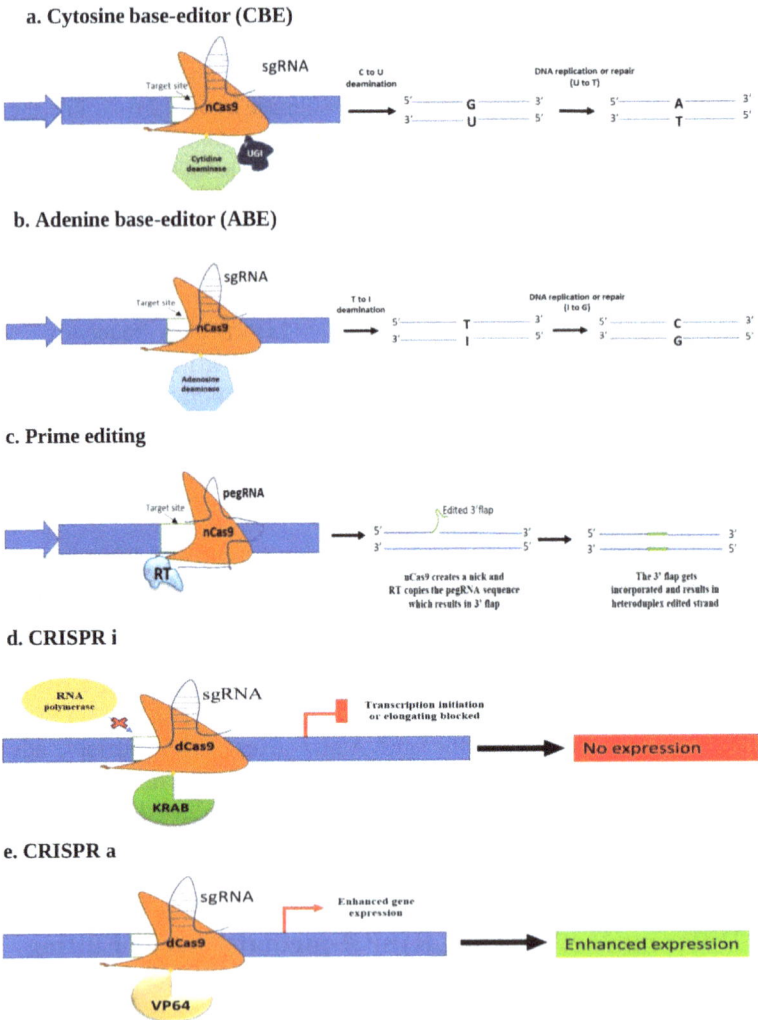

Fig. 14.2. Overview of different CRISPR-based genome editing strategies: a. Cytosine base editor (CBE) - nCas9 fused to cytidine deaminase forms a complex with uracil glycosylase inhibitor (UGI). Cytidine deaminase converts cytosine to uracil, which is further converted to thymine. b. Adenine base-editor (ABE) - nCas9-adenosine deaminase complex converts adenine to guanine with intermediary inosine. c. Prime editors- It consists of nCas9 fused with pegRNA and RT (Reverse Transcriptase). The pegRNA is complementary to the target sequence, which directs nCas9 to make a nick primed by a prime editor (fusion of proteins nCas9 and RT) for reverse transcription. This leads to a 3' DNA flap which hybridizes to form heteroduplex at the target site. The second nick is created in a non-edited strand which is then mutated using the edited strand, creating a fully mutated target site. d. CRISPRi (CRISPR interference) - It can be used to block either transcription initiation or elongation. It is a complex between sgRNA-dCas9 (devoid of nuclease activity) and a repressor such as KRAB (Kruppel-associated box domain) attached to the C-terminus of dCas9. It attaches to the target site and prevents the binding of RNA polymerase, finally blocking gene expression. e. CRISPRa (CRISPR activation) - This system can enhance gene expression. It is a complex between sgRNA-dCas9 (devoid of nuclease activity) and a transcriptional activator such as VP64 (consists of four copies of VP16, a viral protein derived from Herpes Simplex virus) attached to the C-terminus of dCas9. It connects to the target site and recruits specific cofactors which enhance gene expression without editing the target site.

deaminase converts the cytosine from the target site to uracil nucleobase. This uracil then gets converted to thymine nucleobase by the DNA replication process. The uracil glycosylase inhibitor helps block the repair pathway from excising uracil by inhibiting the DNA glycosylase (Gasiunas et al., 2012).

Another base editor called the adenine base-editor (ABE) system converts adenine nucleobase to guanine nucleobase (A to G) (Fig. 14.2b) (Gaudelli et al., 2017). It consists of TadA (transfer RNA adenosine deaminase), which catalyzes the conversion of adenosine to inosine and an nCas9 (D10A) enzyme. The converted inosine in the new strand is then recognized by the polymerase as guanine (G) and introduces the G-C base pair (Kantor et al., 2020).

14.3.2 Prime Editing

The major limitation of base editing is that it cannot go beyond the four transitions and hence cannot create insertions and deletions. Prime editing can overcome this limitation as it does not rely on DSBs and can bring all 12 possible types of point mutations and small targeted indels (Anzalone et al., 2019). This system consists of Cas9 nickase (mutated HNH nuclease) fused with a reverse transcriptase domain and a prime editing guide RNA (pegRNA) (Fig. 14.2c). There are two specifications of pegRNA; first, it contains a complementary sequence to the target site, which directs Cas9. Second, it encodes the primer binding sequence (PBS) at the 3' end of the pegRNA, which hybridizes to the nicked target. This pegRNA targets prime editor protein (PE2- the fusion of nCas9 and reverse transcriptase) to the target site where the Cas9 RuvC nuclease causes a nick at the target site near the PAM site. This nick causes liberation of the 3' end at the targeted DNA site, which is then further used by the prime editor to prime it for reverse transcription by reverse transcriptase. This step results in two DNA flaps: the edited 3' DNA flap obtained from reverse transcription of pegRNA and the original 5' DNA flap. The 5' DNA flap gets excised by cellular endonucleases allowing the 3' DNA flap to hybridize and generate a heteroduplex at the target site. This heteroduplex DNA contains the edited 3' flap and the non-edited strand. The edited strand is used as a template by sgRNA, which directs PE2 to create a nick in the complementary strand, further editing to create a fully mutated complex (Fig. 14.2c) (Anzalone et al., 2020).

14.3.3 Gene Knock-out using CRISPR

CRISPR/Cas has been highly successful in developing plants with superior traits by knocking out the host's susceptible genes. Many single gene knock-outs through the CRISPR system have led to desirable traits; for example, knocking out the *OsGn1a* gene led to an increase in rice grain size (Li et al., 2016). It is known that most of the traits in plants, including yield, quality, and resistance to various stresses, are controlled by a myriad of factors and not just by a single gene. These various factors primarily result from cumulative factors controlled by QTLs (Quantitative Trait Loci). In rice, CRISPR/Cas has successfully knocked out three related genes, GW2, GW5, and TGW6, resulting in increased grain weight of rice (Xu et al., 2016). The action of CRISPR/Cas9 is to generate DSBs, which are repaired through the NHEJ mechanism. This leads to a base frameshift due to the addition/deletion of a few bases. Due to changes in the protein-coding gene sequence, its expression of gene is hindered, resulting in gene silencing/knockout. Hence, this genome editing tool can be used to silence more than one gene at a time with high specificity as compared to T-DNA mutagenesis. This method can be widely used to study gene function or remove undesirable genes to enhance certain traits in plants (Zhang et al., 2021).

14.3.4 Gene Knock-in using CRISPR

Traditionally gene knock-in done through T-DNA insertion has many demerits as it causes silencing of other genes due to position effects. The CRISPR/Cas9 system eliminates position effects by incising at the specific target and incorporating foreign DNA sequences through HDR-mediated repair. Hence, a modified sequence results at the target site without affecting other sites. Gene knock-in through CRISPR/ Cas can be done to develop varieties of crops with enhanced agronomic traits by either substituting a single nucleotide or adding/replacing genes, or modulating gene expression. Knock-in can be used to modulate several genes to enhance elite traits. The *GOS2* promoter was knocked-in *ARGOS8* promoter (ARGOS8 gene-negative regulatory of ethylene response) using CRISPR/Cas, which led to overexpression of the gene increasing drought tolerance in maize (Shi et al., 2017).

14.3.5 Transcriptional Control Models

Genetic modifications such as gene knock-out and knock-in are not always preferred to enhance plant traits. These modifications can cause adverse effects on cells, such as a constitutive expression or repression of genes, leading to cytotoxic effects in the cell; it can also cause improper feedback mechanisms and other phenotypical changes (Donohoue et al., 2018). Hence, unrestrained gene expression can be controlled by specific transcriptional control models such as CRISPR activation (CRISPRa) and CRISPR interference (CRISPRi) (Gilbert et al., 2013; Qi et al., 2013). Another method to regulate transcriptional activity is CRISPR activation (CRISPRa), which can boost expression of a target gene without engineering the promoter sequence or by adding a strong promoter. Both CRISPRi/a provide an alternative method for gene regulation control.

Both the methods require a mutated form of Cas9 known as dCas9 (deactivated Cas9). The dCas9, which results from a mutation in the two nuclease domains RuvC-like and HNH of Cas9, leads to the inactivation of nuclease activity and can only bind to a specific target (Jinek et al., 2012). Since the arrival of dCas9, it has been in use for wide applications such as modulating gene expression, epigenome editing, transcriptional initiation blockage, and many more (Qi et al., 2013; Xu et al., 2016).

14.3.6 CRISPR interference (CRISPRi)

CRISPR interference (CRISPRi) is a transcriptional control module that represses transcription by blocking gene elongation or transcriptional initiation. The presence of a transcriptional repressor fused to dCas9 and sgRNA helps strengthen transcriptional repression. Designing sgRNA that targets promoter sequence or *cis*-acting elements (transcription factor binding sites) can block transcriptional initiation. The above action prevents the binding of RNA polymerase or transcription factors respectively by sterically hindering their binding locus (Larson et al., 2013). Multiple genes can be regulated by designing multiple sgRNAs, or a single gene can be targeted at different sequence positions for a regulated expression of the gene. This system is more efficient than other systems as it doesn't alter the genome sequence and can be tuned accordingly. The dCas9-KRAB module is an effective system to achieve transcriptional repression as dCas9 alone cannot be sometimes enough to repress an endogenous gene expression (Fig. 14.2d). This dCas9-KRAB is a fusion of dCas9 and KRAB (Kruppel-associated box domain). It is involved in epigenome modification through histone methylation and deacetylation by recruiting a heterochromatin-forming complex (Groner et al., 2010; Reynolds et al., 2012).

14.3.7 CRISPR activation (CRISPRa)

In contrast to CRISPRi, CRISPRa (CRISPR activation) is involved in the upregulation of target genes (Gilbert et al., 2013; Qi et al., 2013). This system requires dCas9 fused with a transcriptional activator which increases the targeted endogenous gene expression. This overexpression is beneficial when the native promoter is weak, and the increment in gene expression can improve certain traits or help the plant resist different stresses. Many transcriptional activators such as VP-64, S.A.M. (Synergistic Activation Mediator), VPR, etc., have been found. The first CRISPR transcription activator module dCas9-VP64, derived from Herpes Simplex Virus (HSV), a fusion of four trans-activating domains VP16 to the C-terminus dCas9 (Fig. 14.2e), has been successfully used in enhancing many gene expression. This was the first generation of the transcription activator module, which could increase expression up to two-fold (Chavez et al., 2016). The second-generation activator modules, called VPR, a fusion of VP64 with two other transactivators, the p65 subunit of NF-κB and transactivator domain R from Epstein-Barr virus, are more potent than VP64. While SAM consists of the coat protein of RNA bacteriophage MS2, HSF1 (Heat Shock Transcription Factor 1), and p65 fused with dCas9-VP64. This MS2 modifies the sgRNAs and recruits p65 and HSF1, forming a complex with MS2 known as MPH and facilitating transcription by binding to the modified sgRNA (Chavez et al., 2016).

14.4 Applications of CRISPR/Cas System in Bamboo Improvement

The availability of genome information and efficient genetic transformation methods has a significant role in understanding the gene functions and implications of genome editing. Because of the current advances in whole-genome sequencing, the chromosome-level assembly of Moso bamboo was done in 2018 with the available genome draft (Peng et al., 2013; Zhao et al., 2018). The underlying mechanisms of crucial traits like flowering, lignification, and rapid growth rate in bamboo have been studied by omics-based methods (Ge et al., 2017; Wang et al., 2021; Yang et al., 2021). Many candidate genes have also been found, which can be validated by different molecular tools for developing superior bamboo varieties. But for all this, proper standardization of regeneration and genetic transformation of bamboo in an artificial tissue culture environment is necessary. Generation of stable lines of transgenic plants in bamboo through micropropagation has been tricky, primarily because of the slow regeneration of callus and low transformation frequency. Also, a considerable setback to bamboo tissue culture is a long juvenile phase (taking several decades) and the monocarpic behaviour of bamboos.

Nonetheless, many efforts have been made to transform bamboos and regenerate them in a tissue culture medium. Qiao et al. (2014) successfully transformed Ma bamboo (*Dendrocalamus latiflorus*) using Agrobacterium-mediated transformation to transfer a bacterial gene, *CodA* (*Choline oxidase*), to confer cold tolerance to Ma bamboo (Qiao et al., 2013). Ye et al. (2017) also successfully regenerated transgenic Ma bamboo (*Dendrocalamus latiflorus Munro*) using *Agrobacterium* transforming maize *Lc* gene in Ma bamboo genome, generating anthocyanin over-accumulation phenotype (Ye et al., 2017). Biyun Huang and colleagues have successfully regenerated bamboo plants from immature embryos in recent research (Huang et al., 2022). They also established the CRISPR/Cas9 mediated genome editing method in Moso bamboo, and they targeted *PePDS1* and *PePDS2* with a single conserved sgRNA efficiently derived by PeU3.1 snRNA promoter (Huang et al., 2022). CRISPR/Cas9 system can be employed in proving the various traits of bamboo to enhance its industrial importance.

CRISPR/Cas mediated overexpression of *MADS-box* genes, *PheMADS15* and *PheMADS5*, might lead to early flowering in bamboo, as these genes of Moso bamboo were reported to induce early flowering in *Arabidopsis* (Cheng et al., 2017; Zhang et al., 2018). Similarly, *BtPD1* has been reported to induce early flowering in *Arabidopsis* (Dutta et al., 2021). Another Moso bamboo gene, *PeGRF11*, when expressed in *Nicotiana benthamiana*, enhanced their growth rate (Shi et al., 2019). Growth regulating factors (GRF) has a crucial role during bamboo developmental stages; *PeGRF* genes can be manipulated by CRISPR/Cas system to induce rapid growth in bamboo. Many stress-related genes of bamboo have been found to provide stress resistance in *Arabidopsis*. Overexpression of these genes can help develop stress-tolerant bamboo plants. Widespread application of CRISPR/Cas system in bamboo appears to have a high potential to boost the industrial importance of bamboo species. Therefore improving CRISPR/Cas mediated genome editing in bamboo using various delivery methods and micropropagation techniques is an urgent necessity. A schematics representation of different strategies in genome editing of bamboo using CRISPR/Cas has been presented in Fig. 14.3, where various delivery methods such as *Agrobacterium*, PEG-protoplast, biolistic and nanoparticles has been represented.

14.5 Challenges in Employing CRISPR/Cas System in Bamboo

Within a decade of its application in plant biotechnology, CRISPR/Cas system has become a potent molecular tool for genome editing. The ease of use of this system has made it very popular among plant researchers. CRISPR/Cas system has been rapidly used and is a choice tool for developing valuable traits in plants. Despite all the

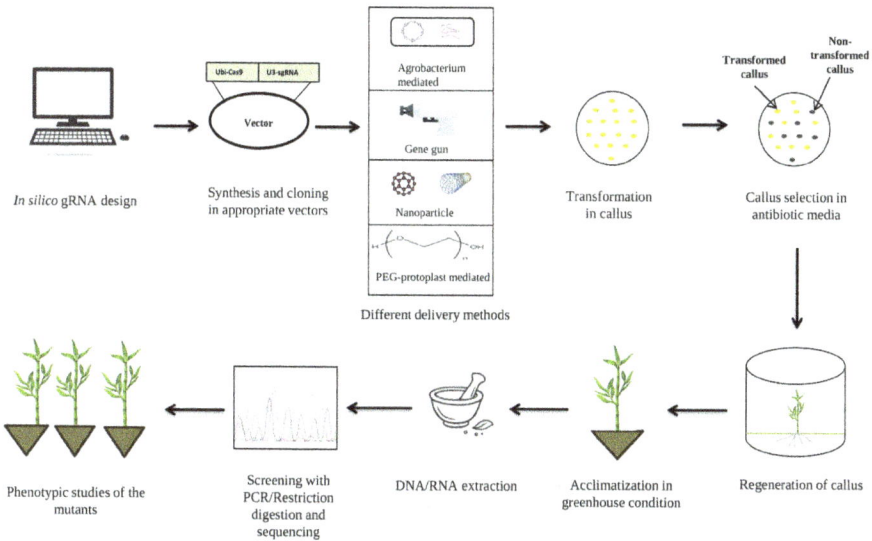

Fig. 14.3. General strategy for CRISPR/Cas mediated genome editing. The first step involves *in silico* designing of gRNA. After this, the gRNA is synthesized and cloned into appropriate DNA vectors. These DNA vectors are then transformed into bamboo callus using various delivery methods such as *Agrobacterium*, PEG-protoplast, biolistic (gene gun), and nanoparticles. The transformed callus is grown into selection media and then transferred to regeneration media, where rooting and shooting of transformed callus occurs. The regenerated plants are then screened for target mutations using PCR and DNA sequencing. Finally, the mutant plants are analysed.

advantages, there are many hurdles to be conquered for the widespread application of CRISPR/Cas in bamboo.

The most important criteria for gene editing using the CRISPR/Cas system is efficient *in vitro* tissue culture-based regeneration of plants. Micropropagation-based regeneration of bamboo has still not been properly standardized and generation of somaclonal variants has rarely been reported. This mostly is because of the very slow regeneration of bamboo callus, where Ye et al. (2017) reported it took around eight months to get healthy callus of ma bamboo. Also, the most widely used method of delivering transgenes such as *Agrobacterium* and biolistic methods has been reported to have a low transformation frequency in bamboo (Ramakrishnan et al., 2020). Therefore poor transformation frequency and long regeneration of callus are serious limitations. The large genome size of bamboo (Moso bamboo-1908 Mb), and their polyploidy nature provide another big hurdle in gene editing using CRISPR/Cas. The presence of multiple alleles and similar repeats makes the design and targeting of genes complicated and increases the chances of off-targets mutations.

CRISPR/Cas system also comes with a few limitations which should be examined before starting the work. These are, firstly, the PAM sequence is required by most of the widely used CRISPR/Cas systems. PAM provides the specificity to genome editing, but it limits the construction of sgRNA of target genes. Secondly, the delivery of CRISPR/Cas cassette into the plant is another major challenge. Mainly in monocots where particle bombardment and *Agrobacterium*-mediated methods are not very successful. However, tissue culture-mediated transformation cannot fulfil the real benefits of genome editing as there will be T-DNA integration events. Thus, other delivery systems like nanoparticles-mediated need to be developed to achieve transgene-free genome editing (Liu et al., 2009). Thirdly, a major concern of the CRISPR/Cas mediated genome editing is off-targeting (Grünewald et al., 2019). These non-specific off-targets can be limited by direct delivery of CRISPR/ ribonucleoprotein (RNP), which does not integrate into the genome, decreasing the non-specific targeting probability. Fourthly, the non-availability of genome sequencing often limits the application of the CRISPR/Cas system, as the selection of a gene for targeting a particular trait is crucial. Lastly, a significant hurdle is the commercialization of CRISPR/Cas mediated modified crop plants because of different government regulatory policies.

14.6 Conclusion

Bamboo is one of the fastest-growing non-timber forest plants. Bamboo has better adaptation to wider environmental conditions and can better cope with increasing climate change. It is one of the versatile plant species on earth with high economic value. It also has an excellent potential for timber, fibre, biofuel, paper, food, and medicines in the industrial sector. Traditional bamboo cultivation faces various adverse effects, such as abiotic and biotic stresses, which must be dealt appropriately. Genetic improvement of bamboo is a tedious task due to its perennial nature, long breeding cycle, and monocarpic behaviour. The availability of whole-genome sequences offers vast potential to improve the quality and quantity of bamboo according to industrial needs. Looking at the future demand, improving bamboo breeding with better traits is the need of the hour. Applying modern techniques for precise genetic modification is a critical necessity to improve bamboo cultivation. Among all these techniques available today, the recently developed CRISPR/Cas9 gene-editing method has been accepted by the larger plant scientist community because of its ease of target design, increased specificity, multiplexing, cost-effectiveness, and mostly undemanding delivery methods. Current day, CRISPR/Cas system has become the first choice of tool for developing valuable traits in plants. Despite a few challenges to using the CRISPR/Cas system, CRISPR/Cas gene-editing technology will be more widely used and undoubtedly play an essential role in bamboo quality and quantity improvement with better growth and development.

References

Anders, C., Niewoehner, O., Duerst, A. and Jinek, M. (2014). Structural basis of PAM-dependent target DNA recognition by the Cas9 endonuclease. *Nature*, 513(7519): 569–73.

Anzalone, A.V., Koblan, L.W. and Liu, D.R. (2020). Genome editing with CRISPR-Cas nucleases, base editors, transposases and prime editors. *Nat. Biotechnol.*, 38(7): 824–844.

Anzalone, A.V., Randolph, P.B., Davis, J.R. et al. (2019). Search-and-replace genome editing without double-strand breaks or donor DNA. *Nature*, 576(7785): 149–157.

Armario Najera, Victoria, Richard M. Twyman, Paul Christou and Changfu Zhu. (2019). Applications of multiplex genome editing in higher plants. *Current Opinion in Biotechnology*, 59: 93–102.

Bandyopadhyay, Anindya, Nagesh Kancharla, Vivek S. Javalkote, Santanu Dasgupta and Thomas P. Brutnell. (2020). CRISPR-Cas12a (Cpf1): A versatile tool in the plant genome editing tool box for agricultural advancement. *Frontiers in Plant Science*, 11.

Barrangou, Rodolphe, Christophe Fremaux, Hélène Deveau et al. (2007). CRISPR provides acquired resistance against viruses in prokaryotes. *Science*, 315(5819): 1709–1712.

Boch, Jens and Ulla Bonas. (2010). Xanthomonas AvrBs3 family-type III effectors: Discovery and function. *Annual Review of Phytopathology*, 48(1): 419–436.

Chavez, A., Tuttle, M., Pruitt, B.W. et al. (2016). Comparison of Cas9 activators in multiple species. *Nat. Methods*, 13(7): 563–567.

Chen, K., Wang, Y., Zhang, R., Zhang, H. and Gao, C. (2019). CRISPR/Cas genome editing and precision plant breeding in agriculture. *Annu. Rev. Plant Biol.*, 70: 667–697.

Cheng, Z., Ge, W., Li, L. et al. (2017). Analysis of MADS-box gene family reveals conservation in floral organ ABCDE model of Moso Bamboo (*Phyllostachys edulis*). *Front. Plant Sci.*, 8: 656.

Cong, L., Ran, F.A., Cox, D. et al. (2013). Multiplex genome engineering using CRISPR/Cas systems. *Science*, 339(6121): 819–23.

Donohoue, P.D., Barrangou, R. and May, A.P. (2018). Advances in industrial biotechnology using CRISPR-Cas systems. *Trends Biotechnol.*, 36(2): 134–146.

Dutta, Smritikana, Anwesha Deb, Prasun Biswas et al. (2021). Identification and functional characterization of two bamboo FD gene homologs having contrasting effects on shoot growth and flowering. *Scientific Reports*, 11(1): 7849.

Endo, Masaki, Masafumi Mikami and Seiichi Toki. (2015). Multigene knockout utilizing off-target mutations of the CRISPR/Cas9 system in rice. *Plant and Cell Physiology*, 56(1): 41–47.

Gaj, T., Gersbach, C.A. and Barbas, C.F. 3rd. (2013). ZFN, TALEN, and CRISPR/Cas-based methods for genome engineering. *Trends Biotechnol.*, 31(7): 397–405.

Gasiunas, G., Barrangou, R., Horvath, P. and Siksnys, V. (2012). Cas9-crRNA ribonucleoprotein complex mediates specific DNA cleavage for adaptive immunity in bacteria. *Proc. Natl. Acad. Sci. USA*, 109(39): E2579–86.

Gaudelli, N.M., Komor, A.C., Rees, H.A. et al. (2017). Programmable base editing of A•T to G•C in genomic DNA without DNA cleavage. *Nature*, 551(7681): 464–471.

Ge, W., Zhang, Y., Cheng, Z., Hou, D., Li, X. and Gao, J. (2017). Main regulatory pathways, key genes and microRNAs involved in flower formation and development of moso bamboo (*Phyllostachys edulis*). *Plant Biotechnol. J.*, 15(1): 82–96.

Gilbert, L.A., Larson, M.H., Morsut, L. et al. (2013). CRISPR-mediated modular RNA-guided regulation of transcription in eukaryotes. *Cell*, 154(2): 442–51.

Groner, A.C., Meylan, S., Ciuffi, A. et al. (2010). KRAB-zinc finger proteins and KAP1 can mediate long-range transcriptional repression through heterochromatin spreading. *PLoS Genet.*, 6(3): e1000869.

Grünewald, J., Zhou, R., Garcia, S.P. et al. (2019). Transcriptome-wide off-target RNA editing induced by CRISPR-guided DNA base editors. *Nature*, 569(7756): 433–437.

Guo, Z.H., Ma, P.F., Yang, G.Q. et al. (2019). Genome sequences provide insights into the reticulate origin and unique traits of woody bamboos. *Mol. Plant*, 12(10): 1353–1365.

Gupta, R.M. and Musunuru, K. 2014. Expanding the genetic editing tool kit: ZFNs, TALENs, and CRISPR-Cas9. *J. Clin. Invest.*, 124(10): 4154–61.

Huang, B., Zhuo, R., Fan, H. et al. (2022). An efficient genetic transformation and CRISPR/Cas9-based genome editing system for Moso Bamboo (*Phyllostachys edulis*). *Front. Plant Sci.*, 13: 822022.

Ishino, Y., Shinagawa, H., Makino, K., Amemura, M. and Nakata, A. (1987). Nucleotide sequence of the iap gene, responsible for alkaline phosphatase isozyme conversion in *Escherichia coli*, and identification of the gene product. *Journal of Bacteriology*, 169(12): 5429–5433.

Jinek, M., Chylinski, K., Fonfara, I., Hauer, M., Doudna, J.A. and Charpentier, E. (2012). A programmable dual-RNA-guided DNA endonuclease in adaptive bacterial immunity. *Science*, 337(6096): 816–21.

Joung, J.K. and Sander, J.D. (2013). TALENs: A widely applicable technology for targeted genome editing. *Nat. Rev. Mol. Cell Biol.*, 14(1): 49–55.

Kantor, Ariel, Michelle E. McClements and Robert E. MacLaren. (2020). CRISPR-Cas9 DNA base-editing and prime-editing. *International Journal of Molecular Sciences*, 21(17).

Komor, A.C., Kim, Y.B., Packer, M.S., Zuris, J.A. and Liu, D.R. (2016). Programmable editing of a target base in genomic DNA without double-stranded DNA cleavage. *Nature*, 533(7603): 420–4.

Larson, M.H., Gilbert, L.A., Wang, X., Lim, W.A., Weissman, J.S. and Qi, L.S. (2013). CRISPR interference (CRISPRi) for sequence-specific control of gene expression. *Nat. Protoc.*, 8(11): 2180–96.

Li, M., Li, X., Zhou, Z. et al. (2016). Reassessment of the four yield-related genes Gn1a, DEP1, GS3, and IPA1 in rice using a CRISPR/Cas9 system. *Front. Plant Sci.*, 7: 377.

Li, Wei, Cong Shi, Kui Li et al. (2020). The draft genome sequence of herbaceous diploid bamboo *Raddia distichophylla*. bioRxiv.

Liu, Qiaoling, Bo Chen, Qinli Wang et al. (2009). Carbon nanotubes as molecular transporters for walled plant cells. *Nano Letters*, 9(3): 1007–1010.

Makarova, Kira S., Yuri I. Wolf, Jaime Iranzo et al. (2020). Evolutionary classification of CRISPR–Cas systems: A burst of class 2 and derived variants. *Nature Reviews Microbiology*, 18(2): 67–83.

Mohanta, T.K., Bashir, T., Hashem, A., Abd Allah, E.F. and Bae, H. (2017). Genome editing tools in plants. *Genes (Basel)*, 8(12).

Mojica, F.J., Díez-Villaseñor, C., García-Martínez, J. and Soria, E. (2005). Intervening sequences of regularly spaced prokaryotic repeats derive from foreign genetic elements. *J. Mol. Evol.*, 60(2): 174–82.

Mojica, F.J., Juez, G. and Rodríguez-Valera, F. (1993). Transcription at different salinities of *Haloferax mediterranei* sequences adjacent to partially modified PstI sites. *Mol. Microbiol.*, 9(3): 613–21.

Montecillo, Jake Adolf V., Luan Luong Chu and Hanhong Bae. (2020). CRISPR-Cas9 system for plant genome editing: Current approaches and emerging developments. *Agronomy*, 10(7): 1033.

Nambiar, T.S., Billon, P., Diedenhofen, G. et al. (2019). Stimulation of CRISPR-mediated homology-directed repair by an engineered RAD18 variant. *Nat. Commun.*, 10(1): 3395.

Nasti, Ryan A. and Daniel F. Voytas. (2021). Attaining the promise of plant gene editing at scale. *Proceedings of the National Academy of Sciences*, 118(22): e2004846117.

Nekrasov, V., Staskawicz, B., Weigel, D., Jones, J.D. and Kamoun, S. (2013). Targeted mutagenesis in the model plant *Nicotiana benthamiana* using Cas9 RNA-guided endonuclease. *Nat. Biotechnol.*, 31(8): 691–3.

Nemudryi, A.A., Valetdinova, K.R., Medvedev, S.P. and Zakian, S.M. (2014). TALEN and CRISPR/Cas genome editing systems: Tools of discovery. *Acta Naturae*, 6(3): 19–40.

Nishimasu, H., Ran, F.A., Hsu, P.D. et al. (2014). Crystal structure of Cas9 in complex with guide RNA and target DNA. *Cell*, 156(5): 935–49.

Peng, Zhenhua, Ying Lu, Lubin Li et al. (2013). The draft genome of the fast-growing non-timber forest species moso bamboo (*Phyllostachys heterocycla*). *Nature Genetics*, 45(4): 456–461.

Petolino, J.F. (2015). Genome editing in plants via designed zinc finger nucleases. *In Vitro Cell Dev. Biol. Plant*, 51(1): 1–8.

Pourcel, C., Salvignol, G. and Vergnaud, G. (2005). CRISPR elements in *Yersinia pestis* acquire new repeats by preferential uptake of bacteriophage DNA, and provide additional tools for evolutionary studies. *Microbiology (Reading)*, 151(Pt 3): 653–663.

Qi, L.S., Larson, M.H., Gilbert, L.A. et al. (2013). Repurposing CRISPR as an RNA-guided platform for sequence-specific control of gene expression. *Cell*, 152(5): 1173–83.

Qiao, Guirong, Huiqing Yang, Ling Zhang et al. (2013). Enhanced cold stress tolerance of transgenic *Dendrocalamus latiflorus* Munro (Ma bamboo) plants expressing a bacterial CodA gene. *In Vitro Cellular & Developmental Biology - Plant*, 50: 385–391.

Ramakrishnan, Muthusamy, Kim Yrjälä, Kunnummal Kurungara Vinod et al. 2020. Genetics and genomics of moso bamboo (*Phyllostachys edulis*): Current status, future challenges, and biotechnological opportunities toward a sustainable bamboo industry. *Food and Energy Security*, 9(4): e229.

Reynolds, N., Salmon-Divon, M., Dvinge, H. et al. (2012). NuRD-mediated deacetylation of H3K27 facilitates recruitment of Polycomb Repressive Complex 2 to direct gene repression. *Embo. J.*, 31(3): 593–605.

Shan, Q., Wang, Y., Li, J. et al. (2013). Targeted genome modification of crop plants using a CRISPR-Cas system. *Nat. Biotechnol.*, 31(8): 686–8.

Shi, J., Gao, H., Wang, H. et al. (2017). ARGOS8 variants generated by CRISPR-Cas9 improve maize grain yield under field drought stress conditions. *Plant Biotechnol. J.*, 15(2): 207–216.

Shi, Y., Liu, H., Gao, Y., Wang, Y., Wu, M. and Xiang, Y. (2019). Genome-wide identification of growth-regulating factors in moso bamboo (*Phyllostachys edulis*): *in silico* and experimental analyses. *Peer J.*, 7: e7510.

Singhal, Poonam, Lalit M. Bal, Santosh Satya, Sudhakar, P. and Naik, S.N. (2013). Bamboo shoots: A novel source of nutrition and medicine. *Critical Reviews in Food Science and Nutrition*, 53: 517–534.

Sun, Y., Shang, L., Zhu, Q.H., Fan, L. and Guo, L. (2022). Twenty years of plant genome sequencing: Achievements and challenges. *Trends Plant Sci.*, 27(4): 391–401.

Wang, Kai-li, Yuanyuan Zhang, Heng-Mu Zhang et al. 2021. MicroRNAs play important roles in regulating the rapid growth of the *Phyllostachys edulis* culm internode. *New Phytologist.*, 231(6): 2215–2230.

Wright, Addison V., James K. Nuñez and Jennifer A. Doudna. (2016). Biology and applications of CRISPR systems: Harnessing nature's toolbox for genome engineering. *Cell*, 164(1): 29–44.

Xie, K. and Yang, Y. (2013). RNA-guided genome editing in plants using a CRISPR-Cas system. *Mol. Plant*, 6(6): 1975–83.

Xu, R., Yang, Y., Qin, R. et al. (2016). Rapid improvement of grain weight via highly efficient CRISPR/Cas9-mediated multiplex genome editing in rice. *J. Genet. Genomics*, 43(8): 529–32.

Yang, K., Li, L., Lou, Y., Zhu, C., Li, X. and Gao, Z. (2021). A regulatory network driving shoot lignification in rapidly growing bamboo. *Plant Physiol.* 187(2): 900–916.

Ye, Shanwen, Changyang Cai, Huibo Ren et al. (2017). An efficient plant regeneration and transformation system of Ma Bamboo (*Dendrocalamus latiflorus Munro*) started from young shoot as explant. *Frontiers in Plant Science*, 8.

Yeasmin, L., Ali, M.N., Gantait, S. and Chakraborty, S. (2015). Bamboo: An overview on its genetic diversity and characterization. *3 Biotech.*, 5(1): 1–11.

Zhang, Dangquan, Zhiyong Zhang, Turgay Unver and Baohong Zhang. (2021). CRISPR/Cas: A powerful tool for gene function study and crop improvement. *Journal of Advanced Research*, 29: 207–221.

Zhang, Y., Tang, D., Lin, X., Ding, M. and Tong, Z. (2018). Genome-wide identification of MADS-box family genes in moso bamboo (*Phyllostachys edulis*) and a functional analysis of PeMADS5 in flowering. *BMC Plant Biol.*, 18(1): 176.

Zhao, H., Gao, Z., Wang, L. et al. (2018). Chromosome-level reference genome and alternative splicing atlas of moso bamboo (*Phyllostachys edulis*). *Gigascience*, 7(10).

Zhu, Haocheng, Chao Li and Caixia Gao. (2020). Applications of CRISPR–Cas in agriculture and plant biotechnology. *Nature Reviews Molecular Cell Biology*, 21(11): 661–677.

Chapter 15

Genome Editing and Future Scope in Bamboos

Muthusamy Ramakrishnan,[1,]* Theivanayagam Maharajan,[2]
Zishan Ahmad,[1] Stanislaus Antony Ceasar[2] and Qiang Wei[1,]*

15.1 Introduction

Genome editing is the process of altering the genetic code of an organism. Genome editing uses enzymes to cut DNA at a specific and targeted site to create a DNA double-standard break (Abdallah et al., 2015). The repair of the double-standard break is accomplished by either non-homologous end joining (NHEJ) or homology-directed repair (HDR). NHEJ creates random mutations (gene knockout), while HDR uses additional DNA to create a desired sequence in the genome (gene knock-in) (Lino et al., 2018). Three types of tools, including zinc finger nucleases (ZFNs), transcription activator-like effector nucleases (TALENs), and clustered regularly interspaced short palindromic repeats (CRISPR)-associated protein (Cas) (CRISPR/Cas) are most used for genome editing technology (Gaj et al., 2013; Arora and Narula, 2017).

15.1.1 ZFNs

ZFNs are tailored and targeted DNA cleavage proteins that are designed to cut DNA sequences at specific sites (Carroll, 2011). They enable targeted gene editing by creating a double-standard break in DNA to replace the target gene through homologous recombination. ZFN contains two domains, a DNA-binding domain,

[1] Co-Innovation Center for Sustainable Forestry in Southern China, Bamboo Research Institute, Key Laboratory of National Forestry and Grassland Administration on Subtropical Forest Biodiversity Conservation, College of Biology and the Environment, Nanjing Forestry University, Nanjing, Jiangsu 210037, China.
[2] Department of Biosciences, Rajagiri College of Social Sciences (Autonomous), Kalamassery, Kochi – 683 104, Kerala, India.
* Corresponding authors: ramky@njfu.edu.cn; weiqiang@njfu.edu.cn

and a DNA-cleaving domain (Carlson et al., 2012; Lee et al., 2016). The DNA-binding domain recognizes a unique 6-base pair in the DNA sequence, while the DNA-cleaving domain consists of a FokI nuclease (Carlson et al., 2012; Gupta et al., 2013). These two domains are linked together to form a zinc finger protein. When fused together, the two domains form a highly specific genomic scissor (Mohanta et al., 2017).

The most important factor for the use of ZFN-based genome editing is the dependence of this technique on the generation of zinc finger proteins that can precisely target a specific DNA sequence in the genome (Mohanta et al., 2017). The most common DNA-binding domain of Cys2His2 (zinc finger) provides the best possible structure for engineering suitable ZFNs with the required sequence specificities (Pabo et al., 2001; Hossain et al., 2015). The Cys2His2 consists of approximately 30 amino acids and has two antiparallel β-sheets (Yusuf et al., 2021). The Cys2His2 binds to the DNA sequence by inserting its α-helix into the major groove of the DNA double helix (Wolfe et al., 2000). Four domains (methylase, FokI cleavage, transcriptional activator, and repressor) are fused to Cys2His2 to form a ZFN (Mani et al., 2005). ZFNs bind specifically to triplet DNA such as 5'-GNN-3', 5'-ANN-3', 5'-CNN-3', and 5'-TNN-3' (Lloyd et al., 2005; Jamieson et al., 2003). In the zinc finger motif, the presence of the Asp residue at the second position of the α-helix promotes cross-strand contact outside the triplet DNA, resulting in overlap of the target sites (Pavletich, 1991). Therefore, when the Asp residue is present in the zinc finger (at the second position of the α-helix), it binds to a DNA target sequence of four base pairs instead of a triplet DNA; hence the design strategy is complex (Mohanta et al., 2017). When the Asp residue is missing from the second position of the α-helix, only two bases on the triplet DNA are recognized, leading to the possibility of detecting degenerate sites (Durai et al., 2005). After recognition of a specific zinc finger protein at a specific DNA site, the restriction endonuclease of Fok1 (type II restriction enzyme) comes into play (Mohanta et al., 2017). FokI recognizes the non-palindromic pentadeoxyribonucleotide sequence (5'-GGATG-3': 5'-CATCC-3') in double-stranded DNA and cleaves the DNA at the recognition site. In general, ZFNs require a recognition site of six base pairs to dimerize the DNA domain, resulting in the formation of a double-strand break (Cathomen and Joung, 2008). In fact, ZFNs have 18 base pair authentication sites, which are sufficient to recognize a unique DNA sequence (Durai et al., 2005). After generating a double-strand break, the corresponding gene in the genome must be targeted. Compared to other genome editing tools, ZFNs allow us to disrupt/integrate arbitrary genomic loci rapidly and randomly in the genome. The mutations caused by ZFNs can be permanent and heritable.

15.1.2 TALENs

TALENs are another type of genome editing tool that has better specificity and performance than ZFNs. They are structurally like the ZFNs containing the endonuclease of FokI but differ in the DNA-binding domains (transcription activator-like effectors (TALEs)) (Joung and Sander, 2013). Like ZFNs, TALENs use FokI (DNA-cleaving domain) to generate double-strand breaks and require dimerization to function (Christian et al., 2010). TALENs contain 33–35 amino acid repeats that are highly conserved except for positions 12 and 13 (Mohanta et al., 2017). Positions 12 and 13 are referred to as variable di-residues of the repeats and show significant correlation with specific nucleotide recognition. The number of amino acid residues between TALENs (DNA-binding domain) and FokI (DNA-cleavage domain) and the number of bases between two separate TALEN-binding sites are key parameters affecting the function of TALENs (Li et al., 2020). When TALENs are constructed, they are transferred to the plasmid vector and then to the target cells (Kusano et al., 2016). The DNA-cleavage domain (FoK1) is fused with the DNA-binding domain (TALE) to generate site-specific double standard breaks, thereby stimulating DNA recombination to complete the TALEN-induced targeted genome modification. TALENs are much larger than ZFNs, so getting them into cells is challenging.

15.1.3 *CRISPR/Cas*

CRISPR is a family of repetitive DNA sequences found in the genomes of archaea (84%) and bacteria (45%) (Ishino et al., 2018). It was first discovered in the downstream of the alkaline phosphatase isozyme gene in *Escherichia coli* (Ishino et al., 1987). It is formerly known as short regularly spaced repeats (SRSRs) and helps to recognize and destroy viral DNA (Richter et al., 2012). In the CRISPR system, small guide RNAs (crRNAs) are used for sequence-specific interference with invading nucleic acids. CRISPR is an array of short, repetitive sequences (repeats) separated by unique sequences (spacers). Both the spacers and the repeats are derived from the nucleic acid of viruses and plasmids (Hille et al., 2016). Some of the proteins involved in the CRIPSR mechanism are called Cas. They can search, cut, and finally transform phage DNA in a specific way (Nadakuduti and Enciso-Rodriguez, 2021). Cas is a protein with enzymatic activity that plays a special role in DNA sequences and CRISPR arrays; therefore, it is also called a nuclease (Makarova and Koonin, 2015). The CRISPR/Cas mechanism can be divided into three steps: (1) insertion of unique sequences into the CRISPR locus (spacer acquisition or adaptation), (2) transcription of the CRISPR locus and processing of crRNA (expression or crRNA biogenesis), and (3) recognition and degradation of nucleic acids by crRNA and Cas proteins (target interference) (Rath et al., 2015; Liu et al., 2018; Hsu et al., 2014). CRISPR/Cas systems have been divided into two classes (Class I and II) (Jinek et al., 2012). Both classes of CRISPR systems can be used for genome editing, but the Class II systems are desirable for genome editing because the methods are much simpler. The Class II CRISPR/Cas systems are categorized by three signature proteins such as Cas1, 2, and 9, which include three subtypes (II-A, II-B, and II-C). Among the three proteins, Cas9 is most used in the CRISPR system (CRISPR/Cas9) for genome editing in various organisms.

15.2 Mechanism and Classifications of CRISPR/Cas System

Compared to the other genome editing systems, CRISPR/Cas system is more suitable and has many advantages. CRISPR/Cas systems are generally classified into two classes and six types with several subtypes (Makarova and Koonin, 2015). Class I systems are considered as evolutionary ancestral systems (Koonin and Makarova, 2019). Class II systems are derived from Class I systems by the insertion of transposable elements encoding different nucleases and are now used as tools for genome editing (Shmakov et al., 2017). Class I systems contain crRNA effector complexes with multiple subunits, whereas Class II systems perform all effector complex functions through a single protein such as Cas9 (Makarova and Koonin, 2015). Due to the unique structure of the effector modules, types I, II, and IV belong to Class I and types II, V, and VI belong to Class II (Makarova et al., 2015). The type 1 system can be further divided into six subtypes (types 1A to 1F). All type 1 systems encode the Cas1, 2, and 3 proteins in a cascade-like complex. The type II systems are further subdivided into three subtypes (type IIA to C) (Rath et al., 2015). Five proteins (Cas1, 2, 4, 9, and Csn2) are encoded by the type IIA and B system, while three proteins (Cas1, 2, and 9) are encoded by the type IIC system (Ka et al., 2018). The type III system can be divided into two subtypes (type IIIA and B). These two subtypes of the type III system contain the Cas10 protein, but their function is not clear (Lin et al., 2020). Most Cas proteins are part of Cascade-like complexes such as Csm (type III-A) and Cmr (type III-B). All type I and II systems target DNA, whereas type III systems target both DNA and RNA. The type II systems were found in bacteria, while the other two types (types I and III) are found in bacteria and archaea.

The CRISPR/Cas9 system is widely used in gene editing technology because it is relatively simple in design and structure. It contains two major components, namely the guide RNA (gRNA) and the Cas9 protein. Of these two components, the Cas9 protein is an RNA-dependent DNA endonuclease that forms a complex with the gRNA (Hsu et al., 2014). The second component of gRNA is a small RNA containing 20 nucleotides that binds to the target DNA (Wu et al., 2014). The protein Cas9 is a large multi-domain DNA endonuclease and was extracted from *Streptococcus pyogenes* (Nishimasu et al., 2014). It is referred to as genetic scissors because it is involved in the cleavage of DNA (formation of double-strand breaks) at specific sites (Mei et al., 2016). The Cas9 protein consists of two lobes, the recognition lobe (REC) and the nuclease (NUC) (Nishimasu et al., 2014). Of these, the REC lobe contains two domains (REC1 and 2) responsible for gRNA binding. The NUC lobe contains three domains (RuvC, HNH, and PAM). Of these, the RuvC and HNH domains are used to cut single-stranded DNA, while the other domain of PAM is responsible for initiating targeted DNA binding (Jiang and Doudna, 2017). The gRNA consists of two parts, crRNA and tracrRNA (Wong et al., 2015). The length of crRNA is 18–20 base pairs, which attaches to the target DNA, while tracrRNA is a long extension of the loops that serves as a binding scaffold for the Cas9 nucleus (Jiang and Doudna, 2017). The CRISPR/Cas9 mechanism can be divided into three steps: Recognition, cleavage, and repair (Asmamaw and Zawdie, 2021). The designed sgRNA activates the Cas9 nuclease and recognizes the target sequence in the gene of interest through its 5'crRNA complement base pair component (Wu et al., 2014). In the absence of sgRNA, the Cas9 protein is inactive. In general, Cas9 recognizes the sequence PAM at 5'-NGG-3' (N can be any nucleotide base) (Gleditzsch et al., 2019). Once Cas9 recognizes a target site with the appropriate PAM, it induces a local DNA fusion that subsequently forms the RNA-DNA hybrid, but it is not yet clear how Cas9 fuses the target DNA sequence (Hsu et al., 2014). Cas9 protein is then activated for DNA cleavage. The complementary and non-complementary strands of the target DNA are cleaved from the HNH and RuvC domains, respectively, forming mainly blunt-ended double-standard breaks. Finally, the double-standard break is repaired by the host cellular machinery.

15.3 Characterization of Various Genes by CRISPR/Cas9 System

The role of different genes has been identified in many crops by the CRISPR/ Cas9 system under biotic and abiotic stresses. Lou et al. (2017) identified the role of the stress activated protein kinase (*SAPK2*) gene in rice under drought stress by the CRISPR/Cas9 system. They reported that *SAPK2* is involved in improving the plant growth under drought stress. In another study, the role of Semi-Rolled Leaf 1 (*SRL1*) and *SRL2* genes was identified in rice under drought stress by the CRISPR/ Cas9 system (Liao et al., 2019). Both genes are involved in increasing survival rate, abscisic acid content, superoxide dismutase (SOD) and catalase (CAT) activities, and the percentage of grain filling in rice under drought stress. In rice, the knockout of the drought and salt tolerance (*DST*) gene improved leaf growth and leaf water retention under drought stress (Kumar et al., 2020). The CRISPR/Cas9 system was used to identify the role of the gene for enhanced response to abscisic acid (*ERA*) under drought stress (Ogata et al., 2020). The silencing of the *ERA1* gene in rice increased stomatal conductance under drought stress. Usman et al. (2020) used the CRISPR/Cas9 system to elucidate the role of the abscisic acid receptor of the pyrabactin resistance gene (*PYL*) in rice under drought stress. Knockout of the *PYL* gene in rice improved plant height, panicle number, panicle length, flag leaf length and width, number of grains per panicle, grain weight, grain length and width, and yield per plant under drought stress. In maize, knock-in of the auxin-regulated gene involved in organ size (*ARGOS*) increased growth and yield under drought stress (Shi et al., 2017). All these results indicate that the CRISPR/Cas9 system helps to identify the accurate role of each gene in plants. Like rice, maize and other cereals, bamboo comes under the family of Poaceae. Recently, the application of the CRISPR/Cas9 system has also been reported in two bamboo species, hexaploid Ma bamboo (*Dendrocalamus latiflorus* Munro) and Moso bamboo (*Phyllostachys edulis*) (Ye et al., 2020; Huang et al., 2022). Various transporters and transcription factors have been identified in bamboo under various stress conditions and during different growth stages (reviewed by Ramakrishnan et al., 2020). However, the role of each transporter and transcription factor has not yet been identified by the

CRISPR/Cas9 system in bamboo compared to other members of the Poaceae family under different environmental stress and at different growth stages. The modification of agronomic traits in bamboo is nearly impossible by traditional breeding due to irregular flowering patterns, and bamboo research is behind due to the lack of efficient genetic transformation.

15.4 Genome Editing and Future Scope in Bamboo

To date, only two studies have reported the functional significance of genome editing in bamboo (Ye et al., 2020; Huang et al., 2022). The first study showed that the silencing of gibberellin-responsive gene 1 (*grg1*) in hexaploid Ma bamboo (2n = 72, AABBCC) altered the plant height using protoplast culture and using rice promoters (*UBI*-Cas9/*OsU6b*-sgRNA). First, they improved the protoplast culture and then optimized the *UBI-Cas9/OsU6b-sgRNA* construct (rice U6 promoters) in the culture. Subsequently, *sgRNA1* and *2* were targeted against three Ma bamboo alleles (*DlmPSY1-A*, *DlmPSY1-B*, and *DlmPSY1-C*). In the results, *sgRNA2* targeted only *DlmPSY1-A1*, but there was no phenotypic change because of the presence of the wild-type alleles *DlmPSY1-B* and *DlmPSY1-C*. Then, the method generated two homozygote *grg1* mutants in Ma bamboo. The mutant enhanced the internode elongation, which increased plant height. The loss of *grg1* function in the transgenic bamboo was confirmed by confirming the putative homozygous mutation, biallelic mutation, and homozygous mutation in sub-genomes A1, B1, and C1, respectively. The results suggest that rice U6 promoter *OsU6b* showed higher editing efficiency and that the generation of homozygous mutations in bamboo species could contribute to study the molecular mechanisms of bamboo growth (Ye et al., 2020).

The second study was conducted on Moso bamboo (Huang et al., 2022). In the study, the plant regeneration system was first established from immature embryos of Moso bamboo, followed by a genetic transformation system with a transformation efficiency of 5%. Subsequently, the promoters of two endogenous U3 small nuclear RNA were used to drive the *sgRNA*. The *PeU3.1* promoter had a higher efficiency and was therefore used for subsequent genome editing. The CRISPR/Cas9 system generated three lines of putative homozygous *pds1pds2* mutants with albino phenotypes, suggesting that the *PeU3.1* promoter is an efficient driver of *sgRNA* for CRISPR/Cas9-based genome editing in Moso bamboo.

These two studies could contribute to further research on the molecular mechanisms behind the rapid growth of bamboo. Several other CRISPR/Cas-mediated genome editing tools such as CRISPR interference (CRISPRi), CRISPR activation (CRISPRa) (Fig. 15.1), base editors, epigenetic engineering, chromatin imaging, and prime editors have been identified and used in crop improvement. All these genome editing tools are becoming increasingly popular in crops due to their accuracy and robustness. Identifying the role of each gene and transcription factor through the currently emerging CRISPR/Cas-mediated tools can help improve the growth of bamboo.

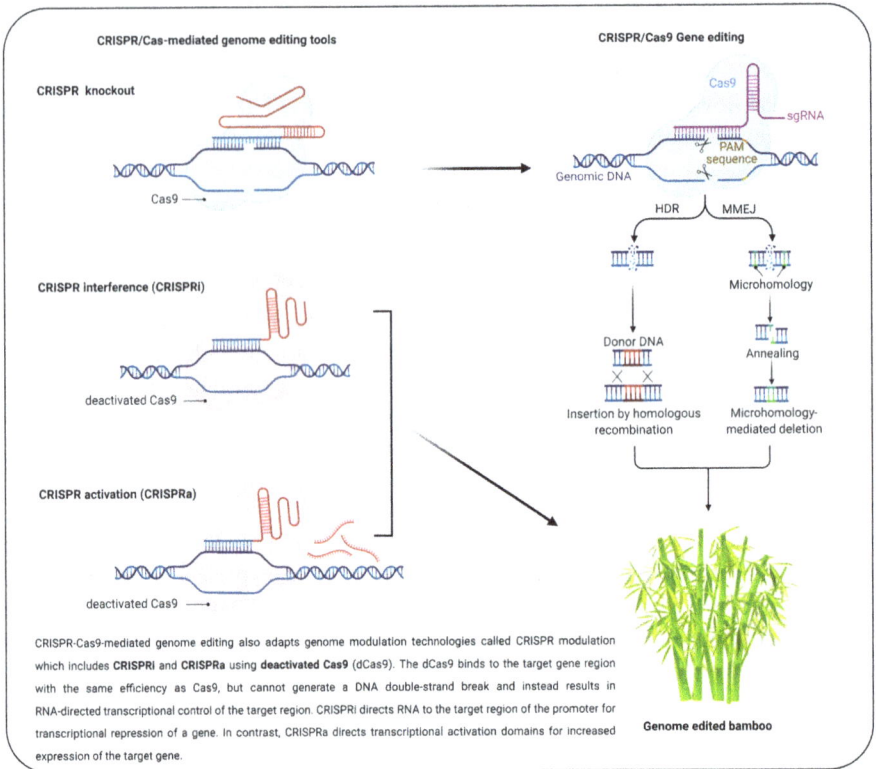

Fig. 15.1. Clustered regularly interspaced short palindromic repeats (CRISPR)-associated protein 9 (Cas9) (CRISPR/Cas9) genome editing system in bamboo. HDR, homology-directed repair; MMEJ, Microhomology-mediated end joining; sgRNA, single guide RNA. Created with BioRender.com.

Acknowledgements and Funding

The authors are grateful for support from the National Natural Science Foundation of China (32071848), a grant from the Jiangxi 'Shuangqian' Program (S2019DQKJ2030), the Natural Science Foundation for Distinguished Young Scholars of Nanjing Forestry University (JC2019004), the Qing Lan Project of Jiangsu Higher Education Institutions, and a project funded by the Priority Academic Program Development of Jiangsu Higher Education Institutions. This work was also supported by the Metasequoia Faculty Research Start-up Funding (grant number 163100028) at Bamboo Research Institute, Nanjing Forestry University for the first author MR.

Author Contributions

MR and TM planned, designed, and wrote the chapter. MR, TM, ZA, SAC, and QW outlined and edited the chapter. All authors edited and revised the chapter.

Declaration of Conflict of Interest

The authors declare that the research was conducted in the absence of any commercial or financial relationships that could be construed as a potential conflict of interest.

References

Abdallah, N.A., Prakash, C.S. and McHughen, A.G. (2015). Genome editing for crop improvement: Challenges and opportunities. *GM Crops & Food*, 6(4): 183–205.

Arora, L. and Narula, A. (2017). Gene editing and crop improvement using CRISPR/Cas9 system. *Frontiers in Plant Science*, 8: 1932.

Asmamaw, M. and Zawdie, B. (2021). Mechanism and applications of CRISPR/Cas-9-mediated genome editing. *Biologics: Targets & Therapy*, 15: 353.

Carlson, D.F., Tan, W., Lillico, S.G., Stverakova, D., Proudfoot, C., Christian, M., Voytas, D.F., Long, C.R., Whitelaw, C.B. and Fahrenkrug, S.C. (2012). Efficient TALEN-mediated gene knockout in livestock. *Proc. Natl. Acad. Sci. USA*, 109(43): 17382–17387.

Carroll, D. (2011). Genome engineering with zinc-finger nucleases. *Genetics*, 188(4): 773–782.

Cathomen, T. and Joung, J.K. (2008). Zinc-finger nucleases: The next generation emerges. *Molecular Therapy*, 16(7): 1200–1207.

Christian, M., Cermak, T., Doyle, E.L., Schmidt, C., Zhang, F., Hummel, A., Bogdanove, A.J. and Voytas, D.F. (2010). Targeting DNA double-strand breaks with TAL effector nucleases. *Genetics*, 186(2): 757–761.

Durai, S., Mani, M., Kandavelou, K., Wu, J., Porteus, M.H. and Chandrasegaran, S. (2005). Zinc finger nucleases: Custom-designed molecular scissors for genome engineering of plant and mammalian cells. *Nucleic Acids Research*, 33(18): 5978–5990.

Gaj, T., Gersbach, C.A. and Barbas III, C.F. (2013). ZFN, TALEN, and CRISPR/Cas-based methods for genome engineering. *Trends in Biotechnology*, 31(7): 397–405.

Gleditzsch, D., Pausch, P., Müller-Esparza, H., Özcan, A., Guo, X., Bange, G. and Randau, L. (2019). PAM identification by CRISPR-Cas effector complexes: Diversified mechanisms and structures. *RNA Biology*, 16(4): 504–517.

Gupta, A., Hall, V.L., Kok, F.O., Shin, M., McNulty, J.C., Lawson, N.D. and Wolfe, S.A. (2013). Targeted chromosomal deletions and inversions in zebrafish. *Genome Research*, 23(6): 1008–1017.

Hille, F. and Charpentier, E. (2016). CRISPR-Cas: Biology, mechanisms, and relevance. *Philosophical Transactions of the Royal Society B: Biological Sciences*, 371(1707): 20150496.

Hossain, M.A., Barrow, J.J., Shen, Y., Haq, M.I. and Bungert, J. (2015). Artificial zinc finger DNA binding domains: Versatile tools for genome engineering and modulation of gene expression. *Journal of Cellular Biochemistry*, 116(11): 2435–2444.

Hsu, P.D., Lander, E.S. and Zhang, F. (2014). Development and applications of CRISPR/Cas9 for genome engineering. *Cell*, 157(6): 1262–1278.

Huang, B., Zhuo, R., Fan, H., Wang, Y., Xu, J., Jin, K. and Qiao, G. (2022). An efficient genetic transformation and CRISPR/Cas9-based genome editing system for Moso bamboo (*Phyllostachys edulis*). *Front. Plant Sci.* https://doi.org/10.3389/fpls.2022.822022.

Ishino, Y., Krupovic, M. and Forterre, P. (2018). History of CRISPR/Cas from encounter with a mysterious repeated sequence to genome editing technology. *Journal of Bacteriology*, 200(7): e00580–17.

Ishino, Y., Shinagawa, H., Makino, K., Amemura, M. and Nakata, A. (1987). Nucleotide sequence of the *iap* gene, responsible for alkaline phosphatase isozyme conversion in *Escherichia coli*, and identification of the gene product. *Journal of Bacteriology*, 169(12): 5429–5433.

Jamieson, A.C., Miller, J.C. and Pabo, C.O. (2003). Drug discovery with engineered zinc-finger proteins. *Nature Reviews Drug Discovery*, 2(5): 361–368.

Jiang, F. and Doudna, J.A. (2017). CRISPR/Cas9 structures and mechanisms. *Annual Review of Biophysics*, 46: 505–529.

Jinek, M., Chylinski, K., Fonfara, I., Hauer, M., Doudna, J.A. and Charpentier, E. (2012). A programmable dual-RNA–guided DNA endonuclease in adaptive bacterial immunity. *Science*, 337(6096): 816–821.

Joung, J.K. and Sander, J.D. (2013). TALENs: A widely applicable technology for targeted genome editing. *Nature Reviews Molecular Cell Biology*, 14(1): 49–55.

Ka, D., Jang, D.M., Han, B.W. and Bae, E. (2018). Molecular organization of the type II-A CRISPR adaptation module and its interaction with Cas9 via Csn2. *Nucleic Acids Research*, 46(18): 9805–9815.

Koonin, E.V. and Makarova, K.S. (2019). Origins and evolution of CRISPR/Cas systems. *Philosophical Transactions of the Royal Society B*, 374(1772): 20180087.

Kumar, V.S., Verma, R.K., Yadav, S.K., Yadav, P., Watts, A., Rao, M.V. and Chinnusamy, V. (2020). CRISPR/Cas9 mediated genome editing of drought and salt tolerance (OsDST) gene in indica mega rice cultivar MTU1010. *Physiology and Molecular Biology of Plants*, 26(6): 1099.

Kusano, H., Onodera, H., Kihira, M., Aoki, H., Matsuzaki, H. and Shimada, H. (2016). A simple gateway-assisted construction system of TALEN genes for plant genome editing. *Scientific Reports*, 6(1): 1–7.

Lee, J., Chung, J.H., Kim, H.M., Kim, D.W. and Kim, H. (2016). Designed nucleases for targeted genome editing. *Plant Biotechnology Journal*, 14(2): 448–462.

Li, H., Yang, Y., Hong, W., Huang, M., Wu, M. and Zhao, X. (2020). Applications of genome editing technology in the targeted therapy of human diseases: Mechanisms, advances, and prospects. *Signal Transduction and Targeted Therapy*, 5(1): 1–23.

Liao, S., Qin, X., Luo, L., Han, Y., Wang, X., Usman, B., Nawaz, G., Zhao, N., Liu, Y. and Li, R. (2019). CRISPR/Cas9-induced mutagenesis of semi-rolled Leaf1, 2 confers curled leaf phenotype and drought tolerance by influencing protein expression patterns and ROS scavenging in rice (*Oryza sativa* L.). *Agronomy*, 9(11): 728.

Lin, J., Feng, M., Zhang, H. and She, Q. (2020). Characterization of a novel type III CRISPR/Cas effector provides new insights into the allosteric activation and suppression of the Cas10 DNase. *Cell Discovery*, 6(1): 1–16.

Lino, C.A., Harper, J.C., Carney, J.P. and Timlin, J.A. (2018). Delivering CRISPR: A review of the challenges and approaches. *Drug Delivery*, 25(1): 1234–1257.

Liu, T., Pan, S., Li, Y., Peng, N. and She, Q. (2018). Type III CRISPR/Cas system: Introduction and its application for genetic manipulations. *Current Issues in Molecular Biology*, 26(1): 1–14.

Lloyd, A., Plaisier, C.L., Carroll, D. and Drews, G.N. (2005). Targeted mutagenesis using zinc-finger nucleases in Arabidopsis. *Proceedings of the National Academy of Sciences*, 102(6): 2232–2237.

Lou, D., Wang, H., Liang, G. and Yu, D. (2017). OsSAPK2 confers abscisic acid sensitivity and tolerance to drought stress in rice. *Frontiers in Plant Science*, 8: 993.

Makarova, K.S. and Koonin, E.V. (2015). Annotation and classification of CRISPR/Cas systems. *CRISPR*, 47–75.

Makarova, K.S., Wolf, Y.I., Alkhnbashi, O.S., Costa, F., Shah, S.A., Saunders, S.J., Barrangou, R., Brouns, S.J., Charpentier, E., Haft, D.H., Horvath, P., Moineau, S., Mojica, F.J., Terns, R.M., Terns, M.P., White, M.F., Yakunin, A.F., Garrett, R.A., van der Oost, J., Backofen, R. and Koonin, E.V. (2015). An updated evolutionary classification of CRISPR/Cas systems. *Nature Reviews Microbiology*, 13(11): 722–736.

Mani, M., Kandavelou, K., Dy, F.J., Durai, S. and Chandrasegaran, S. (2005). Design, engineering, and characterization of zinc finger nucleases. *Biochemical and Biophysical Research Communications*, 335(2): 447–457.

Mei, Y., Zhang, C., Kernodle, B.M., Hill, J.H. and Whitham, S.A. (2016). A foxtail mosaic virus vector for virus-induced gene silencing in maize. *Plant Physiol.*, 171(2): 760–772.

Mohanta, T.K., Bashir, T., Hashem, A., Abd_Allah, E.F. and Bae, H. (2017). Genome editing tools in plants. *Genes*, 8(12): 399.

Nadakuduti, S.S. and Enciso-Rodríguez, F. (2021). Advances in genome editing with CRISPR systems and transformation technologies for plant DNA manipulation. *Frontiers in Plant Science*, 11: 2267.

Nishimasu, H., Ran, F.A., Hsu, P.D., Konermann, S., Shehata, S.I., Dohmae, N., Ishitani, R., Zhang, F. and Nureki, O. (2014). Crystal structure of Cas9 in complex with guide RNA and target DNA. *Cell*, 156(5): 935–949.

Ogata, T., Ishizaki, T., Fujita, M. and Fujita, Y. (2020). CRISPR/Cas9-targeted mutagenesis of OsERA1 confers enhanced responses to abscisic acid and drought stress and increased primary root growth under nonstressed conditions in rice. *PLoS One*, 15(12): e0243376.

Pabo, C.O., Peisach, E. and Grant, R.A. (2001). Design and selection of novel Cys2His2 zinc finger proteins. *Annual Review of Biochemistry*, 70(1): 313–340.

Pavletich, N.P. and Pabo, C.O. (1991). Zinc finger-DNA recognition: Crystal structure of a Zif268-DNA complex at 2.1 Å. *Science*, 252(5007): 809–817.

Ramakrishnan, M., Yrjälä, K., Vinod, K.K., Sharma, A., Cho, J., Satheesh, V. and Zhou, M. (2020). Genetics and genomics of moso bamboo (*Phyllostachys edulis*): Current status, future challenges, and biotechnological opportunities toward a sustainable bamboo industry. *Food Energy Secur.*, 9: e229.

Rath, D., Amlinger, L., Rath, A. and Lundgren, M. (2015). The CRISPR/Cas immune system: Biology, mechanisms, and applications. *Biochimie*, 117: 119–128.

Richter, C., Chang, J.T. and Fineran, P.C. (2012). Function and regulation of clustered regularly interspaced short palindromic repeats (CRISPR)/CRISPR associated (Cas) systems. *Viruses*, 4(10): 2291–2311.

Shi, J., Gao, H., Wang, H., Lafitte, H.R., Archibald, R.L., Yang, M., Hakimi, S.M., Mo, H. and Habben, J.E. (2017). ARGOS 8 variants generated by CRISPR/Cas9 improve maize grain yield under field drought stress conditions. *Plant Biotechnology Journal*, 15(2): 207–216.

Shmakov, S., Smargon, A., Scott, D., Cox, D., Pyzocha, N., Yan, W., Abudayyeh, O.O., Gootenberg, J.S., Makarova, K.S., Wolf, Y.I., Severinov, K., Zhang, F. and Koonin, E.V. (2017). Diversity and evolution of class 2 CRISPR/Cas systems. *Nature Reviews Microbiology*, 15(3): 169–182.

Usman, B., Nawaz, G., Zhao, N., Liao, S., Liu, Y. and Li, R. (2020). Precise editing of the ospyl9 gene by RNA-guided cas9 nuclease confers enhanced drought tolerance and grain yield in rice (*Oryza sativa* l.) by regulating circadian rhythm and abiotic stress responsive proteins. *International Journal of Molecular Sciences*, 21(21): 7854.

Wolfe, S.A., Nekludova, L. and Pabo, C.O. (2000). DNA recognition by Cys2His2 zinc finger proteins. *Annual Review of Biophysics and Biomolecular Structure*, 29(1): 183–212.

Wong, N., Liu, W. and Wang, X. (2015). WU-CRISPR: Characteristics of functional guide RNAs for the CRISPR/Cas9 system. *Genome Biology*, 16(1): 1–8.

Wu, X., Kriz, A.J. and Sharp, P.A. (2014). Target specificity of the CRISPR/Cas9 system. *Quantitative Biology*, 2(2): 59–70.

Ye, S., Chen, G., Kohnen, M.V., Wang, W., Cai, C., Ding, W., Wu, C., Gu, L., Zheng, Y., Ma, X., Lin, C. and Zhu, Q. (2020). Robust CRISPR/Cas9 mediated genome editing and its application in manipulating plant height in the first generation of hexaploid Ma bamboo (*Dendrocalamus latiflorus* Munro). *Plant Biotechnology Journal*, 18(7): 1501.

Yusuf, A.P., Abubakar, M.B., Malami, I., Ibrahim, K.G., Abubakar, B., Bello, M.B., Qusty, N., Elazab, S.T., Imam, M.U., Alexiou, A. and Batiha, G.E.S. (2021). Zinc metalloproteins in epigenetics and their crosstalk. *Life*, 11(3): 186.

Chapter 16

Production of Quality Planting Material in Bamboo

Muralidharan Enarth Maviton

16.1 Introduction

There has been a spurt of interest in recent years for establishing commercial plantations of bamboo around the world. This is the direct outcome of the increasing number of industrial uses that bamboo is being used for, apart from the traditional uses for rural construction, implements, furniture, and food. Bamboo, by virtue of being among the fastest growing plants, is an ideal source of fast-renewable lignocellulosic biomass. Bamboo is a woody perennial grass which has a growth pattern in which new culms are produced every year which can be harvested after they mature in two years. Each plant of bamboo thus consists of a clump of several culms of varying ages interconnected by an underground network of rhizomes. A single clump of bamboo can yield culms continuously until the end of the life cycle which varies from about 30–120 years depending on the species.

Earlier, the entire bamboo consumed, including raw material for the paper and pulp industry, was almost entirely met by collection from forests and bamboo growing in farmland. This mode of supply of raw material is now no longer sustainable since large-scale collection of bamboo from forests is damaging to the ecological balance including loss of biodiversity and wildlife. Cultivation on a large scale of appropriate bamboo species in plantations and agroforestry is the solution to meet the anticipated increase in demand for bamboo in the coming years.

Bamboos, with a few exceptions, are a group of plants that are in the early stage of domestication and scientific studies are relatively recent. The diversity of species and forms in bamboo is large and consist of small herbaceous as well as giant woody

Chief Scientist (Retired), Kerala Forest Research Institute, Peechi, Kerala, 680653 India; Consultant, International Bamboo and Rattan Organization (INBAR), Beijing; Chair, Task Force on Sustainable Bamboo Management, INBAR.
Email: emmurali@gmail.com

species. The sympodial (clump forming) bamboo species are predominantly from the tropical region while the monopodial (running) bamboos are found in the subtropical to temperate areas. The large woody bamboos are the most economically useful species and about a 100 of them are exploited around the world to varying degrees.

While there has been a rapid increase in bamboo plantations around the world, it has not always been accompanied by scientific practices until very recently. There has been an increase in the scientific studies on various aspects of bamboo in some countries, but nevertheless a major lacuna exists in the knowledge on important characteristics of several species that limit their efficient cultivation and utilization. With the increasing interest in large-scale plantations, the issue of quality control of planting material has come into the fore since improved productivity and quality of produce assumes importance as in any agricultural, horticultural, or industrial forestry crop. Bamboo, by virtue of its singular features described below, requires an approach different from other tree crops for genetic improvement, propagation, and production of quality planting material (QPM).

16.2 Flowering Behavior of Bamboo

Bamboos show a flowering behavior that is unique for the plant kingdom in many respects, and this has important implications for its propagation, cultivation, and genetic improvement. Bamboos are typically semelparous (flowering once in its lifetime and dying), mast flowering (synchronous flowering within species), and with a flowering cycle that extends into several decades. Two types of flowering, viz., sporadic flowering and gregarious flowering are encountered in bamboos. Most woody bamboo species, including the economically important ones, flower gregariously and therefore entire populations die at the end of the life cycle. There are bamboo species which are sterile and do not set seeds and die and yet others that flower only rarely.

Dependable records of past flowering events are not common and therefore the intermast period in bamboo has rarely been determined precisely to the year. Moreover, flowering in a locality usually occurs over a period of three years and moves wave-like over the bamboo landscape (Keeley and Bond, 1999). The inability to determine the exact geographical extent of populations that flowered in the past also contributes to this lack of information.

16.3 Bamboo Propagation

16.3.1 Natural Propagation in Bamboo

Except for a few species for which no seed has been recorded, all bamboos reproduce through seeds. Most woody bamboos have a long intermast period ranging from 30 years to 120 years and in most instances the flowering records are not available and therefore the following flowering year is unpredictable with precision. Seeds are produced in great quantities following a gregarious flowering event, but bamboo seeds lose their viability rapidly. When stored at low temperatures in a desiccated condition (4°C and 45% relative humidity), bamboo seeds retain their germinability although much reduced for more than a year (Lakshmi et al., 2018; Somen and Seethalakshmi, 1989). This is therefore a hurdle in planning for propagation in large planting programmes. Bamboos are predominantly outcrossing species and seeds therefore do not breed true and a have a high level of heterogeneity.

16.3.2 Vegetative Propagation Methods

A variety of vegetative propagation methods have been used traditionally by bamboo cultivators since flowering occurred at long and unpredictable intervals and was not a dependable means of propagation. The technique of macro-proliferation has been developed (Kumar, 1991) as a means of clonally propagating bamboo, especially seedlings in the nursery. Vegetative parts of the bamboo clump such as rhizomes offsets, culm and branch cuttings are widely used traditionally for propagation (Banik, 1995). Rhizome offsets taken from 1–3-year-old culms give good success in propagation but due to the large size and difficulty in extraction from the soil without damage, is not considered ideal for a large-scale propagation or transport over long distances. Despite the array of methods available to clonally multiply bamboo, largescale propagation is severely constrained by the quantity of suitable vegetative material that can be collected from a clump and the relatively low multiplication rates that can be achieved. When higher numbers of planting stock are required, rooted culms and branch cuttings are much more suitable, however these methods are not suitable for a large-scale propagation due to limitations of space, time, and labour. Clearly, a method suitable for efficient large-scale clonal propagation is required.

16.3.3 Micropropagation

The limitation faced in large-scale cloning through vegetative propagation is overcome by use of plant tissue culture techniques. Micropropagation, where small plant parts (explants) are used to initiate sterile cultures, high levels of multiplication are achieved through use of plant growth regulators and manipulation of the other components of the nutrient media and the culture environment. The advantages of micropropagation lie in the much higher rates in multiplication, production of uniform, disease and pest free plantlets and the considerable savings in time and space. Micropropagation in bamboo has in the past few decades evolved to a stage where several of the economically important species are routinely propagated and planted around the world. The status of this technology and the latest trends have been reviewed in other chapters in this volume.

16.4 Limitations to Conventional Breeding in Bamboo

Due to the flowering behavior of bamboo, it is difficult for researchers to plan for breeding programmes in bamboo. The largely unpredictable nature of flowering and the rarity of synchronous flowering in clumps that can serve as the putative parents, restricts breeding programmes to chance events.

Nevertheless, there have been reports of interspecific hybrids being produced in China – a region with great diversity of monopodial as well as sympodial species. Hybrids between several bamboo species of *Bambusa* and *Dendrocalamus* have been produced (*Bambusa pervariabilis* with *Dendrocalamus latiflorus*, *D. hamiltonii* with *D. latiflorus*, *B. textilis* with *D. latiflorus*, *B. pervariabilis* with *D. latiflorus* or *B. textilis*) (Zhang and Cheng, 1980, 1986; Zhang, 2000; Wang et al., 2005). Crosses between the monopodial species, *Pleioblastus simonii* with *Phyllostachys violascens*, *Sasa tokugawana* with *S. borealis*, and *Sinobambusa tootsik* with *P. distichus* have also been reported (Lu et al., 2009). The simultaneous flowering in the KFRI Bambusetum in Kerala, India of five species (Sankar and Muralidharan, 2020), allowed for attempts of reciprocal crosses between them and generation of putative hybrids using *in vitro* embryo culture.

The use of molecular methods to rapidly detect true hybrids has facilitated such breeding experiments. Lin et al. (2010) used PCR/ISSR to identify two bamboo hybrids from crossed between *Phyllostachys kwangsiensis* (female parent) and *Phyllostachys bambusoides* (male parent). Another area of research that would benefit bamboo breeding with a good level of success would be long-term pollen storage anticipating flowering in suitable clumps that can serve as the pollen recipient.

16.5 Genetic Improvement in Bamboo

With unsurmountable hurdles in adopting conventional breeding strategies in bamboo, the alternatives for genetic improvement lie in carrying out selections from the extensive germplasm available within the species in the natural populations and under cultivation. When seeds are available from gregarious flowering events, selection at the juvenile stage is not justified unless there is good correlation between juvenile and mature characteristics. Juvenile selection based on seedling height and vigor has been practiced to a limited extent (Banik, 2016). The use of modern molecular methods is sure to hasten and improve our capability in understanding the genetics and enable early selection but for the present this is still in the realm of research. Genetic transformation in bamboo is still in the early stages of development but shows a potentiality to be a significant means of introducing or controlling useful traits in bamboo and has been reviewed in other chapters in this volume. In the meanwhile, bamboo genetic improvement will have to depend on alternative strategies, of which currently the best would involve the following steps:

i. Extensive field surveys to identify populations of each of the economically important species across their natural range or in cultivation.

ii. Assessment of between population and within population genetic variability preferably with molecular tools.

iii. Collection of accessions are made within each species representing the intra-specific genetic variability and efforts made to conserve them in *in situ* as well as *ex situ* germplasm collections for posterity.

iv. Characterization of the accessions is done for the morphological, physico-chemical, mechanical, and genetic parameters relevant to the important end uses.

v. Selection of superior accessions, termed candidate plus clumps (CPC) to match current requirements for various traditional and industrial applications and the newly emerging ones.

vi. Field testing of CPCs is carried out in multilocational trials in all potential agroclimatic zones to assess performance and to enable site-species and site-clone matching.

vii. Mass clonal propagation of the best selections (Plus clumps) for production of QPM for operational planting in commercial bamboo plantations.

The alternative to conventional genetic improvement in bamboo therefore is to clone the Plus clumps that perform best in the multilocational trials for generating planting stock. Generations of farmers in rural communities in bamboo growing areas have traditionally done selections of the best genotypes and propagated them clonally. Selections of candidate plus clumps (CPCs) for traits of economic or silvicultural interest follows the same principles but is carried out by researchers through systematic studies and trials. Certification of the QPM by a recognized agency and large-scale propagation of the certified mother stock under strict quality control is essential to ensure that cultivators have access to authentic QPM of the right species and having the right genetic background. As can be seen in the following section the uniqueness of bamboo imposes certain hurdles for which modern techniques fortunately has solutions which need to be integrated in the quality control system that oversees the production of QPM.

16.6 Hurdles in Species and Clonal Identification

Bamboo taxonomy is heavily dependent on a few vegetative morphological characters in the absence of flowering for most part of the flowering cycle. Thus, the most important morphological features are generally recognized as the culm sheath and new emerging shoot (Sarma and Pathak, 2004) which are vegetative parts that are available for examination only for a few weeks every year. When being clonally propagated in nurseries and tissue culture facilities, the planting stock of different species cannot always be identified from the morphological features, and this can lead to mixing up of batches unless care is taken in keep them segregated. Likewise, within the species, QPM originating from several Plus Clumps will be stocked in the nursery or tissue culture lab and keeping them segregated is essential. Fortunately, there are molecular methods available that help precisely identify the species and clonal identity regardless of the type of tissue, its physiological age or growth environment. Das et al. (2005) had developed species specific SCAR markers for two of the important bamboo species in India, *B. balcooa* and *B. tulda*. Due to the effort and expense involved, these methods are not practicable for routine screening of plants in the nursery; but are essential for certifying the mother clumps intended for mass propagation in nurseries and for sampling by the regulatory agencies of each batch of plants produced by tissue culture labs for genetic fidelity testing or establishing clonal identity.

16.7 DNA Barcoding

DNA barcoding is a molecular technique which uses standard DNA sequence tags for accurate species identification (Hebert et al., 2003). When successfully developed and employed, the DNA barcoding technique can identify a species from any tissue consistently and independent of the developmental stage of the plant and the environment in which it grows. A multi-locus barcoding approach has been suggested as the best strategy for the barcoding of plants (CBOL Plant Working Group, 2009). DNA barcoding is still to be standardized for bamboo as a group but important leads have been obtained (Sijimol et al., 2014; Dev et al., 2020) for many of the economically important species of the Indian subcontinent. DNA barcoding thus promises to be a powerful tool to establish species identity as a prelude to certifying mother stock that is used for QPM production. This is particularly essential since the absence of flowering stages or culms sheath would place hurdles for conventional taxonomists to identify species in field or in the nurseries.

16.8 DNA Fingerprinting

As opposed to DNA barcoding, DNA finger printing, also known as genotyping or identity testing, is used to identify individuals within the same species. In the context of certification of bamboo planting material, DNA fingerprinting is useful to establish that the individuals being tested are of the same genotype or different. This step is essential in two situations, viz., testing for clonal fidelity especially in micropropagated plants where some of the techniques carry the risk of genetic variability or for characterizing a clone and establishing identity when there is a suspicion of mixing up in nursery or dispute on ownership of the commercial variety or clone.

16.9 Modern Analytical Techniques

Several of the modern analytical techniques enable high throughput analysis of the physical and chemical parameters that would be useful in characterizing the germplasm and selecting the desirable ones for QPM production. Confocal laser scanning microscopy (CLSM) based cellulose and lignin assay is a non-invasive method, exploited for the first time for genetic diversity screening in a wild gene pool. This protocol would also be useful for rapid resource screening for various commercial purposes (Bhattacharya et al., 2010). FT NIR is another useful tool for rapid analysis of cellulose/lignin and silica levels in bamboo biomass (Thulasi et al., 2013a, 2013b).

16.10 Certification of Quality Planting Material in Bamboo

Given the complex scenario of difficulties in bamboo species identification, a variety of propagation methods and alternate strategies of genetic improvement, it is imperative that production of quality planting material requires a robust process of certification using the state-of-the-art technology that overcomes the hurdles and ensures that exacting standards of quality are conformed to. An outline of the certification process is given below and summarized in Fig. 16.1.

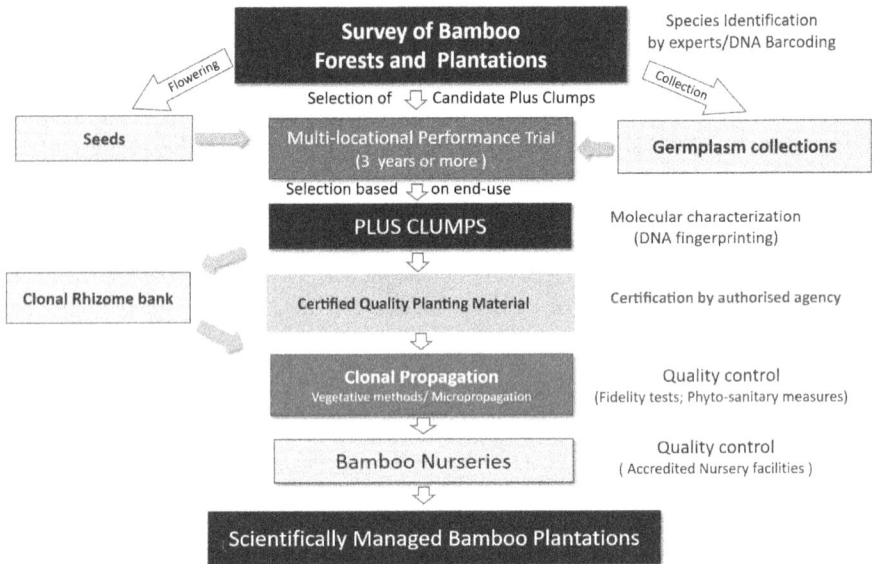

Fig. 16.1. Outline of the scheme for production of QPM in bamboo.

16.11 Identification of Species and Origin of Planting Material

Taxonomic keys are available for the narrowing down identification to the level of species for the major bamboo species, but the process requires considerable specialized expertise. Based on the passport data to be provided with every accession for which certification is sought and supporting data, photographs, geo-coordinates, and herbarium voucher specimens and if required and feasible a field visit to the clumps, the expert will identify the species to which the selected clumps belong to. When bamboo seeds are the starting material, identity is established of the mother clump. Seeds or seedlings can rarely be identified with accuracy based on their morphology and therefore mixing up of planting stock in nurseries should be viewed with concern. Once DNA barcoding is standardised, it will be a more dependable means of identifying the bamboo at any stage of growth.

When certified as originating from seeds from a specific flowering event, the physiological age of the planting material can be confirmed and when the flowering cycle is known, the years remaining for gregarious flowering can be known in advance for many important species to within 2–3 years. Genotyping with DNA fingerprinting will enable identification of individual clones and demonstrate clonal fidelity which is important when tissue culture is the mode of multiplication.

Based on the documentation accompanying each lot of bamboo seeds or vegetative material of known origin, the identity of the species and clonal identity, wherever relevant, will be established by an expert identified by Certification Authority and approved for use in mass propagation and production of certified bamboo QPM in the nurseries or propagation facilities. This certificate is retained as part of the documentation maintained by the nursery for every batch of material used for propagation and production of QPM. Establishment of Clonal Gardens and

Rhizome banks at the bamboo nurseries or research organizations, will ensure that Plus Clumps are available for mass propagation.

16.12 Large Scale Multiplication of Planting Material

Seeds obtained from clumps of correctly identified species or genetically superior clumps of bamboo obtained with the necessary certificates of origin, is mass propagated through micropropagation in tissue culture facilities or vegetative propagation in nurseries and the planting material conforming to the quality standards will be labelled as Certified Bamboo Planting Material by the accredited bamboo nurseries following guidelines issued by the accreditation agencies described in Section 16.15 below.

16.13 Traceability

Appropriate documentation of the chain-of-custody (CoC) of vegetative material used for propagation of planting stocks are maintained by the nurseries through at all stages from the certified mother clump/seeds, through any of the mass propagation procedures, to the point of sale; from where the customers purchase the planting material with labeling that provides essential information on the QPM (Fig. 16.1). In this book, Chapter 3 deals with the germplasm resources of bamboos in detail.

The quality of the propagule used for planting is dependent not only on the genetic superiority of the mother stock, but also on those aspects of the health and vigor of the plants that help it survive and establish well in the field. This is evaluated in terms of the quality of the plant in the nursery at the end of the propagation process and subsequent hardening or acclimatization process.

16.14 Morphological and Plant Health Indicators of Planting Stock Quality

The application of the best propagation practices and nursery management protocol is expected to produce QPM in adequate numbers for use in plantations. Morphological assessment of quality of planting stock (of sympodial bamboo) produced in the nursery is based on the following parameters:

i. Shoot System:
    a. Good Quality: Four sturdy shoots of 50 cm or more in length with healthy leaves.
    b. Acceptable: At least three sturdy shoots which are 50 cm or longer with healthy leaves.
    c. Rejected: Only one to two shoots or leafless/poorly developed shoots.

ii. Rhizome System:
    a. Good Quality: Number of well-developed rhizomes are the same as the shoots or more.
    b. Acceptable: Number of rhizomes at least the same as shoots.
    c. Rejected: Without any well-developed rhizomes/with damaged rhizomes.

iii. Root System:
  a. Good Quality: Each shoot with profuse roots having root hairs.
  b. Acceptable: Planting material with at least two functional roots per shoot.
  c. Rejected: Lacking well developed root system.

16.15 Accreditation of Bamboo Nurseries

It is not sufficient to define the criteria for QPM without looking at the means of producing the propagules routinely in large numbers. Since there are several bamboo species, many methods of propagation in common practice and the planting stock produced differ in their morphology and physiology and their requirement for hardening, etc., quality control in the propagation facility is essential to maintain standards.

The production of QPM in large numbers requires not only infrastructure in terms of space, facilities, and equipment but also trained and qualified technical personnel, a protocol that conforms to good nursery practices, and a system for monitoring that ensures that prescribed norms are being respected. A scheme for accreditation of bamboo nurseries would embody such a system for ensuring that the minimum standard is met and give a measure of confidence to farmers who procure planting material from such nurseries. Preferably the scheme should be implemented by a statutory agency that is mandated to oversee the quality control in agricultural or forestry sectors. The "Guidelines for Accreditation of Bamboo Nurseries, Tissue Culture Laboratories, and Certification of Quality Planting Material" being implemented by the National Bamboo Mission in India is such a scheme (NBM, 2019). The National Certification System for Tissue Culture raised Plants (NCS-TCP) by the Department of Biotechnology in India is an example of the norms set for micropropagation facilities.

16.15.1 Criteria for Accreditation of Bamboo Nurseries

To qualify for accreditation, nurseries should conform to prescribed standards for the following:

  i. Infrastructure
 ii. Planting stock production system
iii. Management practices followed
 iv. Quality parameters of planting stock produced.

Infrastructure: The accreditation agency examines the available infrastructure of the proposed nursery such as the extent and suitability of the land and the facilities such as buildings, stores, polyhouse, shade net nursery, mist chamber, etc., for the propagation and hardening activities. The nursery is expected to have good access to road transport, with fencing, protection against animals, exposure to wind, floods, frost, etc., and year-round availability of good quality water.

Planting stock production system: The propagation system used in the nursery and its ability to produce planting material on large scale is examined. The access to certified mother stock (Plus clumps) and maintenance of rhizome bank to serve as source of vegetative material for propagation is ensured. The labeling system and traceability of planting stock through paper trail is checked.

Management Practices: Good nursery practices are to be adopted by the accredited nurseries. Providing training and regular updating of the technical abilities of the staff and labor employed in the nursery will ensure that the procedures are carried out efficiently and loss at each stage of propagation and hardening is minimized. The nursery is required to maintain a logbook and records pertaining of routine activities, flow chart of production process, the stock position, and details of batches of plants at various stages of propagation. Adoption of quarantine and phyto-sanitary measures to ensure that plants in the nursery do not harbour pest and disease and result in their spread to plantations. Information on the QPM sold from the nursery and advice on aftercare of plants is to be provided to the customers.

16.16 Quality Parameters

The accredited bamboo nurseries are required to conform to the quality standards prescribed for the planting stock and strive to adopt measures to achieve the highest quality with respect to the shoot, rhizome, and root system which have a great influence on the survival and establishment of plants in the field. The nursery is expected to have a proper labeling that indicates the details of the planting material being sold so that the farmer is assured of the suitability of the plants for the selected agroclimatic zone and the end use.

16.17 Conclusions

The unique biological features of bamboo impose hurdles in easy species identification and adoption of conventional genetic improvement, necessitating the use of special methods to produce QPM. Certification of bamboo QPM based on selection of the best genotypes from germplasm collections after multiparameter characterization and field performance trials, should form the basis of mass production of planting material. Accredited bamboo nurseries with adequate infrastructure and technical competence carry out large-scale production of QPM that conforms to prescribed quality standards based on morphological and plant health parameters. Labeled QPM from accredited nurseries will have traceability through chain-of-custody to certified mother stock. Use of modern molecular and analytical methods will aid in the genetic improvement of bamboo and help to evolve a robust system of production of QPM that will result in increasing the productivity and quality of future bamboo plantations and help meet the increasing demand for industrial raw material.

References

Banik, R.L. (1995). *A Manual of Vegetative Propagation of Bamboos. INBAR Tech Report No. 6.* International Network for Bamboo and Rattan, New Delhi, pp. 1–66.

Banik, R.L. (2016). *Silviculture of South Asian Priority Bamboos.* Springer, Singapore. 360 pp. Doi: 10.1007/978-981-10-0569-5.

Bhattacharya, S., Ghosh, J., Sahoo, D., Dey, N. and Pal, A. (2010). Screening of superior fiber-quality traits among wild accessions of *Bambusa balcooa*: Efficient and non-invasive evaluation of fiber developmental stages. *Annals of Forest Science*, 67: 611. Doi: 10.1051/forest/2010024.

CBOL Plant Working Group. (2009). A DNA Barcode for land plants. *Proceedings of the National Academy of Sciences of the United States of America*, 106: 12794–12797. http://dx.doi.org/10.1073/pnas.0905845106.

Das, M., Bhattacharya, S. and Pal, A. (2005). Generation and characterization of SCARs by cloning and sequencing of RAPD products: A strategy for species-specific marker development in bamboo. *Annals of Botany*, 95(5): 835–841. https://doi.org/10.1093/aob/mci088.

Dev, S.A., Sijimol, K., Prathibha, P.S., Sreekumar, V.B. and Muralidharan, E.M. (2020). DNA barcoding as a valuable molecular tool for the certification of planting materials in bamboo. *3 Biotech.*, 10(2): 59. Doi: 10.1007/s13205-019-2018-8. Epub 22 Jan 2020.

Hebert, P.D.N., Cywinska, A., Ball, S.L. and de Waard, J.R. (2003). Biological identifications through DNA barcodes. *Proc. Roy. Soc., London. B*, 270: 313–321.

Keeley, J. and Bond, W.J. (1999). Mast flowering and semelparity in bamboos: The bamboo fire cycle hypothesis. *The American Naturalist*, 154: 383–391. 10.1086/303243.

Lakshmi, C.J., Seethalakshmi, K.K. and Jijeesh, C.M. (2018). Influence of seed storage conditions and germination media on the germination of a priority bamboo species, *Dendrocalamus brandisii* (Munro) Kurz. *J. Tropical Agriculture*, 56 (2): 191–196.

Lin, X.C., Lou, Y.F., Liu, J., Peng, J.S., Liao, G.L. and Fang, W. (2010). Crossbreeding of *Phyllostachys* species (Poaceae) and identification of their hybrids using ISSR markers. *Genet. Mol. Res.*, 9: 1398–404. Doi: 10.4238/vol9-3gmr855.

NBM. (2019). *Guidelines for Accreditation of Bamboo Nurseries, Tissue Culture Laboratories, and Certification of Quality Planting Material.* National Bamboo Mission, India, New Delhi. https://nbm.nic.in/Documents/pdf/Guidelines_accreditation_Nurseries_10062019.pdf (Accessed on 30.03.2022).

Sankar, V.R. and Muralidharan, E.M. (2020). Simultaneous flowering of six bamboo species in Kerala. *Indian Forester*, 146(11): 1081–1083. Doi: 10.36808/if/2020/v146i11/151200.

Sarma, K.K. and Pathak, K.C. (2004). Leaf and culm sheath morphology of some important bamboo species of Assam. *J. Bamboo and Rattan*, 3: 265–281. 10.1163/1569159041765254.

Sijimol, K., Dev, S.A., Muralidharan, E.M. and Sreekumar, V.B. (2014). DNA barcoding: An emerging tool for precise identification and certification of planting stock in taxonomically challenging bamboo species. *J. Bamboo and Rattan*, 13: 29–33.

Somen, C.K. and Seethalakshmi, K.K. (1989). Effect of different storage conditions on the viability of seeds of *Bambusa arundinaceae*. *Seed Sci. and Technol.*, 17: 355–360.

Thulasidas, P.K., Buddhan, S., Muralidharan, E.M. and Pandalai, R.C. (2013a). Silica content in reed bamboo (*Ochlandra travancorica* Gamble) and its rapid prediction using Fourier Transform Near-infrared Spectroscopy. *J. Bamboo and Rattan*, 12(1-4): 43–53.

Thulasidas, P.K., Bhat, K.M., Muralidharan, E.M. et al. (2013b). Evolving elite plants with low lignin and high cellulose reed bamboos (*Ochlandra* spp.) from the Western Ghats of India for pulp and paper industry. *Proceedings of the 2012 IUFRO Conference*, Division 5, 8–13 July 2012, Lisbon, Portugal, OP066, p. 9.

Wang, Y.X., Zhang, G.C. and Li, X.W. (2005). An evaluation on shoot quality of sympodial bamboo species and their hybrids. *J. Bam. Res.*, 24: 39–44.

Zhang, G.C. (2000). The recent situation of bamboo breeding. *J. Bam. Res.*, 19: 13–15.

Zhang, G.C. and Cheng, F.S. (1980). No. 1 of Chengmaqing, a good bamboo hybrid. *Forest Sci. Tech.*, S1: 124–126.

Zhang, G.C. and Cheng, F.S. (1986). Study on bamboo hybridization. *J. Guangdong Forest Sci. Tech.*, 3: 1–5.

Index

A

Abiotic stress 228–231, 233–236, 240, 243–248
Accreditation of nurseries 372
AFLP 121, 122, 126, 132, 134–139
African region 160
Agrobacterium 348, 349
Allozymes 131–134
Alternative polyadenylation 293
Alternative splicing 287, 292, 293
Artificial hybridization 176, 194
Arundinaria 118, 122, 126, 127, 135, 138
Arundinarieae 28, 44–46, 61–63, 71–73, 78–80, 82, 84–90, 92–101, 103–106, 110
Avirulence gene 273

B

Bacteria 256, 266, 271, 274
Bamboo 4–7, 9–21, 27–45, 61, 106–111, 154–160, 162, 166, 168–171, 202–205, 207, 211–220, 228–248, 286–293, 296–299, 301, 304–308, 310–313, 338–340, 347–350, 354, 358–360, 364–373
Bamboo breeding 176, 177
Bamboo flowering 202–205, 211, 219
Bamboo genome 230–232, 234
Bamboo propagation 321–323, 325, 327, 329
Bambuseae 28, 44, 46–51, 53–62, 64–71, 73–78, 80–84, 99, 101–103, 105, 106, 110, 117, 118, 120, 121, 123, 125, 127–131, 135, 140
Bambuseta 27–29, 35, 43–45, 109, 110
Bambusoideae 117–121, 123, 125, 129, 131
Base editing 343, 345
Biochemical defence 272
Botanic gardens 27–37, 39–41, 43, 44, 108–111
Breeding system 176, 177, 185, 191

C

Candidate plus clumps 367, 368
Certification 368–370, 372, 373
Certification of QPM 368

Chain-of-custody 371, 373
Chemical control 274
Chloroplast markers 118, 120, 125–131
Circadian clock 202, 205–209, 211–214, 216–218, 220
Clonal propagation 366, 368
Cold 229, 233, 234, 239–241, 243–245, 247, 248
Conservation 27, 28, 32, 34, 35, 41, 43, 44, 108–111
cpDNA genes 160, 166, 168, 169, 171
CRISPR/Cas9 340, 342, 345, 346, 348, 350, 356–360
Culms 287–293

D

Databases 279, 282
dCas9 344, 346, 347
Delayed flowering 203, 206, 211
Distribution 4–7, 9, 10, 20
Diversity 4, 6, 9, 11, 13, 20
Drought 229, 233–237, 240–248

E

Endophytes 326
Environment 230, 233, 236, 238, 248

F

Flower induction 202, 209, 212, 215, 216
Fruit 185, 192
Functional genetics 293
Fungi 256–259, 266

G

Gene expression 286, 289, 290, 293
Genetic defence 272
Genetic diversity 154, 158–166, 168, 171
Genetic improvement 176, 177, 365, 367–369, 373
Genetic linkage map 193, 194

Genetic map 176, 193, 194
Genetic resistances 228, 229, 231, 236
Genetic transformation 296, 297, 307, 310–313, 320, 321, 323, 330–332
Genome 118, 123, 125, 127, 131, 132
Genome annotation 279–284
Genome editing 354–357, 359, 360
Genomics 229, 230, 233, 234, 236
Germplasm 27, 28, 34, 40, 41, 42, 109, 111
Germplasm collections 367, 373
Germplasm conservation 177
Guide RNAs 356

H

Herbaceous bamboo 117, 118, 120, 125, 130
Hormones 287–291

I

in vitro flowering 324, 328, 329
Insects 256, 258, 259, 266, 270, 271
Irregular flowering 229
ISSR 132–134, 138–141, 143

K

Knock-in/out 343, 345, 346

L

Latent contamination 326, 327
Liquid culture media 330
Living collections 28–30, 37, 40, 108, 109

M

Micropropagation 297, 298, 304, 306, 307, 311, 313, 322, 323, 325, 327, 329, 330, 340, 348, 349
Molecular marker 154, 159, 162, 171
Molecular mechanisms 237, 238
Morphological indicators of QPM 371
Moso bamboo 279, 283, 284

N

Neotropical woody bamboos 118, 120, 127, 129
Northeast region 4–7, 9, 11–15, 17, 19–21
Nuclear genes 166
Nuclear markers 118, 120, 125–128, 131

O

Olyreae 28, 44, 66, 80, 83, 99, 108, 110, 117, 118, 120, 123, 125, 130–132
Organogenesis 297, 304–306

P

Paleotropical woody bamboos 118, 120, 127, 128
Photoautotrophic micropropagation 329, 330
Photoperiodic pathway 205, 207, 212, 216
Photoperiodism 202, 204
Physiology 229, 247, 248
Plant genome 279, 284
Plastid 122, 124, 125, 127
Plus clumps 367, 368, 371, 372
Pollination 177, 178, 190–194
Polymorphism 159, 162
Population structure 158–160, 162, 168, 169, 171
Post-transcriptional gene regulation 291–293

Q

Quality planting material 364, 365, 369, 372

R

RAPD 132, 133, 139–141, 143, 144
Rapid-growth 287–289, 291
Regeneration 297, 298, 303–306, 308, 310, 311, 313
Regulation 286–289, 291–293
Resistance (R)-gene 273
RFLP 132, 133
Rhizome 286, 292, 293
Rhizome bank 109, 110, 371, 372
Rhizome development 286, 292
RNP 350

S

Salinity 234, 238–241, 245, 247, 248
Seed bank 109, 110
Shoot 286, 287, 289, 290, 292, 293
SNP 123, 124, 129, 131, 146, 147
Somatic embryogenesis 304, 307, 308, 311, 313
SSR 132, 140, 141, 144–147
Stable transformation 348
Stress 228–231, 233–248
Symptoms 258–260, 262, 264, 266–268

T

TALEN 339–341, 343
Taxonomy 4–6
Temperate woody bamboos 117, 120, 125–127, 145
Temperature 229, 233, 236, 237, 239, 240, 244
Tools 279, 284
Traceability 371, 373
Transciptome 212–215, 219, 281, 292, 293

Transcription factors 234, 237, 240, 242, 245, 287, 290, 291, 293
Transcriptional gene regulation 287, 288, 291–293
Transcriptional regulation 217, 287
Transcriptomics 229
Transgenics 297, 307
Transposable elements 357

U

Uses 4, 13, 16

V

Virus 256, 267–270, 272, 274, 275

W

Whole genome sequencing 339, 347
Woody (lignified) stems 155

Z

ZFN 339–341, 343

For Product Safety Concerns and Information please contact our EU
representative GPSR@taylorandfrancis.com
Taylor & Francis Verlag GmbH, Kaufingerstraße 24, 80331 München, Germany